Activation of Viruses by Host Proteases

Eva Böttcher-Friebertshäuser
Wolfgang Garten
Hans Dieter Klenk
Editors

Activation of Viruses by Host Proteases

Springer

Editors
Eva Böttcher-Friebertshäuser
Institut für Virologie
Philipps-Universität
Marburg
Germany

Wolfgang Garten
Institut für Virologie
Philipps-Universität
Marburg
Germany

Hans Dieter Klenk
Institut für Virologie
Philipps-Universität
Marburg
Germany

ISBN 978-3-030-09242-9 ISBN 978-3-319-75474-1 (eBook)
https://doi.org/10.1007/978-3-319-75474-1

Printed on acid-free paper

This Springer imprint is published by the registered company Springer International Publishing AG part of Springer Nature
The registered company address is: Gewerbestrasse 11, 6330 Cham, Switzerland

To Purnell W. Choppin, the founder of the field

Preface

More than four decades ago the groups of Purnell Choppin in the United States and of Morio Homma in Japan identified a glycoprotein in paramyxoviruses that induces membrane fusion and initiates infection, and they discovered that its activity depends on posttranslational cleavage by host proteases. At about the same time similar observations were made with the influenza virus hemagglutinin, and proteolytic glycoprotein activation proved to be an important determinant for organ tropism and spread of infection of paramyxoviruses as well as influenza viruses. Thus, proteases were among the first host factors recognized to play a prominent regulatory role in virus infection. Moreover, these studies provided for the first time insight into the molecular mechanisms underlying viral pathogenicity. These early discoveries have been described in detail in previous reviews (Webster RG and Rott R, Cell 50, 665-666, 1987; Nagai Y, Trends Microbiol 1, 81-87, 1993; Klenk HD and Garten G, in: Cellular Receptors for Animal Viruses, Wimmer E, ed, pp241-280, Cold Spring Harbor Press, 1994). Today we know that proteolytic activation of envelope proteins is a characteristic feature of many important viral pathogens including not only influenza virus but also respiratory syncytial virus, Nipah and Hendra viruses, Lassa virus, Marburg and Ebola viruses, SARS and MERS coronaviruses, yellow fever virus, Dengue virus, Zika virus, and tick-borne encephalitis virus, just to name a few. A large number of new activating proteases have also been identified that differ in substrate specificity and expression patterns in tissues and cell compartments, and the concept has been strengthened that these variations have a high impact on the outcome of infection. Of particular interest are activation mutants of viruses that have been used for vaccine design. These and other recent developments will be discussed in this book. Activation by host proteases observed with non-enveloped viruses, such as rotaviruses, will not be addressed.

The first chapters of the book will focus on envelope proteins undergoing proteolytic activation. Most of them induce membrane fusion involving a conformational change that is primed by cleavage of the fusion protein itself or by cleavage of an accessory protein and then triggered by exposure to low pH or interaction with a receptor-binding protein. Based on structural differences, three classes of fusion proteins can be discriminated. Class I fusion proteins will be addressed in contributions by Summer Galloway, Bo Liang, and David Steinhauer on the influenza virus hemagglutinin, by Everett Smith and Rebecca Dutch on the F protein of paramyxoviruses and pneumoviruses, by Antonella Pasquato, Laura Cendron, and Stefan

Kunz on the glycoprotein of arenaviruses, and by Markus Hoffmann, Heike Hofmann-Winkler, and Stefan Pöhlmann on the S protein of SARS and MERS coronaviruses. Activation of the coronavirus S protein is particularly interesting, because it depends on the action of different proteases in concert. As discussed in Chap. 5 the GP protein of filoviruses is also a class I fusion protein undergoing sequential cleavage by two proteases that removes a large carbohydrate-rich segment of GP, thereby exposing the receptor binding region and the fusion loop. The chapter by Franz Heinz and Karin Stiasny illustrates that activation of the class II fusion protein of a flavivirus depends on cleavage of the tightly associated accessory prM protein. Processing by host proteases has also been observed with a few viral proteins that do not induce membrane fusion. Thus, in the chapter on paramyxoviruses the authors will report that, with some strains of Newcastle disease virus, the HN glycoprotein mediating binding to and release from receptors is also activated by proteolytic cleavage.

The section on viral proteins will be followed by reviews of the activating proteases. Obviously, endoproteolytic cleavage is not only necessary for the activation of viruses, but it is also an essential step in the maturation of many cellular proteins with important biological functions, such as peptide hormones, neuropeptides, growth and coagulation factors, cell surface receptors, adhesion molecules, and transcription factors involved in lipid metabolism. Of the large number of vertebrate endoproteases, so far only a fraction has been found to activate viruses. They fall into four major groups, each of which will be reviewed in a separate chapter. The first enzymes identified were soluble trypsin-like serine proteases. As described in the contribution of Hiroshi Kido, these enzymes have several functions in influenza virus pathogenesis of which hemagglutinin activation is only one. In recent years a still increasing number of membrane-anchored serine proteases (MASPs) have been discovered that form the second group (Chap. 8). Proprotein convertases that have the widest spectrum of viral glycoprotein substrates form the third group (Chap. 9). The fourth group of host proteases processing envelope proteins are cathepsins as described in the contribution of Klaudia Brix.

The book will close with a comprehensive overview by Torsten Steinmetzer and Kornelia Hardes on protease inhibitors. Elucidation of the structural details of an increasing number of proteases provides a solid basis for rational inhibitor design. Since host proteases are stable targets, their inhibition should prevent the rapid development of resistance that is observed when viral proteins are addressed. There is experimental evidence for the antiviral potential of such compounds, and there is hope that they may find their way to therapeutic application.

Marburg, Germany — Eva Böttcher-Friebertshäuser
Marburg, Germany — Wolfgang Garten
Marburg, Germany — Hans Dieter Klenk

Contents

Part I

Viral Proteins Activated By Host Proteases

Activation of the Hemagglutinin of Influenza Viruses

1

Summer E. Galloway, Bo Liang, and David A. Steinhauer

Abstract

The hemagglutinin (HA) glycoprotein of influenza viruses is posttranslationally cleaved into the disulfide-linked subunits HA_1 and HA_2, and this proteolytic processing event is critical to the virus life cycle as it is required to activate membrane fusion potential and virus infectivity. High-resolution structures are available for the HA precursor (HA0), the cleaved neutral pH conformation of HA, and the low pH conformation that the HA assumes when triggered by acidification of endosomes to mediate fusion of viral and cellular membrane during virus entry. These structures have provided clues regarding the mechanisms by which proteolytic cleavage activates membrane fusion potential and how subsequent acidification drives the fusion process. It has been known for decades that influenza strains and subtypes can vary with regard to HA cleavage properties and that cleavage site sequences and the proteases that recognize them can represent a major determinant for virus pathogenicity. However, a number of questions remain with respect to the identity and characteristics of the proteases that activate HAs in the various natural hosts and complex ecosystems that constitute the realm of influenza viruses. The continuing study of HA cleavage properties

S. E. Galloway
Influenza Division, National Center for Immunization and Respiratory Diseases, Centers for Disease Control and Prevention, Atlanta, GA, USA
e-mail: nmd3@cdc.gov

B. Liang
Department of Biochemistry, O. Wayne Rollins Research Center, Emory University School of Medicine, Atlanta, GA, USA
e-mail: bo.liang@emory.edu

D. A. Steinhauer (✉)
Department of Microbiology and Immunology, O. Wayne Rollins Research Center, Emory University School of Medicine, Atlanta, GA, USA
e-mail: dsteinh@emory.edu

E. Böttcher-Friebertshäuser et al. (eds.), *Activation of Viruses by Host Proteases*,
https://doi.org/10.1007/978-3-319-75474-1_1

and the proteases involved should illuminate our understanding not only of pathogenicity but other aspects of influenza biology including host range, transmission, and interplay with other microorganisms such as bacteria.

There are three genera of influenza viruses in the *Orthomyxoviridae* family: the influenza A, influenza B, and influenza C. All are human pathogens; however, influenza A viruses are generally of greatest concern as they are responsible for most of the annual seasonal epidemics, and their vast natural reservoir in aquatic avian species provides a fertile source for the unpredictable emergence of novel pandemic strains. For the influenza A and B viruses, the initial stage of entry into host cells is mediated by receptor-binding and membrane fusion functions provided by the hemagglutinin (HA) glycoprotein. A separate viral surface protein, the neuraminidase (NA), is responsible for the receptor-destroying sialidase activity that allows for progeny virus dissemination at the end of the replication cycle. For influenza C viruses, both the receptor-binding function and receptor-destroying esterase functions are provided by a single glycoprotein, termed HEF, for hemagglutinin/esterase/fusion. The HA and HEF proteins of all influenza viruses share a common requirement for protease activation in order to potentiate membrane fusion and activate virus infectivity. This chapter will focus on the proteolytic activation of the HA glycoproteins of influenza A viruses, but many of the concepts outlined herein are equally relevant for the HA and HEF proteins of influenza B and C viruses.

The influenza A viruses are distinguished by their complex ecology and dynamic transmission cycles involving more than 100 avian species and a range of mammalian hosts (Munster et al. 2007; Slusher et al. 2014; Webster et al. 1992). In aquatic birds belonging to the order Anseriformes (ducks, geese, and swans) and Charadriiformes (gulls, terns, and shorebirds) that serve as the "natural" hosts of influenza A virus, 16 antigenically distinct HA subtypes and 9 NA subtypes are known to circulate, though they rarely cause symptomatic disease in these species. Under conducive conditions, viruses maintained within the natural reservoir may transmit to Galliformes (chickens, turkeys, and quail) and passerine birds or mammalian species such as swine, horses, dogs, and humans, causing a range of morbidity and mortality. The outbreaks resulting from cross-species transmission are often limited, but when influenza A virus strains become established in a new host, they can perpetuate for varying lengths of time and have devastating consequences to human health and global economies. Cross-species transmission events of influenza A virus are often characterized by the relatively rapid evolution of the virus to facilitate adaptation to the new host. These adaptive changes may be influenced by a variety of selective pressures, including, but not limited to, the immune response of the host; the availability and structure of host cell receptors; variations in host cell replication machinery; differences in the site of replication, which may involve pH and temperature; and environmental persistence to enable continued transmission. While these concepts are generally understood, we are only at the inception of understanding the complex interplay of viral and host factors involved in host range, transmission, and pathogenicity of these viruses. What we do know is that HA cleavage properties constitute one of the most critical determinants of pathogenicity and the environments and biological niches where activating proteases can be found in nature may play an important role in the complex ecology of influenza A viruses.

Proteolytic cleavage is a fundamental biochemical process that activates a multitude of functions for viruses as well as their hosts. A requirement for proteolytic activation of influenza virus infectivity was first discovered in studies examining influenza virus replication in cell culture. For many years after influenza viruses were first isolated in the 1930s (Shope 1931; Smith et al. 1933), embryonated chicken eggs were the substrate of choice for influenza virus propagation, and even as cell culture systems were developed, only a few strains such as A/WSN/33 (H1N1) were able to replicate and form plaques on cell monolayers. Decades passed before it was reported that virus replication and plaquing efficiency could be enhanced appreciably by the addition of proteases such as trypsin or pancreatin to the cell monolayers (Appleyard and Maber 1974; Came et al. 1968; Tobita and Kilbourne 1974) and that the extent of HA cleavage correlated with cytopathology following infection (Lazarowitz et al. 1973a). These studies were extended to show that HA cleavage activation was required for an early stage in the infection process, but was not required for virion assembly, hemagglutination, or virus adsorption functions (Klenk et al. 1975; Lazarowitz and Choppin 1975). Polyacrylamide gel electrophoresis (PAGE) analyses revealed that cleavage activation resulted in the digestion of the HA precursor protein (HA0) into faster migrating polypeptides known as the HA_1 and HA_2 subunits. N-terminal sequencing of the HA_2 polypeptides derived from various infectious viruses revealed a highly conserved sequence, GLFGAIAGFIE (Skehel and Waterfield 1975), which is the N-terminal portion of the domain now universally referred to as the fusion peptide. Subsequent studies clearly demonstrated that the critical function activated by proteolytic cleavage is HA-mediated membrane fusion of the viral and endosomal membranes, facilitating transfer of the viral genome into host cells during virus entry for replication (Huang et al. 1980, 1981; Maeda and Ohnishi 1980; White and Helenius 1980).

1.1 Structural Basis for Activation of HA Fusion Potential

From a structural perspective, the mechanisms by which influenza HA becomes fusogenic upon protease cleavage and the subsequent acid-induced conformational rearrangements that drive the membrane fusion process in endosomes are quite well developed. For the HA of H3N2 subtype viruses that have circulated in humans since 1968, there is high-resolution structural information for the uncleaved HA0 precursor, the cleaved pre-fusion neutral pH HA present on the surface of infectious virions, and the low pH conformation adopted by HA following the acid-induced structural rearrangements required for fusion (Bizebard et al. 1995; Bullough et al. 1994; Chen et al. 1998; Wilson et al. 1981). Therefore, we base most of our structural discussions on the HA of the 1968 human virus A/Aichi/2/68 (H3N2).

For discussion purposes, we generally assume that infection of a new host is initiated by virions that contain cleaved HAs on their surfaces and are therefore fully activated to facilitate membrane fusion; however, infectious virions can also contain mixed ratios of cleaved and uncleaved HAs on their surfaces, and there may be examples for which cleavage can also occur in endosomes during the early stages of entry. In any case, influenza viruses initiate infection by attaching to sialic acid-containing glycan receptors on host cell surfaces, followed by entry via the endocytic pathway. The trigger for fusion of the viral and endosomal membranes is the acidification of the vesicular compartments by cellular proton pumps. When the endosomal pH reaches a critical threshold, usually between pH 6.0 and 5.0 depending on the viral strain, a number of HA structural rearrangements are triggered that coordinately function to drive the fusion process (Bizebard et al. 1995; Bullough et al. 1994; Chen et al. 1999). The pre- and post-fusion structures of HA are shown in Fig. 1.1, and the conformational changes that take place include detrimerization of the membrane-distal head domains, extrusion of the HA_2 N-terminal fusion peptide domains from the trimer interior, extension of the long HA_2 coiled coil by recruitment of the short HA_2 α-helix and connecting polypeptide to the N-terminal end of the long α-helix of each monomer, and 180° reorientation of the C-terminal domain of the long α-helices to "jackknife," this domain against the central coiled coil. The extension of the central coiled coil directs the fusion peptides to the end of the low pH structure (the top, as shown in Fig. 1.1, right panel), and residues C-terminal to the "jackknifed" helices trace along a groove on the outside of the coiled coil, as can be viewed for the polypeptide chain of one monomer to the left of the coiled coil in Fig. 1.1, and this places the HA_2 C-terminal transmembrane domains at the same end of the rod-shaped structure as the fusion peptides (Fig. 1.1, right panel). Based on these structural changes, a mechanism for initiating the membrane fusion process can be envisaged, in which hydrophobic fusion peptides are "harpooned" into the endosomal membrane to link the viral and cellular membranes via the HA_2 subunit for influenza A virus, and the 180° "jackknife" of the long helices bring the viral and endosomal membranes into proximity as a prerequisite for the fusion process.

Collectively, the acid-induced HA conformational changes that initiate membrane fusion are irreversible, and the observation that the rodlike low pH HA structure is considerably more stable than cleaved neutral pH HA (Carr et al. 1997; Chen et al. 1995, 1999; Ruigrok et al. 1988) indicates that the cleaved neutral pH HA is a metastable molecule that transitions to a more energetically stable form during the fusion process. In contrast, the uncleaved HA is relatively unresponsive to acidification with respect to structural rearrangements, suggesting that proteolytic cleavage primes the HA for fusion by allowing it to adopt the metastable conformation that can be triggered by acidification to induce membrane fusion. A comparison of the structures of the HA0 and cleaved neutral pH HA reveals that only 19 residues relocate upon protease cleavage (Chen et al. 1998; Wilson et al. 1981); however, two critical prerequisites for fusion transpire upon cleavage. First, protease cleavage liberates the hydrophobic fusion peptide from its internal position within the HA0 polypeptide chain, becoming the N-terminal domain of the newly generated HA_2

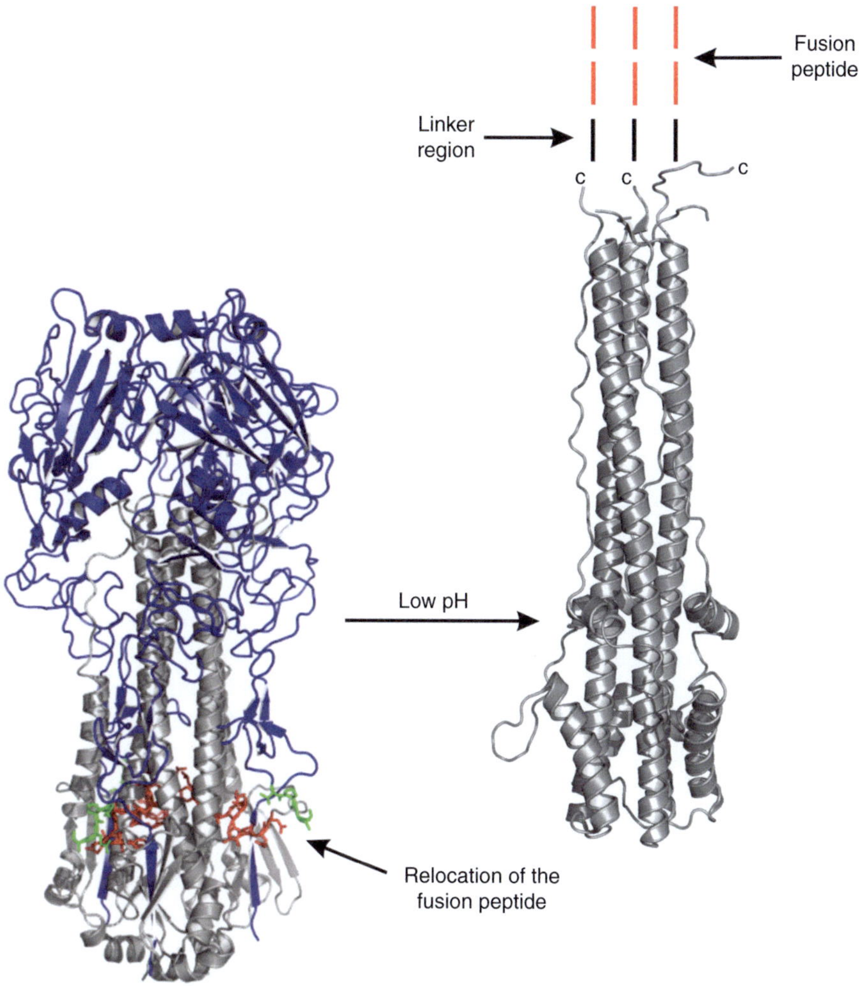

Fig. 1.1 Structures of the neutral pH cleaved HA (left) and the low pH conformation (right) assumed when membrane fusion is triggered by acidification of endosomes. In the neutral pH structure (left), the HA_1 subunit is shown in blue, the HA_2 subunit is in gray, and the C-terminal residues of HA_1 are highlighted in green, and HA_2 N-terminal fusion peptide residues highlighted in red. Keep in mind that the neutral pH structure is derived from HA ectodomains solubilized by proteolytic treatment of virions, which removes the HA transmembrane domains located near the C-terminus of each HA_2 subunit (the viral membrane would be at the bottom of the figure). The low pH rodlike structure (right) is composed entirely of HA_2 residues (gray). Hatched lines indicate regions for which the structure is unknown as they are disordered in the crystal structure (linker region) or were removed by proteolysis to solubilize the protein for crystal preparation (fusion peptide). The rodlike structure illustrates that fusion peptide residues are relocated to the same end of the helical rod as the C-terminal end of the polypeptide chains (labeled C) that are proximal to the membrane anchor domains (removed by protease treatment); therefore, the viral and endosomal membranes are pulled into close proximity with one another by acid-induced conformational changes

subunit. Second, the N-terminal 10 residues of the HA_2 fusion peptide relocate into a "cavity" present in the interior of the trimeric HA0 structure and form new contacts with a number of ionizable residues that line the cavity. These structural changes are depicted in Fig. 1.2, which shows a side view of the HA0 and cleaved HA ectodomains as well as a view down the threefold axes of symmetry. In the uncleaved HA0 structures (panels a and c), the cleavage loop extends out into solution; the conserved arginine residue that serves as the cleavage site of each monomer is indicated by the arrows. The residues colored in green and red are the ones that relocate upon protease cleavage; the relocation of HA_2 N-terminal residues (indicated in red) into the trimer interior is best illustrated when viewed down the threefold axis of cleaved HA (panel d).

A preponderance of evidence suggests that the new contacts formed when the HA_2 N-terminal fusion peptide inserts into the interior of the HA trimer are critical for priming the metastable HA structure for subsequent acid-induced conformational changes. First, let us consider the fusion peptide itself. The N-terminal 11 residues of HA_2 constitute the most highly conserved region of the HA, completely conserved in nearly all strains of the 16 influenza A HA subtype viruses and differing by only one residue in most influenza B strains (Cross et al. 2009; Nobusawa et al. 1991) (Table 1.1). Surprisingly, HAs having mutations in the fusion peptide domain that retain membrane fusion functionality can be selected for in the laboratory and generated easily by reverse genetics and have been documented at nearly every position (Cross et al. 2001, 2009; Daniels et al. 1985; Gething et al. 1986; Korte et al. 2001; Lin et al. 1997; Nobusawa et al. 1995; Orlich and Rott 1994; Qiao et al. 1999; Steinhauer et al. 1995; Yewdell et al. 1993). However, nearly all HA fusion peptide mutants with substitutions in the N-terminal 10 residues mediate membrane fusion at elevated pH relative to wild-type (WT) HA (Cross et al. 2009). Furthermore, when infectious viruses containing such mutants are generated by reverse genetics, minimal passage frequently results in either reversion to the WT HA residue or pseudoreversion at the position that was altered (Cross et al. 2001). These observations suggest an evolutionary pressure on the HA beyond that which operates on functional interactions with target membranes and suggest that fusion peptide contacts in the metastable cleaved neutral pH HA serve as pH "sensors" that are reactive to acidification. As mentioned previously, in the HA0 structure, the cleavage loop is proximal to a cavity lined with ionizable residues, which are subsequently buried by the fusion peptide upon proteolytic cleavage (Chen et al. 1998). Since the relocation of fusion peptide residues constitutes the only structural change that takes place following cleavage, the newly formed contacts are likely to play a role in potentiating fusion activity. An analysis of this region of cleaved HA shows that HA_2 ionizable residues Asp109 and Asp112 are invariantly conserved and form numerous hydrogen bonds with the fusion peptide. Other residues of interest in this region include conserved histidine residues that may serve as potential "trigger" residues for initiating acid-induced conformational changes. The pKa of the histidine side chain is around pH 6.0 in aqueous solution, and therefore, protonation of such residues would occur within the biologically relevant range during endosomal acidification, if accessible to solvent. Of particular interest are HA_1 position 17 and

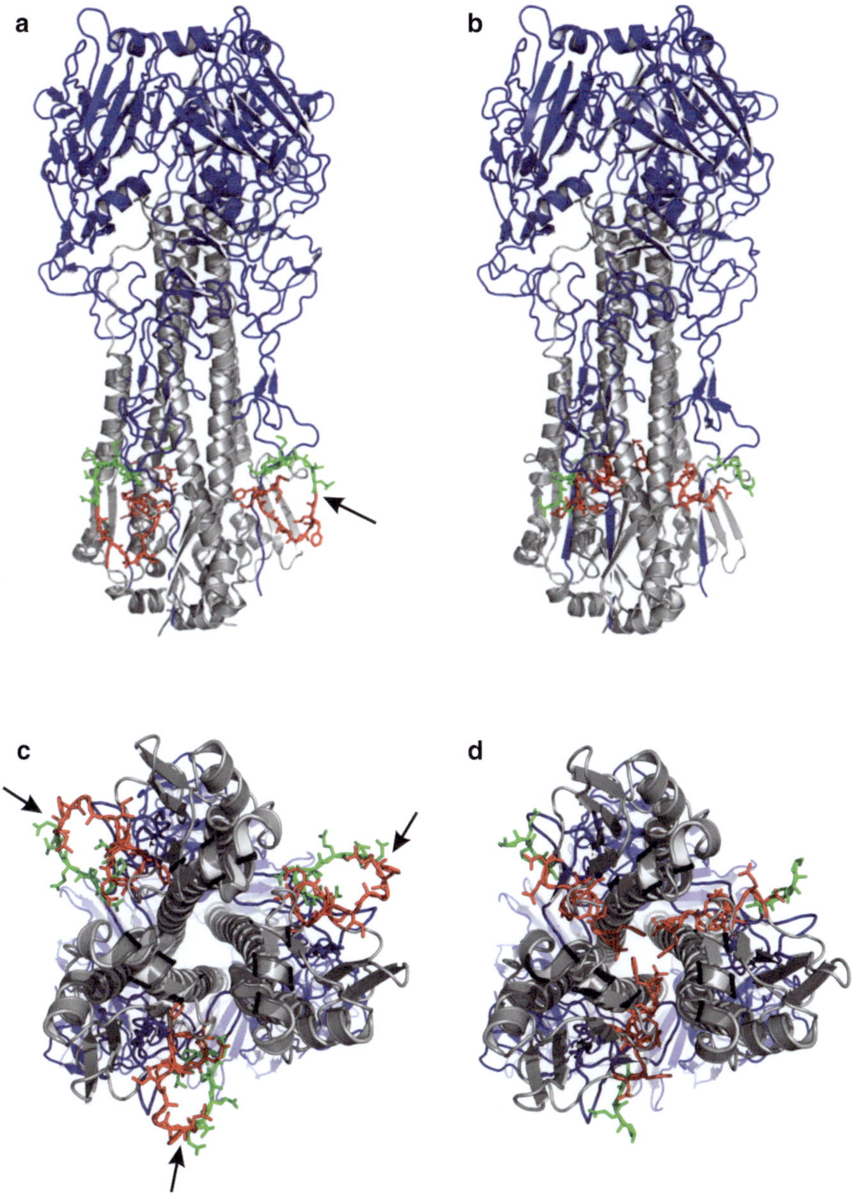

Fig. 1.2 Structures of the HA ectodomains of A/Aichi/2/68 virus (H3N2 subtype). Panels **a** and **c** represent uncleaved HA0 viewed side-on (**a**) or side-down the threefold axis of symmetry (**c**), and panels **b** and **d** represent cleaved neutral pH HA from the same orientations. The residues that constitute the HA_1 subunits are shown in blue and HA_2 in gray, and these residues do not relocate following cleavage (they are superimposable in the two structures). The cleavage loops of HA0 are shown in red and green in panels **a** and **c** with arrows indicating the site of cleavage at residue Arg 329. These are the only residues that relocate following cleavage, with residues that constitute the N-terminus of HA_2 (red) inserting into the trimer interior, as illustrated most clearly in panel **d**

Table 1.1 Cleavage site sequences of HA subtypes known to circulate in wild birds

		HA1									↓	HA2										
		Consensus sequence																				
Subtype	Virus	P						P_4	P_3	P_2	R	G	L	F	G	A	I	A	G	F	I	E
H1	A/duck/Alberta/35/1976	P	S					I	Q	S	R	**G**	-	-	-	-	-	-	-	-	-	-
	A/Puerto Rico/8/1934	P	S					I	Q	S	R	**G**	-	-	-	-	-	-	-	-	-	-
	A/WSN/1933	P	S					I	Q	Y	R	**G**	-	-	-	-	-	-	-	-	-	-
	A/California/04/2009	P	S					I	Q	S	R	**G**	-	-	-	-	-	-	-	-	-	-
H2	A/duck/Germany/1215/1973	P	Q					I	E	S	R	**G**	-	-	-	-	-	-	-	-	-	-
	A/Japan/305/1957	P	Q					I	E	S	R	**G**	-	-	-	-	-	-	-	-	-	-
H3	A/duck/Ukraine/1/1963	P	E					K	Q	T	R	**G**	-	-	-	-	-	-	-	-	-	-
	A/Aichi/2/1968	P	E					K	Q	T	R	**G**	-	-	-	-	-	-	-	-	-	-
H4	A/duck/Czechoslovakia/1/1956	P	E					K	A	S	R	**G**	-	-	-	-	-	-	-	-	-	-
H5	A/tern/South Africa/1961	P	Q	R	E	T	R	R	Q	K	R	**G**	-	-	-	-	-	-	-	-	-	-
	A/Vietnam/1203/2004	P	Q	R	E		R	R	R	K	R	**G**	-	-	-	-	-	-	-	-	-	-
H6	A/shearwater/Australia/1/1972	P	Q					I	E	T	R	**G**	-	-	-	-	-	-	-	-	-	-
H7	A/turkey/England/1963	P	K					R	R	R	R	**G**	-	-	-	-	-	-	-	-	-	-
	A/turkey/Oregon/1971	P	E				N	P	K	T	R	**G**	-	-	-	-	-	-	-	-	-	-
	A/Shanghai/02/2013 (H7N9)	P	E				I	P	K	G	R	**G**	-	-	-	-	-	-	-	-	-	-
H8	A/turkey/Ontario/6118/1968	P	S					V	E	P	R	**G**	-	-	-	-	-	-	-	-	-	-
H9	A/turkey/Wisconsin/1/1966	P	A					V	S	S	R	**G**	-	-	-	-	-	-	-	-	-	-
H10	A/chicken/Germany/N/1949	P	E				V	V	Q	G	R	**G**	-	-	-	-	-	-	-	-	-	-
H11	A/duck/England/1/1956	P	A					I	A	S	R	**G**	-	-	-	-	-	-	-	-	-	-
H12	A/duck/Alberta/60/1976	P	Q					V	Q	D	R	**G**	-	-	-	-	-	-	-	-	-	-
H13	A/gull/Maryland/704/1977	P	A					I	S	N	R	**G**	-	-	-	-	-	-	-	-	-	-
H14	A/duck/Astrakhan/263/1982	P	G					K	Q	A	K	**G**	-	-	-	-	-	-	-	-	-	-
H15	A/duck/Australia/341/1983	P	E				K	I	H	T	R	**G**	-	-	-	-	-	-	-	-	-	-
H16	A/black-headed gull/Sweden/5/1999	P	S					V	G	E	R	**G**	-	-	-	-	-	-	-	-	-	-

Sequences are anchored by the conserved proline (P) N-terminal to the R329 (H3 numbering) that is the site of proteolytic cleavage (denoted with ↓). The P4, P3, P2 positions are labeled such that P1 is the R329 cleavage site. C-terminal to the R329 is the highly conserved HA2 N-terminal fusion peptide (N-GLFGAIAGFIE-C). The first glycine (G) of the fusion peptide is in bold type. The (-) indicates invariant conservation

HA_2 positions 106 and 111. The 16 HA subtypes that circulate in aquatic birds can be phylogenetically and structurally segregated into two groups (Air 1981; Nobusawa et al. 1991; Russell et al. 2004), Group-1 (H1, H2, H5, H6, H8, H9, H11, H12, H13, H16) and Group-2 (H3, H4, H7, H10, H14, H15), and these three positions (HA_1 17 and HA_2 106 and 111) segregate along evolutionary lines and are well conserved within groups. In Group-1, HA_1 17 is a tyrosine, HA_2 106 is an arginine, and HA_2 111 is a histidine; in Group-2, HA_1 17 is a histidine, HA_2 106 is a histidine, and HA_2 111 is a threonine. Extensive mutagenesis studies on these residues of an H3 subtype (Group-2) HA reveals that HA_1 His17 may play a role in triggering structural changes upon acidification (Thoennes et al. 2008). In Group-1 HAs, HA_1 17 is a tyrosine, but nearby HA_2 residue His111 is invariantly conserved and may serve an equivalent role to Group-2 HA_1 His17, and indeed, mutagenesis studies on a number of Group-1 HAs suggest that changes to His111 result in the inactivation of fusion activity (J. Trost, D.A.S., unpublished).

A number of additional studies support the idea that the contacts made between the fusion peptide and the cavity region may be critical in triggering acid-induced structural rearrangements. HA mutants that mediate membrane fusion at elevated pH relative to WT can be found throughout the HA trimer at domain interfaces that rearrange upon acidification (Byrd-Leotis et al. 2015; Cross et al. 2001; Daniels et al. 1985; Doms et al. 1986; Gething et al. 1986; Lin et al. 1997; Qiao et al. 1999; Steinhauer et al. 1995; Thoennes et al. 2008; Zaraket et al. 2013b), whereas stabilizing mutations, or those resulting in an HA that mediates membrane fusion at a lower pH compared to WT, have been identified at fewer positions to date (Byrd-Leotis et al. 2015; Steinhauer et al. 1991; Thoennes et al. 2008; Xu and Wilson 2011; Zaraket et al. 2013a). The identification of mutants that mediate membrane fusion at higher or lower pH relative to WT provided for the rational design of double mutant HAs, in which mutations known to confer high- and low-pH phenotypes were generated in various combinations to examine the cumulative effects. These studies revealed that when the structural locations of the mutations were proximal, the pH phenotype was additive, whereas when the mutations were distal to one another, the phenotype of one mutation was dominant, with changes in the fusion peptide region being most critical in determining the overall fusion phenotype (Steinhauer et al. 1996). These results supported data on the kinetics of conformational changes, as determined using a panel of conformation-specific monoclonal antibodies, which showed that structural alterations in the stem region preceded changes in the HA head domains (White and Wilson 1987). These interpretations were further supported by experiments on the fusion kinetics of single virions with planar lipid bilayers, which indicated that the "withdrawal" of the fusion peptide from the trimer interior is the rate-limiting step (Ivanovic et al. 2013). In addition, another study involving hydrogen-deuterium exchange and mass spectrometry approaches revealed that release of the fusion peptide occurs prior to the structural changes involving coiled-coil rearrangements (Garcia et al. 2015), and studies on the membrane fusion process by Cryo-EM indicate that extrusion of the fusion peptide and insertion into the target membrane as extended intermediates precedes the foldback into helical rods at fusion "dimples" where membrane merger occurs (Calder and Rosenthal 2016). Overall, the data strongly support a mechanism by which HA0 cleavage and coincident conformational changes prime the cleaved neutral pH structure for fusion by generating functional fusion peptides and critically relocating them to structural regions that can respond to acidification.

1.2 Structure of HA0 Cleavage Sites

Not only is cleavage activation required for infectivity, but the available evidence indicates that the sequence and structure of the HA cleavage loop may reveal clues about the ability of HAs to serve as substrates for specific activating proteases and dictate traits such as the pathogenicity of influenza strains and the ecology and host range of influenza viruses. To date, the HA0 precursor structure of three HA subtypes has been determined: the H3 subtype described above from a human 1968 pandemic H3N2 virus (Chen et al. 1998), the H1 subtype representing 1918 human pandemic H1N1 strains (Stevens et al. 2004), and the H16 subtype from an H16N3 avian virus isolated from a black-headed gull (Lu et al. 2012). Figure 1.3 depicts these three structures as a single monomer of the trimer, viewed from equivalent orientation. A visual comparison of these structures reveals major differences in the structural elements of the cleavage loop, including the location of the Arg329 cleavage site (green). Shown in yellow are the conserved electronegative residues discussed above, HA_2 D109 and D112, HA_2 His 17 for H3, and HA_2 His 111 for H1 and H16. The H3 Arg329 cleavage site is located within a relatively standard loop structure that orients away from the trimer surface, leaving it easily accessible at a distal position in the loop. Unlike the H3 HA0, the loop structure of the H1 HA0 packs against the surface of the trimer, and the Arg329 residue is positioned such that it covers the electronegative cavity, though it orients out into solution. The H16 cleavage loop is somewhat similar to the H1 loop in that it packs against the trimer surface, but unlike the other two, it contains a five-residue α-helix that includes positions 325 through the Arg329 that is cleaved by activating proteases. Remarkably, this helix covers the electronegative cavity and orients the side chain of Arg329 into the cavity where it forms a salt bridge with highly conserved HA_2 residue D112. The orientation of Arg329 in H16 HA0 is likely a factor in limiting its accessibility to proteases that can activate other subtypes, but not H16 (see below). It is possible that the differences in cleavage loop structures may exist, in part, due to the location of proximal glycosylation sites, which for H1 and H16 is at HA_1 N20 (H3 numbering), whereas for H3 the glycosylation site is slightly farther away at HA_1 N22 (Chen et al. 1998; Lu et al. 2012; Stevens et al. 2004). It should also be noted that the H3 HA0 structure was derived from protein expressed in mammalian cells, and the H1 and H16 structures were based on proteins expressed using baculovirus expression systems. It will be very interesting to extend our knowledge of HA0 precursor structures to additional subtypes and relate these to data that is accumulating on the range of proteases capable of activating individual HA subtypes.

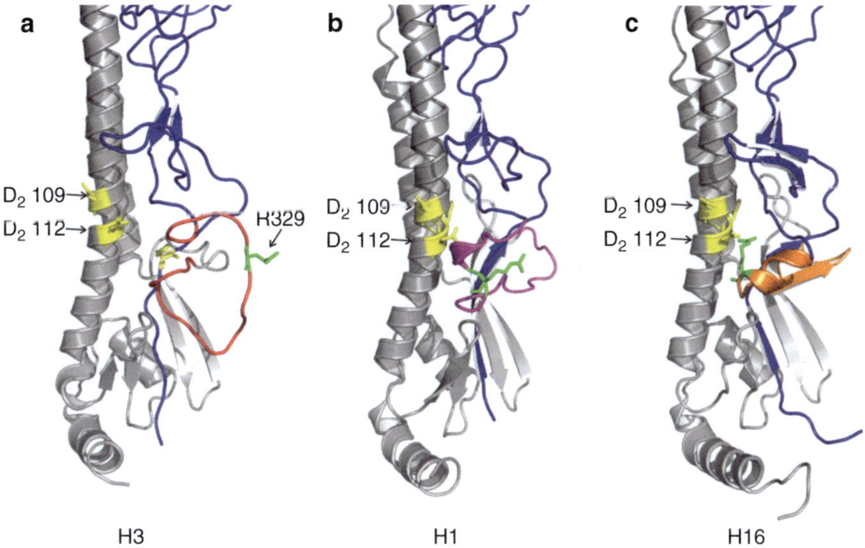

Fig. 1.3 Cleavage site structures showing a single monomer of the HA0 trimer for H3 subtype (**a**), H1 subtype (**b**), and H16 subtype (**c**). (**a**) The cleavage domain of H3 (red) forms a loop structure with R329 (green) oriented out into solution. Shown in yellow are residues that line the electronegative cavity, residues HA_2 D109, HA_2 D112 (subscript denotes the HA_2 subunit in the figure), and HA_1 His 17 (between $D_2$112 and R329). (**b**) The cleavage domain of H1 (magenta) packs more closely against the trimer, but R329 (green) remains oriented out into solution. HA_2 D109, HA_2 D112, and HA_2 H111 are shown in yellow, though H111 is on the opposite face of the long helix and largely hidden in the depicted orientation. (**c**) The cleavage domain of H16 (orange) contains a small helical domain that orients the side chain of R329 (green) into the electronegative cavity, where it forms a salt bridge with HA_2 D112 (again, ionizable residues shown in yellow). In all panels, residues that will constitute the HA_1 and HA_2 subunits following cleavage are shown in blue and gray, respectively

1.3 Activating Proteases

Whereas the residues of the cleavage loop that constitute the N-terminal portion of the fusion peptide are highly conserved, the upstream residues between the highly conserved Pro324 (H3 numbering) and Arg329 cleavage site are quite variable among subtypes and provide important clues regarding cleavage activation and the proteases involved (Tables 1.1 and 1.2). In particular, variations in the number and sequence of arginine and lysine residues at the cleavage site account for major differences in proteolytic activation and proved to be critical determinants of pathogenicity (Bosch et al. 1981; Garten and Klenk 1999; Klenk and Garten 1994; Steinhauer 1999) and host range (Galloway et al. 2013). Other factors that can influence the capacity for particular proteases to activate HAs include the presence or absence of nearby carbohydrates that can alter the accessibility of an activating protease (Deshpande et al. 1987; Kawaoka et al. 1984; Kawaoka and Webster 1989; Ohuchi et al. 1991). Studies on a wide range of influenza strains and subtypes have addressed

Table 1.2 Cleavage site sequences of H7 LPAI and HPAI viruses having insertions derived through recombination

		HA1																									↓	HA2										
		Consensus sequence																																				
		P			P_4	P_3	P_2																				R	**G**	L	F	**G**	A	I	A	**G**	F	I	E
H7 LPAI	A/turkey/ Oregon/1971	P	E	N	P	K	T																				R	**G**	-	-	-	-	-	-	-	-	-	-
	A/Shanghai/ 02/2013	P	E	I	P	K	G																				R	**G**	-	-	-	-	-	-	-	-	-	-
H7 HPAI	A/turkey/ Oregon/1971 (lab)	P	E	N	P	K	T	**S**	**L**	**S**	**P**	**L**	**Y**	**P**	**G**	**R**	**T**	**T**	**D**	**L**	**V**	**Q**	**V**	**P**	**T**	**A**	R	**G**	-	-	-	-	-	-	-	-	-	-
	A/chicken/ Chile/2002	P	E	N	P	K	T	**C**	**S**	**P**	**L**	**S**	**R**	**C**	**R**	**E**	**T**										R	**G**	-	-	-	-	-	-	-	-	-	-
	A/chicken/British Columbia/2004	P	E	N	P	K	**Q**	**A**	**Y**	**Q**	**K**	**R**	**M**	T													R	**G**	-	-	-	-	-	-	-	-	-	-
	A/Mexico/ InDRE7218/2012	P	E	N	P	K	**D**	**R**	**K**	**S**	**R**	**H**	**H**	**R**	T												R	**G**	-	-	-	-	-	-	-	-	-	-
	A/Guangdong/ 17SF003/2016	P	E	V	P	K	R	**K**	**R**	**T**	**A**																R	**G**	-	-	-	-	-	-	-	-	-	-
	A/Taiwan/1/2017	P	E	V	P	K	R	**K**	**R**	**T**	**A**																R	**G**	-	-	-	-	-	-	-	-	-	-

Sequences are anchored by the conserved proline (P) located N-terminal to the R329 (H3 numbering) that is the site of proteolytic cleavage (denoted with ↓). The P4, P3, P2 positions are labeled such that P1 is the R329 cleavage site. HA1 insertions are shown in bold underline type. C-terminal to the R329 is the highly conserved HA2 N-terminal fusion peptide (N-GLFGAIAGFIE-C). The first glycine (G) of the fusion peptide is in bold type. The (-) indicates invariant conservation functional activation of infectivity or fusion potential, and a variety of candidate proteases have been identified that activate low pathogenic avian influenza (LPAI) viruses and mammalian strains at monobasic cleavage sites. They include membrane-anchored serine proteases (MASP) (see Chap. 8) such as human airway tryptase (HAT) (Bottcher et al. 2006; Bertram et al. 2012), transmembrane protease, serine S1 member 2 (TMPRSS2) (Bottcher et al. 2006; Chaipan et al. 2009; Hatesuer et al. 2013; Sakai et al. 2014; Tarnow et al. 2014), TMPRSS4 (Bertram et al. 2010a; Kuhn et al. 2016), and matriptase (Baron et al. 2013; Hamilton et al. 2012), as well as soluble serine proteases (see Chap. 9) such as tryptase Clara (Kido et al. 1992), plasmin (Lazarowitz et al. 1973b), and kallikrein-related peptidases (Hamilton and Whittaker 2013). In considering the numerous studies published to date on an extended panel of proteases, it is apparent that no single universal protease is involved in activating the membrane fusion potential of the HA, as most of the proteases examined activate subsets of viruses in nonoverlapping fashion and with extensive variation in cleavage efficiency. For example, Galloway et al. (2013) characterized cleavage activation from representatives of all 16 avian HA subtypes and several human strains for activation by trypsin, HAT, and TMPRSS2 and observed a variety of phenotypes. Trypsin displayed fairly broad cleavage activation, but showed low or no activity for H12, H13, and H16 subtype HAs, whereas HAT was more limited in its ability to cleave various subtypes, but was able to cleave H9, H11, and H12 HAs. On the other hand, TMPRSS2 was broadly effective but demonstrated no activity against H8 and H12 and was the only one of the three proteases examined in this study that showed specificity for the H13 and H16 HAs. Interestingly, H13 and H16 appear to have a limited host range, generally being isolated only from Charadriiformes, such as gulls and shorebirds, suggesting perhaps that cleavage activation can serve as a host range determinant for particular viruses. Furthermore, a study documenting a repository of Eurasian-lineage reverse genetics vectors and recombinant viruses found that the H13 and H16 recombinant viruses were only able to be recovered when transfected 293T cells were injected into embryonated chicken eggs for further propagation; embryonated chicken eggs are known to express a factor Xa-like serine protease that is capable of cleaving HA0, which may provide the rationale for this observation (Keawcharoen et al. 2010). As mentioned previously, the cleavage loop of the H16 HA appears to be rather structured with the critical Arg329 less exposed than observed in the loop structure of the H3 subtype HA. This may be indicative of host-specific proteases found in Charadriiformes that have greater activity against H13 and H16 HA substrates, but this remains to be determined. Additionally, it was shown that human TMPRSS2 homologues in chicken and swine were capable of activating an influenza virus having a monobasic cleavage site in vitro (Bertram et al. 2012). In another example, studies using matriptase have shown that it can efficiently activate H1 and H9 subtype viruses, but is much less effective against H2 and H3 subtypes. Furthermore, while matriptase was able to cleave H1 subtype HAs, in general, it displayed selective activity against certain H1 subtype strains (Baron et al. 2013; Hamilton et al. 2012). Similarly, studies examining the cleavage activity of the thrombolytic proteases kallikrein (KLK) 5 and 12 with various HA strains belonging to H1, H2, and H3 subtypes found that KLK5 and KLK12 displayed differential activity against HAs within and across subtypes. For example, KLK12, but not KLK5, was able to cleave H2 HA; the opposite was observed with the H3 HAs examined. For the H1 HAs, KLK5 and KLK12 were able to efficiently cleave the HA from A/California/04/2009 (H1N1), but have substantially reduced activity against other H1 strains, such as A/New Caledonia/99, A/South Carolina/18, and the lab-adapted strains A/Puerto Rico/8/34 and A/WSN/33 (Hamilton and Whittaker 2013

Inextricably linked to the recognition of HA as a substrate by activating proteases is the requirement for the protease to generate specific cleavage products containing fusion-competent sequences at the HA_2 N-terminus. As mentioned previously, the HA_2 N-terminal fusion peptide domain that results from cleavage is the most conserved region in HA, with the sequence GLFGAIAGFIE being virtually invariant in natural isolates. Studies have shown that HA digestion by proteases such as thermolysin, which cleaves between the Gly_1 and Leu_2 of authentic HA_2, renders the HA inactive for fusion and results in noninfectious virus (Garten et al. 1981; Lazarowitz and Choppin 1975; Orlich and Rott 1994; Steinhauer et al. 1995). Interestingly, the selection of influenza viruses capable of replicating in the presence of thermolysin revealed HA mutants having a single amino acid insertion just downstream of the Leu_2, effectively restoring the length and spacing of critical residues of the fusion peptide (Orlich and Rott 1994), and fusion peptide length as well as composition has been demonstrated as a requirement for fusion function (Langley et al. 2009; Steinhauer et al. 1995). As mentioned previously, the HAT protease was able to generate HA_1 and HA_2 cleavage products with similar mobility as HAs digested with trypsin, but the HAT-cleaved H12 HA was not capable of mediating membrane fusion (Galloway et al. 2013). It was also shown that a recombinant virus expressing the H12 HA replicated poorly in MDCK cells, but replicated quite well in embryonated chicken eggs (Keawcharoen et al. 2010), which express a factor Xa-like serine protease that is capable of activating HA0 (Gotoh et al. 1990).

An influenza HA cleavage anomaly is observed from studies on the H1N1 strain A/WSN/33. This is a lab-adapted neurovirulent strain selected by passage of the first human influenza A isolate A/WS/33 (Smith et al. 1933) in mouse brain (Francis and Moore 1940) and is efficiently cleaved by the fibrinolytic protease plasmin (Lazarowitz et al. 1973b). The mechanism by which plasmin-mediated HA cleavage occurs was shown to be moderated by the ability of the NA of A/WSN/33 (H1N1) to sequester plasminogen, the zymogen precursor to plasmin, until it is converted to plasmin and able to cleave HA0 into HA_1 and HA_2. The ability of the NA to sequester plasminogen was shown to be dependent on the absence of a carbohydrate at position 146 (N2 numbering) and the presence of a carboxy-terminal lysine (Lys453) (Goto and Kawaoka 1998; Li et al. 1993). This feature has only been observed for the A/WSN/33 NA, but shows specificity for multiple HA subtypes (Goto and Kawaoka 1998). The HA cleavage site sequence of the A/WSN/33 H1 HA contains a serine to tyrosine substitution at the P2 position, resulting in the cleavage site IQY↓R rather than IQS↓R, as is found in most H1 HAs (Table 1.1) (Sun et al. 2010). In addition to the A/WSN/33 H1 HA, a subsequent study reported on the acquisition of a similar cleavage site sequence in a more contemporary seasonal H1N1 strain, A/Beijing/718/2009, that is preferentially cleaved by plasmin

(Tse et al. 2013). These data are consistent with other studies showing that bulky hydrophobic amino acids, such as tyrosine, in the P2 position promote HA cleavage by plasmin. Unlike the A/WSN/33 HA, the ability of the HA to be cleaved by plasmin was found to be independent of the NA. Proteolytic conversion of plasminogen into plasmin is mediated by plasminogen activators. It has recently been shown that interferon stimulates the expression of the plasminogen activator inhibitor PAI-1 which prevents not only activation of plasmin but inhibits also other HA-activating proteases, such as TMPRSS2 and HAT. Inhibition of proteolytic activation of HA may therefore be an important mechanism in innate immunity (Dittmann et al. 2015).

Most studies evaluating the proteolytic activation of HA have been cell-based or in vitro studies; however, following the discovery of the ability of TMPRSS2 to cleave influenza A virus HA (Bottcher et al. 2006), several groups investigated this further in knockout and mutant mice by examining the ability of several different influenza A viruses to replicate and cause disease in *Tmprss2*-deficient mice. These studies showed that H1N1-, H3N2-, and H7N9-infected *Tmprss2*-deficient mice were generally protected from influenza-associated morbidity and mortality (Hatesuer et al. 2013; Sakai et al. 2014; Tarnow et al. 2014). TMPRSS2-mediated influenza virus replication in vivo appears to be limited to influenza A viruses, as a recent study showed that influenza B viruses were able to efficiently replicate and cause disease in *Tmprss2*-deficient mice (Sakai et al. 2016).

Many of the proteases that have been implicated in activating cleavage are secreted and activate HA extracellularly; however, proteases such as TMPRSS2, TMPRSS4, and HAT have membrane-bound forms that may be active during HA transport or at the plasma membrane (Bottcher et al. 2009). In addition, a description of A/WSN/33 HA cleavage in endosomes during virus entry, distinct from the plasmin-directed cleavage described above, has been reported (Boycott et al. 1994). Though this was unique for MDBK cells and the protease involved was not identified, cell-associated proteases have the potential for HA activation during entry as well as during transport following de novo synthesis (Zhirnov et al. 2002). Cumulatively, it appears that among influenza strains, a broad range of proteases have the potential to activate infectivity and that activation can occur in cell-associated fashion at early or late stages of the replication cycle or extracellularly during virus spread in a host or during transmission (see Fig. 1.2 of Bertram et al. (2010b) and Fig. 1.1 of Bottcher-Friebertshauser et al. (2014) for useful illustrations). Proteolytic activation of HA is also an important pathogenic mechanism in viral-bacterial coinfection. It has long been known that various bacteria, including *Staphylococcus aureus*, secrete proteases that cleave HA and promote the development of pneumonia in virus-infected mice (Tashiro et al. 1987; Maeda 1996). Moreover, bacterial proteases may activate host proteases that cleave HA (Scheiblauer et al. 1992).

Perhaps less complicated are the HAs having cleavage site sequences that consist of a stretch of polybasic amino acids, with the general sequence motifs of R-X-(K/R)-R, which can be cleaved in the trans-Golgi by subtilisin-like proteases (see Chap. 10) such as furin and proprotein convertase 5/6 (PC5/6) (Horimoto et al. 1994;

Klenk et al. 1984; Stieneke-Grober et al. 1992). The polybasic motif K-K-K-R that has only been observed on rare occasions with highly pathogenic avian influenza (HPAI) viruses (Kawaoka et al. 1984) is cleaved by TMPRSS13 belonging to the MASP group (Okumura et al. 2010). Notably, furin has been shown to proteolytically activate several other viral glycoproteins, including the respiratory syncytial virus fusion (F) protein, human immunodeficiency virus envelope glycoprotein GP120, Crimean-Congo hemorrhagic fever virus Gn protein, and the Ebola virus GP, among others (Klenk and Garten 1994; Chaps. 2, 5, 9 this volume). Polybasic cleavage sites are only observed with avian influenza viruses of subtypes H5 and H7 (Tables 1.1 and 1.2). Due to the ubiquitous expression of the proteases activating at polybasic cleavage sites, the viruses cause systemic infection with high lethality in poultry. The HPAI viruses evolve from LPAI variants by insertion of the polybasic motif (Garten and Klenk 1999; Klenk and Garten 1994; Steinhauer 1999). As a result of the insertion, the cleavage site loop is extended (Table 1.2), and this extension appears to be essential for high cleavability, since HAs that have acquired the polybasic motif by amino acid substitution are not cleaved by furin (Bottcher-Friebertshauser et al. 2014). However, sensitivity to furin is not always sufficient to confer a HP phenotype. Veits et al. (2012) assessed the capability of multiple HA subtypes to support a HP phenotype by genetically modifying the HA of H1, H2, H3, H4, H6, H8, H10, H11, H14, and H15 subtypes to contain a polybasic cleavage site, followed by the rescue of reassortants after co-transfection with the genes of either a low pathogenic H9N2 or a high pathogenic H5N1. Their results showed that only the reassortants consisting of the polybasic H2, H4, H8, or H14 HA and the remaining genes from the HP H5N1 were capable of supporting the HP phenotype of lethality in chickens. While this study does not rule out the capability of other subtype HAs supporting the HP phenotype, potentially with a different reassorting virus, it does highlight the fact that the polybasic cleavage site alone is not sufficient, and while H5 and H7 subtypes may be predisposed to acquiring a HP phenotype, other HA subtypes are also capable of supporting a HP phenotype.

There is no clear pathway for the emergence of highly pathogenic viruses. Aquatic birds generally display asymptomatic infection with both LPAI and HPAI viruses, despite the fact that they replicate equally well in aquatic birds and poultry, which makes surveillance for HPAI viruses difficult until they present in poultry. In some cases, the emergence of HPAI viruses is through direct transmission from aquatic birds, such as ducks and geese, to poultry, but there have been instances of transmission of HPAI viruses to poultry via aquatic or passerine birds, such as sparrows and starlings, which may display less severe pathogenicity than poultry to infection with HPAI viruses. In other cases, HPAI viruses have evolved from LPAI virus progenitors after introduction to poultry (Steinhauer 1999; Xu et al. 2017).

There are currently two primary mechanisms postulated to be responsible for the insertion of additional nucleotides that encode for basic amino acids. The first is believed to arise through accumulation of point mutations from polymerase slippage of the RNA-dependent RNA polymerase (RdRp) on purine-rich sequences, which are characteristic of the HA cleavage site coding sequences of many H5 and H7 viruses (Perdue et al. 1997). Notably, RdRp slippage on purine-rich or poly(A) sequences is a common feature of many RNA viruses. Briefly, during RNA synthesis of the (+)-sense template, the RdRp may encounter a region of secondary structure near the cleavage site sequence that causes the RdRp to pause and "slip," or release and reprime, on these purine-rich or poly(A) sequences, resulting in the insertion of nucleotides. When these inserted nucleotides maintain the reading frame of the HA and display relative fitness in a given environment, these viruses may be selected for during replication, resulting in the emergence of a novel HPAI virus. The second mechanism is based on RNA recombination of viral or host RNA with that of the influenza A virus HA gene segment. To date, this has only been reported for H7 HPAI viruses. For example, sequence analysis of the viruses responsible for an outbreak of H7 HPAI in Mexico in 2012 revealed the presence of an eight-amino acid insertion (DRKSRHRRTR↓GLF) that resulted from RNA recombination with a host 28S rRNA (Maurer-Stroh et al. 2013); however, the first report of RNA recombination with a eukaryotic 28S rRNA was by Khatchikian et al. (1989) when the authors described the adaptation of A/turkey/Oregon/1971 (H7N3) to chicken embryo cells, which are not permissive for HA cleavage without the addition of exogenous trypsin. Upon sequencing, the authors discovered a 54-nucleotide (18 amino acids) insertion (SLSPLYPGRTTDLQVPTAR↓G) that resulted in an HA that could be cleaved in chicken embryo cells in the absence of exogenous trypsin as well as in MDCK, BHK, MDBK, and Vero cells. The recombination event was postulated to occur by polymerase jumping using the rRNA as a template during (+)-sense RNA synthesis of the (−)-sense RNA template. Notably, the (+)-sense sequence upstream of the insertion site, AAAGACUA, is identical to the 3′ border of the inserted 28S rRNA sequence, yielding a palindromic sequence (AAAGACU) in the adapted HA gene. Evidence for nonhomologous recombination in the laboratory has also been observed between viral HA and NP sequences with A/seal/Mass/1/80 virus (H7N7), resulting in cleavage loop insertions and accompanied by increased pathogenicity (Orlich et al. 1994). Similar recombination events between viral RNA sequences were subsequently observed in two separate outbreaks of H7 HPAI viruses reported in British Columbia in 2004 and Chile in 2002, both of which were the result of H7 LPAI acquiring insertions from an influenza gene segment that yielded a highly pathogenic variant (Pasick et al. 2005; Suarez et al. 2004). The strains isolated in British Columbia had a seven-amino acid insertion (QAYQKR/QMTR↓G) derived through RNA recombination with the M gene segment (M1), and the strains isolated in Chile had a ten-amino acid insertion (CSPLSRCRETR↓G) derived through RNA recombination with the NP gene segment. Analysis of several HPAI H7 HAs revealed the presence of palindromic sequences surrounding the insertions at the HA cleavage site (Maurer-Stroh et al. 2013). More recently, an outbreak of LPAI H7N9 in China is currently in its sixth epidemic wave and has been responsible for more than 1500 laboratory-confirmed human infections since its emergence in 2013 (WHO 2018). In December 2016, the emergence of an HPAI variant of the H7N9 virus was detected in two human patients as well as poultry that has a four-amino acid insertion (KRKRTAR↓G) in the HA cleavage site postulated to have originated from a nonhomologous recombination event (Zhang et al. 2017; Zhu et al. 2017). These observations highlight the need to continue rigorous surveillance to monitor emerging strains for the presence of HA cleavage site characteristics that could lead to HP phenotypes.

Concluding Remarks

The link between HA cleavage properties and pathogenicity has been recognized for decades, beginning with early studies elucidating the effect of exogenous trypsin on influenza A virus replication in cell culture to the continual threat of emerging viruses with modified cleavage site sequences that confer a high pathogenic phenotype. Clearly, a variety of both cell-associated and secreted proteases are capable of cleavage activation of influenza A virus HA, in strain-specific fashion. Despite our existing body of knowledge, many gaps remain that relate to the identity and characteristics of proteases that activate most influenza viruses during replication and transmission in nature, to why certain subtypes are more prone to cleavage site modifications that yield high pathogenic phenotypes, and to what extent host-specific proteases restrict the transmission dynamics of certain influenza A virus HA subtypes. Continued investigation to endeavor to fill these gaps is important to our understanding of influenza virus biology and ecology and has critical implications for the development of novel cleavage-specific therapeutics to decrease influenza morbidity and mortality as well as a better understanding of the parameters contributing the emergence of novel pandemic and/or highly pathogenic influenza viruses. For influenza viruses that infect humans, further identification of candidate proteases as well as more detailed examination of those currently identified could lead to the development of antiviral strategies designed to inhibit cleavage activation. The inhibition of HA-activating proteases would present a novel therapeutic strategy for influenza A virus infection; however, as with any therapy that targets a host protein, there are substantial safety concerns. Interestingly, knockout mice generated for the TMPRSS2 studies discussed in this chapter were found to lack a discernable phenotype (Kim et al. 2006; Sales et al. 2011), providing conservative optimism for further pursuing this overall strategy. For all influenza strains, the virus must encounter an HA-activating protease at some stage of the replication or transmission cycle, be it during transport of newly synthesized HAs, at infected cell surfaces or during spread within infected hosts, or even at appropriate sites of infection in a new host, and this simple biochemical process can potentially have profound implications for a broad range of phenotypes related to virus stability, pathogenicity, host range, and natural ecology.

Acknowledgments The authors acknowledge the support by the US Department of Health and Human Services contract HHSN272201400004C (NIAID Centers of Excellence for Influenza Research and Surveillance).

References

Air GM. Sequence relationships among the hemagglutinin genes of 12 subtypes of influenza A virus. Proc Natl Acad Sci U S A. 1981;78:7639–43.

Appleyard G, Maber HB. Plaque formation by influenza viruses in the presence of trypsin. J Gen Virol. 1974;25:351–7.

Baron J, Tarnow C, Mayoli-Nüssle D, Schilling E, Meyer D, Hammami M, Schwalm F, Steinmetzer T, Guan Y, Garten W, Klenk HD, Böttcher-Friebertshäuser E. Matriptase, HAT, and TMPRSS2 activate the hemagglutinin of H9N2 influenza A viruses. J Virol. 2013;87:1811–20.

Bertram S, Glowacka I, Blazejewska P, Soilleux E, Allen P, Danisch S, Steffen I, Choi SY, Park Y, Schneider H, Schughart K, Pohlmann S. TMPRSS2 and TMPRSS4 facilitate trypsin-independent spread of influenza virus in Caco-2 cells. J Virol. 2010a;84:10016–25.

Bertram S, Glowacka I, Steffen I, Kuhl A, Pohlmann S. Novel insights into proteolytic cleavage of influenza virus hemagglutinin. Rev Med Virol. 2010b;20:298–310.

Bertram S, Heurich A, Lavender H, Gierer S, Danisch S, Perin P, Lucas JM, Nelson PS, Pohlmann S, Soilleux EJ. Influenza and SARS-coronavirus activating proteases TMPRSS2 and HAT are expressed at multiple sites in human respiratory and gastrointestinal tracts. PLoS One. 2012;7:e35876.

Bizebard T, Gigant B, Rigolet P, Rasmussen B, Diat O, Bosecke P, Wharton SA, Skehel JJ, Knossow M. Structure of influenza virus haemagglutinin complexed with a neutralizing antibody. Nature. 1995;376:92–4.

Bosch FX, Garten W, Klenk HD, Rott R. Proteolytic cleavage of influenza virus hemagglutinins: primary structure of the connecting peptide between HA1 and HA2 determines proteolytic cleavability and pathogenicity of Avian influenza viruses. Virology. 1981;113:725–35.

Bottcher E, Freuer C, Steinmetzer T, Klenk HD, Garten W. MDCK cells that express proteases TMPRSS2 and HAT provide a cell system to propagate influenza viruses in the absence of trypsin and to study cleavage of HA and its inhibition. Vaccine. 2009;27:6324–9.

Bottcher E, Matrosovich T, Beyerle M, Klenk HD, Garten W, Matrosovich M. Proteolytic activation of influenza viruses by serine proteases TMPRSS2 and HAT from human airway epithelium. J Virol. 2006;80:9896–8.

Bottcher-Friebertshauser E, Garten W, Matrosovich M, Klenk HD. The hemagglutinin: a determinant of pathogenicity. Curr Top Microbiol Immunol. 2014;385:3–34.

Boycott R, Klenk HD, Ohuchi M. Cell tropism of influenza virus mediated by hemagglutinin activation at the stage of virus entry. Virology. 1994;203:313–9.

Bullough PA, Hughson FM, Skehel JJ, Wiley DC. Structure of influenza haemagglutinin at the pH of membrane fusion. Nature. 1994;371:37–43.

Byrd-Leotis L, Galloway SE, Agbogu E, Steinhauer DA. Influenza hemagglutinin (HA) stem region mutations that stabilize or destabilize the structure of multiple HA subtypes. J Virol. 2015;89:4504–16.

Calder LJ, Rosenthal PB. Cryomicroscopy provides structural snapshots of influenza virus membrane fusion. Nat Struct Mol Biol. 2016;23:853–8.

Came PE, Pascale A, Shimonaski G. Effect of pancreatin on plaque formation by influenza viruses. Arch Gesamte Virusforsch. 1968;23:346–52.

Carr CM, Chaudhry C, Kim PS. Influenza hemagglutinin is spring-loaded by a metastable native conformation. Proc Natl Acad Sci U S A. 1997;94:14306–13.

Chaipan C, Kobasa D, Bertram S, Glowacka I, Steffen I, Tsegaye TS, Takeda M, Bugge TH, Kim S, Park Y, Marzi A, Pohlmann S. Proteolytic activation of the 1918 influenza virus hemagglutinin. J Virol. 2009;83:3200–11.

Chen J, Lee KH, Steinhauer DA, Stevens DJ, Skehel JJ, Wiley DC. Structure of the hemagglutinin precursor cleavage site, a determinant of influenza pathogenicity and the origin of the labile conformation. Cell. 1998;95:409–17.

Chen J, Skehel JJ, Wiley DC. N- and C-terminal residues combine in the fusion-pH influenza hemagglutinin HA(2) subunit to form an N cap that terminates the triple-stranded coiled coil. Proc Natl Acad Sci U S A. 1999;96:8967–72.

Chen J, Wharton SA, Weissenhorn W, Calder LJ, Hughson FM, Skehel JJ, Wiley DC. A soluble domain of the membrane-anchoring chain of influenza virus hemagglutinin (HA2) folds in Escherichia coli into the low-pH-induced conformation. Proc Natl Acad Sci U S A. 1995;92:12205–9.

Cross KJ, Langley WA, Russell RJ, Skehel JJ, Steinhauer DA. Composition and functions of the influenza fusion peptide. Protein Pept Lett. 2009;16:766–78.

Cross KJ, Wharton SA, Skehel JJ, Wiley DC, Steinhauer DA. Studies on influenza haemagglutinin fusion peptide mutants generated by reverse genetics. EMBO J. 2001;20:4432–42.

Daniels RS, Downie JC, Hay AJ, Knossow M, Skehel JJ, Wang ML, Wiley DC. Fusion mutants of the influenza virus hemagglutinin glycoprotein. Cell. 1985;40:431–9.

Deshpande KL, Fried VA, Ando M, Webster RG. Glycosylation affects cleavage of an H5N2 influenza virus hemagglutinin and regulates virulence. Proc Natl Acad Sci U S A. 1987;84:36–40.

Dittmann M, Hoffmann HH, Scull MA, Gilmore RH, Bell KL, Ciancanelli M, Wilson SJ, Crotta S, Yu Y, Flatley B, Xiao JW, Casanova JL, Wack A, Bieniasz PD, Rice CM. A serpin shapes the extracellular environment to prevent influenza A virus maturation. Cell. 2015;160:631–43.

Doms RW, Gething MJ, Henneberry J, White J, Helenius A. Variant influenza virus hemagglutinin that induces fusion at elevated pH. J Virol. 1986;57:603–13.

Francis T, Moore AE. A Study of the Neurotropic Tendency in Strains of the Virus of Epidemic Influenza. J Exp Med. 1940;72:717–28.

Galloway SE, Reed ML, Russell CJ, Steinhauer DA. Influenza HA subtypes demonstrate divergent phenotypes for cleavage activation and pH of fusion: implications for host range and adaptation. PLoS Pathog. 2013;9:e1003151.

Garcia NK, Guttman M, Ebner JL, Lee KK. Dynamic changes during acid-induced activation of influenza hemagglutinin. Structure. 2015;23:665–76.

Garten W, Bosch FX, Linder D, Rott R, Klenk HD. Proteolytic activation of the influenza virus hemagglutinin: the structure of the cleavage site and the enzymes involved in cleavage. Virology. 1981;115:361–74.

Garten W, Klenk HD. Understanding influenza virus pathogenicity. Trends Microbiol. 1999;7:99–100.

Gething MJ, Doms RW, York D, White J. Studies on the mechanism of membrane fusion: site-specific mutagenesis of the hemagglutinin of influenza virus. J Cell Biol. 1986;102:11–23.

Goto H, Kawaoka Y. A novel mechanism for the acquisition of virulence by a human influenza A virus. Proc Natl Acad Sci U S A. 1998;95:10224–8.

Gotoh B, Ogasawara T, Toyoda T, Inocencio NM, Hamaguchi M, Nagai Y. An endoprotease homologous to the blood clotting factor X as a determinant of viral tropism in chick embryo. EMBO J. 1990;9:4189–95.

Hamilton BS, Gludish DW, Whittaker GR. Cleavage activation of the human-adapted influenza virus subtypes by matriptase reveals both subtype and strain specificities. J Virol. 2012;86:10579–86.

Hamilton BS, Whittaker GR. Cleavage activation of human-adapted influenza virus subtypes by kallikrein-related peptidases 5 and 12. J Biol Chem. 2013;288:17399–407.

Hatesuer B, Bertram S, Mehnert N, Bahgat MM, Nelson PS, Pohlmann S, Schughart K. Tmprss2 is essential for influenza H1N1 virus pathogenesis in mice. PLoS Pathog. 2013;9:e1003774.

Horimoto T, Nakayama K, Smeekens SP, Kawaoka Y. Proprotein-processing endoproteases PC6 and furin both activate hemagglutinin of virulent avian influenza viruses. J Virol. 1994;68:6074–8.

Huang RT, Wahn K, Klenk HD, Rott R. Fusion between cell membrane and liposomes containing the glycoproteins of influenza virus. Virology. 1980;104:294–302.

Huang RT, Rott R, Klenk HD. Influenza viruses cause hemolysis and fusion of cells. Virology. 1981;110:243–7.

Ivanovic T, Choi JL, Whelan SP, van Oijen AM, Harrison SC. Influenza-virus membrane fusion by cooperative fold-back of stochastically induced hemagglutinin intermediates. elife. 2013;2:e00333.

Kawaoka Y, Naeve CW, Webster RG. Is virulence of H5N2 influenza viruses in chickens associated with loss of carbohydrate from the hemagglutinin? Virology. 1984;139:303–16.
Kawaoka Y, Webster RG. Interplay between carbohydrate in the stalk and the length of the connecting peptide determines the cleavability of influenza virus hemagglutinin. J Virol. 1989;63:3296–300.
Keawcharoen J, Spronken MI, Vuong O, Bestebroer TM, Munster VJ, Osterhaus AD, Rimmelzwaan GF, Fouchier RA. Repository of Eurasian influenza A virus hemagglutinin and neuraminidase reverse genetics vectors and recombinant viruses. Vaccine. 2010;28:5803–9.
Khatchikian D, Orlich M, Rott R. Increased viral pathogenicity after insertion of a 28S ribosomal RNA sequence into the haemagglutinin gene of an influenza virus. Nature. 1989;340:156–7.
Kido H, Yokogoshi Y, Sakai K, Tashiro M, Kishino Y, Fukutomi A, Katunuma N. Isolation and characterization of a novel trypsin-like protease found in rat bronchiolar epithelial Clara cells. A possible activator of the viral fusion glycoprotein. J Biol Chem. 1992;267:13573–9.
Kim TS, Heinlein C, Hackman RC, Nelson PS. Phenotypic analysis of mice lacking the Tmprss2-encoded protease. Mol Cell Biol. 2006;26:965–75.
Klenk HD, Garten W. Host cell proteases controlling virus pathogenicity. Trends Microbiol. 1994;2:39–43.
Klenk HD, Garten W, Rott R. Inhibition of proteolytic cleavage of the hemagglutinin of influenza virus by the calcium-specific ionophore A23187. EMBO J. 1984;3:2911–5.
Klenk HD, Rott R, Orlich M, Blodorn J. Activation of influenza A viruses by trypsin treatment. Virology. 1975;68:426–39.
Korte T, Epand RF, Epand RM, Blumenthal R. Role of the Glu residues of the influenza hemagglutinin fusion peptide in the pH dependence of fusion activity. Virology. 2001;289:353–61.
Kuhn N, Bergmann S, Kosterke N, Lambertz RL, Keppner A, van den Brand JM, Pohlmann S, Weiss S, Hummler E, Hatesuer B, Schughart K. The proteolytic activation of (H3N2) influenza A virus hemagglutinin is facilitated by different type II transmembrane serine proteases. J Virol. 2016;90:4298–307.
Langley WA, Thoennes S, Bradley KC, Galloway SE, Talekar GR, Cummings SF, Vareckova E, Russell RJ, Steinhauer DA. Single residue deletions along the length of the influenza HA fusion peptide lead to inhibition of membrane fusion function. Virology. 2009;394:321–30.
Lazarowitz SG, Choppin PW. Enhancement of the infectivity of influenza A and B viruses by proteolytic cleavage of the hemagglutinin polypeptide. Virology. 1975;68:440–54.
Lazarowitz SG, Compans RW, Choppin PW. Proteolytic cleavage of the hemagglutinin polypeptide of influenza virus. Function of the uncleaved polypeptide HA. Virology. 1973a;52:199–212.
Lazarowitz SG, Goldberg AR, Choppin PW. Proteolytic cleavage by plasmin of the HA polypeptide of influenza virus: host cell activation of serum plasminogen. Virology. 1973b;56:172–80.
Li S, Schulman J, Itamura S, Palese P. Glycosylation of neuraminidase determines the neurovirulence of influenza A/WSN/33 virus. J Virol. 1993;67:6667–73.
Lin YP, Wharton SA, Martin J, Skehel JJ, Wiley DC, Steinhauer DA. Adaptation of egg-grown and transfectant influenza viruses for growth in mammalian cells: selection of hemagglutinin mutants with elevated pH of membrane fusion. Virology. 1997;233:402–10.
Lu X, Shi Y, Gao F, Xiao H, Wang M, Qi J, Gao GF. Insights into avian influenza virus pathogenicity: the hemagglutinin precursor HA0 of subtype H16 has an alpha-helix structure in its cleavage site with inefficient HA1/HA2 cleavage. J Virol. 2012;86:12861–70.
Maeda H. Role of microbial proteases in pathogenesis. Microbiol Immunol. 1996;40:685–99.
Maeda T, Ohnishi S. Activation of influenza virus by acidic media causes hemolysis and fusion of erythrocytes. FEBS Lett. 1980;122:283–7.
Maurer-Stroh S, Lee RT, Gunalan V, Eisenhaber F. The highly pathogenic H7N3 avian influenza strain from July 2012 in Mexico acquired an extended cleavage site through recombination with host 28S rRNA. Virol J. 2013;10:139.
Munster VJ, Baas C, Lexmond P, Waldenstrom J, Wallensten A, Fransson T, Rimmelzwaan GF, Beyer WE, Schutten M, Olsen B, Osterhaus AD, Fouchier RA. Spatial, temporal, and species variation in prevalence of influenza A viruses in wild migratory birds. PLoS Pathog. 2007;3:e61.

Nobusawa E, Aoyama T, Kato H, Suzuki Y, Tateno Y, Nakajima K. Comparison of complete amino acid sequences and receptor-binding properties among 13 serotypes of hemagglutinins of influenza A viruses. Virology. 1991;182:475–85.

Nobusawa E, Hishida R, Murata M, Kawasaki K, Ohnishi S, Nakajima K. The role of acidic residues in the "fusion segment" of influenza A virus hemagglutinin in low-pH-dependent membrane fusion. Arch Virol. 1995;140:865–75.

Ohuchi R, Ohuchi M, Garten W, Klenk HD. Human influenza virus hemagglutinin with high sensitivity to proteolytic activation. J Virol. 1991;65:3530–7.

Okumura Y, Takahashi E, Yano M, Ohuchi M, Daidoji T, Nakaya T, Bottcher E, Garten W, Klenk HD, Kido H. Novel type II transmembrane serine proteases, MSPL and TMPRSS13, proteolytically activate membrane fusion activity of the hemagglutinin of highly pathogenic avian influenza viruses and induce their multicycle replication. J Virol. 2010;84:5089–96.

Orlich M, Gottwald H, Rott R. Nonhomologous recombination between the hemagglutinin gene and the nucleoprotein gene of an influenza virus. Virology. 1994;204:462–5.

Orlich M, Rott R. Thermolysin activation mutants with changes in the fusogenic region of an influenza virus hemagglutinin. J Virol. 1994;68:7537–9.

Pasick J, Handel K, Robinson J, Copps J, Ridd D, Hills K, Kehler H, Cottam-Birt C, Neufeld J, Berhane Y, Czub S. Intersegmental recombination between the haemagglutinin and matrix genes was responsible for the emergence of a highly pathogenic H7N3 avian influenza virus in British Columbia. J Gen Virol. 2005;86:727–31.

Perdue ML, Garcia M, Senne D, Fraire M. Virulence-associated sequence duplication at the hemagglutinin cleavage site of avian influenza viruses. Virus Res. 1997;49:173–86.

Qiao H, Armstrong RT, Melikyan GB, Cohen FS, White JM. A specific point mutant at position 1 of the influenza hemagglutinin fusion peptide displays a hemifusion phenotype. Mol Biol Cell. 1999;10:2759–69.

Ruigrok RW, Aitken A, Calder LJ, Martin SR, Skehel JJ, Wharton SA, Weis W, Wiley DC. Studies on the structure of the influenza virus haemagglutinin at the pH of membrane fusion. J Gen Virol. 1988;69(Pt 11):2785–95.

Russell RJ, Gamblin SJ, Haire LF, Stevens DJ, Xiao B, Ha Y, Skehel JJ. H1 and H7 influenza haemagglutinin structures extend a structural classification of haemagglutinin subtypes. Virology. 2004;325:287–96.

Sakai K, Ami Y, Nakajima N, Nakajima K, Kitazawa M, Anraku M, Takayama I, Sangsriratanakul N, Komura M, Sato Y, Asanuma H, Takashita E, Komase K, Takehara K, Tashiro M, Hasegawa H, Odagiri T, Takeda M. TMPRSS2 Independency for Haemagglutinin Cleavage In Vivo Differentiates Influenza B Virus from Influenza A Virus. Sci Rep. 2016;6:29430.

Sakai K, Ami Y, Tahara M, Kubota T, Anraku M, Abe M, Nakajima N, Sekizuka T, Shirato K, Suzaki Y, Ainai A, Nakatsu Y, Kanou K, Nakamura K, Suzuki T, Komase K, Nobusawa E, Maenaka K, Kuroda M, Hasegawa H, Kawaoka Y, Tashiro M, Takeda M. The host protease TMPRSS2 plays a major role in in vivo replication of emerging H7N9 and seasonal influenza viruses. J Virol. 2014;88:5608–16.

Sales KU, Hobson JP, Wagenaar-Miller R, Szabo R, Rasmussen AL, Bey A, Shah MF, Molinolo AA, Bugge TH. Expression and genetic loss of function analysis of the HAT/DESC cluster proteases TMPRSS11A and HAT. PLoS One. 2011;6:e23261.

Scheiblauer H, Reinacher M, Tashiro M, Rott R. Interactions between bacteria and influenza A virus in the development of influenza pneumonia. J Infect Dis. 1992;166:783–91.

Shope RE. The etiology of swine influenza. Science. 1931;73:214–5.

Skehel JJ, Waterfield MD. Studies on the primary structure of the influenza virus hemagglutinin. Proc Natl Acad Sci U S A. 1975;72:93–7.

Slusher MJ, Wilcox BR, Lutrell MP, Poulson RL, Brown JD, Yabsley MJ, Stallknecht DE. Are passerine birds reservoirs for influenza A viruses? J Wildl Dis. 2014;50:792–809.

Smith W, Andrewes CH, Laidlaw PP. A virus obtained from influenza patients. Lancet. 1933;2:66–8.

Steinhauer DA. Role of hemagglutinin cleavage for the pathogenicity of influenza virus. Virology. 1999;258:1–20.
Steinhauer DA, Martin J, Lin YP, Wharton SA, Oldstone MB, Skehel JJ, Wiley DC. Studies using double mutants of the conformational transitions in influenza hemagglutinin required for its membrane fusion activity. Proc Natl Acad Sci U S A. 1996;93:12873–8.
Steinhauer DA, Wharton SA, Skehel JJ, Wiley DC. Studies of the membrane fusion activities of fusion peptide mutants of influenza virus hemagglutinin. J Virol. 1995;69:6643–51.
Steinhauer DA, Wharton SA, Skehel JJ, Wiley DC, Hay AJ. Amantadine selection of a mutant influenza virus containing an acid-stable hemagglutinin glycoprotein: evidence for virus-specific regulation of the pH of glycoprotein transport vesicles. Proc Natl Acad Sci U S A. 1991;88:11525–9.
Stevens J, Corper AL, Basler CF, Taubenberger JK, Palese P, Wilson IA. Structure of the uncleaved human H1 hemagglutinin from the extinct 1918 influenza virus. Science. 2004;303:1866–70.
Stieneke-Grober A, Vey M, Angliker H, Shaw E, Thomas G, Roberts C, Klenk HD, Garten W. Influenza virus hemagglutinin with multibasic cleavage site is activated by furin, a subtilisin-like endoprotease. EMBO J. 1992;11:2407–14.
Suarez DL, Senne DA, Banks J, Brown IH, Essen SC, Lee CW, Manvell RJ, Mathieu-Benson C, Moreno V, Pedersen JC, Panigrahy B, Rojas H, Spackman E, Alexander DJ. Recombination resulting in virulence shift in avian influenza outbreak, Chile. Emerg Infect Dis. 2004;10:693–9.
Sun X, Tse LV, Ferguson AD, Whittaker GR. Modifications to the hemagglutinin cleavage site control the virulence of a neurotropic H1N1 influenza virus. J Virol. 2010;84:8683–90.
Tarnow C, Engels G, Arendt A, Schwalm F, Sediri H, Preuss A, Nelson PS, Garten W, Klenk HD, Gabriel G, Bottcher-Friebertshauser E. TMPRSS2 is a host factor that is essential for pneumotropism and pathogenicity of H7N9 influenza A virus in mice. J Virol. 2014;88:4744–51.
Tashiro M, Ciborowski P, Klenk HD, Pulverer G, Rott R. Role of Staphylococcus protease in the development of influenza pneumonia. Nature. 1987;325(6104):536–7.
Thoennes S, Li ZN, Lee BJ, Langley WA, Skehel JJ, Russell RJ, Steinhauer DA. Analysis of residues near the fusion peptide in the influenza hemagglutinin structure for roles in triggering membrane fusion. Virology. 2008;370:403–14.
Tobita K, Kilbourne ED. Genetic recombination for antigenic markers of antigenically different strains of influenza B virus. J Virol. 1974;13:347–52.
Tse LV, Marcano VC, Huang W, Pocwierz MS, Whittaker GR. Plasmin-mediated activation of pandemic H1N1 influenza virus hemagglutinin is independent of the viral neuraminidase. J Virol. 2013;87:5161–9.
Veits J, Weber S, Stech O, Breithaupt A, Graber M, Gohrbandt S, Bogs J, Hundt J, Teifke JP, Mettenleiter TC, Stech J. Avian influenza virus hemagglutinins H2, H4, H8, and H14 support a highly pathogenic phenotype. Proc Natl Acad Sci U S A. 2012;109:2579–84.
Webster RG, Bean WJ, Gorman OT, Chambers TM, Kawaoka Y. Evolution and ecology of influenza A viruses. Microbiol Rev. 1992;56:152–79.
White J, Helenius A. pH-dependent fusion between the Semliki Forest virus membrane and liposomes. Proc Natl Acad Sci U S A. 1980;77:3273–7.
White JM, Wilson IA. Anti-peptide antibodies detect steps in a protein conformational change: low-pH activation of the influenza virus hemagglutinin. J Cell Biol. 1987;105:2887–96.
Wilson IA, Skehel JJ, Wiley DC. Structure of the haemagglutinin membrane glycoprotein of influenza virus at 3 A resolution. Nature. 1981;289:366–73.
World Health Organization. (2018). Influenza at the human-animal interface [PDF file]. Retrieved from http://www.who.int/influenza/human_animal_interface/Influenza_Summary_IRA_HA_interface_25_01_2018_FINAL.pdf?ua=1
Xu R, Wilson IA. Structural characterization of an early fusion intermediate of influenza virus hemagglutinin. J Virol. 2011;85:5172–82.
Xu Y, Ramey AM, Bowman AS, DeLiberto TJ, Killian ML, Krauss S, Nolting JM, Torchetti MK, Reeves AB, Webby RJ, Stallknecht DE, Wan XF. Low-pathogenic influenza A viruses in North

American Diving Ducks contribute to the emergence of a novel highly pathogenic influenza A(H7N8) virus. J Virol. 2017;91:pii: e02208-16.
Yewdell JW, Taylor A, Yellen A, Caton A, Gerhard W, Bachi T. Mutations in or near the fusion peptide of the influenza virus hemagglutinin affect an antigenic site in the globular region. J Virol. 1993;67:933–42.
Zaraket H, Bridges OA, Duan S, Baranovich T, Yoon SW, Reed ML, Salomon R, Webby RJ, Webster RG, Russell CJ. Increased acid stability of the hemagglutinin protein enhances H5N1 influenza virus growth in the upper respiratory tract but is insufficient for transmission in ferrets. J Virol. 2013a;87:9911–22.
Zaraket H, Bridges OA, Russell CJ. The pH of activation of the hemagglutinin protein regulates H5N1 influenza virus replication and pathogenesis in mice. J Virol. 2013b;87:4826–34.
Zhang F, Bi Y, Wang J, Wong G, Shi W, Hu F, Yang Y, Yang L, Deng X, Jiang S, He X, Liu Y, Yin C, Zhong N, Gao GF. Human infections with recently-emerging highly pathogenic H7N9 avian influenza virus in China. J Infect. 2017;75:71–5.
Zhirnov OP, Ikizler MR, Wright PF. Cleavage of influenza a virus hemagglutinin in human respiratory epithelium is cell associated and sensitive to exogenous antiproteases. J Virol. 2002;76:8682–9.
Zhu W, Zhou J, Li Z, Yang L, Li X, Huang W, Zou S, Chen W, Wei H, Tang J, Liu L, Dong J, Wang D, Shu Y. Biological characterisation of the emerged highly pathogenic avian influenza (HPAI) A(H7N9) viruses in humans, in mainland China, 2016 to 2017. Euro Surveill. 2017;22(19):pii: 30533.

Proteolytic Activation of Paramyxoviruses and Pneumoviruses

2

Everett Clinton Smith and Rebecca Ellis Dutch

Abstract

Viruses in the families *Paramyxoviridae* and *Pneumoviridae* infect multiple animal species, and infection can result in varying disease severity. Membrane fusion is an obligate early step during infection and is driven primarily by viral fusion (F) proteins present on the viral envelope. F-mediated membrane fusion begins with insertion of a hydrophobic fusion peptide into the cell membrane and, through a series of conformational changes, culminates in the merger of both the viral and cellular membranes. Proteolytic processing N-terminal to the fusion peptide enables insertion into the cellular membrane, making this cleavage event an essential step in F-promoted membrane fusion. While all F proteins are cleaved by host proteases, the protease utilized and location of F cleavage vary widely among paramyxo- and pneumoviruses. With some paramyxoviruses, proteolytic activation of the hemagglutinin-neuraminidase (HN) glycoprotein has also been observed involving removal of a C-terminal extension from a precursor that blocks the attachment function of this protein. The availability of protein structures and extensive studies on the spatial and temporal processing details have illuminated many important aspects of proteolytic activation of these proteins. However, why such disparate proteolytic cleavage pathways evolved and to what extent they affect pathogenesis are less well understood.

E. C. Smith
Department of Biology, University of the South, Sewanee, TN, USA

R. E. Dutch (✉)
Department of Molecular and Cellular Biochemistry, University of Kentucky, Lexington, KY, USA
e-mail: rebecca.dutch@uky.edu

E. Böttcher-Friebertshäuser et al. (eds.), *Activation of Viruses by Host Proteases*,
https://doi.org/10.1007/978-3-319-75474-1_2

2.1 Paramyxoviruses and Pneumoviruses

The Paramyxovirus and Pneumovirus families include significant human and animal pathogens, along with model viruses which have been used for decades to dissect viral infection processes in laboratory studies (Lamb and Parks 2013). The paramyxoviruses include measles virus (MeV) and mumps virus (MuV), both pathogens which are effectively targeted by vaccines but which have seen a worldwide resurgence, primarily due to low vaccination rates (Phadke et al. 2016). This family also contains the human parainfluenza viruses (HPIV 1–4), which are significant causes of respiratory tract infection, and the zoonotic Hendra and Nipah viruses, categorized as BSL-4 agents due to their high mortality rates and lack of available vaccines or approved antivirals. Paramyxoviruses such as parainfluenza virus 5 (PIV5) and Sendai virus (SeV), while not causing disease in humans, have been extensively used in research. The pneumoviruses, which until 2016 were classified as a subfamily within the paramyxovirus family, include human respiratory syncytial virus (HRSV), a leading cause of hospitalization due to respiratory infection for young children, and human metapneumovirus (HMPV), a respiratory pathogen identified in 2001 (van den Hoogen et al. 2001) that is a major cause of respiratory tract disease worldwide.

The family *Paramyxoviridae* includes seven genera: *Aquaparamyxovirus*, *Avulavirus*, *Ferlavirus*, *Henipavirus*, *Morbillivirus*, *Respirovirus*, and *Rubulavirus*. All members of the family have non-segmented negative-sense RNA genomes which encode six to ten genes critical for virus infection. Despite this variation, a common set of proteins is expressed by every paramyxovirus. These include the fusion (F) protein, responsible for promoting membrane fusion between the viral envelope and a cellular target membrane; the attachment (HN/H/G) protein, required for binding to the host cell receptor; the matrix (M) protein, which is critical for virus assembly; the polymerase (L) protein, which carries out RNA-dependent RNA polymerization to generate mRNA, anti-genomic RNA, and genomic RNA; the nucleocapsid (N) protein, which coats the viral genome; and the phosphoprotein (P), which serves as a polymerase cofactor. The paramyxovirus P gene can also express additional proteins through RNA editing or initiation of translation at alternative sites, though the number of additional proteins expressed varies between viruses (Lamb and Parks 2013). Several paramyxoviruses or pneumoviruses also encode a third glycoprotein, the small hydrophobic (SH) protein, which is dispensable for viral replication but could function as a viroporin (Masante et al. 2014) and is important for pathogenesis in vivo for some viruses (Bukreyev et al. 1997; Li et al. 2011; Ling et al. 2008; Whitehead et al. 1999).

Viruses in the family *Pneumoviridae* were previously classified within the *Paramyxoviridae*. However, a new family was created in 2016, due to the significant differences in nucleocapsid structure, the separation indicated by phylogenetic relationships, and the presence of a unique M2 gene in viruses within the *Pneumoviridae* (Afonso et al. 2016). There are two genera within this family, the *Metapneumovirus* and the *Orthopneumovirus*. Many proteins show significant conservation in sequence and function between the paramyxoviruses and pneumoviruses, including the F protein, which will be the focus of this chapter. However, less homology is seen in proteins such as SH and G across the families, and the unique M2 gene encodes several proteins critical for pneumovirus replication that are not found in the paramyxoviruses.

2.2 The Role of Protease Cleavage in Fusion Protein Activation

Paramyxovirus and pneumovirus particles contain a lipid envelope derived from the host cell membranes where virus assembly occurred. In order for viral infection to initiate, the viral particle must bind to the host cell, and then membrane fusion must occur between the lipid bilayer of the virus and a target membrane of the host cell. This fusion event can occur on the plasma membrane or after internalization via endocytosis, depending on the virus, and fusion is essential for release of the viral genome into the cellular cytosol (Chang and Dutch 2012). The process of membrane fusion for both virus families is promoted by the viral F protein, with the attachment protein involved in triggering F protein function for all paramyxoviruses and some pneumoviruses. In all cases, the function of the F protein is completely dependent on proteolytic processing by host proteases, generating a disulfide-linked $F_1 + F_2$ heterodimer from the uncleaved F_0 precursor.

Membrane fusion is an energetically unfavorable process. Paramyxovirus and pneumovirus F proteins, in a similar manner to other viral fusion proteins such as the influenza HA protein, do not utilize high-energy cofactors to drive this process. Instead, fusion involves a set of dramatic conformational changes in the F protein, and the energetic difference between the metastable pre-fusion state of the F protein and the post-fusion state is thought to provide the energy needed for fusion (Fig. 2.1). F proteins are initially synthesized in the endoplasmic reticulum, where they are co-translationally N-link glycosylated and rapidly folded into trimers. F trimers are then trafficked through the secretory pathway to the plasma membrane, with some F proteins then undergoing endocytic recycling. This pre-fusion F protein trimer is incapable of promoting membrane fusion, however, until it is proteolytically processed by a host protease into a trimer of disulfide-linked heterodimers. Proteolytic cleavage occurs immediately N-terminal to the fusion peptide region of the F protein, a highly hydrophobic region which is capable of inserting into membranes (Fig. 2.1, step c). Cleavage therefore frees the N-terminus of the fusion peptide, and this free domain is crucial for the F protein-promoted fusion process.

For most F proteins, interaction with the attachment protein is sufficient to trigger the conformational changes needed for membrane fusion (Bose et al. 2015; Chang and Dutch 2012). However, a subset of HMPV F proteins are triggered by low pH (Herfst et al. 2008; Schowalter et al. 2006b), while the factors driving the non-low pH triggered HMPV F proteins are still unknown. Early in the cascade of conformational changes, the heptad repeat A regions in the globular head region refold into an extended helical coiled coil (Fig. 2.1, step c). This refolding event propels the fusion peptide toward the target membrane. Insertion of the fusion peptide into

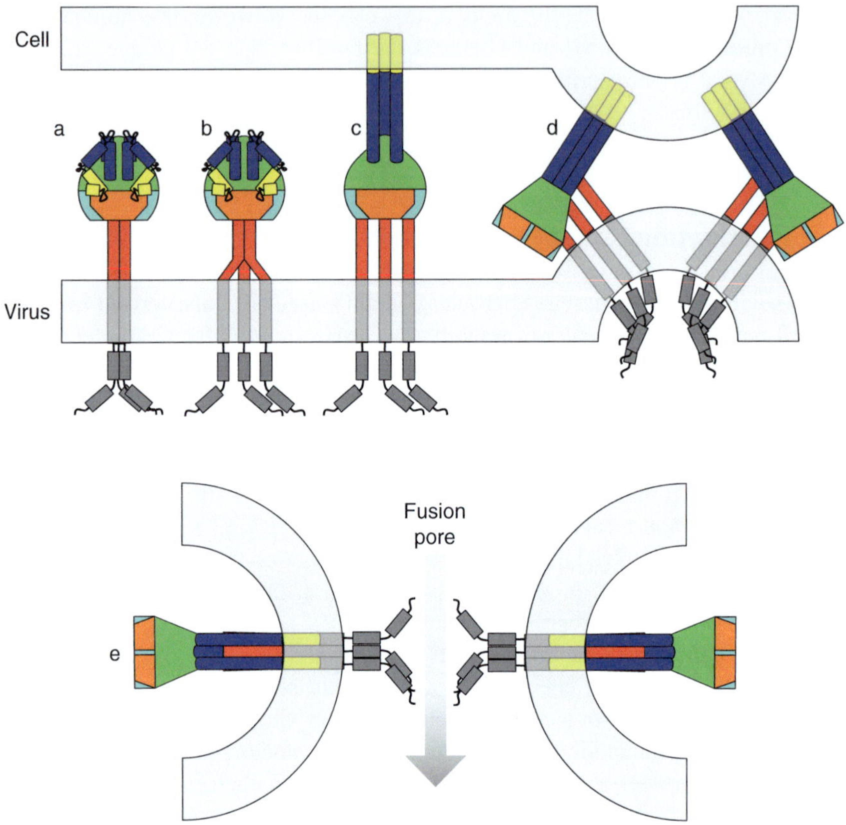

Fig. 2.1 Paramyxovirus and pneumovirus F-promoted membrane fusion. F proteins exist in a metastable, pre-fusion state (step a). Triggering of F results in transmembrane dissociation (TM; step b) followed by refolding of heptad repeat A (HRA; blue) and insertion of the fusion peptide (FP; yellow) into the target cell membrane (step c). Refolding of F brings both the viral and target cell membrane into proximity (step d), and formation of a fusion pore is linked to the presence of a stable six-helix bundle composed of HRA and heptad repeat B (HRB; red; step e)

the target membrane results in an intermediate where the F protein connects the two membranes, a structure that has been visualized using electron microscopy (Kim et al. 2011). Further refolding leads to the formation of a six-helix bundle structure formed by the heptad repeat A coiled coil interacting with the three heptad repeat B helices (Fig. 2.1, step e). Formation of this highly stable bundle structure occurs coincident with the initiation of membrane merger.

Structural studies over the last decade have defined the ectodomain conformation of several F protein forms. The uncleaved pre-fusion structure has been solved for PIV5 F (Yin et al. 2006), Hendra F (Wong et al. 2016; Xu et al. 2015), and RSV F (McLellan et al. 2013) and consists of a globular head domain with a helical stalk, with both the head and stalk regions containing extensive contacts between the three monomers that make up the F protein trimer (Fig. 2.2, left). The site for proteolytic processing is present on an exposed loop N-terminal to the fusion peptide. The structure of the cleaved, pre-fusion form of PIV5 F has also been solved (Welch

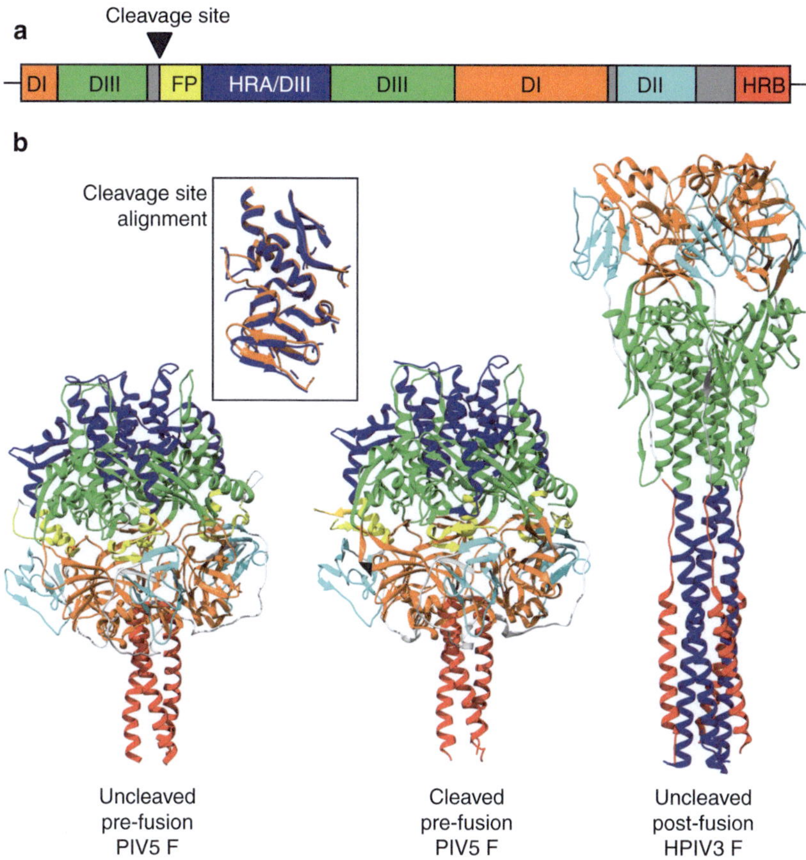

Fig. 2.2 Pre- and post-fusion structures of paramyxovirus F proteins. (**a**) A linear schematic of a general paramyxovirus F without the transmembrane domain (TM) or cytoplasmic tail is shown. The single cleavage site (black triangle) is N-terminal to the FP. Regions in gray do not correspond to specific structural domains. (**b**) The pre-fusion structure of PIV5 F (Yin et al. 2006) is relatively compact (PDB ID: 2B9B), enabling the fusion peptide (FP; yellow) to be partially buried within the globular head domain. Proteolytic cleavage of the PIV5 F_0 form into F_1 + F_2 ((Welch et al. 2012); PDB ID: 4GIP) does not result in significant structural rearrangement, except near the cleavage site (see inset; blue is cleaved F_1 + F_2 and orange is uncleaved F_0). Following membrane fusion, the post-fusion structure of uncleaved HPIV3 ((Yin et al. 2005); PDB ID: 1ZTM) is elongated and stabilized by the formation of a six-helix bundle composed of heptad repeats A and B (HRA and HRB). Both the pre- and post-fusion structures are membrane-anchored by transmembrane domains (TM; not shown) and contain cytoplasmic tails. Colors: HRA, blue; HRB, red; FP, yellow; domain I, orange; domain II, cyan; domain III, green. Molecular graphics images were produced using the UCSF Chimera package from the Computer Graphics Laboratory, University of California, San Francisco (Pettersen et al. 2004)

et al. 2012). Interestingly, the structures of the uncleaved and cleaved PIV5 F proteins are superimposable, with changes observed only in the region near the cleavage site (Fig. 2.2, inset). These results suggest that proteolytic processing itself does not result in significant conformational changes in the F proteins. Previous work analyzing PIV5 F recognition by anti-peptide polyclonal antibodies (Dutch et al. 2001) and a monoclonal antibody (Dutch et al. 2001; Paterson et al. 2000) had indicated more dramatic changes following proteolytic processing, but the differences in antibody recognition may instead reflect the high propensity for cleaved F to trigger to the post-fusion form in the presence of detergents or on biological membranes. Post-fusion forms of a number of F proteins have been determined (McLellan et al. 2011; Swanson et al. 2010; Yin et al. 2005; Zhao et al. 2000) (Fig. 2.2, right). Dramatic rearrangements between the pre- and post-fusion forms are clearly evident, including the formation of a six-helix bundle formed from the heptad repeat A (HRA, blue) and heptad repeat B (HRB, red) regions of the protein. Interestingly, post-fusion F structures were obtained in the absence of proteolytic processing, suggesting that the conformational changes needed to form a six-helix bundle can occur without proteolytic processing when the transmembrane domain and cytoplasmic tail region of the protein are absent. Indeed, studies of a soluble form of the PIV5 F protein confirmed that uncleaved F protein lacking its transmembrane domain region, when exposed to heat, was capable of undergoing conformational changes comparable to those of the full-length cleaved membrane-bound form, though subtle differences between the two were observed (Connolly et al. 2006). However, the uncleaved form was unable to stably associate with liposomes after heat-induced conformational changes, supporting a requirement for cleavage for fusion peptide insertion into a target membrane (Connolly et al. 2006). These studies also demonstrated that cleavage alone was not sufficient to induce conformational changes, as no changes were observed in the uncleaved soluble F protein unless it was exposed to heat (Connolly et al. 2006).

2.3 Proteases Involved in Paramyxovirus F Protein Proteolytic Cleavage

2.3.1 Furin and Its Role in F Protein Cleavage

Many paramyxovirus F proteins contain an R-X-K/R-R sequence immediately N-terminal to their cleavage site, a sequence efficiently recognized by the host cell protease furin. Furin is a ubiquitous secretory pathway protease that belongs to the mammalian subtilisin-like pro-protein convertase family (Klenk and Garten 1994). Furin is a calcium-dependent protease that functions at a wide range of pH values and is known to cleave a large number of precursor proteins within the exocytic or endocytic pathways. Paramyxovirus F proteins containing the furin consensus sequence include those from measles virus (Richardson et al. 1986), mumps virus (Waxham et al. 1987), HPIV3 (Spriggs et al. 1986), PIV5 (Paterson et al. 1984), canine distemper virus (Barrett et al. 1987), reptilian paramyxoviruses (Franke et al.

2006), and some strains of Newcastle disease virus (NDV) (Toyoda et al. 1987). Furin cleavage occurs in the trans-Golgi network as the F protein is trafficked through the secretory pathway (Fig. 2.3, pathway A). The highly efficient cleavage process and the intracellular localization of furin mean that the great majority of F proteins that are processed by furin reach the cell surface in the mature, cleaved $F_1 + F_2$ form. This results in the budding of viral particles containing F proteins predominantly in the fusion-active cleaved form. The presence of fusogenically active F proteins on the cell surface can also lead to efficient cell-cell fusion, termed syncytia formation, between infected cells and neighboring cells.

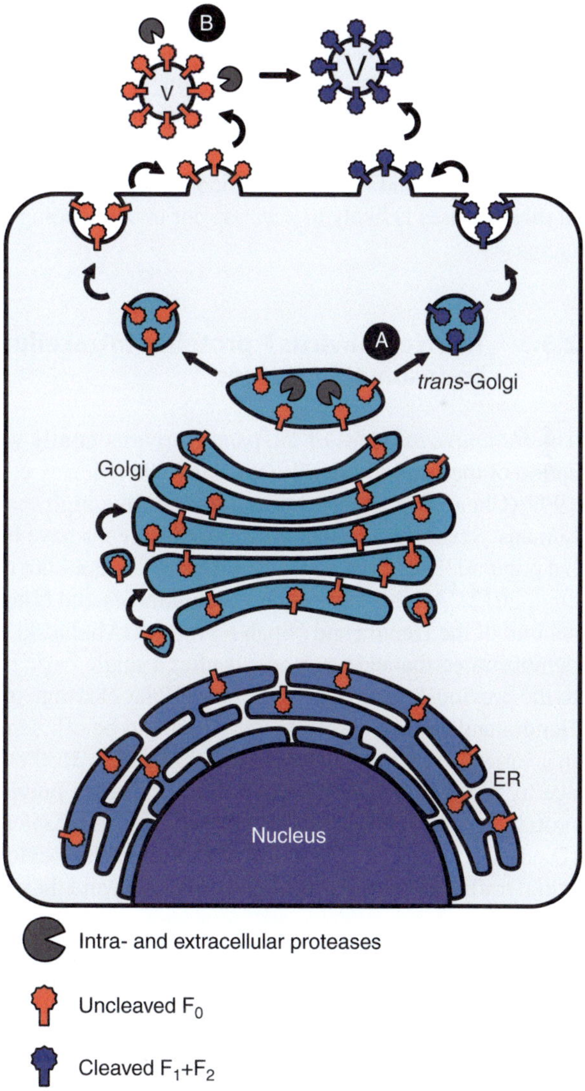

Fig. 2.3 Cleavage of F proteins by intra- and extracellular proteases. (**a**) Most paramyxovirus F proteins are cleaved within the *trans*-Golgi network by the subtilisin-like protease, furin. In this case, the cleaved $F_1 + F_2$ form is incorporated directly into newly formed virus particles. (**b**) Following translation of F into the endoplasmic reticulum (ER), uncleaved (F_0; red) HMPV, HPIV1, or Sendai virus F is incorporated into newly formed virus particles. Extracellular proteases, such as TMPRSS2 and mini-plasmin, then cleave F_0 into the metastable, fusion-competent $F_1 + F_2$ form (blue)

2.3.2 Extracellular Cleavage of Paramyxovirus F Proteins with Monobasic Cleavage Sites

A small number of paramyxoviruses have F proteins containing only a monobasic residue at their cleavage site, and these F proteins are not processed in the exocytic pathway by furin. The F proteins of these viruses therefore arrive at the plasma membrane uncleaved and fusion inactive. F proteins from viruses such as Sendai (Hidaka et al. 1984), HPIV1 (Ambrose et al. 1995), and some strains of Newcastle disease virus (Toyoda et al. 1987) are not processed by any intracellular protease but instead are cleaved by extracellular enzymes present in the respiratory tract (Fig. 2.3, pathway B). Consequently, the infectivity of these viruses is dependent on efficient processing by these extracellular proteases. Several enzymes have been identified that can cleave the F proteins of these viruses, including a secreted trypsin-like protease from Clara cells (Kido et al. 1992), an endoprotease homologous to blood clotting factor X from chick embryo (Gotoh et al. 1990, 1992), and mini-plasmin (Murakami et al. 2001). However, the role of each of these proteases during infection is unclear, and the identification of additional proteases involved in activation of these viruses is likely to occur as our understanding of respiratory tract proteases increases.

2.3.3 The Henipavirus F proteins: Intracellular Cleavage by Endosomal Proteases

The *Henipavirus* genus of the paramyxovirus family was established after identification of the zoonotic Hendra virus in 1994 (Murray et al. 1995) and Nipah virus in 1999 (Chua et al. 2000), both of which give high mortality rates in animals and humans. Several additional viruses in this genus have been identified by analysis of bat populations, including Cedar virus, which does not cause clinical disease (Marsh et al. 2012). Amino acid sequence alignments and N-terminal sequencing of the F_1 subunit of the Hendra and Nipah F proteins (Michalski et al. 2000; Moll et al. 2004) demonstrated that cleavage occurs after a single basic residue. However, in contrast to the previously characterized extracellular cleavage of Sendai F and HPIV1 F, the Hendra and Nipah F proteins were found to be efficiently intracellularly processed in a variety of cell types (Michalski et al. 2000; Moll et al. 2004), suggesting cleavage by a ubiquitous cellular protease. Unlike all previously examined viral fusion proteins, proteolytic processing of both Hendra (Meulendyke et al. 2005) and Nipah (Diederich et al. 2005) was found to require endocytosis of the protein following initial trafficking to the cell surface (Fig. 2.4), and the activating protease was shown to be the endosomal/lysosomal protease cathepsin L (Pager et al. 2006; Pager and Dutch 2005), with cathepsin B also capable of cleaving the F protein in some cell types (Diederich et al. 2012). Cathepsin proteases are involved in the degradation of proteins in the endosome/lysosome, digest a wide variety of substrates, and do not have defined consensus cleavage sequences (Turk and Guncar 2003). Fitting with this, mutation of the cleavage site residues of Hendra F (Craft and Dutch 2005) and

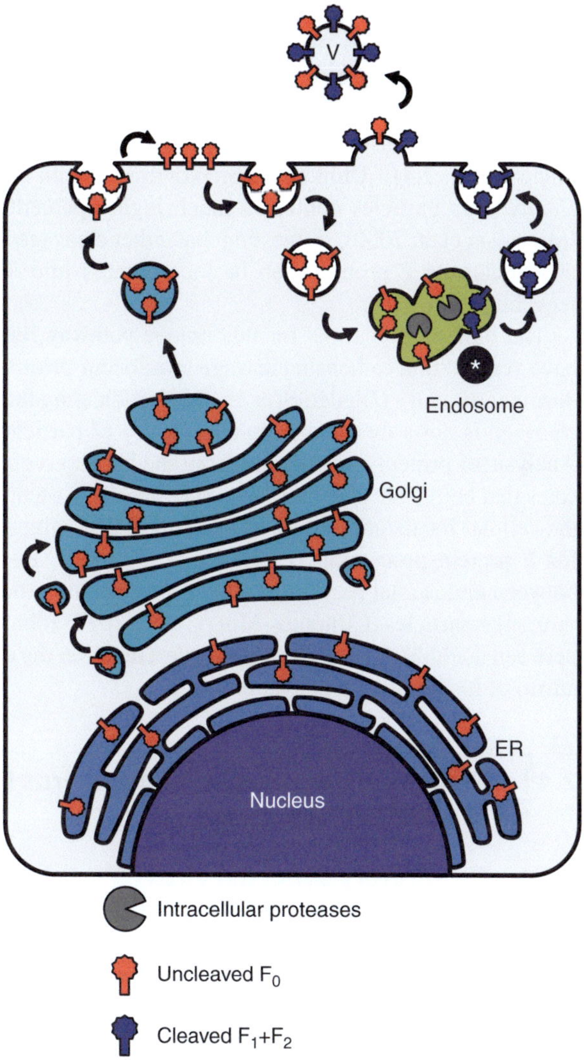

Fig. 2.4 Cleavage of *Henipavirus* F. Following translation of F into the endoplasmic reticulum (ER), uncleaved F (F_0; red) is trafficked to the cell surface. The F_0 form is then endocytosed and cleaved in an endosomal compartment by cathepsin L or B (*). The cleaved, metastable $F_1 + F_2$ form (blue) is then trafficked back to the cell surface and incorporated into newly formed virions along with uncleaved F_0 (red)

Nipah F (Moll et al. 2004) did not ablate proteolytic processing unless the residues corresponding to the cleavage loop were deleted (Moll et al. 2004).

The intriguing question of how a degradative protease can facilitate a specific cleavage event may be partially explained by the finding that Hendra F and Nipah F protein cleavage appears to occur in the recycling endosomal compartments (Diederich et al. 2012; Popa et al. 2012) rather than in the late endosomes/lysosomes which contain the majority of active cathepsins. Proteolysis by cathepsins has been proposed to be affected by both protease concentration and pH (Authier et al. 1996; Pillay et al. 2002), and cathepsin L activity has been shown to be more specific within the higher pH conditions of the early endosomal pathway (Jordans et al. 2009). Thus, the novel mechanism of proteolytic processing for these *Henipavirus* F proteins involves synthesis and transport of the uncleaved F protein to the plasma membrane, endocytosis to bring the F protein into contact with cathepsin L (or B) in the recycling endosomes, and cleavage of the F protein followed by recycling of the fusogenically active cleaved F protein back to the plasma membrane (Fig. 2.4). Unlike paramyxoviruses with furin-processed F proteins, *Henipavirus* particles contain a much higher percentage of uncleaved F protein (Michalski et al. 2000), suggesting that either cleavage is not completely efficient or that uncleaved F proteins can be incorporated into a particle prior to endocytic recycling.

The biological reasons for this unique pathway remain to be clarified. Studies have verified that cathepsin cleavage must occur prior to virus assembly rather than during virus entry (Diederich et al. 2008), indicating that the utilization of cathepsin cleavage is not a mechanism to allow entry of particles with uncleaved F protein. Analysis of proteolytic enzymes from the bat reservoir hosts of these viruses indicates that both cathepsin and furin enzymes are present, though some differences in the cellular localization of furin were suggested, which may impact protease usage for F protein processing (El Najjar et al. 2015). Recent studies indicate a link between endosomal recycling of the F protein and efficient incorporation of F into virus-like particles (Cifuentes-Munoz et al. 2017; Johnston et al. 2017), and this tie between assembly and recycling may be a factor in the evolution of this novel mechanism of fusion protein activation.

2.4 Proteases Involved in Pneumovirus F Protein Proteolytic Cleavage

2.4.1 Respiratory Syncytial Virus F Proteins: A Novel Case of Two Proteolytic Processing Events

The F proteins from HRSV and bovine respiratory syncytial virus (BRSV) are unique among the paramyxo- and pneumovirus family members in their requirement for two proteolytic cleavage events to become fusogenically active (Gonzalez-Reyes et al. 2001; Zimmer et al. 2001). One of the two cleavage sites (site II; after residue 136) is immediately upstream of the fusion peptide region, similar to the cleavage sites observed for all other F proteins from these families (Fig. 2.5a). The second site (site I, after residue 109) is located 26 amino acids closer to the N-terminus of the protein, such that cleavage of both sites results in a disulfide-linked heterodimer plus an additional 27 amino acid peptide, termed p27 (Collins and Crowe 2007). Both cleavage sites have consensus sequences for furin processing (Gonzalez-Reyes et al. 2001; Zimmer et al. 2001), and studies with a soluble form of HRSV suggest that cleavage at site I may facilitate cleavage at site II (Begona Ruiz-Arguello et al. 2002). Structural analysis of the pre-fusion HRSV F bound to a neutralizing antibody (McLellan et al. 2013) shows that the fusion peptide C-terminal to cleavage site II lies buried in the center of the trimer cavity, at

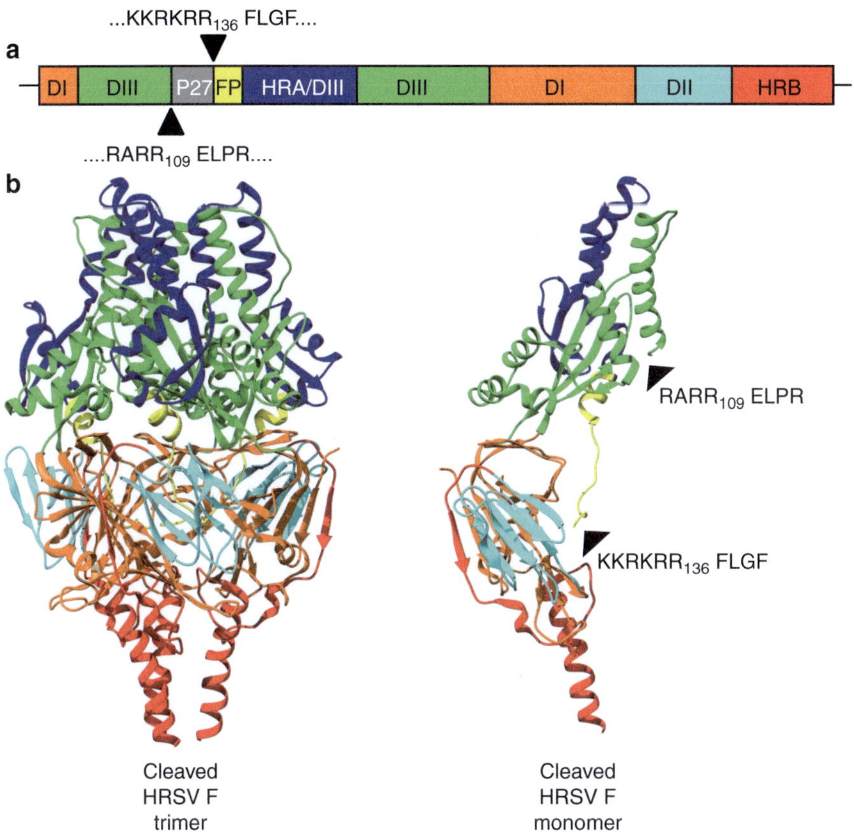

Fig. 2.5 Trimeric and monomeric HRSV F. (**a**) A linear schematic of HRSV F without the transmembrane domain (TM) or cytoplasmic tail is shown. Two cleavage events (black triangles) facilitate the release of a P27 fragment. (**b**) Similar to PIV5 F, the pre-fusion, cleaved structure of HRSV F ((McLellan et al. 2013), PDB ID: 4JHW) is compact, but the fusion peptide (FP; yellow) is buried within the globular head domain. Both cleavage sites are denoted within the monomeric HRSV F structure. Cleavage occurs following R109 and R136. Colors denote the same structural regions as in the PIV5 and HPIV3 structures in Fig. 2.2. Molecular graphic images were produced using the UCSF Chimera package from the Computer Graphics Laboratory, University of California, San Francisco (Pettersen et al. 2004)

least 40 Å away from the last residue of the F_2 subunit (Fig. 2.5b), with p27 no longer present. The buried nature of site II after cleavage suggests a conformational change occurring upon cleavage for HRSV F, in contrast to what is observed for the paramyxovirus F proteins.

The biological advantage for two cleavage sites has been unclear, as studies with BRSV demonstrated that deletion of cleavage site I and the p27 region did not decrease replication in a cell culture model (Zimmer et al. 2002). However, recent work suggests that sequential cleavage at the two sites may play a role in macropinocytosis mediated viral entry (Krzyzaniak et al. 2013). Mass spectrometry analysis of viral particles identified a predominant species corresponding to the F_1 subunit with the p27 still attached, consistent with only site I cleavage occurring prior to virus assembly. However, analysis of the F protein following entry was consistent with cleavage of F_1 at site II occurring during endocytic entry (Krzyzaniak et al. 2013). Infection was blocked when a cell-permeable furin inhibitor was added, but not a cell-impermeable inhibitor. These results support a model where the second cleavage event for HRSV F is promoted by a furin-like protease in an endocytic compartment, and this cleavage is required for fusogenic activity and infection (Fig. 2.6). Interestingly, insertion of the

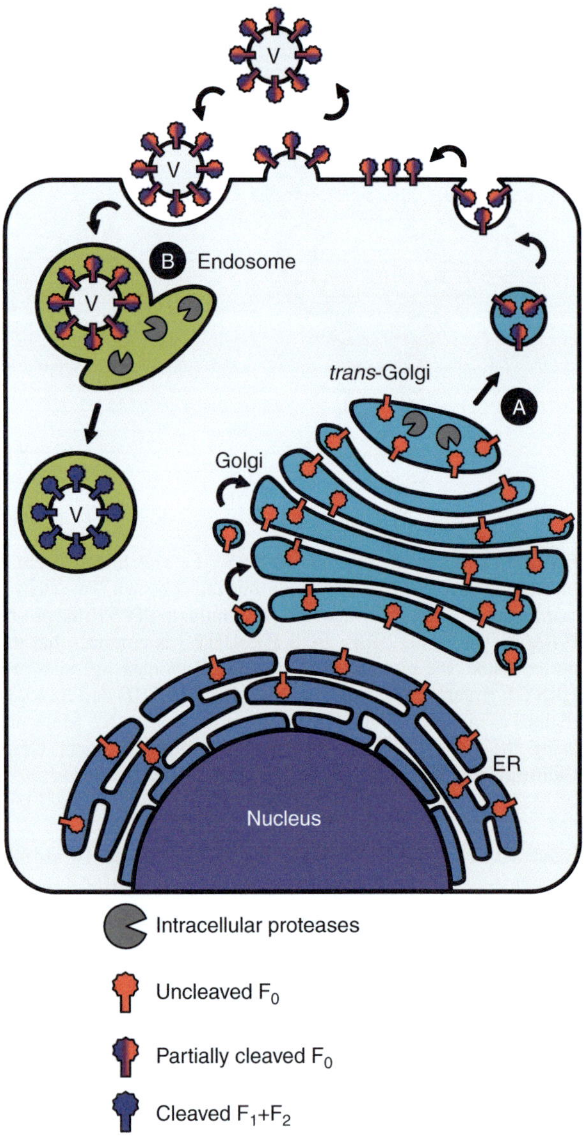

Fig. 2.6 Cleavage of RSV F. Like all paramyxovirus and pneumovirus F proteins, HRSV F is translated into the endoplasmic reticulum (ER) as an uncleaved (F_0; red) form. Unlike other F proteins, HRSV F is cleaved at two sites. (**a**) HRSV F is first cleaved in the *trans*-Golgi network by furin and incorporated into newly formed virus particles (red and blue gradient). (**b**) Following receptor binding and HRSV endocytosis, the second HRSV F cleavage occurs in endosomal compartments, resulting in formation of the cleaved, metastable $F_1 + F_2$ form (blue)

two HRSV F cleavage sites into the Sendai virus F protein resulted in a protein that was no longer dependent on the Sendai attachment protein for fusion (Rawling et al. 2008, 2011), suggesting that sequential cleavage at the two sites may provide a mechanism of F protein triggering for fusion (Bose et al. 2015).

2.4.2 Extracellular Cleavage of F Proteins from the Metapneumoviruses

HMPV F proteins contain a minimal furin consensus sequence at their cleavage site (R-X-X-R), but the lack of a basic residue at the P2 position (second amino acid upstream of the cleavage site) makes these F proteins poor substrates for furin. Additionally, viral growth and F protein cleavage in cell culture models have been shown to be dependent on addition of exogenous proteases such as trypsin (Biacchesi et al. 2004). The host cell protease TMPRSS2 (Shirogane et al. 2008) and proteases from *P. gingivalis* (Pyrc et al. 2011) have been shown to activate HMPV F, but the role of these and other proteases in infection remains to be defined. Some laboratory strains of HMPV have been identified that do not require exogenous proteases for viral growth, and a specific serine to proline change at the P2 residue of the cleavage site was associated with these changes (Schickli et al. 2005). However, a similar mutation in the context of a different strain did not result in trypsin-independent F protein cleavage (Schowalter et al. 2006a), suggesting that more complex factors are involved. A recent study suggests that protease-activated receptor 1-induced increases of cellular furin levels can lead to some intracellular HMPV F cleavage (Aerts et al. 2013).

F proteins from avian metapneumovirus (aMPV) subtypes A and B do not require exogenous trypsin for cleavage or function, indicating that they are cleaved by an endogenous protease (Yun et al. 2015). F proteins from aMPV subtype C show enhanced cleavage when trypsin is added, suggesting that they are less efficiently processed by endogenous proteases. A recent report indicates that TMPRSS12 can cleave the aMPV subtype B F protein (Yun et al. 2016), and this is the first report of this protease cleaving a viral glycoprotein substrate.

2.5 F protein Proteolytic Processing, Pathogenesis, and Antiviral Approaches

F protein proteolytic cleavage is required for promotion of membrane fusion and infection; thus, alterations to the utilized proteases or changes in the efficiency of processing can greatly affect viral pathogenesis. This was first detailed in studies of NDV, where alterations in protease susceptibility were observed between virulent and avirulent strains (Nagai et al. 1976, 1989; Nagai and Klenk 1977). NDV strains with furin-cleaved F proteins were more virulent than those with monobasic cleavage sites, presumably because the ubiquitous nature of furin enables efficient formation of fusogenically active F proteins in any infected cell or organ. In contrast, replication of strains with monobasic cleavage sites is limited to locations where the required extracellular protease is present. However, mutation to create a furin cleavage site did not alter tissue tropism or virulence of other avian paramyxoviruses (Kim et al. 2013; Subbiah et al. 2011; Xiao et al. 2012), supporting the view that, besides proteolytic cleavage, there are additional factors contributing to the pathogenicity of a virus (see Chap. 1). Important questions remain about the role of protease cleavage in pathogenesis for viruses such as HMPV where multiple proteases can promote F cleavage.

As F protein proteolytic processing is essential for viral infectivity, the host cell proteases involved have been examined as antiviral targets. Early work demonstrated that an alpha 1-antitrypsin variant engineered to inhibit furin significantly reduced measles virus titers (Watanabe et al. 1995). Subsequent work has identified a series of potent inhibitors of furin and other pro-protein convertases (Klein-Szanto and Bassi 2017), and a number of these have been shown to have antiviral activity (Shiryaev et al. 2007), including activity against paramyxoviruses (Hardes et al. 2015). Chloroquine, an inhibitor of cathepsin L, has been shown to have antiviral effects on Hendra or Nipah virus in cell culture models (Porotto et al. 2009) but failed to protect in ferret (Pallister et al. 2009) or hamster (Freiberg et al. 2010) models of infection. Natural inhibitors of the airway proteases involved in extracellular activation of F proteins are an attractive potential antiviral, and early studies demonstrated that pulmonary surfactant was a potent inhibitor of Sendai virus activation (Kido et al. 1993), though additional research on natural protease inhibitors targeting paramyxovirus or pneumovirus infections remains to be done. Thus, the experimental results to date provide proof of principle that targeting proteases can inhibit infection, but the challenges of developing an inhibitor that works effectively in animal systems remain.

2.6 Proteolytic Activation of the Hemagglutinin-Neuraminidase Glycoprotein

With two avirulent NDV strains, a precursor HN_0 of the HN glycoprotein, a type II membrane protein, has been observed that is converted by proteolytic removal of a C-terminal extension into the biologically active form (Nagai et al. 1976, 1989; Nagai and Klenk 1977). Crystallographic analysis revealed that the C-terminal extension extends along the outside of the neuraminidase (NA) β-propeller domain and inserts C-terminal residues into the NA domain active site. The C-terminal extension also engages a secondary sialic acid-binding site present in NDV HN proteins, which is located at the NA domain dimer interphase that most likely blocks its attachment function. These results demonstrate that the C-terminal residues lead to an auto-inhibited state of HN, and they explain the requirement for proteolytic activation of HN_0 and associated reduced virulence (Yuan et al. 2011).

Conclusions

Much research has focused on the biophysical and structural aspects of F-mediated membrane fusion, and this process is largely conserved between phylogenetically distinct viruses. However, the widely varying mechanisms utilized for proteolytic activation of paramyxovirus and pneumovirus F proteins suggests that cleavage is not merely a structural requisite but potentially a determinant of viral tropism and pathogenesis. Understanding the temporal and spatial requirements for F cleavage and how they affect pathogenesis could illuminate novel aspects of paramyxo- and pneumovirus biology and be important for the development of efficacious antiviral therapeutics.

References

Aerts L, Hamelin ME, Rheaume C, Lavigne S, Couture C, Kim W, Susan-Resiga D, Prat A, Seidah NG, Vergnolle N, et al. Modulation of protease activated receptor 1 influences human metapneumovirus disease severity in a mouse model. PLoS One. 2013;8:e72529.

Afonso CL, Amarasinghe GK, Banyai K, Bao Y, Basler CF, Bavari S, Bejerman N, Blasdell KR, Briand FX, Briese T, et al. Taxonomy of the order Mononegavirales: update 2016. Arch Virol. 2016;161:2351–60.

Ambrose MR, Hetherington SV, Watson AS, Scroggs RA, Portner A. Molecular evolution of the F glycoprotein of human parainfluenza virus type 1. J Infect Dis. 1995;171:851–6.

Authier F, Posner BI, Bergeron JJ. Endosomal proteolysis of internalized proteins. FEBS Lett. 1996;389:55–60.

Barrett T, Clarke DK, Evans SA, Rima BK. The nucleotide sequence of the gene encoding the F protein of canine distemper virus: a comparison of the deduced amino acid sequence with other paramyxoviruses. Virus Res. 1987;8:373–86.

Begona Ruiz-Arguello M, Gonzalez-Reyes L, Calder LJ, Palomo C, Martin D, Saiz MJ, Garcia-Barreno B, Skehel JJ, Melero JA. Effect of proteolytic processing at two distinct sites on shape and aggregation of an anchorless fusion protein of human respiratory syncytial virus and fate of the intervening segment. Virology. 2002;298:317–26.

Biacchesi S, Skiadopoulos MH, Tran KC, Murphy BR, Collins PL, Buchholz UJ. Recovery of human metapneumovirus from cDNA: optimization of growth in vitro and expression of additional genes. Virology. 2004;321:247–59.

Bose S, Jardetzky TS, Lamb RA. Timing is everything: fine-tuned molecular machines orchestrate paramyxovirus entry. Virology. 2015;479-480:518–31.

Bukreyev A, Whitehead SS, Murphy BR, Collins PL. Recombinant respiratory syncytial virus from which the entire SH gene has been deleted grows efficiently in cell culture and exhibits site-specific attenuation in the respiratory tract of the mouse. J Virol. 1997;71:8973–82.

Chang A, Dutch RE. Paramyxovirus fusion and entry: multiple paths to a common end. Virus. 2012;4:613–36.

Chua KB, Bellini WJ, Rota PA, Harcourt BH, Tamin A, Lam SK, Ksiazek TG, Rollin PE, Zaki SR, Shieh W, et al. Nipah virus: a recently emergent deadly paramyxovirus. Science. 2000;288:1432–5.

Cifuentes-Munoz N, Sun W, Ray G, Schmitt PT, Webb S, Gibson K, Dutch RE, Schmitt AP. Mutations in the transmembrane domain and cytoplasmic tail of hendra virus fusion protein disrupt virus-like-particle assembly. J Virol. 2017;91:pii: e00152-17.

Collins PL, Crowe JE. Respiratory syncytial virus and metapneumovirus. In: Knipe DM, Howley PM, editors. Fields virology. Philadelphia: Lippincott, Williams and Wilkins; 2007. p. 1601–46.

Connolly SA, Leser GP, Yin HS, Jardetzky TS, Lamb RA. Refolding of a paramyxovirus F protein from prefusion to postfusion conformations observed by liposome binding and electron microscopy. Proc Natl Acad Sci U S A. 2006;103:17903–8.

Craft WW Jr, Dutch RE. Sequence motif upstream of the Hendra virus fusion protein cleavage site is not sufficient to promote efficient proteolytic processing. Virology. 2005;341:130–40.

Diederich S, Moll M, Klenk HD, Maisner A. The nipah virus fusion protein is cleaved within the endosomal compartment. J Biol Chem. 2005;280:29899–903.

Diederich S, Sauerhering L, Weis M, Altmeppen H, Schaschke N, Reinheckel T, Erbar S, Maisner A. Activation of the Nipah virus fusion protein in MDCK cells is mediated by cathepsin B within the endosome-recycling compartment. J Virol. 2012;86:3736–45.

Diederich S, Thiel L, Maisner A. Role of endocytosis and cathepsin-mediated activation in Nipah virus entry. Virology. 2008;375:391–400.

Dutch RE, Hagglund RN, Nagel MA, Paterson RG, Lamb RA. Paramyxovirus fusion (F) protein: a conformational change on cleavage activation. Virology. 2001;281:138–50.

El Najjar F, Lampe L, Baker ML, Wang LF, Dutch RE. Analysis of cathepsin and furin proteolytic enzymes involved in viral fusion protein activation in cells of the bat reservoir host. PLoS One. 2015;10:e0115736.

Franke J, Batts WN, Ahne W, Kurath G, Winton JR. Sequence motifs and prokaryotic expression of the reptilian paramyxovirus fusion protein. Arch Virol. 2006;151:449–64.

Freiberg AN, Worthy MN, Lee B, Holbrook MR. Combined chloroquine and ribavirin treatment does not prevent death in a hamster model of Nipah and Hendra virus infection. J Gen Virol. 2010;91:765–72.

Gonzalez-Reyes L, Ruiz-Arguello MB, Garcia-Barreno B, Calder L, Lopez JA, Albar JP, Skehel JJ, Wiley DC, Melero JA. Cleavage of the human respiratory syncytial virus fusion protein at two distinct sites is required for activation of membrane fusion. Proc Natl Acad Sci U S A. 2001;98:9859–64.

Gotoh B, Ogasawara T, Suzuki H, Asaka J, Shimokata K, Rott R, Nagai Y. Expression of factor X and its significance for the determination of paramyxovirus tropism in the chick embryo. EMBO J. 1992;11:2197–202.

Gotoh B, Ogasawara T, Toyoda T, Inocencio M, Hamaguchi M, Nagai Y. An endoprotease homologous to the blood clotting factor X as a determinant of viral tropism in chick embryo. EMBO J. 1990;9:4189–95.

Hardes K, Becker GL, Lu Y, Dahms SO, Kohler S, Beyer W, Sandvig K, Yamamoto H, Lindberg I, Walz L, et al. Novel furin inhibitors with potent anti-infectious activity. ChemMedChem. 2015;10:1218–31.

Herfst S, Mas V, Ver LS, Wierda RJ, Osterhaus AD, Fouchier RA, Melero JA. Low pH induced membrane fusion mediated by human metapneumoviruses F protein is a rare, strain dependent phenomenon. J Virol. 2008;82:8891–5.

Hidaka Y, Kanda T, Iwasaki K, Nomoto A, Shioda T, Shibuta H. Nucleotide sequence of a Sendai virus genome region covering the entire M gene and the 3′ proximal 1013 nucleotides of the F gene. Nucleic Acids Res. 1984;12:7965–73.

Johnston GP, Contreras EM, Dabundo J, Henderson BA, Matz KM, Ortega V, Ramirez A, Park A, Aguilar HC. Cytoplasmic motifs in the nipah virus fusion protein modulate virus particle assembly and egress. J Virol. 2017;91:pii: e02150-16.

Jordans S, Jenko-Kokalj S, Kuhl NM, Tedelind S, Sendt W, Bromme D, Turk D, Brix K. Monitoring compartment-specific substrate cleavage by cathepsins B, K, L, and S at physiological pH and redox conditions. BMC Biochem. 2009;10:23.

Kido H, Sakai K, Kishino Y, Tashiro M. Pulmonary surfactant is a potential endogenous inhibitor of proteolytic activation of Sendai virus and influenza A virus. FEBS Lett. 1993;322:115–9.

Kido H, Yokogoshi Y, Sakai K, Tashiro M, Kishino Y, Fukutomi A, Katunuma N. Isolation and characterization of a novel trypsin-like protease found in rat bronchiolar epithelial Clara cells. J Biol Chem. 1992;267:13573–9.

Kim SH, Xiao S, Shive H, Collins PL, Samal SK. Mutations in the fusion protein cleavage site of avian paramyxovirus serotype 4 confer increased replication and syncytium formation in vitro but not increased replication and pathogenicity in chickens and ducks. PLoS One. 2013;8:e50598.

Kim YH, Donald JE, Grigoryan G, Leser GP, Fadeev AY, Lamb RA, DeGrado WF. Capture and imaging of a prehairpin fusion intermediate of the paramyxovirus PIV5. Proc Natl Acad Sci U S A. 2011;108:20992–7.

Klein-Szanto AJ, Bassi DE. Proprotein convertase inhibition: paralyzing the cell's master switches. Biochem Pharmacol. 2017;140:8–15.

Klenk H-D, Garten W. Host cell proteases controlling virus pathogenicity. Trends Microbiol. 1994;2:39–43.

Krzyzaniak MA, Zumstein MT, Gerez JA, Picotti P, Helenius A. Host cell entry of respiratory syncytial virus involves macropinocytosis followed by proteolytic activation of the F protein. PLoS Pathog. 2013;9:e1003309.

Lamb RA, Parks GD. *Paramyxoviridae:* the viruses and their replication. In: Knipe DM, Howley PM, editors. Fields virology. Philadelphia: Lippincott Williams and Wilkins; 2013. p. 1449–96.

Li Z, Xu J, Patel J, Fuentes S, Lin Y, Anderson D, Sakamoto K, Wang LF, He B. Function of the small hydrophobic protein of J paramyxovirus. J Virol. 2011;85:32–42.

Ling R, Sinkovic S, Toquin D, Guionie O, Eterradossi N, Easton AJ. Deletion of the SH gene from avian metapneumovirus has a greater impact on virus production and immunogenicity in turkeys than deletion of the G gene or M2-2 open reading frame. J Gen Virol. 2008;89:525–33.

Marsh GA, de Jong C, Barr JA, Tachedjian M, Smith C, Middleton D, Yu M, Todd S, Foord AJ, Haring V, et al. Cedar virus: a novel Henipavirus isolated from Australian bats. PLoS Pathog. 2012;8:e1002836.

Masante C, El Najjar F, Chang A, Jones A, Moncman CL, Dutch RE. The human metapneumovirus small hydrophobic protein has properties consistent with those of a viroporin and can modulate viral fusogenic activity. J Virol. 2014;88:6423–33.

McLellan JS, Chen M, Leung S, Graepel KW, Du X, Yang Y, Zhou T, Baxa U, Yasuda E, Beaumont T, et al. Structure of RSV fusion glycoprotein trimer bound to a prefusion-specific neutralizing antibody. Science. 2013;340:1113–7.

McLellan JS, Yang Y, Graham BS, Kwong PD. Structure of respiratory syncytial virus fusion glycoprotein in the postfusion conformation reveals preservation of neutralizing epitopes. J Virol. 2011;85:7788–96.

Meulendyke KA, Wurth MA, McCann RO, Dutch RE. Endocytosis plays a critical role in proteolytic processing of the Hendra virus fusion protein. J Virol. 2005;79:12643–9.

Michalski WP, Crameri G, Wang L, Shiell BJ, Eaton B. The cleavage activation and sites of glycosylation in the fusion protein of hendra virus. Virus Res. 2000;69:83–93.

Moll M, Diederich S, Klenk HD, Czub M, Maisner A. Ubiquitous activation of the Nipah virus fusion protein does not require a basic amino acid at the cleavage site. J Virol. 2004;78:9705–12.

Murakami M, Towatari T, Ohuchi M, Shiota M, Akao M, Okumura Y, Parry MA, Kido H. Mini-plasmin found in the epithelial cells of bronchioles triggers infection by broad-spectrum influenza A viruses and Sendai virus. Eur J Biochem/FEBS. 2001;268:2847–55.

Murray K, Selleck P, Hooper P, Hyatt A, Gould A, Gleeson L, Westbury H, Hiley Lea. A morbillivirus that caused fatal disease in horses and humans. Science. 1995;268:94–7.

Nagai Y, Hamaguchi M, Toyoda T. Molecular biology of Newcastle disease virus. Prog Vet Microbiol Immunol. 1989;5:16–64.

Nagai Y, Klenk H-D. Activation of precursors to both glycoproteins of Newcastle disease virus by proteolytic cleavage. Virology. 1977;77:125–34.

Nagai Y, Klenk H-D, Rott R. Proteolytic cleavage of the viral glycoproteins and its significance for the virulence of Newcastle disease virus. J Virol. 1976;20:501–8.

Pager CT, Craft WW Jr, Patch J, Dutch RE. A mature and fusogenic form of the Nipah virus fusion protein requires proteolytic processing by cathepsin L. Virology. 2006;346:251–7.

Pager CT, Dutch RE. Cathepsin L is involved in proteolytic processing of the Hendra virus fusion protein. J Virol. 2005;79:12714–20.

Pallister J, Middleton D, Crameri G, Yamada M, Klein R, Hancock TJ, Foord A, Shiell B, Michalski W, Broder CC, et al. Chloroquine administration does not prevent Nipah virus infection and disease in ferrets. J Virol. 2009;83:11979–82.

Paterson RG, Harris TJR, Lamb RA. Fusion protein of the paramyxovirus simian virus 5: nucleotide sequence of mRNA predicts a highly hydrophobic glycoprotein. Proc Natl Acad Sci U S A. 1984;81:6706–10.

Paterson RG, Russell CJ, Lamb RA. Fusion protein of the paramyxovirus SV5: destabilizing and stabilizing mutants of fusion activation. Virology. 2000;270:17–30.

Pettersen EF, Goddard TD, Huang CC, Couch GS, Greenblatt DM, Meng EC, Ferrin TE. UCSF Chimera--a visualization system for exploratory research and analysis. J Comput Chem. 2004;25:1605–12.

Phadke VK, Bednarczyk RA, Salmon DA, Omer SB. Association between vaccine refusal and vaccine-preventable diseases in the United States: a review of measles and pertussis. JAMA. 2016;315:1149–58.

Pillay CS, Elliott E, Dennison C. Endolysosomal proteolysis and its regulation. Biochem J. 2002;363:417–29.

Popa A, Carter JR, Smith SE, Hellman L, Fried MG, Dutch RE. Residues in the hendra virus fusion protein transmembrane domain are critical for endocytic recycling. J Virol. 2012;86:3014–26.

Porotto M, Orefice G, Yokoyama CC, Mungall BA, Realubit R, Sganga ML, Aljofan M, Whitt M, Glickman F, Moscona A. Simulating henipavirus multicycle replication in a screening assay leads to identification of a promising candidate for therapy. J Virol. 2009;83:5148–55.

Pyrc K, Strzyz P, Milewska A, Golda A, Schildgen O, Potempa J. Porphyromonas gingivalis enzymes enhance infection with human metapneumovirus in vitro. J Gen Virol. 2011;92:2324–32.

Rawling J, Cano O, Garcin D, Kolakofsky D, Melero JA. Recombinant Sendai viruses expressing fusion proteins with two furin cleavage sites mimic the syncytial and receptor-independent infection properties of respiratory syncytial virus. J Virol. 2011;85:2771–80.

Rawling J, Garcia-Barreno B, Melero JA. Insertion of the two cleavage sites of the respiratory syncytial virus fusion protein in Sendai virus fusion protein leads to enhanced cell-cell fusion and a decreased dependency on the HN attachment protein for activity. J Virol. 2008;82:5986–98.

Richardson C, Hull D, Greer P, Hasel K, Berkovich A, Englund G, Bellini W, Rima B, Lazzarini R. The nucleotide sequence of the mRNA encoding the fusion protein of measles virus (Edmonston strain): a comparison of fusion proteins from several different paramyxoviruses. Virology. 1986;155:508–23.

Schickli JH, Kaur J, Ulbrandt N, Spaete RR, Tang RS. An S101P substitution in the putative cleavage motif of the human metapneumovirus fusion protein is a major determinant for trypsin-independent growth in vero cells and does not alter tissue tropism in hamsters. J Virol. 2005;79:10678–89.

Schowalter RM, Smith SE, Dutch RE. Characterization of human metapneumovirus F protein-promoted membrane fusion: critical roles for proteolytic processing and low pH. J Virol. 2006a;80:10931–41.

Schowalter RM, Wurth MA, Aguilar HC, Lee B, Moncman CL, McCann RO, Dutch RE. Rho GTPase activity modulates paramyxovirus fusion protein-mediated cell-cell fusion. Virology. 2006b;350:323–34.

Shirogane Y, Takeda M, Iwasaki M, Ishiguro N, Takeuchi H, Nakatsu Y, Tahara M, Kikuta H, Yanagi Y. Efficient multiplication of human metapneumovirus in Vero cells expressing the transmembrane serine protease TMPRSS2. J Virol. 2008;82:8942–6.

Shiryaev SA, Remacle AG, Ratnikov BI, Nelson NA, Savinov AY, Wei G, Bottini M, Rega MF, Parent A, Desjardins R, et al. Targeting host cell furin proprotein convertases as a therapeutic strategy against bacterial toxins and viral pathogens. J Biol Chem. 2007;282:20847–53.

Spriggs MK, Olmsted RA, Venkatesan S, Coligan JE, Collins PL. Fusion glycoprotein of human parainfluenza virus type 3: nucleotide sequence of the gene, direct identification of the cleavage-activation site, and comparison with other paramyxoviruses. Virology. 1986;152:241–51.

Subbiah M, Khattar SK, Collins PL, Samal SK. Mutations in the fusion protein cleavage site of avian paramyxovirus serotype 2 increase cleavability and syncytium formation but do not increase viral virulence in chickens. J Virol. 2011;85:5394–405.

Swanson K, Wen X, Leser GP, Paterson RG, Lamb RA, Jardetzky TS. Structure of the Newcastle disease virus F protein in the post-fusion conformation. Virology. 2010;402:372–9.

Toyoda T, Sakaguchi T, Imai K, Inocencio NM, Gotoh B, Hamaguchi M, Nagai Y. Structural comparison of the cleavage-activation site of the fusion glycoprotein between virulent and avirulent strains of Newcastle disease virus. Virology. 1987;158:242–7.

Turk D, Guncar G. Lysosomal cysteine proteases (cathepsins): promising drug targets. Acta Crystallogr D Biol Crystallogr. 2003;59:203–13.

van den Hoogen BG, de Jong JC, Groen J, Kuiken T, de Groot R, Fouchier RA, Osterhaus AD. A newly discovered human pneumovirus isolated from young children with respiratory tract disease. Nat Med. 2001;7:719–24.
Watanabe M, Hirano A, Stenglein S, Nelson J, Thomas G, Wong TC. Engineered serine protease inhibitor prevents furin-catalyzed activation of the fusion glycoprotein and production of infectious measles virus. J Virol. 1995;69:3206–10.
Waxham MN, Server AC, Goodman HM, Wolinsky JS. Cloning and sequencing of the mumps virus fusion protein gene. Virology. 1987;159:381–8.
Welch BD, Liu Y, Kors CA, Leser GP, Jardetzky TS, Lamb RA. Structure of the cleavage-activated prefusion form of the parainfluenza virus 5 fusion protein. Proc Natl Acad Sci U S A. 2012;109:16672–7.
Whitehead SS, Bukreyev A, Teng MN, Firestone CY, St. Claire M, Elkins WR, Collins PL, Murphy BR. Recombinant respiratory syncytial virus bearing a deletion of either the NS2 or SH gene is attenuated in chimpanzees. J Virol. 1999;73:3438–42.
Wong JJ, Paterson RG, Lamb RA, Jardetzky TS. Structure and stabilization of the Hendra virus F glycoprotein in its prefusion form. Proc Natl Acad Sci U S A. 2016;113:1056–61.
Xiao S, Khattar SK, Subbiah M, Collins PL, Samal SK. Mutation of the f-protein cleavage site of avian paramyxovirus type 7 results in furin cleavage, fusion promotion, and increased replication in vitro but not increased replication, tissue tropism, or virulence in chickens. J Virol. 2012;86:3828–38.
Xu K, Chan YP, Bradel-Tretheway B, Akyol-Ataman Z, Zhu Y, Dutta S, Yan L, Feng Y, Wang LF, Skiniotis G, et al. Crystal structure of the pre-fusion nipah virus fusion glycoprotein reveals a novel hexamer-of-trimers assembly. PLoS Pathog. 2015;11:e1005322.
Yin HS, Paterson RG, Wen X, Lamb RA, Jardetzky TS. Structure of the uncleaved ectodomain of the paramyxovirus (hPIV3) fusion protein. Proc Natl Acad Sci U S A. 2005;102:9288–93.
Yin HS, Wen X, Paterson RG, Lamb RA, Jardetzky TS. Structure of the parainfluenza virus 5 F protein in its metastable, prefusion conformation. Nature. 2006;439:38–44.
Yuan P, Swanson KA, Leser GP, Paterson RG, Lamb RA, Jardetzky TS. Structure of the Newcastle disease virus hemagglutinin-neuraminidase (HN) ectodomain reveals a four-helix bundle stalk. Proc Natl Acad Sci U S A. 2011;108:14920–5.
Yun B, Guan X, Liu Y, Gao Y, Wang Y, Qi X, Cui H, Liu C, Zhang Y, Gao L, et al. Trypsin- and low pH-mediated fusogenicity of avian metapneumovirus fusion proteins is determined by residues at positions 100, 101 and 294. Sci Rep. 2015;5:15584.
Yun B, Zhang Y, Liu Y, Guan X, Wang Y, Qi X, Cui H, Liu C, Zhang Y, Gao H, et al. TMPRSS12 Is an Activating Protease for Subtype B Avian Metapneumovirus. J Virol. 2016;90:11231–46.
Zhao X, Singh M, Malashkevich VN, Kim PS. Structural characterization of the human respiratory syncytial virus fusion protein core. Proc Natl Acad Sci U S A. 2000;97:14172–7.
Zimmer G, Budz L, Herrler G. Proteolytic activation of respiratory syncytial virus fusion protein. Cleavage at two furin consensus sequences. J Biol Chem. 2001;276:31642–50.
Zimmer G, Conzelmann KK, Herrler G. Cleavage at the furin consensus sequence RAR/KR(109) and presence of the intervening peptide of the respiratory syncytial virus fusion protein are dispensable for virus replication in cell culture. J Virol. 2002;76:9218–24.

Cleavage of the Glycoprotein of Arenaviruses

3

Antonella Pasquato, Laura Cendron, and Stefan Kunz

Abstract

The arenaviruses are a large family of emerging negative-stranded RNA viruses that include several severe human pathogens causing hemorrhagic fevers with high mortality. During the arenavirus life cycle, processing of the viral envelope glycoprotein precursor (GPC) by the cellular subtilisin kexin isozyme-1 (SKI-1)/site-1 protease (S1P) is crucial for productive infection. The ability of newly emerging arenaviruses to hijack human SKI-1/S1P is a key factor for zoonotic transmission and human disease potential. Apart from being an essential host factor for arenavirus infection, SKI-1/S1P is involved in the regulation of important physiological processes and linked to major human diseases. This chapter provides an overview of the mechanisms of arenavirus GPC processing by SKI-1/S1P including recent findings. We will highlight to what extent the molecular mechanisms of SKI-1/S1P cleavage of viral GPC differ from processing of SKI-1/S1P's cellular substrates and discuss the implications for virus-host interaction and coevolution. Moreover, we will show how the use of the viral GPC as a "molecular probe" uncovered novel and unusual aspects of SKI-1/S1P biosynthesis and maturation. The crucial role of SKI-1/S1P in arenavirus infection and other major human diseases combined with its nature as an enzyme makes SKI-1/S1P further an attractive target for therapeutic intervention. In the last part, we will therefore cover past and present efforts to identify specific SKI-1/S1P inhibitors.

A. Pasquato · S. Kunz (✉)
Lausanne University Hospital, Institute of Microbiology, Lausanne, Switzerland
e-mail: Antonella.Pasquato@chuv.ch; Stefan.Kunz@chuv.ch

L. Cendron
Department of Biology, University of Padova, Padova, Italy
e-mail: laura.cendron@unipd.it

E. Böttcher-Friebertshäuser et al. (eds.), *Activation of Viruses by Host Proteases*,
https://doi.org/10.1007/978-3-319-75474-1_3

3.1 Arenavirus Structure, Genome Organization, and Basic Virology

The arenaviruses are a large and diverse family of emerging enveloped negative-stranded viruses that include several severe human pathogens (Buchmeier et al. 2007). The *Arenaviridae* family has been recently separated by the International Committee on Taxonomy of Viruses into the genus *Mammarenavirus* and the genus *Reptarenavirus* (Radoshitzky et al. 2015). Based on phylogenetic and serological data, the mammarenaviruses are divided into two major groups: the Old World and the New World complex. For simplicity, we will henceforth use the term "arenaviruses" synonymous for the entire family and members of the *Mammarenavirus* genus, whereas viruses of the *Reptarenavirus* genus will be specifically referred to. The Old World arenavirus lineage contains the prototypic arenavirus lymphocytic choriomeningitis virus (LCMV) with worldwide distribution. The infection of LCMV in the mouse represents one of the most powerful models in experimental virology and immunology (Oldstone 2002). LCMV is further a relevant human pathogen in pediatric and transplantation medicine (Bonthius 2009; Palacios et al. 2008). The highly pathogenic Lassa virus (LASV) is endemic in Western Africa (McCormick and Fisher-Hoch 2002), and Lujo virus (LUJV) recently emerged in Southern Africa associated with a cluster of fatal infections (Briese et al. 2009). The African arenaviruses Mopeia, Mobala, and Ippy virus have so far not been associated with human disease. The New World arenaviruses are divided into Clades, A, B, C, and D, the latter corresponding to former Clade A/B or A/rec (Radoshitzky et al. 2015). Clade B contains the human pathogenic Junin (JUNV), Machupo (MACV), Guanarito (GTOV), Sabia (SABV), and Chapare (CHAV) virus, together with the nonpathogenic Tacaribe (TCRV), Amapari, and Cupixi virus. In nature each arenavirus species has one or a limited number of closely related rodent species as reservoirs that are persistently infected, with the exception of TCRV that was isolated from bats (Buchmeier et al. 2007) and recently detected in host-seeking *Amblyomma americanum* ticks (Sayler et al. 2014). The current phylogenetic diversity of arenaviruses is likely the result of long-term coevolution between viruses and their host species, involving vertical and horizontal transfer of viruses within and between populations (Emonet et al. 2009).

Arenaviruses are enveloped negative-stranded RNA viruses, whose non-lytic life cycle is confined to the cytoplasm (De La Torre 2009). In electron microscopy, viral particles appear spherical to pleomorphic, with diameters of 50–300 nm. The arenavirus genome is comprised of two RNA segments, L (*c.* 7.3 kb) and S (*c.* 3.5 kb), containing two open reading frames in opposite orientation, separated by a noncoding intergenic region with a predicted hairpin structure. The viral S RNA encodes the nucleoprotein (NP) and the envelope glycoprotein precursor (GPC), whereas the L RNA encodes the viral RNA-dependent RNA polymerase L and the viral matrix protein Z. Synthesized as a single polypeptide chain, the viral GPC is posttranslationally cleaved by the cellular protease subtilisin kexin isozyme-1/site-1 protease (SKI-1/S1P) to yield the mature virion glycoproteins GP1 and GP2 (Fig. 3.1).

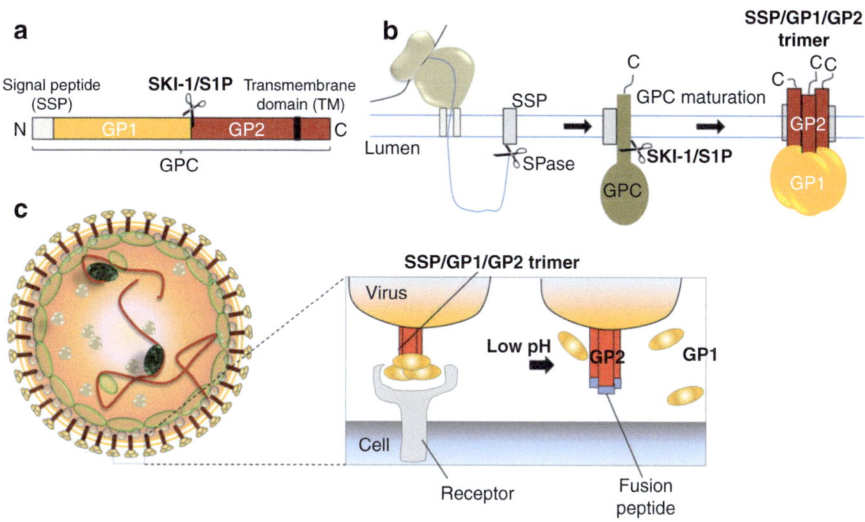

Fig. 3.1 Processing of arenavirus GPC by SKI-1/S1P. (**a**) The arenavirus GPC precursor is comprised of the stable signal peptide (SSP), GP1, and GP2. The transmembrane domain and the site of SKI-1/S1P cleavage are indicated (scissors). (**b**) Sequential processing of arenavirus GPC in the secretory pathway by signal peptidase (SPase) and SKI-1/S1P. The mature tripartite complex SSP/GP1/GP2 forms the mature trimeric GP spike. (**c**) The mature GP trimer decorates the virion surface and engages cellular receptors. Under acidic pH, GP1 dissociates and liberates the fusion peptide of GP2, triggering fusion between the viral and the cellular membrane

Several excellent reviews cover different steps of the arenavirus life cycle (Fehling et al. 2012; Grant et al. 2012; Urata and de la Torre 2011; Emonet et al. 2011; Torriani et al. 2017; Nunberg and York 2012; Wolff et al. 2013; Loureiro et al. 2012), and only a short summary will be given here. The first step of arenavirus infection requires attachment of the viral particle to cellular receptor(s). Most Old World and Clade C New World arenaviruses use dystroglycan, a ubiquitously expressed receptor for proteins of the extracellular matrix (ECM) as a high-affinity receptor (Cao et al. 1998; Oldstone and Campbell 2011). The cellular receptor for the pathogenic Clade B New World arenaviruses was identified as human transferrin receptor 1 (TfR1) (Radoshitzky et al. 2007), a highly conserved cargo receptor involved in iron metabolism. The ability of a Clade B New World arenavirus to use human TfR1 is crucial for its potential to cause zoonotic infection and hemorrhagic fever in man, whereas nonpathogenic viruses use TfR1 orthologues from other species (Helguera et al. 2012; Radoshitzky et al. 2011). More recently, the Tyro3/Axl/Mer (TAM) receptor tyrosine kinases Axl and Tyro3/Dtk, T cell immunoglobulin mucin (TIM) proteins 1 and 4, as well as the C-type lectins DC-specific ICAM-3-grabbing nonintegrin (DC-SIGN) and LSECtin have been identified as novel candidate receptors for arenaviruses (Shimojima and Kawaoka 2012; Shimojima et al. 2012; Jemielity et al. 2013; Goncalves et al. 2013; Martinez et al. 2013). Upon initial attachment to the target cell, arenavirus particles are taken up by receptor-mediated endocytosis. Consistent with the use of TfR1 as a receptor, Clade B New World viruses enter via clathrin-mediated endocytosis (Martinez et al. 2007), whereas Old World arenaviruses use a pathway resembling macropinocytosis (Iwasaki et al. 2014; Oppliger et al. 2016; Torriani et al. 2017). The virus passes through the multivesicular endosome and reaches the late endosome (Pasqual et al. 2011b), where low pH triggers fusion of the viral membrane with the limiting membrane of the late endosome by the fusion-active GP2, creating a "fusion pore." At the late endosome, LASV GP1 undergoes a unique "receptor switch" and engages the late endosomal/lysosomal resident protein LAMP1 for efficient fusion (Jae et al. 2014). The dependence of LASV, but not other arenaviruses including the closely related LCMV, on LAMP1 as a late endosomal entry factor represents an interesting analogy to the filoviruses Ebola virus, whose fusion depends on the late endosomal protein Niemann-Pick C1 (Jae and Brummelkamp 2015).

By an unknown mechanism of "uncoating," the arenavirus ribonucleoprotein (RNP) comprised of viral RNA, NP, and L is released into the cytosol. Viral transcription is initiated at the incoming polymerase complex, resulting in expression of NP and L. As NP accumulates, the viral polymerase shifts to a replicase mode, generating full-length antigenomic RNAs serving as templates for the transcription of GPC and Z as well as synthesis of genomic RNA. Newly synthesized NP assembles the viral replication-transcription complexes that are membrane-associated structures that contain cellular lipids and proteins (Baird et al. 2012; Knopp et al. 2015). In the final stages of the arenavirus life cycle, progeny particles assemble and are released by budding from the plasma membrane. The key factor in the budding process is the small RING finger Z protein that functions as a bona fide matrix protein in arenavirus particle assembly (Urata and De La Torre 2011; Perez et al. 2003). As with other matrix proteins of enveloped viruses, arenavirus Z interacts with the cytosolic tail of GP2 (Capul et al. 2007) and specific cellular factors of the endosomal/multiple vesicle body pathway to drive the budding of viral particles from "budding zones" (Wolff et al. 2013; Urata and De La Torre 2011; Fehling et al. 2012; Perez et al. 2003).

Recent studies isolated, identified, and characterized novel and highly divergent arenaviruses from snakes associated with boid inclusion body disease (Bodewes et al. 2013; Hetzel et al. 2013; Stenglein et al. 2012). The genome organization of these viruses corresponds to arenaviruses, and they show a high degree of divergence. Notably, the GPC of the viral envelope seems more related to filoviruses. No cases of infections in other species have been reported so far, though reptarenaviruses are capable of infecting mammalian and arthropod cells *in vitro* (Hepojoki et al. 2015). Interestingly, reptarenavirus infection of mammalian cells occurred efficiently at 30 °C but was markedly reduced at mammalian body temperature, likely highlighting adaptation to their reptile hosts (Hepojoki et al. 2015).

The most prevalent human pathogen among the arenaviruses is the Old World arenavirus LASV that causes a severe viral hemorrhagic fever with high mortality in humans. Every year LASV causes over 300,000 infections in Western Africa (Mccormick and Fisher-Hoch 2002) and has been declared one of the eight top emerging pathogens by WHO in 2015 (Sweileh 2017). There is currently neither an efficient cure nor a licensed vaccine, resulting in case fatality rates of 15–30% (Yun and Walker 2012). In the USA, JUNV, MACV, GTOV, and SABV have emerged as causative agents of hemorrhagic fevers with high case-fatality rates with JUNV representing the most important public health problem (Grant et al. 2012). Novel arenaviruses emerge on the average every 2 years and can be associated with severe diseases (Briese et al. 2009; Delgado et al. 2008). Highly pathogenic arenaviruses have been included in the list of Category A pathogens by the Centers for Disease Control and Prevention. Global climate changes may influence population dynamics of natural rodent host populations, likely increasing exposure of humans to these pathogens.

The pathophysiology of fatal arenavirus infection is not well understood and involves viral and host immune factors (Yun and Walker 2012; Prescott et al. 2017). A highly predictive parameter for disease outcome is the viral load, indicating a close competition between viral spread and replication and the patient's immune system (Prescott et al. 2017). Drugs targeting specific steps of the viral life cycle can reduce multiplication and spread of the virus. This may provide the patient's immune system a window of opportunity to develop an antiviral immune response. An in-depth understanding of the molecular mechanisms underlying arenavirus multiplication and virus-host cell interaction is therefore of great importance to develop novel and efficacious strategies for antiviral therapeutic intervention.

3.2 Arenavirus GP Structure and Function

The arenavirus GPC is synthesized initially as a single polypeptide precursor that is sequentially cleaved by cellular signal peptidases and then by SKI-1/S1P (Lenz et al. 2001; Rojek et al. 2008; Beyer et al. 2003) (Fig. 3.1a, b). Processing of GPC by SKI-1/S1P yields the N-terminal GP1, which is implicated in binding to the cellular receptors (Borrow and Oldstone 1992) and the transmembrane GP2 that mediates fusion and resembles class I viral fusion proteins (Eschli et al. 2006; Igonet et al. 2011; Parsy et al. 2013). Arenavirus GPC contains a remarkably stable signal peptide (SSP) of 58 amino acids that contains two hydrophobic domains and undergoes myristoylation at its N-terminus (Eichler et al. 2003a, b; York et al. 2004; Froeschke et al. 2003). The SSP becomes part of a mature tripartite complex SSP/GP1/GP2 where it interacts with the GP2 subunit (Fig. 3.1b). Recent electron cryomicroscopy combined with tomography revealed that SSP/GP1/GP2 complexes of LASV assemble into a trimeric spike that is 9 nm high and 10 nm wide and undergoes significant changes when exposed to low pH (Li et al. 2016). Structural studies on the GP1 of LASV, MACV, and JUNV revealed a similar compact α/β fold, despite significant sequence deviation (Bowden et al. 2009; Cohen-Dvashi et al. 2015; Mahmutovic et al. 2015). Structural studies on the complex of MACV GP1 with its cellular receptor hTfR1 revealed that the GP1 monomer represents the functional unit of receptor recognition and that trimerization is not required for receptor binding (Abraham et al. 2010; Radoshitzky et al. 2011). Notably, MACV GP1 binds to the apical surface of hTfR1 without competing with transferrin binding.

More recent crystallographic studies resolved the structure of the pre-fusion conformation of the mature envelope GP of the prototypic Old World arenavirus LCMV (Hastie et al. 2016). Within the pre-fusion trimer, LCMV GP1 and GP2 undergo extensive interactions, involving ionic bonds. In contrast to the New World arenaviruses, monomeric LCMV GP1 is unable to bind the receptor dystroglycan with high affinity, suggesting that either avidity or the quaternary structure of the pre-fusion trimer is required. Once delivered to the late endosome, low pH sets off a series of conformational changes leading to shedding of GP1 and triggering of fusion of the viral and cellular membrane mediated by GP2 (Fig. 3.1c). The post-fusion conformation of arenavirus GP2 is similar to the six-helix bundle conformation common to a number of class I fusion proteins of enveloped viruses (Igonet et al. 2011; Parsy et al. 2013). The SSP is crucial for transport and processing of arenavirus GPC (Messina et al. 2012; York and Nunberg 2007; York et al. 2004; Eichler et al. 2003a, b). Both N- and C-termini of SSP are located in the cytosol (Agnihothram et al. 2007), and SSP associates non-covalently with a zinc-binding domain within the cytoplasmic tail of GP2 (Agnihothram et al. 2006; Briknarova et al. 2011). The SSP-GP2 interactions critically modulate pH-induced activation of membrane fusion (York and Nunberg 2006, 2009) and are targeted by a range of potent arenavirus fusion inhibitors (Shankar et al. 2016; York et al. 2008), pinpointing this unique feature of arenavirus fusion as a target for the development of antiviral therapeutics.

3.3 The Proprotein Convertase SKI-1/S1P Cleaves Arenavirus GPC

A crucial step of arenavirus infection is the maturation of the envelope glycoprotein precursor GPC. With the exception of the Crimean-Congo hemorrhagic fever virus which belongs to the *Bunyavirus* family, mammarenaviridae are the only viral pathogens known to hijack the proprotein convertase SKI-1/S1P to process their envelope GP. Proprotein convertases (PC) are a family of nine conserved calcium-dependent serine endoproteases and include the basic PCs PC1/3, PC2, furin, PC4, PACE4, PC5/6, and PC7, as well as the nonbasic PCs SKI-1/S1P and PCSK9 (Seidah and Prat 2007, 2012). The PCs share homology to the kexin subfamily of subtilases with a distinctive "Ser/His/Asp" catalytic triad that mediates peptide bond scission (Seidah and Prat 2002). Furin, PC5/6B, PC7, and SKI-1/S1P are membrane-anchored, while the remaining enzymes are secreted (PC4, PC5/6A, PACE4, and PCSK9) or retained in granules (PC1/3, PC2) (Seidah 2011). Basic PCs have similar but not identical *consensus* sequences K/RXnR↓ that may result in overlapping patterns of substrate cleavage. In contrast, SKI-1/S1P and PCSK9 cleave after hydrophobic or small residues, BX(hydrophobic)X↓ (Pasquato et al. 2006) and VFAQ↓, respectively (Benjannet et al. 2004). Processing by PC is essential for the proper function of a plethora of cellular proteins, including prohormones, growth factor precursors, transcription factors, proteases, and adhesion molecules.

The convertase SKI-1/S1P has been co-discovered by the laboratory of Nabil Seidah, working on the biosynthesis of brain-derived neurotrophic factor (Seidah et al. 1999), and the group of Brown and Goldstein, investigating the regulation of cholesterol metabolism (Sakai et al. 1998). SKI-1/S1P is a type I membrane protein synthesized as an inactive precursor of 1052 amino acids, comprised of a signal peptide, an N-terminal prodomain, and a catalytic domain (Fig. 3.2a). The transmembrane domain of SKI-1/S1P is followed by a basic cytosolic tail (amino acids 1023–1052). As all PCs, SKI-1/S1P activation requires removal of an N-terminal prodomain that assists the correct folding of the protease (Fig. 3.2a). Upon translocation into the ER, SKI-1/S1P undergoes autocatalytic maturation by sequential cleavages of the N-terminal prodomain first at sites B′/B (**R**K**V**F↓**R**S**L**K$_{137}$↓), followed by site C (**R**R**L**L$_{186}$↓) and the newly described site C′ (**R**R**A**S$_{166}$↓) (Da Palma et al. 2014). The end product, the C form of SKI-1/S1P, represents the fully mature enzyme (Toure et al. 2000; Elagoz et al. 2002). In contrast to basic PCs, maturation of SKI-1/S1P is unique because fragments of the truncated prodomain remain attached to the catalytic subunit of the protease (Fig. 3.2b). While retention of the prodomain prevents catalytic activation of basic PCs, the complexes of SKI-1/S1P with the attached prodomain fragments are enzymatically active and may differently interact with cellular and viral substrates, as detailed below (Da Palma et al. 2014) (Fig. 3.2b). Using the GPC of arenaviruses as "molecular probes," recent studies revealed that the prodomain of SKI-1/S1P has a modular structure. Specifically, the N-terminal AB fragment represents an autonomous structural and functional unit that is necessary and sufficient for SKI-1/S1P folding and partial activation (Da Palma et al. 2016). In contrast, the C-terminal BC fragment of the prodomain lacks a defined structure but seems crucial for autoprocessing and full activation. The AB sequence of the prodomain is evolutionary highly conserved, whereas the BC fragment shows considerable variation and is even missing in some species. Phylogenetic and functional studies suggest that primordial SKI-1/S1P may have contained a simpler prodomain consisting of the conserved AB fragment, whereas the BC region appears as a later evolutionary acquisition possibly allowing subtle regulation of the maturation process (Da Palma et al. 2016).

SKI-1/S1P plays a key role in regulation of lipid metabolism and other physiological processes and is linked to a wide range of human disorders, including hypercholesterolemia, vascular diseases, cancer, and viral infections (Fig. 3.3a). Its proven role in human diseases and the nature as an enzyme make SKI-1/S1P an interesting target for therapeutic intervention. The activity of SKI-1/S1P was first linked to cholesterol and fatty acid biosynthesis, where it was implicated in the activation of the sterol regulating protein factors (SREBP) (Sakai et al. 1998). Other transcription factors were then shown to be activated in a SKI-1/S1P-dependent manner, including activating transcription factor (ATF) 6 that senses ER stress (Ye et al. 2000) and members of the cAMP response element-binding proteins (CREB) family (Kondo et al. 2005). All these substrates share a similar mechanism of activation which involves processing by SKI-1/S1P followed by cleavage by site-2 protease (S2P). The initial, rate-limiting SKI-1/S1P processing occurs in the lumen of the Golgi compartment of the secretory pathway. The SKI-1/S1P cleavage unmasks a

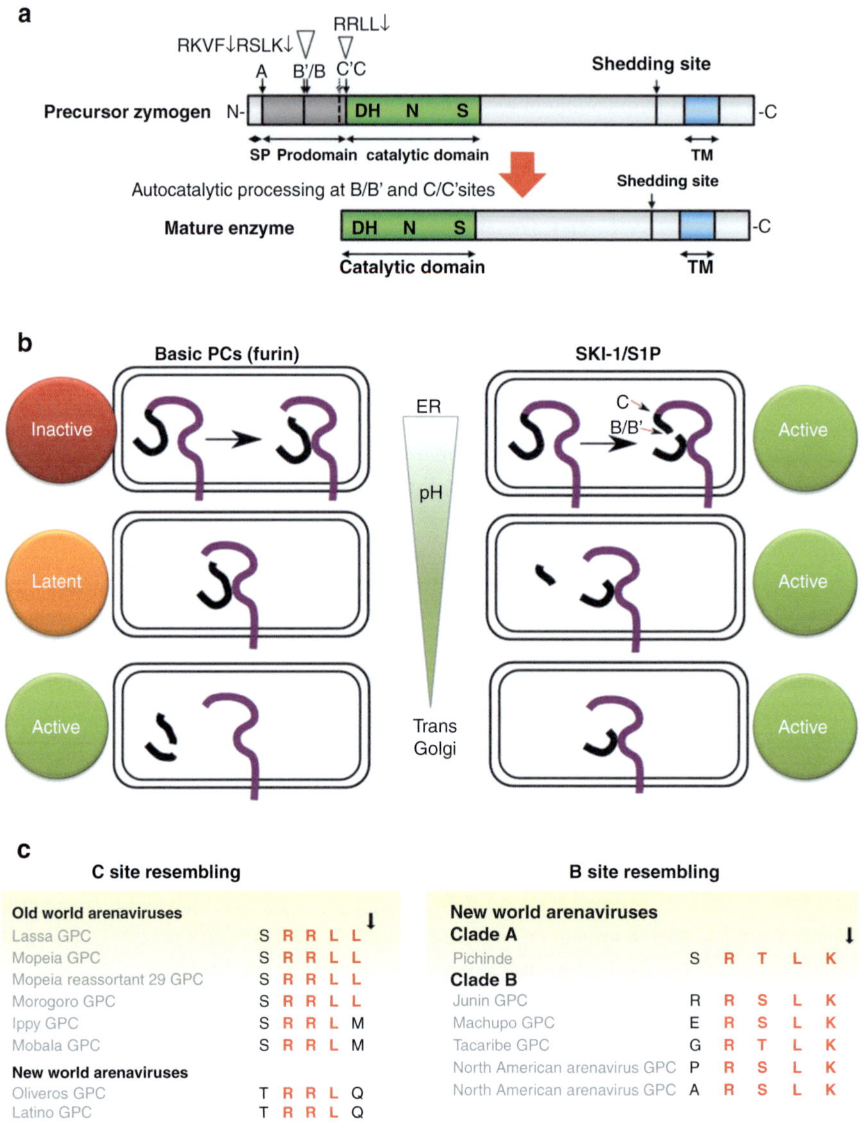

Fig. 3.2 The unusual mechanism of SKI-1/S1P maturation. (**a**) SKI-1/S1P is synthesized as an inactive zymogen precursor that undergoes autocatalytic cleavage to remove the prodomain. The signal peptide (SP), prodomain, catalytic domain, shedding site, and transmembrane domain, as well as autoprocessing sites A, B′/B, and C′/C and their corresponding amino acid sequences, are indicated. Autoprocessing at site C generates the mature enzyme. (**b**) Schematic representation of the maturation of the prototypic basic PC furin (left) and SKI-1/S1P (right). Initial autoprocessing of the furin prodomain results in a catalytically inactive latent complex between prodomain and enzyme. Complete removal of the prodomain is required to liberate the active enzyme late in the secretory pathway. In contrast, immature forms of SKI-1/S1P containing prodomain fragments of different lengths present all along the secretory pathway are catalytically active. (**c**) The GP1/GP2 SKI-1/S1P processing sites resemble sites B and C of autoprocessing. For details, please see text

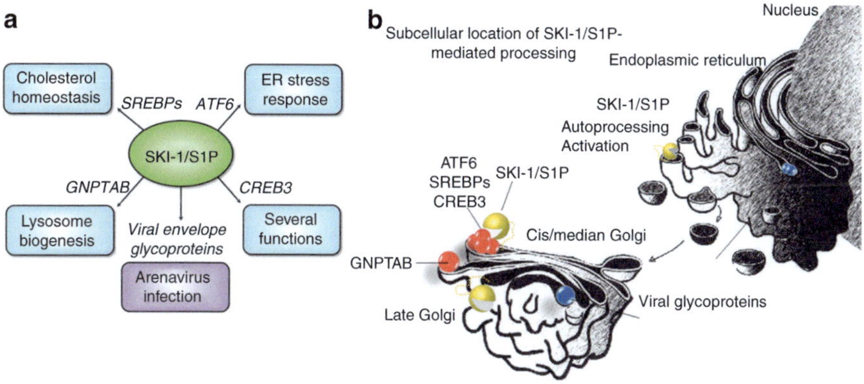

Fig. 3.3 Cellular substrates of SKI-1/S1P. (**a**) Cellular substrates of SKI-1/S1P are linked to major human disorders. (**b**) Subcellular location of SKI-1/S1P autocatalytic activation and SKI-1/S1P-mediated processing of the major cellular substrates. For details, please see text

second site of processing located close to the cytosolic face of the membrane, which is subsequently cleaved by the metalloprotease S2P. Processing by S2P releases a soluble fragment into the cytosol that subsequently enters the nucleus and acts in transcriptional regulation. In addition to transcription factors, SKI-1/S1P is implicated in processing of pro-brain-derived neurotrophic factor (Seidah et al. 1999); *N*-acetylglucosamine-1-phosphotransferase (GNPTAB), which is responsible for the correct sorting of lysosomal proteins (Marschner et al. 2011); and the renin receptor (Nakagawa et al. 2016). SKI-1/S1P further plays a role in bone and muscle formation (Gorski et al. 2011, 2016), ECM signaling, and axial development (Achilleos et al. 2015), as well as fur pigmentation (Rutschmann et al. 2012). However, the exact SKI-1/S1P substrates involved in these latter processes have not yet been clearly identified.

3.4 The Mechanism of SKI-1/S1P Processing of Arenavirus GPC Differs from Cellular Substrates

Alignment of the putative GP1/GP2 cleavage sites in arenavirus GPC reveals extensive sequence variation (Table 3.1). Due to this unusual residue pattern at the processing site, the protease responsible for GPC cleavage was found long after the identification of a *consensus* motif. So far, SKI-1/S1P has been implicated in processing of all mammarenavirus GPCs tested, including the Old World viruses LASV (Lenz et al. 2000, 2001), LCMV (Pinschewer et al. 2003; Beyer et al. 2003), the distantly related LUJV (Oppliger et al. 2015; Urata et al. 2015), as well as New World arenaviruses of different Clades (Rojek et al. 2008; Pasquato et al. 2011; Oppliger et al. 2015). Alignment of the sequences surrounding the putative cleavage site (P10-P10′ positions) reveals the presence of a highly conserved Arg and hydrophobic residue at P4 and P2 positions, respectively (Table 3.1). Interestingly, for the Old World and Clade C New World viruses, the sequences upstream from the scissile bond are of hydrophobic character.

Table 3.1 Amino acid sequences at the GP1/GP2 cleavage sites of arenavirus GPC

ID	Name	Sequence at cleavage site
>YP_009141003.1	Mariental	RSIYISRRIL GTFTWTLSDS
>ADX32840.1	Menekre	KSIYISRRLL GTFSWTLSDN
>AFU54705.1	Middle Pease River	RQAVGIRKLQ AFFSWTLSDN
>YP_516226.1	Mobala	REIYISRRLM GTFTWTLSDS
>AEO89355.1	Mopeia	RNFYISRRLL GLFTWTLSDS
>ABC71134.1	Mopeia	RSSYISRRLL GLFTWTLSDS
>AFY05576.1	Mopeia Lassa reassortant 29	RDIYISRRLL GTFTWTLSDS
>AFY05594.1	Mopeia Lassa reassortant 29	RDMYISRRLL GTFTWTLSDS
>AFY05619.1	Mopeia virus AN20410	RNFYISRRLL GLFTWTLSDS
>YP_003090214.	Morogoro	KNFYISRRLL GLFTWTLSDS
>AGT56422.1	Natoeduori	RTRFIARKLA GTFSWTLSDD
>AFU54675.1	North America	GQTAGIRKLQ AFFSWTLSDN
>ABW96598.1	North America	KQVIKVRKLL AFFTWSLSDA
>ABW96600.1	North America	KQMIKPRSLK SFFSWSLSDA
>ABW96602.1	North America	KQMIGARSLK AFFTWSLSDA
>AFD98839.1	Ocozocoautla de Espinosa	KNMFTRRTLK AFFSWSLTDS
>AJZ76770.1	Okahandja	RSIYLSRRLR SVFSWTLTDA
>AKG54821.1	Oliveros	GQSFITRRLQ AFLTWTLSDS
>YP_001936017.1	Parana	AYSSVSRKLL GFFTWDISDS
>AIN76883.1	Patawa	AYSSVSRKLM GFFTWDISDS
>AER45493.1	Pichinde	AYSSVSRKLL GFFTWDLSDS
>ACD71458.1	Pirital	AYSSVSRKLL GFFTWDISDS
>AAT88084.1	Pirital	AYGSVSRKLL GFFTWDISDS
>YP_089665.1	Sabia	GRSSGSRRPL GIFSWTITDA
>ABW96596.1	Skinner Tank	SQIVRARKLH AFFTWSLTDS
>AHW46355.1	Tacaribe	KSIAVGRTLK AFFSWSLTDP
>ACC99352.1	Tamiami	TQVVRARRIL SFFTWSLSDA
>ABU94341.1	Tonto creek	NQVIRARKLH AFFTWSLTDS
>AFU54702.1	Whitewater Arroyo	SQMIKARRLQ NFFSWSLSDA
>YP_001911113.1	Whitewater Arroyo	KQMIKSRTLK SFFAWSLSDA

The GenBank ID, name of virus species, and amino acids at the GP1/GP2 cleavage site are displayed. The conserved R residues in position 4 and hydrophobic residues in P2 position are highlighted

Maturation of GPC by SKI-1/S1P is strictly required for the production of infectious particles and viral cell-to-cell spread (Beyer et al. 2003; Lenz et al. 2001; Rojek et al. 2008). A crucial role for SKI-1/S1P for arenavirus dissemination *in vivo* is further suggested by the observation that mice bearing the "wood rat" mutation Y496C in SKI-1/S1P show enhanced resistance to infection with LCMV due to impairment of GPC processing (Popkin et al. 2011). Accordingly, proof-of-concept studies with protein- and peptide-based SKI-1/S1P inhibitors revealed that targeting GPC maturation represents a novel and promising antiviral strategy (Maisa et al. 2009; Rojek et al. 2010) as will be further developed below. Inhibition of SKI-1/S1P in infected cells results in the formation of noninfectious "naked" particles that contain viral RNP but lack GP (Lenz et al. 2001; Kunz et al. 2003; Rojek et al. 2008), indicating specific incorporation of fully mature, processed GP. How the arenavirus budding machinery is capable to achieve this specificity is currently unknown. Notably, SKI-1/S1P processing is not required for cell-surface transport of arenavirus GPC (Kunz et al. 2003; Schlie et al. 2010a), and small amounts of uncleaved GPC can be detected at the surface of infected cells (Kunz et al. 2003). However, in contrast to trimeric mature SSP/GP1/GP2 complexes, the uncleaved GPC forms monomers and oligomers spanning a wide size range, indicating that SKI-1/S1P processing is critical for the correct oligomeric state (Schlie et al. 2010a). Moreover, mutations in the cytosolic tail of LCMV and LASV GP2 affect SKI-1/S1P processing of the ectodomain (Schlie et al. 2010b; Kunz et al. 2003), suggesting some sort of transmission of structural information through the membrane. Since viral budding requires interactions of the cytosolic domain of GP2 with the matrix protein Z (Capul et al. 2007), processing by SKI-1/S1P may be required for targeting mature GP to putative "budding domains" and/or unmasking GP2 binding domains to Z.

The cleavage sites of arenavirus GPC differ from cellular substrates and resemble the B and C autoprocessing motifs of SKI-1/S1P (Fig. 3.2c). The GPC of JUNV contains the sequence RSLK↓ (B site), whereas LASV and LCMV GPCs are cleaved at the motifs RRLL↓ and RRLA↓ (C site). LASV GPC with the recognition sequence RRLL undergoes SKI-1/S1P processing early in the secretory pathway (Lenz et al. 2001), whereas LCMV GPC containing RRLA is processed in late Golgi or post-Golgi compartments (Wright et al. 1990; Beyer et al. 2003). Membrane-associated SKI-1/S1P is found predominantly in the early Golgi where cellular SKI-1/S1P substrates are cleaved (Pullikotil et al. 2007). Thus, the data at hand indicate that SKI-1/S1P is active in at least three different sub-compartments of the secretory pathway, ER/*cis*-Golgi (LASV GPC), median Golgi (SREBPs, ATF6, CREBs, GNPTAB), and late Golgi (LCMV GPC) (Fig. 3.3b). How arenaviruses selected specific subcellular compartments for SKI-1/S1P-mediated GPC maturation is still not fully understood, but recent studies gave some hints. Subtle changes of the sequence at the cleavage site can have drastic effects on the location and efficiency of GPC maturation, despite maintaining the RXLX↓ *consensus* motif. As an example, processing of an LCMV GPC mutant containing the cleavage site RRLL derived from LASV GPC is redirected from late Golgi to the ER/*cis*-Golgi. In contrast, introduction of the LCMV GPC cleavage motif RRLA into the

LASV GPC backbone results in an uncleavable protein (Burri et al. 2012). The unusual mechanism of zymogen activation and maturation of SKI-1/S1P described above (Fig. 3.2b) may contribute to this phenomenon. As mentioned above, autocatalytic processing is necessary but not sufficient to remove the SKI-1/S1P prosegment, resulting in different already active forms of the enzyme still bearing prodomain fragments of distinct lengths (Fig. 3.2b). It is conceivable that such enzyme/prodomain complexes located in defined sub-compartments of the secretory pathway may show differential specificity for viral and cellular substrates. Further evidence for differential recognition of viral and cellular substrates by SKI-1/S1P comes from the observation that the mutations R130E and R134E within the B′/B autoprocessing site result in selective impairment of viral GPC processing, but not cleavage of cellular substrates (Burri et al. 2012).

Due to their non-lytic strategy of replication, arenaviruses can establish persistent infections *in vitro* and *in vivo* without causing overt signs of pathology. Considering the multiple roles of SKI-1/S1P in maintaining cellular functions, the largely nonoverlapping subcellular localization of viral vs. cellular substrates (Fig. 3.3b) may be a consequence of the extensive coevolution of the viruses with their reservoir hosts. Accordingly, high expression levels of GPC during acute arenavirus infection do not interfere with the SKI-1/S1P-mediated processing of ATF6 involved in the host cell's ER stress response (Pasqual et al. 2011a). As a consequence, arenavirus infection results in specific and transient activation of the ATF6-regulated branch of the cellular ER stress response that includes upregulation of chaperones, adjusting the folding capacity to the increased demand. The differential subcellular location of SKI-1/S1P processing of viral and cellular substrates may therefore allow extensive viral replication and gene expression without causing overt cytopathic effects (Oldstone 2002).

3.5 Optimized Recognition of Arenavirus GPC by SKI-1/S1P: Viral Advantage and Achilles' Heel

Comparison of the currently known sequences of arenavirus GPCs revealed the presence of a highly conserved aromatic residue at position P7 relative to the SKI-1/S1P recognition sites in Old World and Clade C New World arenaviruses, but not in New World viruses of Clades A and B or cellular substrates (Burri et al. 2013) (Table 3.1). Early experimental evidence already supported the notion that an aromatic amino acid at P7 somehow promotes SKI-1/S1P cleavage of both LASV (Pasquato et al. 2006) and LCMV GPC (Beyer et al. 2003). Subsequent molecular modeling allowed docking of the LASV GPC-derived peptides into the putative catalytic pocket of SKI-1/S1P (Burri et al. 2013). These studies revealed that the aromatic "signature residue" in position P7 of some viral GPC recognition sequences interacts with residue Y285 located in the extended substrate-binding pocket of SKI-1/S1P (Burri et al. 2013). Indeed, introduction of the mutation Y285A into SKI-1/S1P gives an enzyme that is markedly impaired in processing of LASV GPC, but not GPC of New World arenaviruses or cellular substrates. During coevolution with their mammalian hosts, the GPCs of Old World and Clade C New World viruses apparently expanded the molecular contacts with SKI-1/S1P beyond the classical four amino acid recognition sequences, resulting in an enlarged binding surface. The concept that critical residues flanking the classical recognition sites can modulate PC processing is also supported by findings with basic PCs and their cellular substrates. A comparative study showed the influence of surrounding amino acids on the relative PC cleavage efficiency (Remacle et al. 2008). The presence of an N at position P1′ after the scissile bond in the substrate growth differentiation factor-11 markedly reduces cleavage by PACE4, furin, and PC7 while being selectively permissive to PC5/6 (Essalmani et al. 2008). Moreover, Constam and colleagues demonstrated that L and G at P2′ and P3′ positions in nodal, a regulator of the fate of pluripotent cells, dramatically enhance basic PC processing, further supporting this concept (Constam and Robertson 1999). The specificity of the interaction between Y285 of SKI-1/S1P and aromatic P7 residues for LASV GPC processing, but not cleavage of cellular substrates, makes this interaction a promising target for the development of specific antiviral drugs against this important human pathogen. Perturbation of the interaction of the highly conserved aromatic side chain in position P7 of LASV GPC with the contact residue Y285 of SKI-1/S1P, e.g., by a small molecule, is not expected to affect the SKI-1/S1P catalytic triad, limiting unwanted side effects. The markedly reduced processing of LASV GPC bearing a Y to A mutation in position P7 (Burri et al. 2013) suggests that viral escape variants lacking the aromatic signature residue at P7 may have impaired fitness, making the P7/Y285 interaction a true “Achilles’ heel” of the virus.

3.6 The Processing of Reptarenavirus GPC Is Largely Unknown

For the mammarenaviridae, several lines of evidence suggest that the maturation of the viral GPC by SKI-1/S1P is a crucial step in the virus life cycle. In contrast, little is known about the biosynthesis, maturation, and processing of reptarenavirus GPC. The sequences of reptarenavirus GPC seem to deviate from the canonical SSP/GP1/GP2 composition of mammarenaviruses and bear some resemblance to filovirus GP (Li et al. 2016; Stenglein et al. 2012). In contrast to arenavirus GPC, whose processing critically depends on SKI-1/S1P, the GP of the filovirus Ebola undergoes processing by furin, the prototypic member of the basic PC family, although this cleavage seems dispensable for virus infection and propagation (Neumann et al. 2002) (see also Chap. 5). Further studies are required to identify the specific cellular protease(s) involved in reptarenavirus GPC processing and to see if this step is crucial for functional maturation. A very different mechanism cannot be ruled out, including the use of another class of host-derived proteases that may function in the secretory pathway during biosynthesis, at the cell surface, or during the entry process, as illustrated by the use of endosomal cathepsins by filoviruses (Hunt et al. 2012) or coronaviruses that use multiple proteases, including cathepsins, cell surface transmembrane protease/serine proteases, furin, and trypsin (Millet and Whittaker 2015, see also Chap. 4).

3.7 Targeting SKI-1/S1P-Mediated GPC Processing as an Anti-arenaviral Strategy

A major challenge for the development of drugs against human pathogenic arenaviruses is the limited structural information available on the pathogens. As all viruses, arenaviruses critically depend on the molecular machinery of the host cell for their multiplication. This is particularly true for the biosynthesis of the envelope GP that involves a complex interplay with cellular factors, including SKI-1/S1P. So far, all human pathogenic arenaviruses seem to be SKI-1/S1P-dependent, and SKI-1/S1P activity is crucial for productive infection. In the absence of SKI-1/S1P activity, infected cells produce noninfectious naked particles devoid of GP, due to a yet-unknown mechanism of selective incorporation of cleaved GP. Studies have been carried out to investigate the ability of arenaviruses to escape SKI-1/S1P. Recombinant LCMV containing the furin cleavage site RRRR↓ instead of RRLA↓ was found to be replication competent and behaved similarly to the wild-type virus in cell culture (Rojek et al. 2010). Although this suggested that arenaviruses may, at least in principle, use other proteases than SKI-1/S1P for GPC processing, persistent virus infection of SKI-1/S1P null cells or inhibitor treatment so far never resulted in the emergence of SKI-1/S1P-independent viral escape variants (Rojek et al. 2008, 2010; Pasquato et al. 2012b). Moreover, complete inhibition of SKI-1/S1P seems not required to restrict arenavirus infection *in vivo*, since the partially active "wood rat" variant of the enzyme conferred significant protection and prevented persistent viral infection with LCMV (Popkin et al. 2011). In sum, inhibition of SKI-1/S1P appears as a promising therapeutic approach to combat arenavirus infection (Pasquato et al. 2012a).

Direct inhibition of the mature protease is a widely used approach to interfere with the normal enzymatic activity. Although this approach often gives excellent results, a catalytically dead enzyme results in a general loss of activity toward all substrates, cellular and viral alike. Unwanted side effects must be carefully taken into consideration. Considering the crucial role of SKI-1/S1P in major physiological processes and disorders, including infection with highly pathogenic arena- and bunyaviruses, efforts have been made to develop a panel of protein-based and small molecule inhibitors that will be covered below.

3.7.1 Protein-Based Strategies

The naturally occurring serpin α_1-antitrypsin (AT) had been long known as a potent suicide inhibitor for trypsin-like proteases, entrapping the enzyme in a stable complex following cleavage at the reactive site loop. Subsequently, mutations have been inserted into the reactive site loop of α_1-AT to introduce the $B(X)_nB$↓ motif of basic PCs, yielding a potent protein-based furin inhibitor (α_1-PDX) (Anderson et al. 1993). Based on the homology of the catalytic sites of basic and nonbasic PCs, the reactive site loop of α_1-AT was further mutated to introduce the SKI-1/S1P BX(hydrophobic)X↓ motif RRVL. In cell culture experiments, α_1-AT RRVL efficiently blocked SKI-1/S1P-mediated processing of SREBPs. Overexpression of α_1-AT RRVL inhibited LASV GPC maturation and had a strong antiviral effect by markedly reducing cell-to-cell spread and infectious viral particle production (Maisa et al. 2009), providing proof of concept.

3.7.2 Peptide-Based Compounds and Small Molecules

Analogous to peptide chloromethyl ketones (CMK) developed to inhibit furin-like proteases, two CMK peptides were designed, containing the IYISRRLL and RRLL motifs derived from LASV GPC. Both CMK peptides act as irreversible inhibitors of SKI-1/S1P at the level of substrate maturation (Pasquato et al. 2006). Proof-of-principle studies showed that these small molecules potently block infection of LCMV (Rojek et al. 2010). However, due to their toxicity, CMK inhibitors are not suitable as therapeutic agents.

The aminopyrrolidine amide compound PF-429242 is a reversible, competitive inhibitor of SKI-1/S1P discovered by Pfizer Inc. that efficiently blocks processing of endogenous cellular substrates (Hay et al. 2007). Following PF-429242 administration *in vivo*, SREBP-2 activation in mice is dramatically reduced resulting in a marked drop of plasma cholesterol levels (Hawkins et al. 2008). Pharmacological inhibition of SKI-1/S1P activity by PF-429242 also blocks LASV and LCMV GPC maturation reducing cell-to-cell propagation with only mild off-target effects (Urata et al. 2011; Pasquato et al. 2012b). Cells persistently infected with LCMV were efficiently cleared by treatment with PF-429242 without emergence of drug-resistant viral escape variants (Pasquato et al. 2012b). Initial evaluation of PF-429242 *in vivo* in a murine model raised concerns about applications against chronic diseases, such as familial hypercholesterolemia (Hawkins et al. 2008). However, human pathogenic arenaviruses cause acute diseases, limiting antiviral treatment to a time window of a few weeks. Considering these relatively short periods of treatment, the toxicological and pharmacokinetic profile of PF-429242 makes it still an interesting experimental drug candidate (Hawkins et al. 2008).

The current standard of care for treatment of human arenavirus infection is an off-label use of the nucleoside analogue ribavirin (1-β-D-ribofuranosyl-1,2,4-triazole-3-carboxamide) (Parker 2005). Early administration of ribavirin reduces the fatality in human Lassa fever (Mccormick et al. 1986) and experimental infections with MACV (Kilgore et al. 1995) and JUNV (Weissenbacher et al. 1987) in animals. However, to achieve high efficacy, ribavirin needs to be administered early during infection intravenously and is often associated with side effects. Novel anti-arenaviral drugs may be used individually or in combination with therapy and ribavirin to combat human pathogenic arenaviruses, allowing lower doses of ribavirin. Indeed, the combination of PF-429242 with ribavirin revealed stronger than additive effect of the two drugs (Pasquato et al. 2012b). The basis for this apparent synergism may lie in the distinct underlying antiviral mechanisms of the two drugs. Depending on the concentration used, ribavirin inhibits arenavirus infection at the level of replication (Ruiz-Jarabo et al. 2003) and shows drug action as a mutagen (Moreno et al. 2011). In contrast, PF-429242 affects the biosynthesis of the viral GP, blocking the formation of infectious progeny virus from infected cells.

3.7.3 A Novel Cell-Based Sensor for SKI-1/S1P as a Platform for High-Throughput Drug Screening

As outlined above, the currently available SKI-1/S1P inhibitors are potent and specific and represent invaluable experimental drugs for proof-of-concept studies. However, they face considerable restrictions regarding therapeutic use in clinical medicine. The protein nature of α_1-AT RRVL makes cell permeability and *in vivo* drug delivery challenging. The smaller SKI-1/S1P-specific decanoylated CMK peptides show good cell permeability, but due to their toxicity profile and short half-life, their use is restricted. PF-429242 has low cytotoxicity and shows *in vitro* IC_{50} in the low micromolar range. However, the compound has an unfavorable pharmacokinetic profile *in vivo* and has to the best of our knowledge not yet entered the clinical test phase. Considering the promise of SKI-1/S1P as a drug target and the limitations of the candidate inhibitors at hand, the identification of novel small molecule inhibitors for SKI-1/S1P is of high priority.

Conventional approaches to study substrate processing by SKI-1/S1P use homogeneous biochemical assays including synthetic chromogenic peptides and purified soluble enzyme (Pasquato et al. 2006). These systems have greatly contributed to our current understanding of the biochemistry of SKI-1/S1P and lead to the discovery of candidate drugs like PF-429242. However, as mentioned above, evidence is accumulating that the interaction of SKI-1/S1P with its substrates is more complex and may also be regulated at the level of subcellular location. Robust and quantifiable cell-based assays are therefore needed to screen for inhibitors of SKI-1/S1P processing of specific substrates in the authentic cellular context. To close this gap, a novel reliable cell-based assay allowing the quantitative detection of the enzymatic activity of endogenous SKI-1/S1P has been developed, exploiting key findings on the processing of viral GPC (Da Palma et al. 2014). The assay is based on a chimeric protein composed of a *Gaussia* luciferase (GLuc) reporter anchored to the membrane by the stump region of SKI-1/S1P through a virus-derived cleavable peptide sequence (Fig. 3.4). The SKI-1/S1P-cleavable 9mer sequence IYIS**RRLL**↓G used in the prototypic sensor is derived from LASV GPC, which is one of the best substrates currently known (Pasquato et al. 2006, Lenz et al. 2001), assuring optimal sensitivity and specificity of the sensor. The membrane anchor of the sensor mimics that of SKI-1/S1P, allowing correct cellular targeting and favoring optimal substrate-enzyme recognition. Upon processing, this sensor releases the soluble reporter GLuc to the medium, where it can be easily detected using a sensitive and cost-efficient luciferase assay. The sensor recapitulates the key features of the viral substrate from which the processing site has been derived, both in terms of subcellular localization and efficiency of cleavage. The robust and reliable nature of this novel cell-based sensor assay allows implementation in high-throughput screening (HTS).

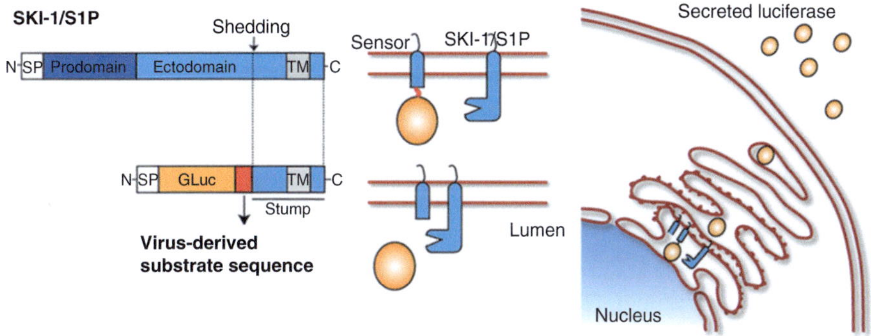

Fig. 3.4 A cell-based sensor for the detection of endogenous SKI-1/S1P activity. Schematic of the SKI-1/S1P sensor. The SKI-1/S1P-derived stump region and the GLuc reporter, as well as the virus-derived peptide comprising the cleavage motif, are indicated. Processing of the sensor by endogenous SKI-1/S1P releases the GLuc reporter (sphere) that is secreted into the tissue culture supernatant, where it can be detected *via* luminescence assay. For details, please see text

PCs are essential for normal cell functions during development and in adults (Seidah and Prat 2012); nonetheless several studies showed that specific inhibition of a definite PC is not detrimental in case of *in vivo* short-term treatment (Hay et al. 2007; Shiryaev et al. 2007). Thus, targeting viral GP cleavage is a novel promising therapeutic approach to fight against virulent pathogens such as hemorrhagic arenaviruses. Our modern globalized world increasingly faces the threat of emerging viruses due to human migration, rapidly progressing urbanization, almost free global trade, exposure to animals, and climatic changes. The development of novel broadly specific antivirals is therefore important to help meeting these unmatched medical problems.

3.7.4 Use of the SKI-1/S1P Sensor to Predict Protease Use of Newly Emerging Arenaviruses

New arenaviruses are rapidly emerging and are in some cases associated with severe human diseases. With the advent of powerful next-generation sequencing approaches, we expect to see accelerated discovery of many new arenavirus species, pathogenic or not, in the years to come. However, in many cases viruses may not be isolated, and only genetic information will become available. Considering the extensive variation at the known and putative SKI-1/S1P recognition sequences in known arenavirus GPC, defined *consensus* sequences cannot easily be found (Burri et al. 2013; Pasquato et al. 2011). Data at hand indicate that knowledge of the sequence P1–8 and residue P1′ of the putative GP1/GP2 cleavage site of a novel arenavirus GPC would be necessary and sufficient to test if the new virus can hijack human SKI-1/S1P, which is a prerequisite for productive infection and hence disease potential in man. In a recent study, the cell-based SKI-1/S1P sensor was applied to make a first prediction if the recently emerged LUJV GPC is processed by human

SKI-1/S1P. Phylogenetic analysis identified LUJV as an outlier within the Old World arenaviruses as the sequence RKLM↓K at the putative GP1/GP2 border differed significantly from the known *consensus* (Burri et al. 2013). Despite these important differences, a sensor containing a 9mer peptide derived from LUJV GPC underwent efficient processing by human SKI-1/S1P, which was then validated with authentic full-length GPC (Oppliger et al. 2015).

Considering the promise of SKI-1/S1P as therapeutic target for novel antiviral drugs to combat human pathogenic arenaviruses, SKI-1/S1P dependence of a newly emerging pathogenic arenavirus assessed by this new sensor may open the possibility for rapid intervention. The SKI-1/S1P sensor can further be used to assess the processing of arenavirus GPCs by SKI-1/S1P orthologues derived from other species, shedding light on the complex ecology of arenaviruses. Conceptually, the sensor platform developed for SKI-1/S1P based on lessons learned from arenavirus GPC cleavage may be applicable for other human proteases that are responsible for processing of a plethora of viral envelope GPs in a wide range of species. For the virology research community such a platform may serve as a rapid and cost-effective evaluation of viral GP processing by human proteases that may contribute to our preparedness against the threat of emerging viruses.

Acknowledgments The authors would like to apologize to all those colleagues whose excellent work could not be covered due to space limitations. This work was supported by Swiss National Science Foundation grant 310030_170108 to S.K. and funds to S.K. from the University of Lausanne.

References

Abraham J, corbett KD, farzan M, Choe H, Harrison SC. Structural basis for receptor recognition by New World hemorrhagic fever arenaviruses. Nat Struct Mol Biol. 2010;17:438–44.

Achilleos A, Huffman NT, Marcinkiewicyz E, Seidah NG, Chen Q, Dallas SL, Trainor PA, Gorski JP. Mbtps1/Ski-1/S1p proprotein convertase is required for Ecm signaling and axial elongation during somitogenesis and vertebral development. Hum Mol Genet. 2015;24:2884–98.

Agnihothram SS, York J, Nunberg JH. Role of the stable signal peptide and cytoplasmic domain of G2 in regulating intracellular transport of the Junin virus envelope glycoprotein complex. J Virol. 2006;80:5189–98.

Agnihothram SS, York J, Trahey M, Nunberg JH. Bitopic membrane topology of the stable signal peptide in the tripartite Junin virus Gp-C envelope glycoprotein complex. J Virol. 2007;81:4331–7.

Anderson ED, Thomas L, Hayflick JS, Thomas G. Inhibition of Hiv-1 Gp160-dependent membrane fusion by a furin-directed alpha 1-antitrypsin variant. J Biol Chem. 1993;268:24887–91.

Baird NL, York J, Nunberg JH. Arenavirus infection induces discrete cytosolic structures for Rna replication. J Virol. 2012;86:11301–10.

Benjannet S, Rhainds D, Essalmani R, Mayne J, Wickham L, Jin W, Asselin MC, Hamelin J, Varret M, Allard D, Trillard M, Abifadel M, Tebon A, Attie AD, Rader DJ, Boileau C, Brissette L, Chretien M, Prat A, Seidah NG. Narc-1/Pcsk9 and its natural mutants: zymogen cleavage and effects on the low density lipoprotein (Ldl) receptor and Ldl cholesterol. J Biol Chem. 2004;279:48865–75.

Beyer WR, Popplau D, Garten W, Von Laer D, Lenz O. Endoproteolytic processing of the lymphocytic choriomeningitis virus glycoprotein by the subtilase Ski-1/S1p. J Virol. 2003;77:2866–72.

Bodewes R, Kik MJ, Raj VS, Schapendonk CM, Haagmans BL, Smits SL, Osterhaus AD. Detection of novel divergent arenaviruses in boid snakes with inclusion body disease in the Netherlands. J Gen Virol. 2013;94:1206–10.

Bonthius DJ. Lymphocytic choriomeningitis virus: a prenatal and postnatal threat. Adv Pediatr. 2009;56:75–86.
Borrow P, Oldstone MB. Characterization of lymphocytic choriomeningitis virus-binding protein(s): a candidate cellular receptor for the virus. J Virol. 1992;66:7270–81.
Bowden TA, Crispin M, Graham SC, Harvey DJ, Grimes JM, Jones EY, Stuart DI. Unusual molecular architecture of the machupo virus attachment glycoprotein. J Virol. 2009;83:8259–65.
Briese T, Paweska JT, Mcmullan LK, Hutchison SK, Street C, Palacios G, Khristova ML, Weyer J, Swanepoel R, Egholm M, Nichol ST, Lipkin WI. Genetic detection and characterization of lujo virus, a new hemorrhagic fever-associated arenavirus from Southern Africa. PLoS Pathog. 2009;5:E1000455.
Briknarova K, Thomas CJ, York J, Nunberg JH. Structure Of a zinc-binding domain in the Junin virus envelope glycoprotein. J Biol Chem. 2011;286:1528–36.
Buchmeier MJ, De La Torre JC, Peters CJ. Arenaviridae: the viruses and their replication. In: Knipe DL, Howley PM, editors. Fields Virology. 4th ed. Philadelphia: Lippincott-Raven; 2007.
Burri DJ, Da Palma JR, Seidah NG, Zanotti G, Cendron L, Pasquato A, Kunz S. Differential recognition of old world and New World arenavirus envelope glycoproteins by subtilisin kexin isozyme 1 (Ski-1)/site 1 protease (S1p). J Virol. 2013;87:6406–14.
Burri DJ, Pasqual G, Rochat C, Seidah NG, Pasquato A, Kunz S. Molecular characterization of the processing of arenavirus envelope glycoprotein precursors by subtilisin kexin isozyme-1/site-1 protease. J Virol. 2012;86:4935–46.
Cao W, Henry MD, Borrow P, Yamada H, Elder JH, Ravkov EV, Nichol ST, Compans RW, Campbell KP, Oldstone MB. Identification of alpha-dystroglycan as a receptor for lymphocytic choriomeningitis virus and Lassa fever virus [see comments]. Science. 1998;282:2079–81.
Capul AA, Perez M, Burke E, Kunz S, Buchmeier MJ, De La Torre JC. Arenavirus Z-glycoprotein association requires Z myristoylation but not functional ring or late domains. J Virol. 2007;81:9451–60.
Cohen-Dvashi H, Cohen N, Israeli H, Diskin R. Molecular mechanism for Lamp1 recognition by Lassa virus. J Virol. 2015;89:7584–92.
Constam DB, Robertson EJ. Regulation of bone morphogenetic protein activity by Pro domains and proprotein convertases. J Cell Biol. 1999;144:139–49.
Da Palma JR, Burri DJ, Oppliger J, Salamina M, Cendron L, De Laureto PP, Seidah NG, Kunz S, Pasquato A. Zymogen activation and subcellular activity of subtilisin kexin isozyme 1/site 1 protease. J Biol Chem. 2014;289:35743–56.
Da Palma JR, Cendron L, Seidah NG, Pasquato A, Kunz S. Mechanism of folding and activation of subtilisin kexin isozyme-1 (Ski-1)/site-1 protease (S1p). J Biol Chem. 2016;291:2055–66.
De La Torre JC. Molecular and cell biology of the prototypic arenavirus lcmv: implications for understanding and combating hemorrhagic fever arenaviruses. Ann N Y Acad Sci. 2009;1171 Suppl 1:E57–64.
Delgado S, Erickson BR, Agudo R, Blair PJ, Vallejo E, Albarino CG, Vargas J, Comer JA, Rollin PE, Ksiazek TG, Olson JG, Nichol ST. Chapare virus, a newly discovered arenavirus isolated from a fatal hemorrhagic fever case In Bolivia. PLoS Pathog. 2008;4:E1000047.
Eichler R, Lenz O, Strecker T, Eickmann M, Klenk HD, Garten W. Identification of Lassa virus glycoprotein signal peptide as a trans-acting maturation factor. EMBO Rep. 2003a;4:1084–8.
Eichler R, Lenz O, Strecker T, Garten W. Signal peptide of Lassa virus glycoprotein Gp-C exhibits an unusual length. FEBS Lett. 2003b;538:203–6.
Elagoz A, Benjannet S, Mammarbassi A, Wickham L, Seidah NG. Biosynthesis and cellular trafficking of the convertase Ski-1/S1p: ectodomain shedding requires Ski-1 activity. J Biol Chem. 2002;277:11265–75.
Emonet SE, Urata S, De La Torre JC. Arenavirus reverse genetics: new approaches for the investigation of arenavirus biology and development of antiviral strategies. Virology. 2011;411:416–25.
Emonet SF, De La Torre JC, Domingo E, Sevilla N. Arenavirus genetic diversity and its biological implications. Infect Genet Evol. 2009;9:417–29.
Eschli B, Quirin K, Wepf A, Weber J, Zinkernagel R, Hengartner H. Identification of an N-terminal trimeric coiled-coil core within arenavirus glycoprotein 2 permits assignment to class I viral fusion proteins. J Virol. 2006;80:5897–907.

Essalmani R, Zaid A, Marcinkiewicz J, Chamberland A, Pasquato A, Seidah NG, Prat A. In vivo functions of the proprotein convertase Pc5/6 during mouse development: Gdf11 is a likely substrate. Proc Natl Acad Sci U S A. 2008;105:5750–5.

Fehling SK, Lennartz F, Strecker T. Multifunctional nature of the arenavirus ring finger protein Z. Virus. 2012;4:2973–3011.

Froeschke M, Basler M, Groettrup M, Dobberstein B. Long-lived signal peptide of lymphocytic choriomeningitis virus glycoprotein Pgp-C. J Biol Chem. 2003;278:41914–20.

Goncalves AR, Moraz ML, Pasquato A, Helenius A, Lozach PY, Kunz S. Role of Dc-sign in Lassa virus entry into human dendritic cells. J Virol. 2013;87:11504–15.

Gorski JP, Huffman NT, Chittur S, Midura RJ, Black C, Oxford J, Seidah NG. Inhibition of proprotein convertase Ski-1 blocks transcription of key extracellular matrix genes regulating osteoblastic mineralization. J Biol Chem. 2011;286:1836–49.

Gorski JP, Huffman NT, Vallejo J, Brotto L, Chittur SV, Breggia A, Stern A, Huang J, Mo C, Seidah NG, Bonewald L, Brotto M. Deletion of Mbtps1 (Pcsk8, S1p, Ski-1) gene in osteocytes stimulates soleus muscle regeneration and increased size and contractile force with age. J Biol Chem. 2016;291:4308–22.

Grant A, Seregin A, Huang C, Kolokoltsova O, Brasier A, Peters C, Paessler S. Junin virus pathogenesis and virus replication. Virus. 2012;4:2317–39.

Hastie KM, Igonet S, Sullivan BM, Legrand P, Zandonatti MA, Robinson JE, Garry RF, Rey FA, Oldstone MB, Saphire EO. Crystal structure of the prefusion surface glycoprotein of the prototypic arenavirus Lcmv. Nat Struct Mol Biol. 2016;23:513–21.

Hawkins JL, Robbins MD, Warren LC, Xia D, Petras SF, Valentine JJ, Varghese AH, Wang IK, Subashi TA, Shelly LD, Hay BA, Landschulz KT, Geoghegan KF, Harwood HJ Jr. Pharmacologic inhibition of site 1 protease activity inhibits sterol regulatory element-binding protein processing and reduces lipogenic enzyme gene expression and lipid synthesis in cultured cells and experimental animals. J Pharmacol Exp Ther. 2008;326:801–8.

Hay BA, Abrams B, Zumbrunn AY, Valentine JJ, Warren LC, Petras SF, Shelly LD, Xia A, Varghese AH, Hawkins JL, Van Camp JA, Robbins MD, Landschulz K, Harwood HJ Jr. Aminopyrrolidineamide Inhibitors Of Site-1 Protease. Bioorg Med Chem Lett. 2007;17:4411–4.

Helguera G, Jemielity S, Abraham J, Cordo SM, Martinez MG, Rodriguez JA, Bregni C, Wang JJ, Farzan M, Penichet ML, Candurra NA, Choe H. An antibody recognizing the apical domain of human transferrin receptor 1 efficiently inhibits the entry of all New World hemorrhagic fever arenaviruses. J Virol. 2012;86:4024–8.

Hepojoki J, Kipar A, Korzyukov Y, Bell-Sakyi L, Vapalahti O, Hetzelb U. Replication of boid inclusion body disease-associated arenaviruses is temperature sensitive in both boid and mammalian cells. J Virol. 2015;89:1119–28.

Hetzel U, Sironen T, Laurinmaki P, Liljeroos L, Patjas A, Henttonen H, Vaheri A, Artelt A, Kipar A, Butcher SJ, Vapalahti O, Hepojoki J. Isolation, identification, and characterization of novel arenaviruses, the etiological agents of boid inclusion body disease. J Virol. 2013;87:10918–35.

Hunt CL, Lennemann NJ, Maury W. Filovirus entry: a novelty in the viral fusion world. Virus. 2012;4:258–75.

Igonet S, Vaney MC, Vonhrein C, Bricogne G, Stura EA, Hengartner H, Eschli B, Rey FA. X-ray structure of the arenavirus glycoprotein Gp2 in its postfusion hairpin conformation. Proc Natl Acad Sci U S A. 2011;108:19967–72.

Iwasaki M, Urata S, Cho Y, Ngo N, De La Torre JC. Cell entry of lymphocytic choriomeningitis virus is restricted in myotubes. Virology. 2014;458-459:22–32.

Jae LT, Brummelkamp TR. Emerging intracellular receptors for hemorrhagic fever viruses. Trends Microbiol. 2015;23:392–400.

Jae LT, Raaben M, Herbert AS, Kuehne AI, Wirchnianski AS, Soh TK, Stubbs SH, Janssen H, Damme M, Saftig P, Whelan SP, Dye JM, Brummelkamp TR. Virus entry. Lassa virus entry requires a trigger-induced receptor switch. Science. 2014;344:1506–10.

Jemielity S, Wang JJ, Chan YK, Ahmed AA, Li W, Monahan S, Bu X, Farzan M, Freeman GJ, Umetsu DT, Dekruyff RH, Choe H. Tim-family proteins promote infection of multiple enveloped viruses through virion-associated phosphatidylserine. PLoS Pathog. 2013;9:E1003232.

Kilgore PE, Peters CJ, Mills JN, Rollin PE, Armstrong L, Khan AS, Ksiazek TG. Prospects for the control of bolivian hemorrhagic fever. Emerg Infect Dis. 1995;1:97–100.

Knopp KA, Ngo T, Gershon PD, Buchmeier MJ. Single nucleoprotein residue modulates arenavirus replication complex formation. MBio. 2015;6:E00524-15.

Kondo S, Murakami T, Tatsumi K, Ogata M, Kanemoto S, Otori K, Iseki K, Wanaka A, Imaizumi K. Oasis, a Creb/Atf-family member, modulates Upr signalling in astrocytes. Nat Cell Biol. 2005;7:186–94.

Kunz S, Edelmann KH, De La Torre JC, Gorney R, Oldstone MB. Mechanisms for lymphocytic choriomeningitis virus glycoprotein cleavage, transport, and incorporation into virions. Virology. 2003;314:168–78.

Lenz O, Ter Meulen J, Feldmann H, Klenk HD, Garten W. Identification of a novel consensus sequence at the cleavage site of the Lassa virus glycoprotein. J Virol. 2000;74:11418–21.

Lenz O, Ter Meulen J, Klenk HD, Seidah NG, Garten W. The Lassa virus glycoprotein precursor Gp-C is proteolytically processed by subtilase Ski-1/S1p. Proc Natl Acad Sci U S A. 2001;98:12701–5.

Li S, Sun Z, Pryce R, Parsy ML, Fehling SK, Schlie K, Siebert CA, Garten W, Bowden TA, Strecker T, Huiskonen JT. Acidic Ph-induced conformations and Lamp1 binding of the Lassa virus glycoprotein spike. PLoS Pathog. 2016;12:E1005418.

Loureiro ME, D'antuono A, Levingston Macleod JM, Lopez N. Uncovering viral protein-protein interactions and their role in arenavirus life cycle. Virus. 2012;4:1651–67.

Mahmutovic S, Clark L, Levis SC, Briggiler AM, Enria DA, Harrison SC, Abraham J. Molecular basis for antibody-mediated neutralization of New World hemorrhagic fever mammarenaviruses. Cell Host Microbe. 2015;18:705–13.

Maisa A, Stroher U, Klenk HD, Garten W, Strecker T. Inhibition of Lassa virus glycoprotein cleavage and multicycle replication by site 1 protease-adapted alpha(1)-antitrypsin variants. PLoS Negl Trop Dis. 2009;3:E446.

Marschner K, Kollmann K, Schweizer M, Braulke T, Pohl S. A key enzyme in the biogenesis of lysosomes is a protease that regulates cholesterol metabolism. Science. 2011;333:87–90.

Martinez MG, Bialecki MA, Belouzard S, Cordo SM, Candurra NA, Whittaker GR. Utilization of human Dc-Sign and L-Sign for entry and infection of host cells by the New World arenavirus, Junin virus. Biochem Biophys Res Commun. 2013;441:612–7.

Martinez MG, Cordo SM, Candurra NA. Characterization of Junin arenavirus cell entry. J Gen Virol. 2007;88:1776–84.

Mccormick JB, Fisher-Hoch SP. Lassa fever. Curr Top Microbiol Immunol. 2002;262:75–109.

Mccormick JB, King IJ, Webb PA, Scribner CL, Craven RB, Johnson KM, Elliott LH, Belmont-Williams R. Lassa fever. Effective therapy with ribavirin. N Engl J Med. 1986;314:20–6.

Messina EL, York J, Nunberg JH. Dissection of the role of the stable signal peptide of the arenavirus envelope glycoprotein in membrane fusion. J Virol. 2012;86:6138–45.

Millet JK, Whittaker GR. Host cell proteases: critical determinants of coronavirus tropism and pathogenesis. Virus Res. 2015;202:120–34.

Moreno H, Gallego I, Sevilla N, De La Torre JC, Domingo E, Martin V. Ribavirin can be mutagenic for arenaviruses. J Virol. 2011;85:7246–55.

Nakagawa T, Suzuki-Nakagawa C, Watanabe A, Asami E, Matsumoto M, Nakano M, Ebihara A, Uddin MN, Suzuki F. Site-1 protease is required for the generation of soluble (pro)renin receptor. J Biochem. 2016;161(4):369–79.

Neumann G, Feldmann H, Watanabe S, Lukashevich I, Kawaoka Y. Reverse genetics demonstrates that proteolytic processing of the ebola virus glycoprotein is not essential for replication in cell culture. J Virol. 2002;76:406–10.

Nunberg JH, York J. The curious case of arenavirus entry, and its inhibition. Virus. 2012;4:83–101.

Oldstone MB. Biology and pathogenesis of lymphocytic choriomeningitis virus infection. In: Oldstone MB, editor. Arenaviruses. New York: Springer; 2002.

Oldstone MB, Campbell KP. Decoding arenavirus pathogenesis: essential roles for alpha-dystroglycan-virus interactions and the immune response. Virology. 2011;411:170–9.

Oppliger J, Da Palma JR, Burri DJ, Bergeron E, Khatib AM, Spiropoulou CF, Pasquato A, Kunz S. A molecular sensor to characterize arenavirus envelope glycoprotein cleavage by subtilisin kexin isozyme 1/site 1 protease. J Virol. 2015;90:705–14.

Oppliger J, Torriani G, Herrador A, Kunz S. Lassa virus cell entry via dystroglycan involves an unusual pathway of macropinocytosis. J Virol. 2016;90:6412–29.

Palacios G, Druce J, Du L, Tran T, Birch C, Briese T, Conlan S, Quan PL, Hui J, Marshall J, Simons JF, Egholm M, Paddock CD, Shieh WJ, Goldsmith CS, Zaki SR, Catton M, Lipkin WI. A new arenavirus in a cluster of fatal transplant-associated diseases. N Engl J Med. 2008;358:991–8.

Parker WB. Metabolism and antiviral activity of ribavirin. Virus Res. 2005;107:165–71.

Parsy ML, Harlos K, Huiskonen JT, Bowden TA. Crystal structure of venezuelan hemorrhagic fever virus fusion glycoprotein reveals a class 1 Postfusion architecture with extensive glycosylation. J Virol. 2013;87:13070–5.

Pasqual G, Burri DJ, Pasquato A, De La Torre JC, Kunz S. Role of the host cell's unfolded protein response in arenavirus infection. J Virol. 2011a;85:1662–70.

Pasqual G, Rojek JM, Masin M, Chatton JY, Kunz S. Old world arenaviruses enter the host cell via the multivesicular body and depend on the endosomal sorting complex required for transport. PLoS Pathog. 2011b;7:E1002232.

Pasquato A, Burri DJ, Kunz S. Current drug discovery strategies against arenavirus infections. Expert Rev Anti-Infect Ther. 2012a;10:1297–309.

Pasquato A, Burri DJ, Traba EG, Hanna-El-Daher L, Seidah NG, Kunz S. Arenavirus envelope glycoproteins mimic autoprocessing sites of the cellular proprotein convertase subtilisin kexin isozyme-1/site-1 protease. Virology. 2011;417:18–26.

Pasquato A, Pullikotil P, Asselin MC, Vacatello M, Paolillo L, Ghezzo F, Basso F, Di Bello C, Dettin M, Seidah NG. The proprotein convertase Ski-1/S1p. In vitro analysis of Lassa virus glycoprotein-derived substrates and ex vivo validation of irreversible peptide inhibitors. J Biol Chem. 2006;281:23471–81.

Pasquato A, Rochat C, Burri DJ, Pasqual G, De La Torre JC, Kunz S. Evaluation of the anti-arenaviral activity of the subtilisin kexin isozyme-1/site-1 protease inhibitor Pf-429242. Virology. 2012b;423:14–22.

Perez M, Craven RC, De La Torre JC. The small ring finger protein Z drives arenavirus budding: implications for antiviral strategies. Proc Natl Acad Sci U S A. 2003;100:12978–83.

Pinschewer DD, Perez M, Sanchez AB, De La Torre JC. Recombinant lymphocytic choriomeningitis virus expressing vesicular stomatitis virus glycoprotein. Proc Natl Acad Sci U S A. 2003;100:7895–900.

Popkin DL, Teijaro JR, Sullivan BM, Urata S, Rutschmann S, De La Torre JC, Kunz S, Beutler B, Oldstone M. Hypomorphic mutation in the site-1 protease Mbtps1 endows resistance to persistent viral infection in a cell-specific manner. Cell Host Microbe. 2011;9:212–22.

Prescott JB, Marzi A, Safronetz D, Robertson SJ, Feldmann H, Best SM. Immunobiology of Ebola and Lassa virus infections. Nat Rev Immunol. 2017;17:195–207.

Pullikotil P, Benjannet S, Mayne J, Seidah NG. The proprotein convertase Ski-1/S1p: alternate translation and subcellular localization. J Biol Chem. 2007;282:27402–13.

Radoshitzky SR, Abraham J, Spiropoulou CF, Kuhn JH, Nguyen D, Li W, Nagel J, Schmidt PJ, Nunberg JH, Andrews NC, Farzan M, Choe H. Transferrin receptor 1 is a cellular receptor for New World haemorrhagic fever arenaviruses. Nature. 2007;446:92–6. Epub 2007 Feb 7

Radoshitzky SR, Bao Y, Buchmeier MJ, Charrel RN, Clawson AN, Clegg CS, Derisi JL, Emonet S, Gonzalez JP, Kuhn JH, Lukashevich IS, Peters CJ, Romanowski V, Salvato MS, Stenglein MD, De La Torre JC. Past, present, and future of arenavirus taxonomy. Arch Virol. 2015;160:1851–74.

Radoshitzky SR, Longobardi LE, Kuhn JH, Retterer C, Dong L, Clester JC, Kota K, Carra J, Bavari S. Machupo virus glycoprotein determinants for human transferrin receptor 1 binding and cell entry. PLoS One. 2011;6:E21398.

Remacle AG, Shiryaev SA, Oh ES, Cieplak P, Srinivasan A, Wei G, Liddington RC, Ratnikov BI, Parent A, Desjardins R, Day R, Smith JW, Lebl M, Strongin AY. Substrate cleavage analysis of furin and related proprotein convertases. A Comparative Study. J Biol Chem. 2008;283:20897–906.

Rojek JM, Lee AM, Nguyen N, Spiropoulou CF, Kunz S. Site 1 protease is required for proteolytic processing of the glycoproteins of the South American hemorrhagic fever viruses Junin, Machupo, and Guanarito. J Virol. 2008;82:6045–51.

Rojek JM, Pasqual G, Sanchez AB, Nguyen NT, De La Torre JC, Kunz S. Targeting the proteolytic processing of the viral glycoprotein precursor is a promising novel antiviral strategy against arenaviruses. J Virol. 2010;84:573–84.

Ruiz-Jarabo CM, Ly C, Domingo E, De La Torre JC. Lethal mutagenesis of the prototypic arenavirus lymphocytic choriomeningitis virus (Lcmv). Virology. 2003;308:37–47.

Rutschmann S, Crozat K, Li X, Du X, Hanselman JC, Shigeoka AA, Brandl K, Popkin DL, Mckay DB, Xia Y, Moresco EM, Beutler B. Hypopigmentation and maternal-zygotic embryonic lethality caused by a hypomorphic Mbtps1 mutation in mice. G3 (Bethesda). 2012;2:499–504.

Sakai J, Rawson RB, Espenshade PJ, Cheng D, Seegmiller AC, Goldstein JL, Brown MS. Molecular identification of the sterol-regulated luminal protease that cleaves Srebps and controls lipid composition of animal cells. Mol Cell. 1998;2:505–14.

Sayler KA, Barbet AF, Chamberlain C, Clapp WL, Alleman R, Loeb JC, Lednicky JA. Isolation of Tacaribe virus, a Caribbean arenavirus, from host-seeking Amblyomma americanum ticks in Florida. PLoS One. 2014;9:E115769.

Schlie K, Maisa A, Lennartz F, Stroher U, Garten W, Strecker T. Characterization of Lassa virus glycoprotein oligomerization and influence of cholesterol on virus replication. J Virol. 2010a;84:983–92.

Schlie K, Strecker T, Garten W. Maturation cleavage within the ectodomain of Lassa virus glycoprotein relies on stabilization by the cytoplasmic tail. FEBS Lett. 2010b;584:4379–82.

Seidah NG. The proprotein convertases, 20 years later. Methods Mol Biol. 2011;768:23–57.

Seidah NG, Mowla SJ, Hamelin J, Mamarbachi AM, Benjannet S, Toure BB, Basak A, Munzer JS, Marcinkiewicz J, Zhong M, Barale JC, Lazure C, Murphy RA, Chretien M, Marcinkiewicz M. Mammalian subtilisin/kexin isozyme Ski-1: a widely expressed proprotein convertase with a unique cleavage specificity and cellular localization. Proc Natl Acad Sci U S A. 1999;96:1321–6.

Seidah NG, Prat A. Precursor convertases in the secretory pathway, cytosol and extracellular milieu. Essays Biochem. 2002;38:79–94.

Seidah NG, Prat A. The proprotein convertases are potential targets in the treatment of dyslipidemia. J Mol Med (Berl). 2007;85:685–96.

Seidah NG, Prat A. The biology and therapeutic targeting of the proprotein convertases. Nat Rev Drug Discov. 2012;11:367–83.

Shankar S, Whitby LR, Casquilho-Gray HE, York J, Boger DL, Nunberg JH. Small-molecule fusion inhibitors bind the Ph-sensing stable signal peptide-Gp2 subunit interface of the Lassa virus envelope glycoprotein. J Virol. 2016;90:6799–807.

Shimojima M, Kawaoka Y. Cell surface molecules involved in infection mediated by lymphocytic choriomeningitis virus glycoprotein. J Vet Med Sci. 2012;74:1363–6.

Shimojima M, Stroher U, Ebihara H, Feldmann H, Kawaoka Y. Identification of cell surface molecules involved in dystroglycan-independent Lassa virus cell entry. J Virol. 2012;86:2067–78.

Shiryaev SA, Remacle AG, Ratnikov BI, Nelson NA, Savinov AY, Wei G, Bottini M, Rega MF, Parent A, Desjardins R, Fugere M, Day R, Sabet M, Pellecchia M, Liddington RC, Smith JW, Mustelin T, Guiney DG, Lebl M, Strongin AY. Targeting host cell furin proprotein convertases as a therapeutic strategy against bacterial toxins and viral pathogens. J Biol Chem. 2007;282:20847–53.

Stenglein MD, Sanders C, Kistler AL, Ruby JG, Franco JY, Reavill DR, Dunker F, Derisi JL. Identification, characterization, and in vitro culture of highly divergent arenaviruses from boa constrictors and annulated tree boas: candidate etiological agents for snake inclusion body disease. MBio. 2012;3:E00180-12.

Sweileh WM. Global research trends of World Health Organization's top eight emerging pathogens. Glob Health. 2017;13:9.

Torriani G, Galan-Navarro C, Kunz S. Lassa virus cell entry reveals new aspects of virus-host cell interaction. J Virol. 2017;91:pii: e01902-16.

Toure BB, Munzer JS, Basak A, Benjannet S, Rochemont J, Lazure C, Chretien M, Seidah NG. Biosynthesis and enzymatic characterization of human Ski-1/S1p and the processing of its inhibitory prosegment. J Biol Chem. 2000;275:2349–58.

Urata S, De La Torre JC. Arenavirus Budding. Adv Virol. 2011;2011:180326.

Urata S, Weyer J, Storm N, Miyazaki Y, Van Vuren PJ, Paweska JT, Yasuda J. Analysis of assembly and budding of Lujo virus. J Virol. 2015;90:3257–61.

Urata S, Yun N, Pasquato A, Paessler S, Kunz S, De La Torre JC. Antiviral activity of a small-molecule inhibitor of arenavirus glycoprotein processing by the cellular site 1 protease. J Virol. 2011;85:795–803.

Weissenbacher MC, Laguens RP, Coto CE. Argentine hemorrhagic fever. Curr Top Microbiol Immunol. 1987;134:79–116.

Wolff S, Ebihara H, Groseth A. Arenavirus budding: a common pathway with mechanistic differences. Virus. 2013;5:528–49.

Wright KE, Spiro RC, Burns JW, Buchmeier MJ. Post-translational processing of the glycoproteins of lymphocytic choriomeningitis virus. Virology. 1990;177:175–83.

Ye J, Rawson RB, Komuro R, Chen X, Dave UP, Prywes R, Brown MS, Goldstein JL. Er stress induces cleavage of membrane-bound Atf6 by the same proteases that process Srebps. Mol Cell. 2000;6:1355–64.

York J, Dai D, Amberg SM, Nunberg JH. Ph-induced activation of arenavirus membrane fusion is antagonized by small-molecule inhibitors. J Virol. 2008;82:10932–9.

York J, Nunberg JH. Role of the stable signal peptide of Junin arenavirus envelope glycoprotein in Ph-dependent membrane fusion. J Virol. 2006;80:7775–80.

York J, Nunberg JH. Distinct requirements for signal peptidase processing and function in the stable signal peptide subunit of the Junin virus envelope glycoprotein. Virology. 2007;359:72–81.

York J, Nunberg JH. Intersubunit interactions modulate Ph-induced activation of membrane fusion by the Junin virus envelope glycoprotein Gpc. J Virol. 2009;83:4121–6.

York J, Romanowski V, Lu M, Nunberg JH. The signal peptide of the Junin arenavirus envelope glycoprotein is myristoylated and forms an essential subunit of the mature G1-G2 complex. J Virol. 2004;78:10783–92.

Yun NE, Walker DH. Pathogenesis of Lassa fever. Virus. 2012;4:2031–48.

Priming Time: How Cellular Proteases Arm Coronavirus Spike Proteins

4

Markus Hoffmann*, Heike Hofmann-Winkler*, and Stefan Pöhlmann

Abstract

Coronaviruses are enveloped RNA viruses that infect mammals and birds. Infection of humans with globally circulating human coronaviruses is associated with the common cold. In contrast, transmission of animal coronaviruses to humans can result in severe disease: The severe acute respiratory syndrome (SARS) and the Middle East respiratory syndrome (MERS) are responsible for hundreds of deaths in Asia and the Middle East, respectively, and are both caused by members of the genus *Betacoronavirus*, SARS-CoV, and MERS-CoV that were zoonotically transmitted from an animal host to humans. At present, neither vaccines nor specific treatment is available to combat coronavirus infection in humans, and novel antiviral strategies are urgently sought. The viral spike protein (S) mediates the first essential step in coronavirus infection, viral entry into target cells. For this, the S protein critically depends on priming by host cell proteases, and the responsible enzymes are potential targets for antiviral intervention. Recent studies revealed that the endosomal cysteine protease cathepsin L and the serine proteases furin and TMPRSS2 prime the S proteins of SARS-CoV and MERS-CoV and provided evidence that successive S protein cleavage at two sites is required for S protein priming. Moreover, mechanisms that control protease choice were unraveled, and insights were obtained into which enzyme promotes viral spread in the host. Here, we will provide basic information on S protein function and proteolytic priming, and we will then discuss recent progress in our understanding of the priming of the S proteins of SARS-CoV and MERS-CoV.

*M.H. and H.H-.W. contributed equally to this work.

M. Hoffmann · H. Hofmann-Winkler · S. Pöhlmann (✉)
Infection Biology Unit, German Primate Center, Göttingen, Germany
e-mail: spoehlmann@dpz.eu

E. Böttcher-Friebertshäuser et al. (eds.), *Activation of Viruses by Host Proteases*,
https://doi.org/10.1007/978-3-319-75474-1_4

4.1 Introduction

Coronaviruses (CoV) belong to the *Coronavirinae* subfamily that forms along with the subfamily *Torovirinae* the virus family *Coronaviridae* within the order *Nidovirales*. The *Coronavirinae* subfamily harbors four genera (Fig. 4.1): *Alpha-*, *Beta-*, *Gamma-*, and *Deltacoronavirus* (Adams and Carstens 2012; Woo et al. 2012). Coronaviruses are enveloped viruses that contain a single-stranded RNA genome of positive polarity comprising roughly 30 kilobases. The virus particles are spherical and with a diameter

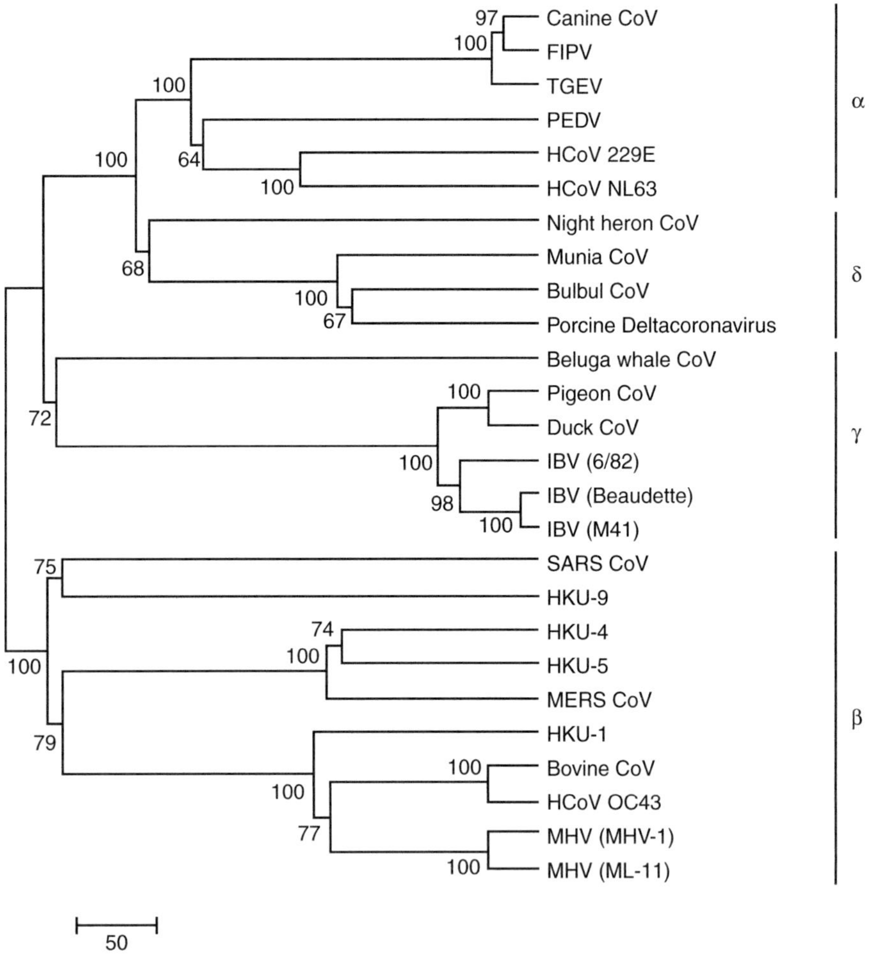

Fig. 4.1 Phylogenetic relationship among coronaviruses based on their spike glycoproteins. The amino acid sequences of coronavirus spike glycoproteins representing all four genera (*Alpha-*, α; *Beta-*, β; *Gamma-*, γ; *Deltacoronavirus*, δ) within the *Coronavirinae* subfamily were aligned and utilized to generate a phylogenetic tree (neighbor-joining method). Italicized numbers at the nodes indicate bootstrap values

of 80–120 nm (Belouzard et al. 2012). They contain the genome, which is associated with the nucleoprotein (NP), forming a ribonucleoprotein complex (RNP) (Belouzard et al. 2012). Depending on the virus, three or four viral proteins are embedded in the viral envelope: Membrane protein (M), envelope protein (E), and spike glycoprotein (S) are present in all coronaviruses, while some members of the genus *Betacoronavirus* additionally contain a hemagglutinin-esterase protein (HE). M and E are required for viral assembly (Belouzard et al. 2012), HE promotes release of viruses from infected cells (Vlasak et al. 1988), and the S protein, which is in the focus of this review, facilitates viral entry into target cells. The S protein is also responsible for the corona-like shape of these viruses in electron micrographs, on the basis of which the name *coronavirus* was coined (Berry and Almeida 1968; Du et al. 2009).

Coronaviruses infect a broad range of vertebrate hosts with alpha- and betacoronaviruses targeting different mammals, while gamma- and deltacoronaviruses mainly infect birds (Breslin et al. 1999; Cavanagh et al. 2001; Jonassen et al. 2005). It is believed that coronaviruses of the genera *Alpha-* and *Betacoronavirus* have emerged from bats, while gamma- and deltacoronaviruses seem to originate from birds (Graham and Baric 2010; Woo et al. 2012). Coronavirus infection is mainly associated with respiratory and enteric diseases but, depending on the virus, can also lead to hepatic (Lane and Hosking 2010) and neurologic manifestations (Foley and Leutenegger 2001).

Human coronaviruses (HCoVs) are known since 1965 when they were identified in patients suffering from the common cold (Tyrrell and Bynoe 1965). Most of HCoVs known today (HCoV-229E, HCoV-NL63, HCoV-OC43, and HCoV-HKU1) infect ciliated epithelia cells of the nasopharynx (Afzelius 1994; Weiss and Navas-Martin 2005) and cause self-limiting upper respiratory tract diseases in immunocompetent individuals, with symptoms like headache, sore throat, and malaise being frequently observed. In rare events, infection can spread to the lower respiratory tract, causing bronchiolitis, bronchitis, and pneumonia, particularly in infants, the elderly, and immunocompromised individuals (Masters and Perlman 2013).

Within the last 20 years, two novel HCoVs emerged that cause severe and frequently fatal infections in humans (Drosten et al. 2003; Lu et al. 2015; Reusken et al. 2016; Zaki et al. 2012). In 2002, the outbreak of severe acute respiratory syndrome coronavirus (SARS-CoV) in Southern China and its subsequent worldwide spread was associated with roughly 8100 infections of which 10% took a fatal course, with the elderly being mainly affected (Peiris et al. 2003). In the aftermath of the SARS pandemic, it has been revealed that bats harbor numerous SARS-CoV-related viruses as well as other coronaviruses that may be zoonotically transmitted to humans via intermediate hosts (Hu et al. 2015; Lu et al. 2015). In 2012, the Middle East respiratory syndrome coronavirus (MERS-CoV), another novel, highly pathogenic coronavirus emerged in Saudi Arabia, causing a SARS-like disease (Zaki et al. 2012). MERS-CoV infection is associated with a case-fatality rate of 35% (WHO Health Organisation 2017), and comorbidities like diabetes mellitus, chronic renal disease, and hypertension constitute major risk factors for a lethal outcome of the disease (Assiri et al. 2013). Like SARS-CoV, MERS-CoV is a zoonotic virus originating from an animal reservoir, dromedary camels (Mohd et al. 2016). As the MERS epidemic is still ongoing, there are concerns that human-to-human transmission, which is very infrequent at present (Alsolamy and Arabi 2015), might become more efficient due to adaptive mutations in the viral genome (Dudas and Rambaut 2016; Reusken et al. 2016).

Coronaviruses also constitute a severe threat to animal health. For instance, porcine epidemic diarrhea coronavirus (PEDV) infects the epithelia of the small intestine and causes villous atrophy, resulting in diarrhea and severe dehydration (Debouck and Pensaert 1980; Jung et al. 2006). The virus was first described in Europe in the 1970s and was originally not perceived as a major threat to animal health (Debouck and Pensaert 1980; Pensaert and de 1978). Recently, however, highly virulent PEDV strains emerged that cause lethal infection in 80–100% of piglets and weight loss in adult pigs (Debouck and Pensaert 1980; Lee 2015). PEDV spread can have severe consequences: The introduction of PEDV in the USA resulted in major economic losses among pig farmers and a 10% decline in the American pig population (Lee 2015; Li et al. 2012; Liu et al. 2016; Stevenson et al. 2013). As there are no effective vaccines or specific treatments available, current containment strategies are mainly limited to rigorous disinfection routines.

Coronaviruses constitute a severe threat to animal and human health, as discussed above, and the development of antivirals is an important task. Host cell factors required for coronavirus spread but dispensable for cellular survival are attractive targets, since their blockade might suppress infection by several coronaviruses and might be associated with a high barrier against resistance development. The viral S protein mediates the first step in coronavirus spread, viral entry into target cells. However, the S protein is synthesized as an inactive precursor and requires cleavage by host cell proteases for conversion into an active form. The cellular enzymes responsible constitute targets for antiviral intervention, and recent studies provided important insights into their identity, expression, and target sites in the viral S protein. Moreover, novel mechanisms governing protease choice by coronaviruses have been uncovered. The present manuscript will review and discuss these findings, focusing on SARS-CoV and MERS-CoV.

4.2 The Coronavirus Spike Protein: Viral Key for Entry into the Target Cell

Domain organization. The S protein of coronaviruses contains an N-terminal signal peptide which primes the nascent polyprotein for import into the ER. In the ER, the S protein is extensively modified with N-linked glycans, which may provide protection against neutralizing antibodies (Walls et al. 2016b). After passing the quality control mechanisms of the ER, the S protein is transported to the site of viral budding, the endoplasmic reticulum/Golgi intermediate compartment (ERGIC). Homotrimers of the S protein, for which atomic structures have recently been reported (Kirchdoerfer et al. 2016; Walls et al. 2016a), are incorporated into the viral membrane and mediate viral entry into target cells. For this, the S protein combines two biological functions: First, its surface unit, S1, binds to a specific receptor located at the surface of host cells and thereby determines cellular tropism and, as a consequence, viral pathogenesis. Second, the transmembrane unit, S2, mediates fusion between the viral envelope and a target cell membrane (Fig. 4.2).

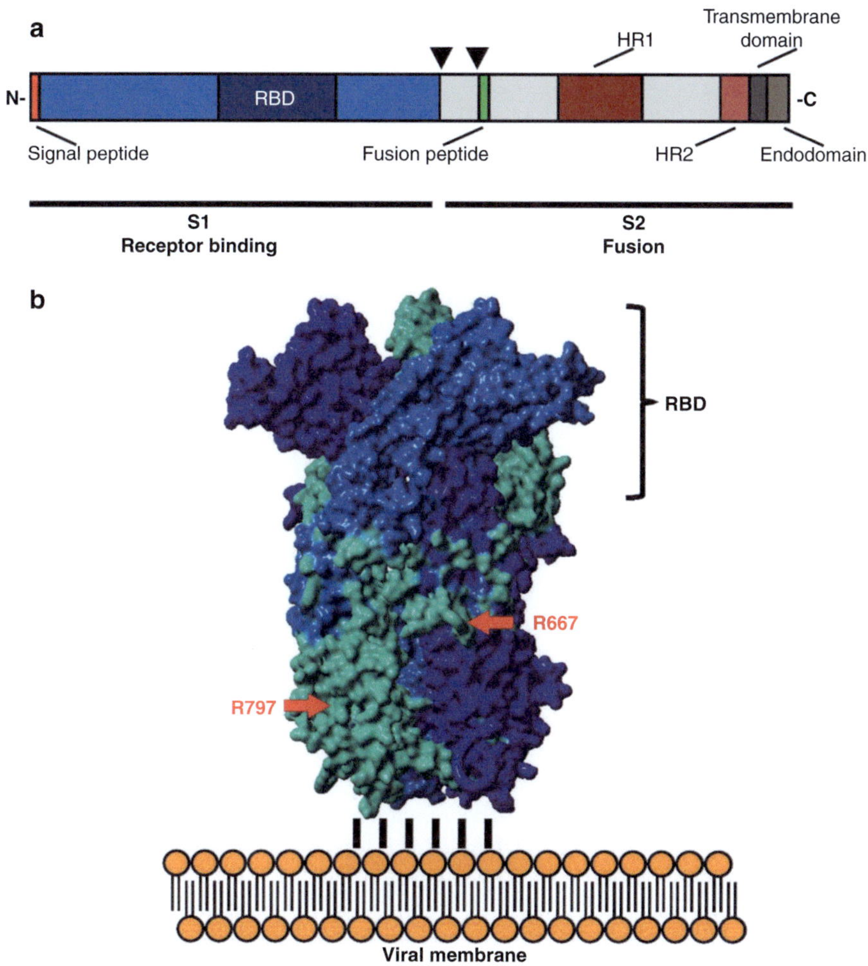

Fig. 4.2 Domain organization and structure of the coronavirus spike glycoprotein. (**a**) Schematic illustration of a coronavirus spike (S) glycoprotein consisting of the subdomains S1 and S2. At the N-terminus of the S1 subdomain resides the signal peptide that allows for introduction of nascent S proteins into the host cells' secretory pathway. Additionally, this subdomain harbors amino acid residues responsible for virus attachment to target cells (receptor-binding domain, RBD). The S2 subdomain contains the structural components of the membrane fusion machinery (fusion peptide, heptad repeats (HR) 1 and 2), anchors the S protein in the lipid envelope via the transmembrane domain, and interacts with the viral ribonucleoprotein complex through its endodomain. Location of the S1/S2 border and the S2′ position is indicated by black triangles. (**b**) 3D-model of trimeric SARS-CoV S protein (amino acid residues 261–1058) schematically positioned on the outside of the viral envelope. The protein structure ID, 5WRG, (Gui et al. 2017) was downloaded from the RCSB Protein Data Bank and analyzed using the YASARA software (www.yasara.org, Krieger and Vriend 2014). Each S protein monomer is colored individually, and the position of the RBD is indicated. Further, the locations of the arginines at the S1/S2 border (R667) and S2′ position (R797) are highlighted

Table 4.1 Host cell receptors of selected alpha- and betacoronaviruses

Genus	Virus	Receptor
Alphacoronavirus	FIPV	CD13
	TGEV	CD13
	HCoV-NL63	Angiotensin-converting enzyme 2 (ACE2)
	HCoV-229E	CD13
Betacoronavirus	SARS-CoV	Angiotensin-converting enzyme 2 (ACE2)
	MERS-CoV	Dipeptidyl peptidase 4 (DPP4)
	HCoV-OC43	*N*-Acetyl-9-*O*-acetyl neuraminic acid (Neu5,9Ac2)
	MHV	Carcinoembryonic antigen-related cell adhesion molecule 1a (CEACAM1a)

Cellular receptors. Coronaviruses use a broad range of receptors for entry into target cells (Table 4.1). Alphacoronaviruses like HCoV-229E, transmissible gastroenteritis coronavirus (TGEV), and porcine respiratory coronavirus (PRCV) engage the zinc metalloproteinase CD13 from their natural host as well as feline CD13 (feCD13) as entry receptor (Tresnan and Holmes 1998), with different residues in feCD13 being required for recognition by the respective coronaviral S proteins (Tusell et al. 2007). Despite high amino acid sequence similarity within the S1 subunit, the S proteins of HCoV-229E and -NL63 interact with different host cell receptors, namely, CD13 (Yeager et al. 1992) and angiotensin-converting enzyme 2 (ACE2) (Hofmann et al. 2005). Notably, ACE2 is also employed by SARS-CoV for entry (Li et al. 2003; Wang et al. 2004), although the S protein of this betacoronavirus and NL63-S share little sequence similarity. Other members of the betacoronaviruses use different entry receptors: MERS-CoV uses human dipeptidyl peptidase 4 (DPP4), mouse hepatitis virus (MHV) interacts with carcinoembryonic antigen-related cell adhesion molecule 1 (CEACAM1) (Dveksler et al. 1991; Williams et al. 1991), and neuraminic acid is used by bovine CoV and HCoV-OC43 for attachment to cells (Kunkel and Herrler 1993; Schultze et al. 1991). Similarly, sialic acid-containing surface molecules serve as attachment factors or receptors for TGEV, PEDV, and avian infectious bronchitis virus (IBV) (Cavanagh and Davis 1986; Deng et al. 2016; Krempl et al. 1997; Liu et al. 2015; Schultze et al. 1992).

Structural insights into receptor choice. The proteolytic priming of the viral S proteins is in the center of this review. However, priming and receptor binding can be intimately connected, and structural analyses provide valuable explanations for coronavirus receptor specificity. Therefore, structural aspects of S protein binding to its receptor will be briefly discussed. Binding to a receptor is mediated by a receptor-binding domain (RBD), which is located in the surface unit S1. The S1 subunit generally consists of an N-terminal (NTD) and a C-terminal domain (CTD) (Li 2012), which can serve as RBD either alone or in combination. For most coronavirus analyzed, the S1-NTD is responsible for binding to host cell glycans (Krempl et al. 1997; Liu et al. 2015; Peng et al. 2012; Promkuntod et al. 2014), whereas the S1-CTD targets the a proteinaceous receptor (Du et al. 2013; Godet et al. 1994; Hofmann et al. 2006; Lin et al. 2008; Liu et al. 2015; Mou et al. 2013; Wong et al. 2004). All S1-CTD investigated so far are characterized by a core domain overlaid by an external region, which directly contacts the receptor (Li 2016). The S1-CTD of SARS-S comprises a core of five ß-sheets in antiparallel orientation, headed by a rather globular external region (Li et al. 2005a) in which amino acids N479 and T487 mediate high affinity binding to ACE2 (Li et al. 2005b). The S protein of SARS-CoV from palm civets, a potential intermediate host (Guan et al. 2003; Ksiazek et al. 2003; Rota et al. 2003; Song et al. 2005; Wu et al. 2005), harbors amino acids at positions 479 and 487 which preclude efficient binding to human ACE2 (Li 2008), and acquisition of mutations at these positions was sufficient for cross-species transmission during the SARS epidemic (Li 2008; Li et al. 2005b; Qu et al. 2005; Song et al. 2005; Wu et al. 2011, 2012). Within human ACE2, two lysine residues (K31 and K353) are critical for SARS-S binding (Li 2008; Wu et al. 2011, 2012), and an exchange to histidine at position 353 present in murine ACE2 renders this protein unsuitable for efficient SARS-S binding (Li et al. 2004, 2005b). Similarly, the rat homologue of ACE2 contains a glycosylated asparagine at position 82 which sterically blocks S protein interaction (Frieman et al. 2012; Li et al. 2004). These findings show that subtle variations within the S protein and its receptor can dramatically impact cross-species transmission of coronaviruses.

The core domain of the S1-CTD in MERS-S structurally resembles that of SARS-S (Chen et al. 2013; Lu et al. 2013; Wang et al. 2013; Yuan et al. 2017), but the extended core domains are different, with the MERS-S extended core consisting of antiparallel ß-sheets forming a flat surface which targets DPP4 (Raj et al. 2013). The MERS-S binding site on DPP4 is located within a propeller-like structure conserved in bat, camel, and human DPP4 (Barlan et al. 2014; van et al. 2014), and MERS-related CoV have been isolated from both bats and camels (Alagaili et al. 2014; Annan et al. 2013; Haagmans et al. 2014; Lau et al. 2013). In contrast, rodent DPP4 homologues are nonfunctional as MERS-CoV receptors (Cockrell et al. 2014; Coleman et al. 2014; Fukuma et al. 2015; Peck et al. 2015; Raj et al. 2014), probably due to steric hindrance due to a glycosylation in rodent DPP4 (Peck et al. 2015).

In a recent publication, Yuan and colleagues analyzed trimeric MERS- and SARS-S proteins in their pre-fusion conformation using single-particle cryo-electron microscopy (Yuan et al. 2017). Their results revealed an unexpected flexibility of the respective RBDs: in the "lying state," the RBDs are buried inside the trimer, whereas in the "standing state" the RBDs are exposed for receptor interaction (Yuan et al. 2017). Hereby, MERS-S1/S2 trimers appeared with one or two of the RBDs in the standing conformation, thus being able to contact DPP4, whereas SARS-S trimers showed two or all three RBDs in the lying state, thus being incapable of receptor binding without further conformational change. The flexibility of the RBDs might therefore alleviate receptor interaction for subsequent virus entry (Yuan et al. 2017).

Finally, it should be noted that the RBD constitutes the most important target for neutralizing antibodies (Bonavia et al. 2003; Breslin et al. 2003; Godet et al. 1994; He et al. 2004; Kubo et al. 1994). Additionally, sequence comparison of six HCoV S2 domains suggests that also the fusion peptide, the HR1 domain, and the central helix, which are exposed at the surface of the stem region of S protein trimers, can be targeted by neutralizing antibodies (Yuan et al. 2017). Therefore, the structural information discussed above not only provides insights into S protein receptor interactions but also helps to understand how they can be inhibited by antibodies (Du et al. 2008; Lan et al. 2015; Oh et al. 2014; Tai et al. 2017; Walls et al. 2016a).

Membrane fusion. The transmembrane unit S2 harbors domains required for fusion between viral and host cell membrane, including a fusion peptide and two heptad repeats (HR1 and HR2). These elements are followed by a transmembrane (TM) domain and a C-terminal intracytoplasmic tail (Fig. 4.2), which plays a role in S protein sorting. The HR domains consist of α-helices, and their position and amino acid sequences are conserved among all groups within the coronavirus family (de Groot et al. 1987). Membrane fusion commences with the insertion of the fusion peptide into the target cell membrane. Subsequently, the HR regions fold back onto each other, resulting in the formation of a thermostable six-helix bundle structure (Bosch et al. 2003; Duquerroy et al. 2005; Lu et al. 2014; White and Whittaker 2016). As a consequence, the membranes are pulled into close contact and ultimately fuse. Several unrelated viral glycoproteins exhibit the same domain organization and membrane fusion mechanism as CoV S proteins (Dimitrov 2004; White and Whittaker 2016). These proteins are collectively termed class I membrane fusion proteins and contain α-helices as the predominant structural element (Belouzard et al. 2012; Bosch et al. 2003; Tripet et al. 2004; White and Whittaker 2016). All viral class I membrane fusion proteins require a trigger to overcome the energy barrier associated with membrane fusion reaction, low pH, and/or potentially receptor binding. Moreover, viral class I membrane fusion proteins are invariably synthesized as inactive precursors and depend on priming by host cell proteases to transit into an active form, and the general aspects of CoV S protein priming will be discussed in the next section.

4.3 Proteolytic Priming of Coronavirus Spike Proteins: Basic Concepts

The proteolytic separation of the S1 and S2 subunits, termed priming, provides the CoV S protein with the structural flexibility required for the membrane fusion reaction. Initial studies, conducted with the envelope protein of human immunodeficiency virus (HIV) and the hemagglutinin of highly pathogenic avian influenza A viruses (FLUAV), indicated that cleavage occurs in the constitutive secretory pathway of infected cells and is carried out by furin or related subtilisin-like proteases (Hallenberger et al. 1992; Stieneke-Gröber et al. 1992). Moreover, cleavage was shown to occur at the border between the surface and transmembrane units of these glycoproteins (Hallenberger et al. 1992; Stieneke-Gröber et al. 1992). However, subsequent studies, many of which were conducted in recent years, showed that priming of CoV S proteins is substantially more complex and can impact the cellular localization of membrane fusion. The major advances of our understanding of S protein priming relative to early studies will be briefly outlined below and will then be discussed in detail in the context of SARS-CoV and MERS-CoV infection.

Two cleavage sites. Initial studies reported cleavage of viral glycoproteins at the border between surface and transmembrane unit, but more than one cleavage event

		S1/S2		S2'	
α	Canine CoV (1-71)	772-TTTPNFYY	YSIY-783	955-HNSKRKYR	SAIE-966
	FIPV	771-TTTPNFYY	YSIY-782	954-HNSKRKYG	SAIE-965
	TGEV (Miller)	768-TTTPNFYY	YSIY-779	951-DNSKRKYR	SAIE-962
	HCoV 229E	526-EMPKFFYA	SNGT-537	679-SGSRVAGR	SAIE-690
	HCoV NL63	708-QLPNFYYV	SNGG-721	863-HSSRIAGR	SALE-874
	PEDV	727-ELPGFFYH	SNDG-738	884-SGRVVQKR	SVIE-895
β	SARS CoV	660-YHTVSLLR	STSQ-671	790-DPLKPTKR	SFIE-801
	MERS CoV	744-TLTPRSVR	SVPG-755	880-STGSRSAR	SAIE-891
	Bovine CoV	761-STKRRSRR	SITT-772	906-DCNKVSSR	SAIE-917
	HCoV OC43	756-SKNRRSRG	AITT-767	896-ECSKASSR	SAIE-907
	MHV (MHV-1)	750-STSHRARR	SIST-761	901-TTAALKGR	SVIE-912
	MHV (ML-11)	750-STSHRARS	SVST-761	900-TMAAQTGR	SAIE-911
	HCoV HKU-1	753-SSSRRKRR	SISA-764	897-PHCGSSSR	SFFE-908
	Bat CoV HKU-4	742-VPPVSTFR	SYSA-753	879-GGSSSSYR	SAIE-890
	Bat CoV HKU-5	738-TTSSRVRR	ATSG-749	877-TTGERKYR	STIE-888
	Bat CoV HKU-9	669-ANADSLPR	LQLV-670	802-NCGATTYR	SAFS-813
γ	IBV (Beaudette)	530-NGTRRFRR	SITE-541	683-NPSSRRKR	SLIE-694
	IBV (6/82)	531-NGTRRSRR	SITG-542	684-PPSSASGR	SFIE-695
	IBV (M41)	530-NGTRRFRR	SITE-541	683-TPSSPRRR	SFIE-694
	Pigeon CoV	535-KETRRFRR	STSE-546	688-PSSSPRGR	SFIE-688
	Duck CoV	533-NVSRRHRR	SISE-544	686-PHNSPSGR	SFIE-697
δ	Bulbul CoV	520-IAASTFYY	STNA-531	670-ITSKAGGR	SAIE-681
	Night Heron CoV	564-QETPNFYI	ANNA-575	733-QNNSPQKR	SVIE-744
	Munia CoV	513-IPTPTFYY	TTNA-524	662-LSNKIGEK	SVIE-673
	Porcine DeltaCoV	515-IVTPTFFY	STNA-526	665-LTPRVGGR	SAIE-676

Fig. 4.3 Amino acid residues at the S1/S2 interphase and S2′ position among different coronavirus spike proteins. Partial sequence alignment of amino acid residues of coronavirus spike glycoproteins from all four genera located at sites used for S protein activation, S1/S2 border, and the S2′ position (numbers indicate the respective regions of the respective full length S proteins). Basic amino acid residues upstream of the S1/S2 border and the S2′ position are written in bold letters. Moreover, mono- and multibasic motifs suitable for host cell protease-mediated S protein activation are highlighted (gray boxes)

might be required for S protein activation (Belouzard et al. 2009; Millet and Whittaker 2014). Thus, it is now appreciated that several S proteins are cleaved at the interface between the S1 and S2 subunits, termed S1/S2 site, and at a site located near the N-terminus of the fusion peptide, termed S2′ site (Fig. 4.3). The latter cleavage might be of particular importance since it generates the mature N-terminus of the fusion peptide, which is required for insertion into the target cell membrane and thus the successful execution of the membrane fusion reaction (Belouzard et al. 2009; Millet and Whittaker 2014).

Multiple priming enzymes, multiple cellular locations for priming. Several enzymes, pertaining to different protease families, can be hijacked by CoV S proteins for priming. The pH-dependent cysteine protease cathepsin L, TMPRSS2, and other members of the type II transmembrane serine protease (TTSP) family as well as the serine protease furin can prime S proteins during viral entry into target cells (Bertram et al. 2012, 2013; Gierer et al. 2013; Glowacka et al. 2011; Matsuyama et al. 2010; Millet and Whittaker 2014; Shirato et al. 2013; Simmons et al. 2005). In addition, furin can cleave CoV S proteins in infected cells (Bergeron et al. 2005; Millet and Whittaker 2015; Yamada and Liu 2009). These proteases are expressed at different sites in cells, and their intracellular localization determines the cellular location of S protein-driven membrane fusion. For instance, cathepsin L is expressed in endosomes and cleaves S proteins upon viral uptake into these vesicles (Burkard et al. 2014; Huang et al. 2006; Qiu et al. 2006; Simmons et al. 2005; White and Whittaker 2016), while TTSPs process their ligands at the cell surface and are believed to cleave S proteins at this site (Glowacka et al. 2011; Matsuyama et al. 2010; Shulla et al. 2011). Finally, S protein processing in infected cells can determine which proteases can be engaged for priming during viral entry into target cells, suggesting an intricate connection between proteolysis events (Park et al. 2016).

Link between receptor binding and priming. Receptor binding and priming are frequently viewed as separate events. For instance, the FLUAV hemagglutinin is primed by proteases in infected cells and uses sialic acid modified proteins or lipids on the surface of target cells as entry receptor (Hamilton et al. 2012). In contrast, receptor engagement and priming can be intimately connected for CoV S proteins. Thus, SARS-S on cell-free virions is inactivated by trypsin cleavage, while trypsin cleavage of virion-associated SARS-S bound to its receptor ACE2 primes the S protein for membrane fusion (Belouzard et al. 2009; Matsuyama et al. 2005; Simmons et al. 2004, 2005). Similarly, DPP4 binding of MERS-S, precleaved at the S1/S2 site, is believed to be required for subsequent priming by TMPRSS2, as discussed above (Millet and Whittaker 2014; Park et al. 2016). On the basis of these findings, it has been postulated that receptor binding can induce conformational changes in S proteins that expose cleavage sites for priming proteases.

Priming and triggering of S proteins: Distinction without a difference? Viral glycoproteins are usually triggered by protonation and/or receptor binding, which allow the proteins to overcome the energy barrier associated with membrane fusion. However, neither binding to receptor nor exposure to low pH is sufficient to trigger the S proteins of MERS-CoV and SARS-CoV (Li et al. 2006; Sha et al. 2006; Simmons et al. 2004). Therefore, it is conceivable that proteolytic processing of these S proteins may suffice for triggering. In order to reflect this finding, we will replace "priming" by "activating" in the remainder of this discussion.

4.4 Proteolytic Activation of the Spike Proteins of SARS-CoV and MERS-CoV

4.4.1 Cathepsin L: Endosomal Activator of the Spike Protein

The role of cathepsin L in coronavirus entry has been discovered in the context of SARS-CoV infection. Initial studies showed that SARS-S-driven entry is pH-dependent (Hofmann et al. 2004; Huang et al. 2006; Simmons et al. 2004, 2005) but also discovered that exposure to low pH fails to trigger the membrane fusion activity of the S protein (Simmons et al. 2004), arguing that protons might indirectly promote SARS-CoV entry. Simmons and coworkers provided an explanation for this at first sight paradoxical finding: They showed that inhibitors of cathepsin L activity block SARS-S-driven entry into host cells, while recombinant cathepsin L can activate S protein-driven membrane fusion (Simmons et al. 2005), indicating that the pH-dependency of SARS-S-driven entry stems from protons being required for cathepsin L activity rather than from protonation of SARS-S triggering the membrane fusion activity. Moreover, they demonstrated that trypsin treatment of cell-bound viruses allows SARS-S-driven entry into cells pretreated with a cathepsin L inhibitor while trypsin treatment of cell-free particles abrogated infectivity (Simmons et al. 2005). Thus, activation of the S protein at the cell surface can override the need for endosomal cathepsin L activity for SARS-S-driven entry and is likely promoted by S protein interactions with ACE2. These findings established cathepsin L as CoV-activating protease and are in keeping with previous reports demonstrating a role of cathepsin L in reovirus uncoating (Ebert et al. 2002) and Ebola virus glycoprotein (EBOV-GP) activation (Chandran et al. 2005).

What is known about cathepsin L expression and physiological functions? The cathepsin family encompasses serine (cathepsins A and E), aspartic (cathepsins D and E), and cysteine proteases (cathepsins B, C, F, H, K, L, O, S, V, X, and W in humans) (Turk et al. 2012). The cysteine cathepsins are localized to lysosomes and cleave a variety of extra- and intracellular substrates, preferentially after basic or hydrophobic residues. The cathepsins B, H, L, C, X, F, O, and V are ubiquitously expressed and seem to be required for protein degradation and turnover in a cell type-independent fashion. In contrast, expression of cathepsins K, W, and S is cell type- or tissue-specific, suggesting more specialized functions (Lecaille et al. 2002). Although a slightly acidic pH is required for activity of cysteine cathepsins and exposure to neutral pH may irreversibly abrogate enzymatic activity (Turk et al. 1995), these enzymes may also be localized and active in compartments other than lysosomes. For instance, cleavage of histones by cathepsin L in the nucleus has been reported and may regulate the cell cycle (Goulet et al. 2004), while activity of cathepsins in the extracellular space may contribute to degradation of the extracellular matrix and the resulting pathologies (Fonovic and Turk 2014; Obermajer et al. 2008). Cysteine cathepsins are generated as preproenzymes: An N-terminal signal peptide facilitates ER import and is removed cotranslationally; the remaining propeptide is required for proper folding of the protein and for transport into endo- and lysosomes in a mannose-6-phosphate receptor (M6PR)-dependent fashion (Hasilik et al. 2009; Saftig and Klumperman 2009; Turk et al. 2012). Moreover, the propeptide blocks the substrate binding site and thereby prevents premature activity of the enzyme. Finally, the propeptide is removed either by autocatalytic cleavage or by other proteases, resulting in the generation of mature, proteolytically active enzymes, which may be present as single-chain or double-chain (attached by a disulfide bond) forms (Turk et al. 2012).

The demonstration that cathepsin L can activate SARS-S, at least in cell lines (Huang et al. 2006; Simmons et al. 2005), raises the question at which site the S protein is processed by this protease. Bosch and colleagues have demonstrated with recombinant proteins that cathepsin L cleaves SARS-S at T678 (Bosch et al. 2008), which represents a region in which furin cleavage occurs in other CoV S proteins. However, it remains to be investigated whether this residue is indeed required for SARS-S activation by cathepsin L during viral entry. The S1/S2 cleavage site, defined by R667 (Belouzard et al. 2009; Follis et al. 2006; Simmons et al. 2011), and the S2′ site, defined by R797 (Belouzard et al. 2009), are required for SARS-S activation by trypsin. However, both sites are dispensable for cathepsin L-dependent, SARS-S-driven host cell entry (Belouzard et al. 2009; Simmons et al. 2011). It remains to be determined whether both sites are indeed not recognized by this protease or whether cathepsin L can activate SARS-S at surrogate sites, in case R667 and R797 are not available. The latter possibility would be in keeping with the low substrate specificity of cathepsin L.

Several CoVs other than SARS-CoV can use cathepsin L for S protein activation, including PEDV (Liu et al. 2016), MHV (Burkard et al. 2014; Qiu et al. 2006), HCoV-229E (Kawase et al. 2009), and MERS-CoV (Gierer et al. 2013; Qian et al. 2013; Shirato et al. 2013; Yang et al. 2015). Although MERS-S activation by cathepsin L has not been observed by all studies (Burkard et al. 2014), these results indicate that inhibitors targeting this protease might display broad anti-CoV activity. A notable exception is HCoV-NL63, which was reported to enter target cells in a pH-dependent but cathepsin L-independent fashion (Huang et al. 2006). Although these results are not undisputed (Hofmann et al. 2006), they suggest that NL63-S might exploit endosomal proteases other than cathepsin L for entry and cysteine cathepsins with substrate specificity and expression similar to cathepsin L are potential candidates. In sum, cathepsin L can activate diverse CoV upon endosomal entry (Fig. 4.4). The mechanisms controlling choice of cathepsin L versus other CoV-activating proteases as well as their role in CoV spread in vivo have only recently been discovered and will be discussed in the next section.

4.4.2 Activation of the Spike Protein by Type II Transmembrane Serine Proteases at the Cell Surface

Type II transmembrane serine proteases (TTSPs) have been identified as activators of viral infection by Böttcher and coworkers, who showed that the TTSPs TMPRSS2 and HAT cleaved and thereby activated FLUAV-HA, at least upon directed expression in cell lines (Böttcher et al. 2006). Subsequent studies showed that TMPRSS2 can also activate HA upon endogenous expression in cell lines (Bertram et al. 2010b; Böttcher-Friebertshäuser et al. 2011) and provided evidence that this protease is expressed in FLUAV target cell in the human respiratory tract (Bertram et al. 2012), suggesting that TMPRSS2 could promote FLUAV spread in the infected host. Indeed, Hatesuer and colleagues (Hatesuer et al. 2013) as well as subsequent studies (Sakai et al. 2014; Tarnow et al. 2014) demonstrated that mice lacking *tmprss2* are largely resistant to spread and pathogenesis of several FLUAV subtypes and could link this finding to absence of HA activation. Moreover, polymorphisms in the TMPRSS2 gene in humans which increase TMPRSS2 expression were shown

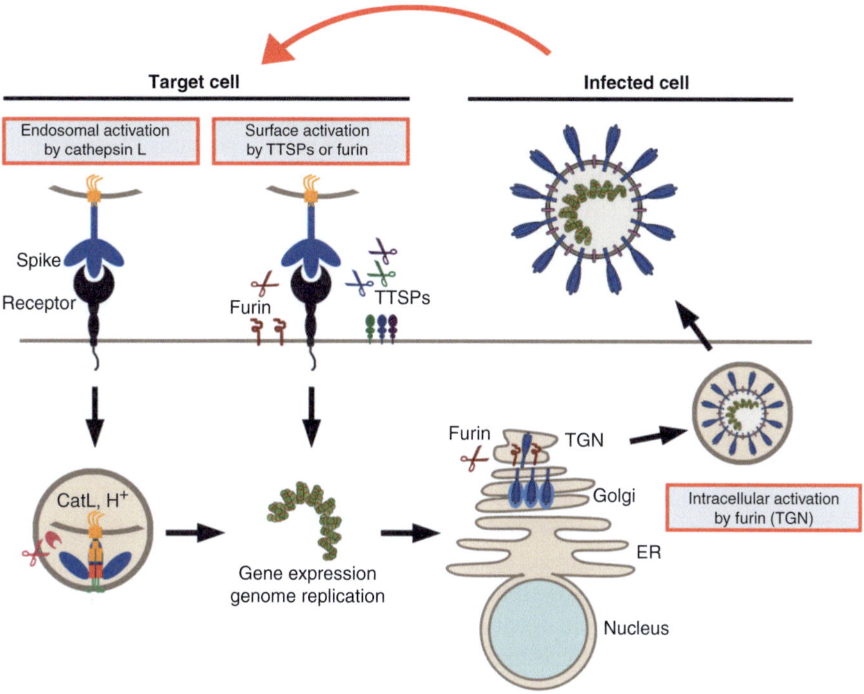

Fig. 4.4 Activation of coronavirus spike proteins by host cell proteases occurs at different stages in the viral life cycle. Binding of the viral spike (S) protein to a cellular receptor can induce endocytosis of virions. In the endosome, the pH-dependent cysteine protease cathepsin L (CatL) can activate the S protein for fusion within the endosomal membrane. Alternatively, receptor binding may expose a protease cleavage site and may thus promote S protein activation at the plasma membrane by type II transmembrane serine proteases (TTSPs) or furin. Membrane fusion allows the release of the viral genome into the cytoplasm, the site of viral genome replication and protein translation. The S protein is synthesized in the constitutive secretory pathway, where some S proteins can be cleaved by furin or other pro-protein convertases during passage through the *trans*-Golgi network (TGN). Finally, nascent virions are assembled at the endoplasmic reticulum/Golgi intermediate compartment and are released from infected cells through exocytosis

to be associated with severe influenza, suggesting that this protease might also promote FLUAV spread in humans (Cheng et al. 2015). Finally, it is noteworthy that several other TTSPs can activate HA upon directed expression in cell culture, including TMPRSS4, DESC1, MSPL, and matriptase (Baron et al. 2013; Beaulieu et al. 2013; Bertram et al. 2010b; Chaipan et al. 2009; Hamilton et al. 2012; Zmora et al. 2014) and that TMPRSS4 has recently been shown to promote spread of a H3N2 FLUAV in mice that showed partial TMPRSS2-independence (Kuhn et al. 2016).

TTSP are membrane-anchored serine proteases that play an important role in several physiological processes, including maintenance of homeostasis (Antalis et al. 2010, 2011; Szabo and Bugge 2011). They exhibit a characteristic domain organization: An N-terminal cytoplasmic tail is followed by transmembrane domain, a stem region, and a C-terminal protease domain (Bugge et al. 2009; Hooper et al. 2001). The cytoplasmic tail might be involved in targeting the protease to the cellular membrane, while the transmembrane domain anchors the proteins in the plasma membrane (Bugge et al. 2009). The stem region has a strictly modular organization and may be composed of up to 11 different protein domains (Antalis et al. 2011; Hooper et al. 2001). The number and configuration of these domains is characteristic for specific TTSPs, and the stem region is known to function in protein-protein interactions or protein-ligand interactions (Hooper et al. 2001). Finally, the protease domain harbors a conserved catalytic triad of histidine, aspartate, and serine, which is essential for enzymatic activity. TTSPs are synthesized as inactive pro-proteins, zymogens, and are either autoactivated or activated by another protease. Activation requires cleavage at a site located at the interface between stem region and protease domain and may result in shedding of the enzymatically active protease domain into the extracellular space (Antalis et al. 2010; Bugge et al. 2009; Hooper et al. 2001).

Three studies independently demonstrated that TMPRSS2 does not only activate FLUAV-HA but also cleaves and activates SARS-S (Glowacka et al. 2011; Matsuyama et al. 2010; Shulla et al. 2011). They found that directed expression of TMPRSS2 in target cells allowed SARS-S-driven entry, despite previous treatment of cells with lysosomotropic agents (i.e., elevated endosomal pH) or cathepsin L inhibitors. These results indicate that TMPRSS2 activates virion-associated SARS-S early during viral entry and thereby renders entry independent of cathepsin L activity (Fig. 4.4). Activation is believed to occur at the plasma membrane, likely after S protein binding to ACE2. Notably, ACE2 and TMPRSS2 interact (Shulla et al. 2011), and it is conceivable that conformational changes in SARS-S that are induced upon SARS-S binding might expose the TMPRSS2 cleavage site in the S protein. SARS-S activation by TMPRSS2 was only observed when S protein and protease were located in different membranes (i.e., viral and cellular membranes, respectively) and thus depends on SARS-S cleavage in *trans* (Glowacka et al. 2011; Matsuyama et al. 2010). However, TMPRSS2 can also cleave SARS-S when both proteins are localized in the same membrane (*cis*-cleavage), at least upon directed expression, and this may result in shedding of soluble SARS-S into the extracellular space, where the S protein can serve as decoy for neutralizing antibodies (Glowacka et al. 2011).

The cleavage site of TMPRSS2 in SARS-S is largely unclear, although one can speculate that R797 might be involved. One report suggested that R667 might be dispensable for SARS-S cleavage by TMPRSS2, but a quantitative analysis was not provided (Bertram et al. 2011). In contrast, R667 was found to be essential for SARS-S processing by HAT (Bertram et al. 2011), although it was not determined if this residue is also required for SARS-S activation by this protease. In this context, differences in SARS-S activation by TMPRSS2 and HAT should be noted: TMPRSS2 can activate SARS-S for cell-cell and virus-cell fusion in *trans* (Bertram et al. 2011; Matsuyama et al. 2010). In contrast, HAT can only activate SARS-S for cell-cell but not virus-cell fusion, and activation is observed in both the *cis* and the *trans* setting (Bertram et al. 2011). Whether these observations reflect general differences in the activation reaction or simply mirror the somewhat less efficient expression of HAT as compared to TMPRSS2 in transfected cells remains to be investigated. Finally, it is noteworthy that besides TMPRSS2 and HAT, several other TTSPs can cleave and activate SARS-S. Thus, DESC1 and MSPL can cleave and activate SARS-S in *trans* (Zmora et al. 2014), at least upon directed expression, and thus seem to function in a TMPRSS2-like fashion. Cleavage of SARS-S by recombinant TMPRSS11A has also been demonstrated, and both R667 and R797 were identified as cleavage sites (Kam et al. 2009). Moreover, exposure of SARS-S-bearing particles to recombinant TMPRSS11A augmented entry into cultured respiratory epithelium, and mutation of R797 reduced entry into these cells to background levels. In contrast, mutation of R667 had only a modest effect (Kam et al. 2009). Finally, it should be highlighted that the exploitation of TTSPs for S protein activation is not limited to SARS-CoV: TMPRSS2 expression facilitates cathepsin L-independent 229E-S- and MERS-S-driven entry into target cells (Bertram et al. 2013; Gierer et al. 2013; Kawase et al. 2012; Shirato et al. 2013, 2017). Moreover, acquisition of use of human proteases, including TMPRSS2, for S protein activation has been suggested to be a determinant of zoonotic transmission of MERS-CoV-related viruses from bats to humans (Yang et al. 2014, 2015). TMPRSS2 has also been reported to play a role in PEDV infection. In this context, protease activity seems to be required for efficient release of progeny virions from infected cells (Shirato et al. 2011). The underlying mechanism is unknown, but one can speculate that either modulation of S protein glycosylation (Bertram et al. 2010a) or, more plausible, TMPRSS2-mediated inactivation of an antiviral host cell factor might be responsible.

The findings discussed above indicate that S protein activation by TTSPs is a complex process and is governed, among other factors, by the localization of the S protein in joint or in opposite membranes. Another layer of complexity is added by the observation that TMPRSS2 cannot only cleave the SARS-S protein but can also process its entry receptor ACE2 (Heurich et al. 2014). Thus, TMPRSS2 and the metalloprotease a disintegrin and metalloproteinase domain 17 (ADAM17) cleave ACE2 close to its transmembrane domain, and cleavage may result in ACE2 shedding (Haga et al. 2008; Heurich et al. 2014). Moreover, it has been proposed that ACE2 cleavage by ADAM17 is required for efficient SARS-S-driven entry (Haga et al. 2008), while ACE2 processing by TMPRSS2 seems to account for the augmentation of viral infectivity observed upon directed expression of TMPRSS2 in target cells (Heurich et al. 2014). These findings suggest that TTSPs and other proteases can impact S protein-driven entry by ways other than S protein activation, but the underlying mechanism remains to be investigated.

What is the evidence that TMPRSS2 and potentially other TTSPs promote coronavirus spread in the infected host? For HAT, DESC1, MSPL, and TMPRSS11A, the evidence is limited to the demonstration of mRNA and/or protein expression in the lung and to S protein activation upon directed expression of protease or addition of recombinant protease (Bertram et al. 2012; Kam et al. 2009; Zmora et al. 2014). In contrast, a constantly accumulating body of evidence suggests an important contribution of TMPRSS2 to SARS-CoV spread in the host: TMPRSS2 is coexpressed with ACE2 in the human lung (Bertram et al. 2012), and TMPRSS2-positive cells were found to harbor SARS-CoV antigen in experimentally infected cynomolgus macaques (Matsuyama et al. 2010). Moreover, blockade of TMPRSS2 activity reduced SARS-CoV infection of respiratory epithelium (Kawase et al. 2012). Finally, a serine protease inhibitor active against TMPRSS2 reduced viral spread and pathogenesis in a rodent mode of SARS-CoV infection, while blockade of cathepsin L activity had no appreciable effect (Zhou et al. 2015). These findings suggest that SARS-CoV, like FLUAV, might depend on TMPRSS2 for spread within hosts. Such a scenario would be in keeping with the findings that cathepsin L is not expressed in respiratory epithelium at levels sufficient for MERS-S activation (Park et al. 2016) and that HCoV-229E isolated from patients uses TMPRSS2 while viral variants adapted to growth in cell culture employ cathepsin L (Kawase et al. 2009; Shirato et al. 2017), suggesting that S protein activation by cathepsin L might be the result of cell culture adaptation. In sum, several lines of evidence suggest that TMPRSS2 but not cathepsin L activity is important for CoV spread in the infected host. This raises the question which determinants control whether cathepsin L or TMPRSS2 is used for S protein activation. Recent insights obtained for MERS-S activation provide interesting answers and will be discussed below.

4.4.3 Furin Can Activate Coronavirus Spike Proteins in the Constitutive Secretory Pathway of Infected Cells and During Viral Entry into Target Cells

Furin, a subtilisin-like serine protease, belongs to the family of pro-protein convertases (PPCs), which comprises nine members (Seidah and Prat 2012). Seven of these enzymes process substrates at basic residues and are required for activation of various cellular proteins, including hormones, growth factors, and adhesion molecules. Cleavage occurs at single or paired basic residues, which fit the following rule: (R/K)Xn(R/K)↓ (Nakayama 1997; Seidah et al. 2013; Seidah and Prat 2012), with the arrow indicating the cleavage site, X indicating any amino acid, and n corresponding to a 0, 2, 4, or 6, respectively. Furin is ubiquitously expressed and found in the *trans*-Golgi-network (TGN) from where it can be transported to the cell surface and back again via the endosomal compartment (Bosshart et al. 1994; Molloy et al. 1994). Two PPCs, SKI-1 and PCSK9, play a role in cholesterol/lipid homeostasis and cleave substrates at nonbasic residues (Seidah et al. 2013; Seidah and Prat 2012). Like cathepsin L and TTSPs, PCCs are synthesized as zymogens and the presence of a prosegment, which is removed by autocatalytic activation but remains non-covalently associated with the protease, prevents premature activity.

Many CoV S proteins harbor a furin motif at the S1/S2 site, and processing of, for example, the S protein of MHV, strain A59 (de Haan et al. 2004), and IBV (Yamada and Liu 2009) by furin has been demonstrated. Moreover, the insertion of a furin motif in the S protein of PEDV allows for trypsin-independent viral spread in cell culture (Li et al. 2015). The contribution of furin to SARS-S activation is less clear. It has been documented that a pro-protein convertase inhibitor blocks

SARS-CoV spread in cell culture, but mutational analysis failed to demonstrate robust processing of the S protein by this protease (Bergeron et al. 2005). Moreover, insertion of a furin motif at the S1/S2 site augmented SARS-S-driven cell-cell but not virus-cell fusion (Follis et al. 2006). In contrast, a prominent role of furin in MERS-S activation has been documented. Thus, it has been shown that furin can cleave MERS-S at the S1/S2 site during S protein biogenesis in the constitutive secretory pathway of infected cells and at the S2′ site during S protein-driven entry into target cells (Millet and Whittaker 2014). Blockade of furin expression or activity reduced MERS-S-driven entry (Burkard et al. 2014; Millet and Whittaker 2014), indicating that furin is an activator of MERS-S (Fig. 4.4), although the requirement of furin activity for MERS-S-driven entry has not been observed by a separate study (Gierer et al. 2015) and thus might be cell type specific to some extent (Millet and Whittaker 2014).

The observation that furin is an activator of certain CoV S proteins raises the question which determinants control whether S proteins are activated by furin, cathepsin L, or TMPRSS2. An intriguing answer has been provided by a recent study by Park and colleagues. They showed that cleavage of MERS-S at the S1/S2 site in infected cells determines whether MERS-S is activated by cathepsin L or TMPRSS2 during viral entry into target cells (Park et al. 2016). Thus, only precleaved MERS-S seems to be able to undergo the conformational changes upon receptor binding that are required for activation by TMPRSS2. If the S protein is uncleaved and thus conformationally rigid, binding to receptor results in viral uptake into endosomes, where the S protein is activated by cathepsin L (Park et al. 2016). However, this activation pathway seems to be less robust as compared to activation by TMPRSS2 and does not allow efficient entry into cells within respiratory epithelium, due to expression of insufficient amounts of cathepsin L (Park et al. 2016). In sum, cleavage at the S1/S2 site in infected cells can determine protease choice during entry into target cells, with only cleaved S proteins exhibiting sufficient conformational flexibility for activation at the cell surface by TMPRSS2.

Conclusions

The cleavage activation of the S protein of coronaviruses by host cell proteases is required for viral infectivity, and the responsible enzymes constitute potential targets for antiviral intervention. Studies within the recent years provided interesting insights regarding the nature of the S protein-activating proteases, the mechanisms that control protease choice, and the contribution of specific enzymes to viral spread in the infected host (Fig. 4.4). The pH-dependent cysteine protease cathepsin L can activate the S proteins of SARS-CoV, MERS-CoV, PEDV, and other pathogenic coronaviruses upon viral uptake into endosomes. However, cathepsin L might not be sufficiently expressed in respiratory epithelium to support viral spread in this important target tissue, and, at least for some CoVs, efficient S protein activation by cathepsin L might be the result of viral passaging in cell lines. Such a scenario would be compatible with the finding that EBOV-GP is activated by cathepsin B and L for entry into cell lines (Chandran et al. 2005), while expression of these proteases is dispensable for efficient viral spread in mice (Marzi et al. 2012). The type II transmembrane serine protease TMPRSS2 activates FLUAV-HA and is essential for spread of diverse FLUAV in rodent and likely also human hosts. Similarly, TMPRSS2 activates the S proteins of SARS-CoV and MERS-CoV and is expressed in cells in the human respiratory epithelium that also express the SARS-CoV receptor, ACE2. Moreover, TMPRSS2 activity is required for efficient SARS-S- and MERS-S-driven entry into cultured respiratory epithelium, and a protease inhibitor active against TMPRSS2 suppresses SARS-CoV spread and pathogenesis in a rodent model. Finally, pre-cleavage of MERS-S by furin in infected cells is essential for subsequent S protein activation by TMPRSS2 for entry into target cells, potentially by providing the S protein with increased conformational flexibility. Thus, TMPRSS2 is an attractive antiviral target, and specific inhibitors of this enzyme might exert activity against a broad spectrum of respiratory viruses. Initial efforts to generate such inhibitors have been documented (Meyer et al. 2013), and compounds with high specificity for TMPRSS2 can be expected to suppress viral spread without inducing unwanted side effects, since tmprss2 is dispensable for normal development and homeostasis in mice (Kim et al. 2006).

References

Adams MJ, Carstens EB. Ratification vote on taxonomic proposals to the International Committee on Taxonomy of Viruses (2012). Arch Virol. 2012;157(7):1411–22.

Afzelius BA. Ultrastructure of human nasal epithelium during an episode of coronavirus infection. Virchows Arch. 1994;424(3):295–300.

Alagaili AN, Briese T, Mishra N, Kapoor V, Sameroff SC, Burbelo PD, de WE, Munster VJ, Hensley LE, Zalmout IS, Kapoor A, Epstein JH, Karesh WB, Daszak P, Mohammed OB, Lipkin WI. Middle East respiratory syndrome coronavirus infection in dromedary camels in Saudi Arabia. MBio. 2014;5(2):e00884-14.

Alsolamy S, Arabi YM. Infection with Middle East respiratory syndrome coronavirus. Can J Respir Ther. 2015;51(4):102.

Annan A, Baldwin HJ, Corman VM, Klose SM, Owusu M, Nkrumah EE, Badu EK, Anti P, Agbenyega O, Meyer B, Oppong S, Sarkodie YA, Kalko EK, Lina PH, Godlevska EV, Reusken C, Seebens A, Gloza-Rausch F, Vallo P, Tschapka M, Drosten C, Drexler JF. Human betacoronavirus 2c EMC/2012-related viruses in bats, Ghana and Europe. Emerg Infect Dis. 2013;19(3):456–9.

Antalis TM, Bugge TH, Wu Q. Membrane-anchored serine proteases in health and disease. Prog Mol Biol Transl Sci. 2011;99:1–50.

Antalis TM, Buzza MS, Hodge KM, Hooper JD, Netzel-Arnett S. The cutting edge: membrane-anchored serine protease activities in the pericellular microenvironment. Biochem J. 2010;428(3):325–46.

Assiri A, Al-Tawfiq JA, Al-Rabeeah AA, Al-Rabiah FA, Al-Hajjar S, Al-Barrak A, Flemban H, Al-Nassir WN, Balkhy HH, Al-Hakeem RF, Makhdoom HQ, Zumla AI, Memish ZA. Epidemiological, demographic, and clinical characteristics of 47 cases of Middle East respiratory syndrome coronavirus disease from Saudi Arabia: a descriptive study. Lancet Infect Dis. 2013;13(9):752–61.

Barlan A, Zhao J, Sarkar MK, Li K, PB MC Jr, Perlman S, Gallagher T. Receptor variation and susceptibility to Middle East respiratory syndrome coronavirus infection. J Virol. 2014;88(9):4953–61.

Baron J, Tarnow C, Mayoli-Nussle D, Schilling E, Meyer D, Hammami M, Schwalm F, Steinmetzer T, Guan Y, Garten W, Klenk HD, Böttcher-Friebertshäuser E. Matriptase, HAT, and TMPRSS2 activate the hemagglutinin of H9N2 influenza A viruses. J Virol. 2013;87(3):1811–20.

Beaulieu A, Gravel E, Cloutier A, Marois I, Colombo E, Desilets A, Verreault C, Leduc R, Marsault E, Richter MV. Matriptase proteolytically activates influenza virus and promotes multicycle replication in the human airway epithelium. J Virol. 2013;87(8):4237–51.

Belouzard S, Chu VC, Whittaker GR. Activation of the SARS coronavirus spike protein via sequential proteolytic cleavage at two distinct sites. Proc Natl Acad Sci U S A. 2009;106(14):5871–6.

Belouzard S, Millet JK, Licitra BN, Whittaker GR. Mechanisms of coronavirus cell entry mediated by the viral spike protein. Virus. 2012;4(6):1011–33.

Bergeron E, Vincent MJ, Wickham L, Hamelin J, Basak A, Nichol ST, Chretien M, Seidah NG. Implication of proprotein convertases in the processing and spread of severe acute respiratory syndrome coronavirus. Biochem Biophys Res Commun. 2005;326(3):554–63.

Berry DM, Almeida JD. The morphological and biological effects of various antisera on avian infectious bronchitis virus. J Gen Virol. 1968;3(1):97–102.

Bertram S, Dijkman R, Habjan M, Heurich A, Gierer S, Glowacka I, Welsch K, Winkler M, Schneider H, Hofmann-Winkler H, Thiel V, Pöhlmann S. TMPRSS2 activates the human coronavirus 229E for cathepsin-independent host cell entry and is expressed in viral target cells in the respiratory epithelium. J Virol. 2013;87(11):6150–60.

Bertram S, Glowacka I, Blazejewska P, Soilleux E, Allen P, Danisch S, Steffen I, Choi SY, Park Y, Schneider H, Schughart K, Pöhlmann S. TMPRSS2 and TMPRSS4 facilitate trypsin-independent spread of influenza virus in Caco-2 cells. J Virol. 2010a;84(19):10016–25.

Bertram S, Glowacka I, Muller MA, Lavender H, Gnirss K, Nehlmeier I, Niemeyer D, He Y, Simmons G, Drosten C, Soilleux EJ, Jahn O, Steffen I, Pöhlmann S. Cleavage and activation of the severe acute respiratory syndrome coronavirus spike protein by human airway trypsin-like protease. J Virol. 2011;85(24):13363–72.

Bertram S, Glowacka I, Steffen I, Kuhl A, Pöhlmann S. Novel insights into proteolytic cleavage of influenza virus hemagglutinin. Rev Med Virol. 2010b;20(5):298–310.

Bertram S, Heurich A, Lavender H, Gierer S, Danisch S, Perin P, Lucas JM, Nelson PS, Pöhlmann S, Soilleux EJ. Influenza and SARS-coronavirus activating proteases TMPRSS2 and HAT are expressed at multiple sites in human respiratory and gastrointestinal tracts. PLoS One. 2012;7(4):e35876.

Bonavia A, Zelus BD, Wentworth DE, Talbot PJ, Holmes KV. Identification of a receptor-binding domain of the spike glycoprotein of human coronavirus HCoV-229E. J Virol. 2003;77(4):2530–8.

Bosch BJ, Bartelink W, Rottier PJ. Cathepsin L functionally cleaves the severe acute respiratory syndrome coronavirus class I fusion protein upstream of rather than adjacent to the fusion peptide. J Virol. 2008;82(17):8887–90.

Bosch BJ, van der Zee R, de Haan CA, Rottier PJ. The coronavirus spike protein is a class I virus fusion protein: structural and functional characterization of the fusion core complex. J Virol. 2003;77(16):8801–11.

Bosshart H, Humphrey J, Deignan E, Davidson J, Drazba J, Yuan L, Oorschot V, Peters PJ, Bonifacino JS. The cytoplasmic domain mediates localization of furin to the trans-Golgi network en route to the endosomal/lysosomal system. J Cell Biol. 1994;126(5):1157–72.

Böttcher E, Matrosovich T, Beyerle M, Klenk HD, Garten W, Matrosovich M. Proteolytic activation of influenza viruses by serine proteases TMPRSS2 and HAT from human airway epithelium. J Virol. 2006;80(19):9896–8.

Böttcher-Friebertshäuser E, Stein DA, Klenk HD, Garten W. Inhibition of influenza virus infection in human airway cell cultures by an antisense peptide-conjugated morpholino oligomer targeting the hemagglutinin-activating protease TMPRSS2. J Virol. 2011;85(4):1554–62.

Breslin JJ, Mork I, Smith MK, Vogel LK, Hemmila EM, Bonavia A, Talbot PJ, Sjostrom H, Noren O, Holmes KV. Human coronavirus 229E: receptor binding domain and neutralization by soluble receptor at 37 degrees C. J Virol. 2003;77(7):4435–8.

Breslin JJ, Smith LG, Fuller FJ, Guy JS. Sequence analysis of the turkey coronavirus nucleocapsid protein gene and 3′ untranslated region identifies the virus as a close relative of infectious bronchitis virus. Virus Res. 1999;65(2):187–93.
Bugge TH, Antalis TM, Wu Q. Type II transmembrane serine proteases. J Biol Chem. 2009;284(35):23177–81.
Burkard C, Verheije MH, Wicht O, van Kasteren SI, van Kuppeveld FJ, Haagmans BL, Pelkmans L, Rottier PJ, Bosch BJ, de Haan CA. Coronavirus cell entry occurs through the endo–/lysosomal pathway in a proteolysis-dependent manner. PLoS Pathog. 2014;10(11):e1004502.
Cavanagh D, Davis PJ. Coronavirus IBV: removal of spike glycopolypeptide S1 by urea abolishes infectivity and haemagglutination but not attachment to cells. J Gen Virol. 1986;67(Pt 7):1443–8.
Cavanagh D, Mawditt K, Sharma M, Drury SE, Ainsworth HL, Britton P, Gough RE. Detection of a coronavirus from turkey poults in Europe genetically related to infectious bronchitis virus of chickens. Avian Pathol. 2001;30(4):355–68.
Chaipan C, Kobasa D, Bertram S, Glowacka I, Steffen I, Tsegaye TS, Takeda M, Bugge TH, Kim S, Park Y, Marzi A, Pöhlmann S. Proteolytic activation of the 1918 influenza virus hemagglutinin. J Virol. 2009;83(7):3200–11.
Chandran K, Sullivan NJ, Felbor U, Whelan SP, Cunningham JM. Endosomal proteolysis of the Ebola virus glycoprotein is necessary for infection. Science. 2005;308(5728):1643–5.
Chen Y, Rajashankar KR, Yang Y, Agnihothram SS, Liu C, Lin YL, Baric RS, Li F. Crystal structure of the receptor-binding domain from newly emerged Middle East respiratory syndrome coronavirus. J Virol. 2013;87(19):10777–83.
Cheng Z, Zhou J, To KK, Chu H, Li C, Wang D, Yang D, Zheng S, Hao K, Bosse Y, Obeidat M, Brandsma CA, Song YQ, Chen Y, Zheng BJ, Li L, Yuen KY. Identification of TMPRSS2 as a susceptibility gene for severe 2009 pandemic A(H1N1) influenza and A(H7N9) influenza. J Infect Dis. 2015;212(8):1214–21.
Cockrell AS, Peck KM, Yount BL, Agnihothram SS, Scobey T, Curnes NR, Baric RS, Heise MT. Mouse dipeptidyl peptidase 4 is not a functional receptor for Middle East respiratory syndrome coronavirus infection. J Virol. 2014;88(9):5195–9.
Coleman CM, Matthews KL, Goicochea L, Frieman MB. Wild-type and innate immune-deficient mice are not susceptible to the Middle East respiratory syndrome coronavirus. J Gen Virol. 2014;95(Pt 2):408–12.
de Groot RJ, Luytjes W, Horzinek MC, van der Zeijst BA, Spaan WJ, Lenstra JA. Evidence for a coiled-coil structure in the spike proteins of coronaviruses. J Mol Biol. 1987;196(4):963–6.
de Haan CA, Stadler K, Godeke GJ, Bosch BJ, Rottier PJ. Cleavage inhibition of the murine coronavirus spike protein by a furin-like enzyme affects cell-cell but not virus-cell fusion. J Virol. 2004;78(11):6048–54.
Debouck P, Pensaert M. Experimental infection of pigs with a new porcine enteric coronavirus, CV 777. Am J Vet Res. 1980;41(2):219–23.
Deng F, Ye G, Liu Q, Navid MT, Zhong X, Li Y, Wan C, Xiao S, He Q, Fu ZF, Peng G. Identification and comparison of receptor binding characteristics of the spike protein of two porcine epidemic diarrhea virus strains. Virus. 2016;8(3):55.
Dimitrov DS. Virus entry: molecular mechanisms and biomedical applications. Nat Rev Microbiol. 2004;2(2):109–22.
Drosten C, Gunther S, Preiser W, van der Werf S, Brodt HR, Becker S, Rabenau H, Panning M, Kolesnikova L, Fouchier RA, Berger A, Burguiere AM, Cinatl J, Eickmann M, Escriou N, Grywna K, Kramme S, Manuguerra JC, Muller S, Rickerts V, Sturmer M, Vieth S, Klenk HD, Osterhaus AD, Schmitz H, Doerr HW. Identification of a novel coronavirus in patients with severe acute respiratory syndrome. N Engl J Med. 2003;348(20):1967–76.
Du L, He Y, Zhou Y, Liu S, Zheng BJ, Jiang S. The spike protein of SARS-CoV—a target for vaccine and therapeutic development. Nat Rev Microbiol. 2009;7(3):226–36.
Du L, Zhao G, Kou Z, Ma C, Sun S, Poon VK, Lu L, Wang L, Debnath AK, Zheng BJ, Zhou Y, Jiang S. Identification of a receptor-binding domain in the S protein of the novel human coronavirus Middle East respiratory syndrome coronavirus as an essential target for vaccine development. J Virol. 2013;87(17):9939–42.

Du L, Zhao G, Lin Y, Sui H, Chan C, Ma S, He Y, Jiang S, Wu C, Yuen KY, Jin DY, Zhou Y, Zheng BJ. Intranasal vaccination of recombinant adeno-associated virus encoding receptor-binding domain of severe acute respiratory syndrome coronavirus (SARS-CoV) spike protein induces strong mucosal immune responses and provides long-term protection against SARS-CoV infection. J Immunol. 2008;180(2):948–56.
Dudas G, Rambaut A. MERS-CoV recombination: implications about the reservoir and potential for adaptation. Virus Evol. 2016;2(1):vev023.
Duquerroy S, Vigouroux A, Rottier PJ, Rey FA, Bosch BJ. Central ions and lateral asparagine/glutamine zippers stabilize the post-fusion hairpin conformation of the SARS coronavirus spike glycoprotein. Virology. 2005;335(2):276–85.
Dveksler GS, Pensiero MN, Cardellichio CB, Williams RK, Jiang GS, Holmes KV, Dieffenbach CW. Cloning of the mouse hepatitis virus (MHV) receptor: expression in human and hamster cell lines confers susceptibility to MHV. J Virol. 1991;65(12):6881–91.
Ebert DH, Deussing J, Peters C, Dermody TS. Cathepsin L and cathepsin B mediate reovirus disassembly in murine fibroblast cells. J Biol Chem. 2002;277(27):24609–17.
Foley JE, Leutenegger C. A review of coronavirus infection in the central nervous system of cats and mice. J Vet Intern Med. 2001;15(5):438–44.
Follis KE, York J, Nunberg JH. Furin cleavage of the SARS coronavirus spike glycoprotein enhances cell-cell fusion but does not affect virion entry. Virology. 2006;350(2):358–69.
Fonovic M, Turk B. Cysteine cathepsins and extracellular matrix degradation. Biochim Biophys Acta. 2014;1840(8):2560–70.
Frieman M, Yount B, Agnihothram S, Page C, Donaldson E, Roberts A, Vogel L, Woodruff B, Scorpio D, Subbarao K, Baric RS. Molecular determinants of severe acute respiratory syndrome coronavirus pathogenesis and virulence in young and aged mouse models of human disease. J Virol. 2012;86(2):884–97.
Fukuma A, Tani H, Taniguchi S, Shimojima M, Saijo M, Fukushi S. Inability of rat DPP4 to allow MERS-CoV infection revealed by using a VSV pseudotype bearing truncated MERS-CoV spike protein. Arch Virol. 2015;160(9):2293–300.
Gierer S, Bertram S, Kaup F, Wrensch F, Heurich A, Kramer-Kuhl A, Welsch K, Winkler M, Meyer B, Drosten C, Dittmer U, von HT, Simmons G, Hofmann H, Pöhlmann S. The spike protein of the emerging betacoronavirus EMC uses a novel coronavirus receptor for entry, can be activated by TMPRSS2, and is targeted by neutralizing antibodies. J Virol. 2013;87(10):5502–11.
Gierer S, Muller MA, Heurich A, Ritz D, Springstein BL, Karsten CB, Schendzielorz A, Gnirss K, Drosten C, Pöhlmann S. Inhibition of proprotein convertases abrogates processing of the middle eastern respiratory syndrome coronavirus spike protein in infected cells but does not reduce viral infectivity. J Infect Dis. 2015;211(6):889–97.
Glowacka I, Bertram S, Muller MA, Allen P, Soilleux E, Pfefferle S, Steffen I, Tsegaye TS, He Y, Gnirss K, Niemeyer D, Schneider H, Drosten C, Pöhlmann S. Evidence that TMPRSS2 activates the severe acute respiratory syndrome coronavirus spike protein for membrane fusion and reduces viral control by the humoral immune response. J Virol. 2011;85(9):4122–34.
Godet M, Grosclaude J, Delmas B, Laude H. Major receptor-binding and neutralization determinants are located within the same domain of the transmissible gastroenteritis virus (coronavirus) spike protein. J Virol. 1994;68(12):8008–16.
Goulet B, Baruch A, Moon NS, Poirier M, Sansregret LL, Erickson A, Bogyo M, Nepveu A. A cathepsin L isoform that is devoid of a signal peptide localizes to the nucleus in S phase and processes the CDP/Cux transcription factor. Mol Cell. 2004;14(2):207–19.
Graham RL, Baric RS. Recombination, reservoirs, and the modular spike: mechanisms of coronavirus cross-species transmission. J Virol. 2010;84(7):3134–46.
Guan Y, Zheng BJ, He YQ, Liu XL, Zhuang ZX, Cheung CL, Luo SW, Li PH, Zhang LJ, Guan YJ, Butt KM, Wong KL, Chan KW, Lim W, Shortridge KF, Yuen KY, Peiris JS, Poon LL. Isolation and characterization of viruses related to the SARS coronavirus from animals in southern China. Science. 2003;302(5643):276–8.
Gui M, Song W, Zhou H, Xu J, Chen S, Xiang Y, Wang X. Cryo-electron microscopy structures of the SARS-CoV spike glycoprotein reveal a prerequisite conformational state for receptor binding. Cell Res. 2017;27(1):119–29.

Haagmans BL, Al Dhahiry SH, Reusken CB, Raj VS, Galiano M, Myers R, Godeke GJ, Jonges M, Farag E, Diab A, Ghobashy H, Alhajri F, Al-Thani M, Al-Marri SA, Al Romaihi HE, Al KA, Bermingham A, Osterhaus AD, AlHajri MM, Koopmans MP. Middle East respiratory syndrome coronavirus in dromedary camels: an outbreak investigation. Lancet Infect Dis. 2014;14(2):140–5.

Haga S, Yamamoto N, Nakai-Murakami C, Osawa Y, Tokunaga K, Sata T, Yamamoto N, Sasazuki T, Ishizaka Y. Modulation of TNF-alpha-converting enzyme by the spike protein of SARS-CoV and ACE2 induces TNF-alpha production and facilitates viral entry. Proc Natl Acad Sci U S A. 2008;105(22):7809–14.

Hallenberger S, Bosch V, Angliker H, Shaw E, Klenk HD, Garten W. Inhibition of furin-mediated cleavage activation of HIV-1 glycoprotein gp160. Nature. 1992;360(6402):358–61.

Hamilton BS, Whittaker GR, Daniel S. Influenza virus-mediated membrane fusion: determinants of hemagglutinin fusogenic activity and experimental approaches for assessing virus fusion. Virus. 2012;4(7):1144–68.

Hasilik A, Wrocklage C, Schroder B. Intracellular trafficking of lysosomal proteins and lysosomes. Int J Clin Pharmacol Ther. 2009;47(Suppl 1):S18–33.

Hatesuer B, Bertram S, Mehnert N, Bahgat MM, Nelson PS, Pöhlmann S, Schughart K. Tmprss2 is essential for influenza H1N1 virus pathogenesis in mice. PLoS Pathog. 2013;9(12):e1003774.

He Y, Zhou Y, Liu S, Kou Z, Li W, Farzan M, Jiang S. Receptor-binding domain of SARS-CoV spike protein induces highly potent neutralizing antibodies: implication for developing subunit vaccine. Biochem Biophys Res Commun. 2004;324(2):773–81.

Heurich A, Hofmann-Winkler H, Gierer S, Liepold T, Jahn O, Pöhlmann S. TMPRSS2 and ADAM17 cleave ACE2 differentially and only proteolysis by TMPRSS2 augments entry driven by the severe acute respiratory syndrome coronavirus spike protein. J Virol. 2014;88(2):1293–307.

Hofmann H, Hattermann K, Marzi A, Gramberg T, Geier M, Krumbiegel M, Kuate S, Uberla K, Niedrig M, Pöhlmann S. S protein of severe acute respiratory syndrome-associated coronavirus mediates entry into hepatoma cell lines and is targeted by neutralizing antibodies in infected patients. J Virol. 2004;78(12):6134–42.

Hofmann H, Pyrc K, van der Hoek L, Geier M, Berkhout B, Pöhlmann S. Human coronavirus NL63 employs the severe acute respiratory syndrome coronavirus receptor for cellular entry. Proc Natl Acad Sci U S A. 2005;102(22):7988–93.

Hofmann H, Simmons G, Rennekamp AJ, Chaipan C, Gramberg T, Heck E, Geier M, Wegele A, Marzi A, Bates P, Pöhlmann S. Highly conserved regions within the spike proteins of human coronaviruses 229E and NL63 determine recognition of their respective cellular receptors. J Virol. 2006;80(17):8639–52.

Hooper JD, Clements JA, Quigley JP, Antalis TM. Type II transmembrane serine proteases. Insights into an emerging class of cell surface proteolytic enzymes. J Biol Chem. 2001;276(2):857–60.

Hu B, Ge X, Wang LF, Shi Z. Bat origin of human coronaviruses. Virol J. 2015;12:221.

Huang IC, Bosch BJ, Li F, Li W, Lee KH, Ghiran S, Vasilieva N, Dermody TS, Harrison SC, Dormitzer PR, Farzan M, Rottier PJ, Choe H. SARS coronavirus, but not human coronavirus NL63, utilizes cathepsin L to infect ACE2-expressing cells. J Biol Chem. 2006;281(6):3198–203.

Jonassen CM, Kofstad T, Larsen IL, Lovland A, Handeland K, Follestad A, Lillehaug A. Molecular identification and characterization of novel coronaviruses infecting graylag geese (Anser anser), feral pigeons (Columbia livia) and mallards (Anas platyrhynchos). J Gen Virol. 2005;86(Pt 6):1597–607.

Jung K, Ahn K, Chae C. Decreased activity of brush border membrane-bound digestive enzymes in small intestines from pigs experimentally infected with porcine epidemic diarrhea virus. Res Vet Sci. 2006;81(3):310–5.

Kam YW, Okumura Y, Kido H, Ng LF, Bruzzone R, Altmeyer R. Cleavage of the SARS coronavirus spike glycoprotein by airway proteases enhances virus entry into human bronchial epithelial cells in vitro. PLoS One. 2009;4(11):e7870.

Kawase M, Shirato K, Matsuyama S, Taguchi F. Protease-mediated entry via the endosome of human coronavirus 229E. J Virol. 2009;83(2):712–21.

Kawase M, Shirato K, van der Hoek L, Taguchi F, Matsuyama S. Simultaneous treatment of human bronchial epithelial cells with serine and cysteine protease inhibitors prevents severe acute respiratory syndrome coronavirus entry. J Virol. 2012;86(12):6537–45.

Kim TS, Heinlein C, Hackman RC, Nelson PS. Phenotypic analysis of mice lacking the Tmprss2-encoded protease. Mol Cell Biol. 2006;26(3):965–75.

Kirchdoerfer RN, Cottrell CA, Wang N, Pallesen J, Yassine HM, Turner HL, Corbett KS, Graham BS, McLellan JS, Ward AB. Pre-fusion structure of a human coronavirus spike protein. Nature. 2016;531(7592):118–21.

Krempl C, Schultze B, Laude H, Herrler G. Point mutations in the S protein connect the sialic acid binding activity with the enteropathogenicity of transmissible gastroenteritis coronavirus. J Virol. 1997;71(4):3285–7.

Krieger E, Vriend G. YASARA view—molecular graphics for all devices—from smartphones to workstations. Bioinformatics. 2014;30(20):2981–2.

Ksiazek TG, Erdman D, Goldsmith CS, Zaki SR, Peret T, Emery S, Tong S, Urbani C, Comer JA, Lim W, Rollin PE, Dowell SF, Ling AE, Humphrey CD, Shieh WJ, Guarner J, Paddock CD, Rota P, Fields B, DeRisi J, Yang JY, Cox N, Hughes JM, LeDuc JW, Bellini WJ, Anderson LJ. A novel coronavirus associated with severe acute respiratory syndrome. N Engl J Med. 2003;348(20):1953–66.

Kubo H, Yamada YK, Taguchi F. Localization of neutralizing epitopes and the receptor-binding site within the amino-terminal 330 amino acids of the murine coronavirus spike protein. J Virol. 1994;68(9):5403–10.

Kuhn N, Bergmann S, Kosterke N, Lambertz RL, Keppner A, van den Brand JM, Pöhlmann S, Weiss S, Hummler E, Hatesuer B, Schughart K. The proteolytic activation of (H3N2) influenza A virus hemagglutinin is facilitated by different type II transmembrane serine proteases. J Virol. 2016;90(9):4298–307.

Kunkel F, Herrler G. Structural and functional analysis of the surface protein of human coronavirus OC43. Virology. 1993;195(1):195–202.

Lan J, Yao Y, Deng Y, Chen H, Lu G, Wang W, Bao L, Deng W, Wei Q, Gao GF, Qin C, Tan W. Recombinant receptor binding domain protein induces partial protective immunity in Rhesus Macaques against Middle East respiratory syndrome coronavirus challenge. EBioMedicine. 2015;2(10):1438–46.

Lane TE, Hosking MP. The pathogenesis of murine coronavirus infection of the central nervous system. Crit Rev Immunol. 2010;30(2):119–30.

Lau SK, Li KS, Tsang AK, Lam CS, Ahmed S, Chen H, Chan KH, Woo PC, Yuen KY. Genetic characterization of Betacoronavirus lineage C viruses in bats reveals marked sequence divergence in the spike protein of pipistrellus bat coronavirus HKU5 in Japanese pipistrelle: implications for the origin of the novel Middle East respiratory syndrome coronavirus. J Virol. 2013;87(15):8638–50.

Lecaille F, Kaleta J, Bromme D. Human and parasitic papain-like cysteine proteases: their role in physiology and pathology and recent developments in inhibitor design. Chem Rev. 2002;102(12):4459–88.

Lee C. Porcine epidemic diarrhea virus: An emerging and re-emerging epizootic swine virus. Virol J. 2015;12:193.

Li F. Structural analysis of major species barriers between humans and palm civets for severe acute respiratory syndrome coronavirus infections. J Virol. 2008;82(14):6984–91.

Li F. Evidence for a common evolutionary origin of coronavirus spike protein receptor-binding subunits. J Virol. 2012;86(5):2856–8.

Li F. Structure, function, and evolution of coronavirus spike proteins. Annu Rev Virol. 2016;3(1):237–61.

Li F, Berardi M, Li W, Farzan M, Dormitzer PR, Harrison SC. Conformational states of the severe acute respiratory syndrome coronavirus spike protein ectodomain. J Virol. 2006;80(14):6794–800.

Li F, Li W, Farzan M, Harrison SC. Structure of SARS coronavirus spike receptor-binding domain complexed with receptor. Science. 2005a;309(5742):1864–8.

Li W, Greenough TC, Moore MJ, Vasilieva N, Somasundaran M, Sullivan JL, Farzan M, Choe H. Efficient replication of severe acute respiratory syndrome coronavirus in mouse cells is limited by murine angiotensin-converting enzyme 2. J Virol. 2004;78(20):11429–33.

Li W, Li H, Liu Y, Pan Y, Deng F, Song Y, Tang X, He Q. New variants of porcine epidemic diarrhea virus, China, 2011. Emerg Infect Dis. 2012;18(8):1350–3.

Li W, Moore MJ, Vasilieva N, Sui J, Wong SK, Berne MA, Somasundaran M, Sullivan JL, Luzuriaga K, Greenough TC, Choe H, Farzan M. Angiotensin-converting enzyme 2 is a functional receptor for the SARS coronavirus. Nature. 2003;426(6965):450–4.

Li W, Wicht O, van Kuppeveld FJ, He Q, Rottier PJ, Bosch BJ. A single point mutation creating a furin cleavage site in the spike protein renders porcine epidemic diarrhea coronavirus trypsin independent for cell entry and fusion. J Virol. 2015;89(15):8077–81.

Li W, Zhang C, Sui J, Kuhn JH, Moore MJ, Luo S, Wong SK, Huang IC, Xu K, Vasilieva N, Murakami A, He Y, Marasco WA, Guan Y, Choe H, Farzan M. Receptor and viral determinants of SARS-coronavirus adaptation to human ACE2. EMBO J. 2005b;24(8):1634–43.

Lin HX, Feng Y, Wong G, Wang L, Li B, Zhao X, Li Y, Smaill F, Zhang C. Identification of residues in the receptor-binding domain (RBD) of the spike protein of human coronavirus NL63 that are critical for the RBD-ACE2 receptor interaction. J Gen Virol. 2008;89(Pt 4):1015–24.

Liu C, Ma Y, Yang Y, Zheng Y, Shang J, Zhou Y, Jiang S, Du L, Li J, Li F. Cell entry of porcine epidemic diarrhea coronavirus is activated by lysosomal proteases. J Biol Chem. 2016;291(47):24779–86.

Liu C, Tang J, Ma Y, Liang X, Yang Y, Peng G, Qi Q, Jiang S, Li J, Du L, Li F. Receptor usage and cell entry of porcine epidemic diarrhea coronavirus. J Virol. 2015;89(11):6121–5.

Lu G, Hu Y, Wang Q, Qi J, Gao F, Li Y, Zhang Y, Zhang W, Yuan Y, Bao J, Zhang B, Shi Y, Yan J, Gao GF. Molecular basis of binding between novel human coronavirus MERS-CoV and its receptor CD26. Nature. 2013;500(7461):227–31.

Lu G, Wang Q, Gao GF. Bat-to-human: spike features determining 'host jump' of coronaviruses SARS-CoV, MERS-CoV, and beyond. Trends Microbiol. 2015;23(8):468–78.

Lu L, Liu Q, Zhu Y, Chan KH, Qin L, Li Y, Wang Q, Chan JF, Du L, Yu F, Ma C, Ye S, Yuen KY, Zhang R, Jiang S. Structure-based discovery of Middle East respiratory syndrome coronavirus fusion inhibitor. Nat Commun. 2014;5:3067.

Marzi A, Reinheckel T, Feldmann H. Cathepsin B & L are not required for ebola virus replication. PLoS Negl Trop Dis. 2012;6(12):e1923.

Masters PS, Perlman S. Coronaviridae. In: Knipe DM, Howley PM, editors. Fields virology. 1st ed. Philadelphia: Wolters Kluwer Health/Lippincott Williams & Wilkins; 2013.

Matsuyama S, Nagata N, Shirato K, Kawase M, Takeda M, Taguchi F. Efficient activation of the severe acute respiratory syndrome coronavirus spike protein by the transmembrane protease TMPRSS2. J Virol. 2010;84(24):12658–64.

Matsuyama S, Ujike M, Morikawa S, Tashiro M, Taguchi F. Protease-mediated enhancement of severe acute respiratory syndrome coronavirus infection. Proc Natl Acad Sci U S A. 2005;102(35):12543–7.

Meyer D, Sielaff F, Hammami M, Böttcher-Friebertshäuser E, Garten W, Steinmetzer T. Identification of the first synthetic inhibitors of the type II transmembrane serine protease TMPRSS2 suitable for inhibition of influenza virus activation. Biochem J. 2013;452(2):331–43.

Millet JK, Whittaker GR. Host cell entry of Middle East respiratory syndrome coronavirus after two-step, furin-mediated activation of the spike protein. Proc Natl Acad Sci U S A. 2014;111(42):15214–9.

Millet JK, Whittaker GR. Host cell proteases: critical determinants of coronavirus tropism and pathogenesis. Virus Res. 2015;202:120–34.

Mohd HA, Al-Tawfiq JA, Memish ZA. Middle East Respiratory Syndrome Coronavirus (MERS-CoV) origin and animal reservoir. Virol J. 2016;13:87.

Molloy SS, Thomas L, VanSlyke JK, Stenberg PE, Thomas G. Intracellular trafficking and activation of the furin proprotein convertase: localization to the TGN and recycling from the cell surface. EMBO J. 1994;13(1):18–33.

Mou H, Raj VS, van Kuppeveld FJ, Rottier PJ, Haagmans BL, Bosch BJ. The receptor binding domain of the new Middle East respiratory syndrome coronavirus maps to a 231-residue region in the spike protein that efficiently elicits neutralizing antibodies. J Virol. 2013;87(16):9379–83.

Nakayama K. Furin: a mammalian subtilisin/Kex2p-like endoprotease involved in processing of a wide variety of precursor proteins. Biochem J. 1997;327(Pt 3):625–35.

Obermajer N, Jevnikar Z, Doljak B, Kos J. Role of cysteine cathepsins in matrix degradation and cell signalling. Connect Tissue Res. 2008;49(3):193–6.

Oh J, Lee KW, Choi HW, Lee C. Immunogenicity and protective efficacy of recombinant S1 domain of the porcine epidemic diarrhea virus spike protein. Arch Virol. 2014;159(11):2977–87.
Park JE, Li K, Barlan A, Fehr AR, Perlman S, PB MC Jr, Gallagher T. Proteolytic processing of Middle East respiratory syndrome coronavirus spikes expands virus tropism. Proc Natl Acad Sci U S A. 2016;113(43):12262–7.
Peck KM, Cockrell AS, Yount BL, Scobey T, Baric RS, Heise MT. Glycosylation of mouse DPP4 plays a role in inhibiting Middle East respiratory syndrome coronavirus infection. J Virol. 2015;89(8):4696–9.
Peiris JS, Chu CM, Cheng VC, Chan KS, Hung IF, Poon LL, Law KI, Tang BS, Hon TY, Chan CS, Chan KH, Ng JS, Zheng BJ, Ng WL, Lai RW, Guan Y, Yuen KY, Hku Uch Sars Study Group. Clinical progression and viral load in a community outbreak of coronavirus-associated SARS pneumonia: a prospective study. Lancet. 2003;361(9371):1767–72.
Peng G, Xu L, Lin YL, Chen L, Pasquarella JR, Holmes KV, Li F. Crystal structure of bovine coronavirus spike protein lectin domain. J Biol Chem. 2012;287(50):41931–8.
Pensaert MB, de BP. A new coronavirus-like particle associated with diarrhea in swine. Arch Virol. 1978;58(3):243–7.
Promkuntod N, van Eijndhoven RE, de VG, Grone A, Verheije MH. Mapping of the receptor-binding domain and amino acids critical for attachment in the spike protein of avian coronavirus infectious bronchitis virus. Virology. 2014;448:26–32.
Qian Z, Dominguez SR, Holmes KV. Role of the spike glycoprotein of human Middle East respiratory syndrome coronavirus (MERS-CoV) in virus entry and syncytia formation. PLoS One. 2013;8(10):e76469.
Qiu Z, Hingley ST, Simmons G, Yu C, Das SJ, Bates P, Weiss SR. Endosomal proteolysis by cathepsins is necessary for murine coronavirus mouse hepatitis virus type 2 spike-mediated entry. J Virol. 2006;80(12):5768–76.
Qu XX, Hao P, Song XJ, Jiang SM, Liu YX, Wang PG, Rao X, Song HD, Wang SY, Zuo Y, Zheng AH, Luo M, Wang HL, Deng F, Wang HZ, Hu ZH, Ding MX, Zhao GP, Deng HK. Identification of two critical amino acid residues of the severe acute respiratory syndrome coronavirus spike protein for its variation in zoonotic tropism transition via a double substitution strategy. J Biol Chem. 2005;280(33):29588–95.
Raj VS, Mou H, Smits SL, Dekkers DH, Muller MA, Dijkman R, Muth D, Demmers JA, Zaki A, Fouchier RA, Thiel V, Drosten C, Rottier PJ, Osterhaus AD, Bosch BJ, Haagmans BL. Dipeptidyl peptidase 4 is a functional receptor for the emerging human coronavirus-EMC. Nature. 2013;495(7440):251–4.
Raj VS, Smits SL, Provacia LB, van den Brand JM, Wiersma L, Ouwendijk WJ, Bestebroer TM, Spronken MI, van AG, Rottier PJ, Fouchier RA, Bosch BJ, Osterhaus AD, Haagmans BL. Adenosine deaminase acts as a natural antagonist for dipeptidyl peptidase 4-mediated entry of the Middle East respiratory syndrome coronavirus. J Virol. 2014;88(3):1834–8.
Reusken CB, Raj VS, Koopmans MP, Haagmans BL. Cross host transmission in the emergence of MERS coronavirus. Curr Opin Virol. 2016;16:55–62.
Rota PA, Oberste MS, Monroe SS, Nix WA, Campagnoli R, Icenogle JP, Penaranda S, Bankamp B, Maher K, Chen MH, Tong S, Tamin A, Lowe L, Frace M, DeRisi JL, Chen Q, Wang D, Erdman DD, Peret TC, Burns C, Ksiazek TG, Rollin PE, Sanchez A, Liffick S, Holloway B, Limor J, McCaustland K, Olsen-Rasmussen M, Fouchier R, Gunther S, Osterhaus AD, Drosten C, Pallansch MA, Anderson LJ, Bellini WJ. Characterization of a novel coronavirus associated with severe acute respiratory syndrome. Science. 2003;300(5624):1394–9.
Saftig P, Klumperman J. Lysosome biogenesis and lysosomal membrane proteins: trafficking meets function. Nat Rev Mol Cell Biol. 2009;10(9):623–35.
Sakai K, Ami Y, Tahara M, Kubota T, Anraku M, Abe M, Nakajima N, Sekizuka T, Shirato K, Suzaki Y, Ainai A, Nakatsu Y, Kanou K, Nakamura K, Suzuki T, Komase K, Nobusawa E, Maenaka K, Kuroda M, Hasegawa H, Kawaoka Y, Tashiro M, Takeda M. The host protease TMPRSS2 plays a major role in in vivo replication of emerging H7N9 and seasonal influenza viruses. J Virol. 2014;88(10):5608–16.

Schultze B, Cavanagh D, Herrler G. Neuraminidase treatment of avian infectious bronchitis coronavirus reveals a hemagglutinating activity that is dependent on sialic acid-containing receptors on erythrocytes. Virology. 1992;189(2):792–4.

Schultze B, Gross HJ, Brossmer R, Herrler G. The S protein of bovine coronavirus is a hemagglutinin recognizing 9-O-acetylated sialic acid as a receptor determinant. J Virol. 1991;65(11):6232–7.

Seidah NG, Prat A. The biology and therapeutic targeting of the proprotein convertases. Nat Rev Drug Discov. 2012;11(5):367–83.

Seidah NG, Sadr MS, Chretien M, Mbikay M. The multifaceted proprotein convertases: their unique, redundant, complementary, and opposite functions. J Biol Chem. 2013;288(30):21473–81.

Sha Y, Wu Y, Cao Z, Xu X, Wu W, Jiang D, Mao X, Liu H, Zhu Y, Gong R, Li W. A convenient cell fusion assay for the study of SARS-CoV entry and inhibition. IUBMB Life. 2006;58(8):480–6.

Shirato K, Kanou K, Kawase M, Matsuyama S. Clinical isolates of human coronavirus 229E bypass the endosome for cell entry. J Virol. 2017;91(1):e01387-16.

Shirato K, Kawase M, Matsuyama S. Middle East respiratory syndrome coronavirus infection mediated by the transmembrane serine protease TMPRSS2. J Virol. 2013;87(23):12552–61.

Shirato K, Matsuyama S, Ujike M, Taguchi F. Role of proteases in the release of porcine epidemic diarrhea virus from infected cells. J Virol. 2011;85(15):7872–80.

Shulla A, Heald-Sargent T, Subramanya G, Zhao J, Perlman S, Gallagher T. A transmembrane serine protease is linked to the severe acute respiratory syndrome coronavirus receptor and activates virus entry. J Virol. 2011;85(2):873–82.

Simmons G, Bertram S, Glowacka I, Steffen I, Chaipan C, Agudelo J, Lu K, Rennekamp AJ, Hofmann H, Bates P, Pöhlmann S. Different host cell proteases activate the SARS-coronavirus spike-protein for cell-cell and virus-cell fusion. Virology. 2011;413(2):265–74.

Simmons G, Gosalia DN, Rennekamp AJ, Reeves JD, Diamond SL, Bates P. Inhibitors of cathepsin L prevent severe acute respiratory syndrome coronavirus entry. Proc Natl Acad Sci U S A. 2005;102(33):11876–81.

Simmons G, Reeves JD, Rennekamp AJ, Amberg SM, Piefer AJ, Bates P. Characterization of severe acute respiratory syndrome-associated coronavirus (SARS-CoV) spike glycoprotein-mediated viral entry. Proc Natl Acad Sci U S A. 2004;101(12):4240–5.

Song HD, Tu CC, Zhang GW, Wang SY, Zheng K, Lei LC, Chen QX, Gao YW, Zhou HQ, Xiang H, Zheng HJ, Chern SW, Cheng F, Pan CM, Xuan H, Chen SJ, Luo HM, Zhou DH, Liu YF, He JF, Qin PZ, Li LH, Ren YQ, Liang WJ, Yu YD, Anderson L, Wang M, Xu RH, Wu XW, Zheng HY, Chen JD, Liang G, Gao Y, Liao M, Fang L, Jiang LY, Li H, Chen F, Di B, He LJ, Lin JY, Tong S, Kong X, Du L, Hao P, Tang H, Bernini A, Yu XJ, Spiga O, Guo ZM, Pan HY, He WZ, Manuguerra JC, Fontanet A, Danchin A, Niccolai N, Li YX, Wu CI, Zhao GP. Cross-host evolution of severe acute respiratory syndrome coronavirus in palm civet and human. Proc Natl Acad Sci U S A. 2005;102(7):2430–5.

Stevenson GW, Hoang H, Schwartz KJ, Burrough ER, Sun D, Madson D, Cooper VL, Pillatzki A, Gauger P, Schmitt BJ, Koster LG, Killian ML, Yoon KJ. Emergence of Porcine epidemic diarrhea virus in the United States: clinical signs, lesions, and viral genomic sequences. J Vet Diagn Investig. 2013;25(5):649–54.

Stieneke-Gröber A, Vey M, Angliker H, Shaw E, Thomas G, Roberts C, Klenk HD, Garten W. Influenza virus hemagglutinin with multibasic cleavage site is activated by furin, a subtilisin-like endoprotease. EMBO J. 1992;11(7):2407–14.

Szabo R, Bugge TH. Membrane-anchored serine proteases in vertebrate cell and developmental biology. Annu Rev Cell Dev Biol. 2011;27:213–35.

Tai W, Wang Y, Fett CA, Zhao G, Li F, Perlman S, Jiang S, Zhou Y, Du L. Recombinant receptor-binding domains of multiple Middle East respiratory syndrome coronaviruses (MERS-CoVs) induce cross-neutralizing antibodies against divergent human and camel MERS-CoVs and antibody escape mutants. J Virol. 2017;91(1):e01651-16.

Tarnow C, Engels G, Arendt A, Schwalm F, Sediri H, Preuss A, Nelson PS, Garten W, Klenk HD, Gabriel G, Böttcher-Friebertshäuser E. TMPRSS2 is a host factor that is essential for pneumotropism and pathogenicity of H7N9 influenza A virus in mice. J Virol. 2014;88(9):4744–51.

Tresnan DB, Holmes KV. Feline aminopeptidase N is a receptor for all group I coronaviruses. Adv Exp Med Biol. 1998;440:69–75.
Tripet B, Howard MW, Jobling M, Holmes RK, Holmes KV, Hodges RS. Structural characterization of the SARS-coronavirus spike S fusion protein core. J Biol Chem. 2004;279(20):20836–49.
Turk B, Bieth JG, Bjork I, Dolenc I, Turk D, Cimerman N, Kos J, Colic A, Stoka V, Turk V. Regulation of the activity of lysosomal cysteine proteinases by pH-induced inactivation and/or endogenous protein inhibitors, cystatins. Biol Chem Hoppe Seyler. 1995;376(4):225–30.
Turk V, Stoka V, Vasiljeva O, Renko M, Sun T, Turk B, Turk D. Cysteine cathepsins: from structure, function and regulation to new frontiers. Biochim Biophys Acta. 2012;1824(1):68–88.
Tusell SM, Schittone SA, Holmes KV. Mutational analysis of aminopeptidase N, a receptor for several group 1 coronaviruses, identifies key determinants of viral host range. J Virol. 2007;81(3):1261–73.
Tyrrell DA, Bynoe ML. Cultivation of a novel type of common-cold virus in organ cultures. Br Med J. 1965;1(5448):1467–70.
van DN, Miazgowicz KL, Milne-Price S, Bushmaker T, Robertson S, Scott D, Kinne J, McLellan JS, Zhu J, Munster VJ. Host species restriction of Middle East respiratory syndrome coronavirus through its receptor, dipeptidyl peptidase 4. J Virol. 2014;88(16):9220–32.
Vlasak R, Luytjes W, Leider J, Spaan W, Palese P. The E3 protein of bovine coronavirus is a receptor-destroying enzyme with acetylesterase activity. J Virol. 1988;62(12):4686–90.
Walls AC, Tortorici MA, Bosch BJ, Frenz B, Rottier PJ, DiMaio F, Rey FA, Veesler D. Cryo-electron microscopy structure of a coronavirus spike glycoprotein trimer. Nature. 2016a;531(7592):114–7.
Walls AC, Tortorici MA, Frenz B, Snijder J, Li W, Rey FA, DiMaio F, Bosch BJ, Veesler D. Glycan shield and epitope masking of a coronavirus spike protein observed by cryo-electron microscopy. Nat Struct Mol Biol. 2016b;23(10):899–905.
Wang N, Shi X, Jiang L, Zhang S, Wang D, Tong P, Guo D, Fu L, Cui Y, Liu X, Arledge KC, Chen YH, Zhang L, Wang X. Structure of MERS-CoV spike receptor-binding domain complexed with human receptor DPP4. Cell Res. 2013;23(8):986–93.
Wang P, Chen J, Zheng A, Nie Y, Shi X, Wang W, Wang G, Luo M, Liu H, Tan L, Song X, Wang Z, Yin X, Qu X, Wang X, Qing T, Ding M, Deng H. Expression cloning of functional receptor used by SARS coronavirus. Biochem Biophys Res Commun. 2004;315(2):439–44.
Weiss SR, Navas-Martin S. Coronavirus pathogenesis and the emerging pathogen severe acute respiratory syndrome coronavirus. Microbiol Mol Biol Rev. 2005;69(4):635–64.
White JM, Whittaker GR. Fusion of enveloped viruses in endosomes. Traffic. 2016;17(6):593–614.
WHO Health Organisation. Middle East respiratory syndrome coronavirus (MERS-CoV). 2017. http://www.who.int/emergencies/mers-cov/en/. Accessed 27 April 2017.
Williams RK, Jiang GS, Holmes KV. Receptor for mouse hepatitis virus is a member of the carcinoembryonic antigen family of glycoproteins. Proc Natl Acad Sci U S A. 1991;88(13):5533–6.
Wong SK, Li W, Moore MJ, Choe H, Farzan M. A 193-amino acid fragment of the SARS coronavirus S protein efficiently binds angiotensin-converting enzyme 2. J Biol Chem. 2004;279(5):3197–201.
Woo PC, Lau SK, Lam CS, Lau CC, Tsang AK, Lau JH, Bai R, Teng JL, Tsang CC, Wang M, Zheng BJ, Chan KH, Yuen KY. Discovery of seven novel Mammalian and avian coronaviruses in the genus deltacoronavirus supports bat coronaviruses as the gene source of alphacoronavirus and betacoronavirus and avian coronaviruses as the gene source of gammacoronavirus and deltacoronavirus. J Virol. 2012;86(7):3995–4008.
Wu D, Tu C, Xin C, Xuan H, Meng Q, Liu Y, Yu Y, Guan Y, Jiang Y, Yin X, Crameri G, Wang M, Li C, Liu S, Liao M, Feng L, Xiang H, Sun J, Chen J, Sun Y, Gu S, Liu N, Fu D, Eaton BT, Wang LF, Kong X. Civets are equally susceptible to experimental infection by two different severe acute respiratory syndrome coronavirus isolates. J Virol. 2005;79(4):2620–5.
Wu K, Chen L, Peng G, Zhou W, Pennell CA, Mansky LM, Geraghty RJ, Li F. A virus-binding hot spot on human angiotensin-converting enzyme 2 is critical for binding of two different coronaviruses. J Virol. 2011;85(11):5331–7.

Wu K, Peng G, Wilken M, Geraghty RJ, Li F. Mechanisms of host receptor adaptation by severe acute respiratory syndrome coronavirus. J Biol Chem. 2012;287(12):8904–11.

Yamada Y, Liu DX. Proteolytic activation of the spike protein at a novel RRRR/S motif is implicated in furin-dependent entry, syncytium formation, and infectivity of coronavirus infectious bronchitis virus in cultured cells. J Virol. 2009;83(17):8744–58.

Yang Y, Du L, Liu C, Wang L, Ma C, Tang J, Baric RS, Jiang S, Li F. Receptor usage and cell entry of bat coronavirus HKU4 provide insight into bat-to-human transmission of MERS coronavirus. Proc Natl Acad Sci U S A. 2014;111(34):12516–21.

Yang Y, Liu C, Du L, Jiang S, Shi Z, Baric RS, Li F. Two mutations were critical for bat-to-human transmission of middle east respiratory syndrome coronavirus. J Virol. 2015;89(17):9119–23.

Yeager CL, Ashmun RA, Williams RK, Cardellichio CB, Shapiro LH, Look AT, Holmes KV. Human aminopeptidase N is a receptor for human coronavirus 229E. Nature. 1992;357(6377):420–2.

Yuan Y, Cao D, Zhang Y, Ma J, Qi J, Wang Q, Lu G, Wu Y, Yan J, Shi Y, Zhang X, Gao GF. Cryo-EM structures of MERS-CoV and SARS-CoV spike glycoproteins reveal the dynamic receptor binding domains. Nat Commun. 2017;8:15092.

Zaki AM, van BS, Bestebroer TM, Osterhaus AD, Fouchier RA. Isolation of a novel coronavirus from a man with pneumonia in Saudi Arabia. N Engl J Med. 2012;367(19):1814–20.

Zhou Y, Vedantham P, Lu K, Agudelo J, Carrion R Jr, Nunneley JW, Barnard D, Pöhlmann S, McKerrow JH, Renslo AR, Simmons G. Protease inhibitors targeting coronavirus and filovirus entry. Antivir Res. 2015;116:76–84.

Zmora P, Blazejewska P, Moldenhauer AS, Welsch K, Nehlmeier I, Wu Q, Schneider H, Pöhlmann S, Bertram S. DESC1 and MSPL activate influenza A viruses and emerging coronaviruses for host cell entry. J Virol. 2014;88(20):12087–97.

Proteolytic Processing of Filovirus Glycoproteins

5

Viktor Volchkov and Hans Dieter Klenk

Abstract

Filoviruses (Marburg virus and Ebola virus) have a single envelope glycoprotein (GP) that initiates infection. GP is a class I fusion protein that forms trimeric spikes composed of heterodimers of the subunits GP1 and GP2. GP1 and GP2 are derived from the precursor pre-GP by furin cleavage during exocytosis. GP1 contains a receptor-binding core topped by a glycan cap and a heavily glycosylated mucin-like domain, while GP2 contains a fusion loop and a membrane anchor. After entering cells by macropinocytosis, the glycan cap and the mucin-like domain are removed from GP1 by endosomal cathepsins B and L exposing the binding site for the Niemann-Pick C1 receptor. It appears that there is no strict requirement for specific proteases involved in GP processing. Thus, furin is not indispensible for GP1-2 cleavage, and GP1 may be trimmed not only by cathepsins B and L but also by other endosomal proteases.

Two soluble glycoproteins of Ebola virus are also processed by host proteases. A significant amount of GP1,2 is cleaved by the metalloprotease TACE and shed from the surface of infected cells (GP1,2 delta). The secreted protein sGP is derived from the precursor pre-sGP by furin cleavage.

V. Volchkov
CIRI, INSERM U111, Université Lyon, Lyon, France
e-mail: viktor.volchkov@inserm.fr

H. D. Klenk (✉)
Institute of Virology, Philipps-University Marburg, Marburg, Germany
e-mail: klenk@staff.uni-marburg.de

E. Böttcher-Friebertshäuser et al. (eds.), *Activation of Viruses by Host Proteases*,
https://doi.org/10.1007/978-3-319-75474-1_5

5.1 Introduction

Filoviruses comprising Marburg virus (MARV) and 5 Ebola virus (EBOV) species (Zaire, Sudan, Reston, Bundibugyo, and Tai Forest virus) cause fulminant hemorrhagic fevers in man and nonhuman primates. MARV and EBOV have a zoonotic background and, except for Reston virus, are endemic in sub-Saharan Africa. Since the discovery of MARV in 1967 and EBOV in 1976, the viruses re-emerged with increasing frequency. Most of the outbreaks were dramatic but confined to relatively short time periods and small geographic areas. Between 2013 and 2015, however, an unprecedented EBOV outbreak occurred in West Africa with almost 30,000 human infections and more than 11,000 deaths.

The non-segmented negative-stranded RNA genome of filoviruses contains seven genes: NP, VP35, VP40, GP, VP30, VP24, and L. The GP gene of EBOV has two overlapping reading frames from which three glycoproteins are expressed by transcriptional editing: the envelope glycoprotein GP and two nonstructural glycoproteins, sGP and ssGP. In contrast, the envelope glycoprotein of MARV is expressed as the only gene product from a single open reading frame (Volchkov et al. 1995, 2005; Sanchez et al. 1996).

5.2 Biosynthesis and Maturation of Filovirus Glycoproteins

GP is a type I membrane glycoprotein that matures during export through the exocytotic transport route to the cell surface. ER-associated GP, designated pre-GPer, contains oligomannosidic *N*-glycans and shows sensitivity to endoglycosidase H treatment. Oligomerization of GP occurs already within the ER early after pre-GPer synthesis (V. Volchkov, unpublished results). Pre-GPer lacks the signal peptide sequence which is co-translationally cleaved by cellular signal peptidase. The second precursor identified, designated pre-GP, represents the Golgi-associated form of GP. This precursor contains mature *N*-glycans and is O-glycosylated. Still within the Golgi apparatus, pre-GP is processed by proteolytic cleavage into GP1,2 consisting of the amino-terminal fragment GP1 and the carboxy-terminal fragment GP2 linked by a disulfide bond (Volchkov et al. 1998a; Sanchez et al. 1998) (Fig. 5.1). GP1,2 complexes are present at the surface of EBOV-infected cells and build up trimeric spikes on virions. Proteolytic processing of the envelope glycoprotein of filoviruses has been unnoticed for a rather long period of time, largely due to the fact that pre-GP, mature GP1,2, and the GP1 subunit have similar migration rates on polyacrylamide gels and that GP2 tends to escape detection because it partly co-migrates with the VP24 protein. We know now, however, that cleavage of GP is remarkably efficient and that unprocessed GP is not present on Ebola virions in any significant amount.

EBOV GP is cleaved into subunits GP1 and GP2 by furin at the motif R-T-R-R501 (Volchkov et al. 1998a). Furin cleavage was assessed by the observation that cleavage efficiency was dramatically reduced when GP was expressed in the furin-deficient LoVo cell line but was fully restored in these cells by

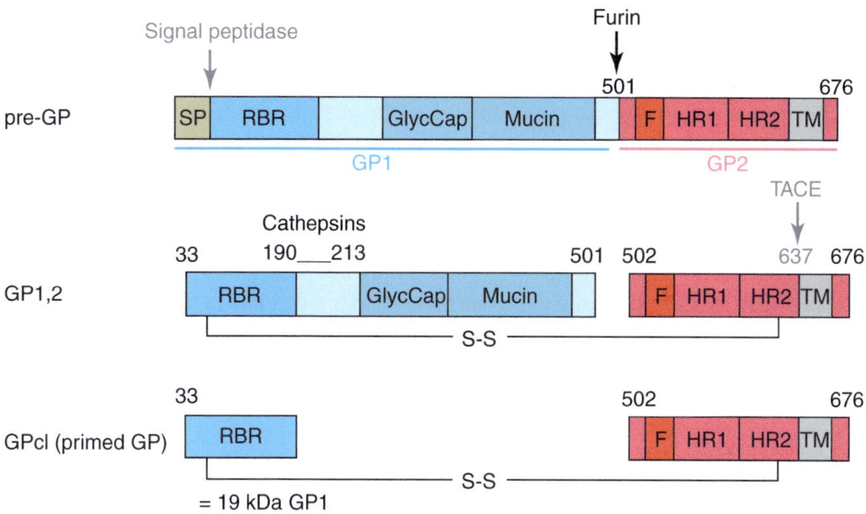

Fig. 5.1 Processing of EBOV GP. GP is a type I membrane protein that matures on the exocytotic transport route. After co-translational removal of the amino-terminal signal by the signal peptidase, pre-GP is cleaved by furin into GP1 and GP2. GP1 contains the receptor-binding region (RBR), the glycan cap (GlycCap), and the mucin-like domain (Mucin). GP2 contains the fusion loop (F), two heptad repeats (HR1, HR2), and the transmembrane anchor. GP1,2 forms trimeric spikes that are incorporated into virions or shed from the cell surface by the metalloprotease TACE. After cell entry by macropinocytosis, glycan cap and mucin-like domain are removed in endosomes by cathepsins yielding GPcl. GPcl contains GP2 linked by a disulfide bond to a 19 kDa fragment of GP1 with the receptor-binding region

vector-expressed furin. The finding that cleavage was effectively inhibited by peptidyl-chloromethylketone containing a furin motif or by site-directed mutagenesis of the furin site further supported this concept.

The surface glycoprotein of MARV is proteolytically processed in a similar way as that of EBOV; two precursor molecules and mature GP1,2 consisting of the disulfide-linked cleavage products GP1 and GP2 were identified in cells expressing MARV GP and in Marburg virions (Volchkov et al. 1998a, 2000). Interestingly, MARV GP contains two sites suitable for furin cleavage: R-R-K-R435 and R-L-R-R561. It appears that the second site is not used for protein processing, possibly due to conformational constraints. Site-directed mutagenesis revealed that MARV GP is indeed proteolytically processed at the first furin site (Volchkov et al. 2000). Mutations introduced at the multibasic site revealed the consensus sequence recognized by furin or the related proprotein convertase PC5/6 which contains Arg at positions −1 and −4 as a minimal requirement and Arg/Lys at position −2 for cleavage optimization (see Chap. 9). Thus, substitution R435L at position −1 resulted in a dramatic loss of cleavage, whereas mutation K434M at position −2 showed a reduction in cleavage efficiency (Volchkov et al. 2000).

A fraction of EBOV GP1,2 that is not incorporated into virions is released from the cell surface after removal of the membrane anchor by the metalloprotease

TACE (tumor necrosis factor-alpha-converting enzyme) (Dolnik et al. 2004) (Fig. 5.1). Released GP1,2, designated GP1,2 delta, is present in the trimeric form which, however, is more labile than GP1,2 trimers, indicating that the membrane anchor has a stabilizing function. GP1,2 delta released from virus-infected cells activates non-infected dendritic cells and macrophages causing the massive secretion of pro- and anti-inflammatory cytokines and increased vascular permeability. These activities may be instrumental for the excessive and dysregulated inflammatory host reactions to infection and, thus, contribute to the high pathogenicity of the virus (Escudero-Pérez et al. 2014). There is also evidence that fine-tuning of the levels of EBOV GP expressed at the surface of infected cells via GP shedding plays an important role in EBOV replication by orchestrating the balance between optimal virion GP content and cytotoxicity caused by GP (Dolnik et al. 2015). TACE, also designated ADAM17, is a member of the ADAM (a disintegrin and metalloprotease) family, a large group of zinc-dependent cell surface proteases. TACE mediates shedding of many membrane proteins and has therefore been proposed to have the function of a common sheddase. Most, but not all, substrates are cleaved between two hydrophobic residues, but neither a specific recognition sequence nor a specific secondary structure at the cleavage site appears to be required (Althoff et al. 2001).

The secreted glycoprotein (sGP) of EBOV is derived from a precursor (pre-sGP) that has a length of 364 amino acids and shares the amino-terminal 295 amino acids with the membrane glycoprotein GP. Like pre-GP, pre-sGP undergoes several co- and posttranslational processing events, such as signal peptide cleavage, N- and O-glycosylation, oligomerization, and proteolytic cleavage by furin to sGP and a small peptide, designated delta-peptide (Volchkova et al. 1998, 1999). sGP, like GP1,2 delta (Dolnik et al. 2004), may have a decoy function by binding EBOV-specific neutralizing antibodies (Sanchez et al. 1996; Volchkov et al. 1998b). There is also evidence that the cytotoxicity caused by GP is down-regulated through the expression of sGP (Volchkov et al. 2001).

5.3 The Role of GP in Host Cell Entry

The mature envelope glycoprotein of filoviruses is a class I fusion protein that forms trimeric spikes composed of disulfide-linked GP1,2 heterodimers. The structure of the EBOV glycoprotein has been analyzed in detail. Early studies gave insight into the post-fusion structure of GP2 (Gallaher 1996; Malashkevich et al. 1999; Weissenhorn et al. 1998a, b; Volchkov et al. 1992). More recently, the structure of GP1,2 trimers in the pre-fusion state has been elucidated (Lee et al. 2008; Lee and Saphire 2009). According to these studies, the trimeric spike is shaped like a chalice. The bowl of the chalice is assembled by the three GP1 subunits, and the base is formed by the GP2 subunits that cradle and encircle the GP1 trimer (Fig. 5.2). The bowl which is formed by discontinuous sections of the amino-terminal region of GP1 (residues 33–226) contains residues required for binding to an endosomal receptor and is covered by a glycan cap with a cluster of N-linked oligosaccharides (residues 227–310). Between the glycan cap and the carboxy-terminal end of GP1

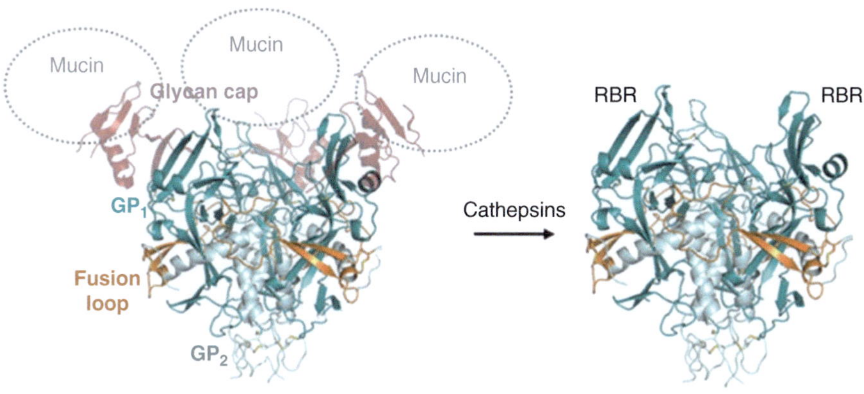

Fig. 5.2 The structure of the EBOV spike. The structure of the trimeric spike before and after removal of glycan caps and mucin-like domains exposing the receptor binding regions (RBR) is shown. GP1 and GP2 are colored in teal and light blue, respectively. Glycan cap (magenta) and fusion loop (yellow) are also indicated. The mucin-like domains were deleted for crystallization and have been modeled here as not-to-scale circles. For crystallization of GPcl, glycan caps have been removed from mucin-deleted GP1,2 by thermolysin treatment (modified from Bornholdt et al. 2016)

stretches a mucin-like domain that is about 150 amino acids long and heavily loaded with *O*-glycans. The GP2 subunit contains the hydrophobic fusion loop, two heptad repeats typical for class I fusion proteins, and the membrane anchor (Fig. 5.1).

GP presumably initiates infection by binding to the cell surface. A number of cell surface receptors have been implicated, but none of them proved to be necessary and sufficient for viral entry. It is widely accepted, however, that filoviruses are internalized after surface attachment by macropinocytosis and transported to endosomes (Saeed et al. 2010; Nanbo et al. 2010; Aleksandrowicz et al. 2011). Within endosomes, EBOV GP1,2 is cleaved by cathepsins B and/or L which is an important step in the infection process (Chandran et al. 2005; Kaletsky et al. 2007; Sanchez 2007; Schornberg et al. 2006). Cathepsin trims EBOV GP1 from its original size (ca. 130 kDa) to an initial 50-kDa fragment, followed by further cleavage to an approximately 19-kDa species of GP1 bound to GP2 by non-covalent linkages and a disulfide bridge between C53 and C609 (Jeffers et al. 2002; Volchkova et al. 1998) (Fig. 5.1). The crystal structure suggests that the site of the final cathepsin cleavage is a loop reaching from residues 189 to 214 (Lee and Saphire 2009). This concept is supported by biochemical studies indicating that the cleavage site is located at amino acid 190 (Dube et al. 2009). Thus, the entire glycan cap and the mucin-like domain are removed yielding a glycoprotein called GPcl that contains the receptor-binding site exposed on the truncated GP1 subunit and the fusion loop on GP2 (Fig. 5.2). The endosomal receptor has been identified as the cholesterol transporter Niemann-Pick C1 (NPC1) (Carette et al. 2011; Côté et al. 2011). NPC1 is a ubiquitously expressed endosomal membrane protein involved in the fusion and fission of endosomes and lysosomes (Goldman and Krise 2010). After cathepsin cleavage and receptor binding, the GP2 subunit unwinds from its GP1 clamp and rearranges irreversibly into a six-helix bundle to drive fusion of viral and endosomal membrane (Bornholdt et al. 2016;Wang et al. 2016). It has also been suggested that cathepsins are required for a step in genome delivery following fusion triggering (Spence et al. 2016).

Like EBOV, MARV enters cells by macropinocytosis and endosomal fusion, but there are some differences in the structure and in endosomal processing of the glycoproteins. Structural analysis by crystallography and small angle X-ray scattering in solution indicated that the mucin-like domains of EBOV GP project upward, whereas with MARV GP they have a more equatorial orientation. Furthermore, the glycan cap is more flexible with MARV GP than with EBOV GP. Thus, the receptor-binding site appears to be tightly masked on the surface of EBOV spikes but more exposed on the surface of MARV spikes prior to endosomal cleavage (Hashiguchi et al. 2015). This study showed also other structural differences, particularly at the putative cleavage site, which may explain previous observations indicating that, unlike EBOV, MARV does not depend on cathepsin B for endosomal GP processing (Gnirss et al. 2012; Misasi et al. 2012).

5.4 Proteases Responsible for GP Processing

The data presented so far strongly support the concept that removal of the glycan cap and the mucin-like domain which is essential for filovirus infectivity depends on cleavage of GP at the GP1-GP2 interphase followed by endosomal processing of GP1,2 to GPcl. The nature of the proteases responsible for cleavage, however, has been and still is a matter of debate.

The finding that EBOV GP is cleaved into GP1 and GP2 by furin (Volchkov et al. 1998a) did not come as a surprise, since this protease is responsible for the activation of many viral glycoproteins. The role of furin in the EBOV life cycle became a mystery, however, when several groups reported that substitution of all basic amino acids at the furin cleavage site did not significantly affect virus infectivity. Initially, these unexpected data were obtained, when pseudotype systems based on murine leukemia virus (Wool-Lewis and Bates 1999) and vesicular stomatitis virus (Ito et al. 2001) were used which allowed generation of surrogate virions carrying mutated EBOV GP. The mutated glycoprotein was shown to be transported to the plasma membrane and to be incorporated into virions, predominantly in the uncleaved form, and the pseudotyped viruses infected a wide range of cell types from diverse origins. Subsequently, it was reported that recombinant EBOV carrying GP in which the multibasic cleavage site was replaced by nonbasic amino acids was able to replicate in Vero E6 cells (Neumann et al. 2002) and to cause lethal infection in nonhuman primates (Neumann et al. 2007). These findings are frequently used as arguments against an essential function of furin cleavage in EBOV replication. There is evidence, however, that does not fully support this conclusion. Close inspection of the data obtained with the pseudotypes reveals that small amounts of GP1,2 were present. Likewise, a minor, but clearly detectable, fraction of GP was present in the cleaved form in the recombinant EBOV (Volchkov et al. 2005). Furthermore, recombinant EBOV replicated with significantly reduced growth kinetics, when the furin cleavage site was replaced by nonbasic amino acids (Neumann et al. 2002). It therefore appears that cleavage of GP into subunits GP1 and GP2 is accomplished not only by furin but also, yet with lower efficiency, by other proteases that still have to be identified. It is also conceivable that only a fraction of GP has to be present in cleaved form to allow infection as has been observed with other viruses (see Chap. 6). In any case, there appears to be a preference for furin cleavage, since this is the most efficient processing form. This concept is underlined by the high conservation of the multibasic cleavage site with filoviruses. The only exception is Reston EBOV. Here, the consensus sequence of a typical furin cleavage site is missing which has been suspected to account, at least in part, for the low human pathogenicity of this virus (Volchkov et al. 1998a).

As has been pointed out above, endosomal processing of EBOV GP1,2 is mediated by the cysteine proteases cathepsin B and L. There is evidence, however, that, again, both enzymes are not indispensible for this process. It could be shown that Zaire EBOV entry was reduced in cell culture upon selective inhibition of cathepsin B, but not cathepsin L. Interestingly, all other EBOV species entered the cells efficiently when cathepsin B and/or L activity was blocked. Moreover, cathepsin B and cathepsin L knockout mice were equally susceptible to a lethal dose of mouse-adapted Zaire EBOV as wild-type animals, with no difference in virus replication and time of death (Marzi et al. 2012). Thus, it appears that, like cleavage of GP into subunits GP1 and GP2, endosomal trimming to GPcl is mediated by an array of proteases. This concept is also supported by the observation that cathepsin can be replaced by thermolysin to convert GP1,2 into structurally and functionally competent GPcl (Brecher et al. 2012). EBOV may therefore not be a very suitable target for therapeutic approaches based on protease inhibitors (Marzi et al. 2012), quite in contrast to other viruses, such as influenza virus, where this strategy is more promising because of the high specificity of the proteases required for activation (see Chaps. 8, 9, 11).

Conclusions

Proteolytic processing of the envelope glycoprotein of filoviruses is complex involving a sequence of cleavage steps at different stages of the viral life cycle. In this respect it resembles proteolytic activation of other envelope proteins, such as the F protein of Respiratory Syncytial Virus (see Chap. 2) and presumably the S protein of coronaviruses (see Chap. 4), that are also cleaved first during exocytosis by one and subsequently upon virus entry by another enzyme. Although cleavage of pre-GP to GP1,2 and triming of GP1,2 to GPcl play essential roles in the processing of the filovirus envelope protein, it is not clear whether there is a strict requirement for furin and cathepsins, respectively. The specificity of the cleavage reactions and the proteases involved will have to be analysed in more detail in future studies. It is well known that cleavage primes a viral fusion protein for the conformational change required for activity, but it has never been shown before that fusion activity depends on removal of a large carbohydrate shield from the top of the spike as is the case with filoviruses. Another unique feature is the high amount of virus-encoded glycoproteins that are secreted or shed by proteolytic cleavage from EBOV-infected cells and may play important roles in the course of infection and in pathogenesis.

References

Aleksandrowicz P, Marzi A, Biedenkopf N, Beimforde N, Becker S, Hoenen T, Feldmann H, Schnittler HJ. Ebola virus enters host cells by macropinocytosis and clathrin-mediated endocytosis. J Infect Dis. 2011;204(Suppl 3):S957–67.

Althoff K, Müllberg J, Aasland D, Voltz N, Kallen K, Grötzinger J, Rose-John S. Recognition sequences and structural elements contribute to shedding susceptibility of membrane proteins. Biochem J. 2001;353(Pt 3):663–72.

Bornholdt ZA, Ndungo E, Fusco ML, Bale S, Flyak AI, Crowe JE Jr, Chandran K, Saphire EO. Host-primed Ebola virus GP exposes a hydrophobic NPC1 receptor-binding pocket, revealing a target for broadly neutralizing antibodies. MBio. 2016;7(1):e02154-15.

Brecher M, Schornberg KL, Delos SE, Fusco ML, Saphire EO, White JM. Cathepsin cleavage potentiates the Ebola virus glycoprotein to undergo a subsequent fusion-relevant conformational change. J Virol. 2012;86(1):364–72.

Carette JE, Raaben M, Wong AC, Herbert AS, Obernosterer G, Mulherkar N, Kuehne AI, Kranzusch PJ, Griffin AM, Ruthel G, Dal Cin P, Dye JM, Whelan SP, Chandran K, Brummelkamp TR. Ebola virus entry requires the cholesterol transporter Niemann-Pick C1. Nature. 2011;477(7364):340–3.

Chandran K, Sullivan NJ, Felbor U, Whelan SP, Cunningham JM. Endosomal proteolysis of the Ebola virus glycoprotein is necessary for infection. Science. 2005;308(5728):1643–5.

Côté M, Misasi J, Ren T, Bruchez A, Lee K, Filone CM, Hensley L, Li Q, Ory D, Chandran K, Cunningham J. Small molecule inhibitors reveal Niemann-Pick C1 is essential for Ebola virus infection. Nature. 2011;477(7364):344–8.

Dolnik O, Volchkova V, Garten W, Carbonnelle C, Becker S, Kahnt J, Ströher U, Klenk HD, Volchkov V. Ectodomain shedding of the glycoprotein GP of Ebola virus. EMBO J. 2004; 23(10):2175–84.

Dolnik O, Volchkova VA, Escudero-Perez B, Lawrence P, Klenk HD, Volchkov VE. Shedding of Ebola virus surface glycoprotein is a mechanism of self-regulation of cellular cytotoxicity and has a direct effect on virus infectivity. J Infect Dis. 2015;212(Suppl 2):S322–8.

Dube D, Brecher MB, Delos SE, Rose SC, Park EW, Schornberg KL, Kuhn JH, White JM. The primed ebolavirus glycoprotein (19-kilodalton GP1,2): sequence and residues critical for host cell binding. J Virol. 2009;83(7):2883–91.

Escudero-Pérez B, Volchkova VA, Dolnik O, Lawrence P, Volchkov VE. Shed GP of Ebola virus triggers immune activation and increased vascular permeability. PLoS Pathog. 2014;10(11): e1004509.

Gallaher WR. Similar structural models of the transmembrane proteins of Ebola and avian sarcoma viruses. Cell. 1996;85(4):477–8.

Gnirss K, Kühl A, Karsten C, Glowacka I, Bertram S, Kaup F, Hofmann H, Pöhlmann S. Cathepsins B and L activate Ebola but not Marburg virus glycoproteins for efficient entry into cell lines and macrophages independent of TMPRSS2 expression. Virology. 2012;424(1):3–10.

Goldman SD, Krise JP. Niemann-Pick C1 functions independently of Niemann-Pick C2 in the initial stage of retrograde transport of membrane-impermeable lysosomal cargo. J Biol Chem. 2010;285(7):4983–94.

Hashiguchi T, Fusco ML, Bornholdt ZA, Lee JE, Flyak AI, Matsuoka R, Kohda D, Yanagi Y, Hammel M, Crowe JE Jr, Saphire EO. Structural basis for Marburg virus neutralization by a cross-reactive human antibody. Cell. 2015;160(5):904–12.

Ito H, Watanabe S, Takada A, Kawaoka Y. Ebola virus glycoprotein: proteolytic processing, acylation, cell tropism, and detection of neutralizing antibodies. J Virol. 2001;75(3):1576–80.

Jeffers SA, Sanders DA, Sanchez A. Covalent modifications of the Ebola virus glycoprotein. J Virol. 2002;76(24):12463–72.

Kaletsky RL, Simmons G, Bates P. Proteolysis of the Ebola virus glycoproteins enhances virus binding and infectivity. J Virol. 2007;81(24):13378–84.

Lee JE, Fusco ML, Hessell AJ, Oswald WB, Burton DR, Saphire EO. Structure of the Ebola virus glycoprotein bound to an antibody from a human survivor. Nature. 2008;454(7201):177–82.

Lee JE, Saphire EO. Ebolavirus glycoprotein structure and mechanism of entry. Future Virol. 2009;4(6):621–35.
Malashkevich VN, Schneider BJ, McNally ML, Milhollen MA, Pang JX, Kim PS. Core structure of the envelope glycoprotein GP2 from Ebola virus at 1.9-A resolution. Proc Natl Acad Sci U S A. 1999;96(6):2662–7.
Marzi A, Reinheckel T, Feldmann H. Cathepsin B & L are not required for Ebola virus replication. PLoS Negl Trop Dis. 2012;6(12):e1923.
Misasi J, Chandran K, Yang JY, Considine B, Filone CM, Côté M, Sullivan N, Fabozzi G, Hensley L, Cunningham J. Filoviruses require endosomal cysteine proteases for entry but exhibit distinct protease preferences. J Virol. 2012;86(6):3284–92.
Nanbo A, Imai M, Watanabe S, Noda T, Takahashi K, Neumann G, Halfmann P, Kawaoka Y. Ebolavirus is internalized into host cells via macropinocytosis in a viral glycoprotein-dependent manner. PLoS Pathog. 2010;6(9):e1001121.
Neumann G, Feldmann H, Watanabe S, Lukashevich I, Kawaoka Y. Reverse genetics demonstrates that proteolytic processing of the Ebola virus glycoprotein is not essential for replication in cell culture. J Virol. 2002;76(1):406–10.
Neumann G, Geisbert TW, Ebihara H, Geisbert JB, Daddario-DiCaprio KM, Feldmann H, Kawaoka Y. Proteolytic processing of the Ebola virus glycoprotein is not critical for Ebola virus replication in nonhuman primates. J Virol. 2007;81(6):2995–8.
Saeed MF, Kolokoltsov AA, Albrecht T, Davey RA. Cellular entry of Ebola virus involves uptake by a macropinocytosis-like mechanism and subsequent trafficking through early and late endosomes. PLoS Pathog. 2010;6(9):e1001110.
Sanchez A, Trappier SG, Mahy BW, Peters CJ, Nichol ST. The virion glycoproteins of Ebola viruses are encoded in two reading frames and are expressed through transcriptional editing. Proc Natl Acad Sci U S A. 1996;93(8):3602–7.
Sanchez A, Yang ZY, Xu L, Nabel GJ, Crews T, Peters CJ. Biochemical analysis of the secreted and virion glycoproteins of Ebola virus. J Virol. 1998;72(8):6442–7.
Sanchez A. Analysis of filovirus entry into vero e6 cells, using inhibitors of endocytosis, endosomal acidification, structural integrity, and cathepsin (B and L) activity. J Infect Dis. 2007;196(Suppl 2):S251–8.
Schornberg K, Matsuyama S, Kabsch K, Delos S, Bouton A, White J. Role of endosomal cathepsins in entry mediated by the Ebola virus glycoprotein. J Virol. 2006;80(8):4174–8.
Spence JS, Krause TB, Mittler E, Jangra RK, Chandran K. Direct visualization of Ebola virus fusion triggering in the endocytic pathway. MBio. 2016;7(1):e01857-15.
Volchkov VE, Becker S, Volchkova VA, Ternovoj VA, Kotov AN, Netesov SV, Klenk HD. GP mRNA of Ebola virus is edited by the Ebola virus polymerase and by T7 and vaccinia virus polymerases. Virology. 1995;214(2):421–30.
Volchkov VE, Blinov VM, Netesov SV. The envelope glycoprotein of Ebola virus contains an immunosuppressive-like domain similar to oncogenic retroviruses. FEBS Lett. 1992;305(3):181–4.
Volchkov VE, Feldmann H, Volchkova VA, Klenk HD. Processing of the Ebola virus glycoprotein by the proprotein convertase furin. Proc Natl Acad Sci U S A. 1998a;95(10):5762–7.
Volchkov VE, Volchkova VA, Dolnik O, Feldmann H, Klenk HD. Polymorphism of filovirus glycoproteins. Adv Virus Res. 2005;64:359–81.
Volchkov VE, Volchkova VA, Mühlberger E, Kolesnikova LV, Weik M, Dolnik O, Klenk HD. Recovery of infectious Ebola virus from complementary DNA:RNA editing of the GP gene and viral cytotoxicity. Science. 2001;291:1965–9.
Volchkov VE, Volchkova VA, Slenczka W, Klenk HD, Feldmann H. Release of viral glycoproteins during Ebola virus infection. Virology. 1998b;245(1):110–9.
Volchkov VE, Volchkova VA, Ströher U, Becker S, Dolnik O, Cieplik M, Garten W, Klenk HD, Feldmann H. Proteolytic processing of Marburg virus glycoprotein. Virology. 2000;268(1):1–6.
Volchkova VA, Feldmann H, Klenk HD, Volchkov VE. The nonstructural small glycoprotein sGP of Ebola virus is secreted as an antiparallel-orientated homodimer. Virology. 1998;250(2):408–14.
Volchkova VA, Klenk HD, Volchkov VE. Delta-peptide is the carboxy-terminal cleavage fragment of the nonstructural small glycoprotein sGP of Ebola virus. Virology. 1999;265(1):164–71.

Wang H, Shi Y, Song J, Qi J, Lu G, Yan J, Gao GF. Ebola viral glycoprotein bound to its endosomal receptor Niemann-Pick C1. Cell. 2016;164(1–2):258–68.

Weissenhorn W, Calder LJ, Wharton SA, Skehel JJ, Wiley DC. The central structural feature of the membrane fusion protein subunit from the Ebola virus glycoprotein is a long triple-stranded coiled coil. Proc Natl Acad Sci U S A. 1998a;95(11):6032–6.

Weissenhorn W, Carfí A, Lee KH, Skehel JJ, Wiley DC. Crystal structure of the Ebola virus membrane fusion subunit, GP2, from the envelope glycoprotein ectodomain. Mol Cell. 1998b;2(5):605–16.

Wool-Lewis RJ, Bates P. Endoproteolytic processing of the Ebola virus envelope glycoprotein: cleavage is not required for function. J Virol. 1999;73(2):1419–26.

Proteolytic Activation of Flavivirus Envelope Proteins

6

Franz X. Heinz and Karin Stiasny

Abstract

Flaviviruses comprise a number of important human vector-borne pathogens, including yellow fever, dengue, Zika, West Nile, Japanese encephalitis, and tick-borne encephalitis viruses. New technologies for determining high-resolution structures of viral particles have provided unprecedented insights into the molecular organization of this group of enveloped, icosahedral viruses in different stages of assembly and maturation. The viral fusion protein E forms a metastable herringbone-like array at the surface of mature viruses, spring-loaded to mediate membrane fusion upon encountering the acidic pH in endosomes. The E protein does not require proteolytic cleavage for activation, but an accessory protein (prM), associated tightly with E in the initially assembled noninfectious immature viruses, has to be cleaved by furin in the trans-Golgi network during virus release, thus priming E for fusion. A complex interplay of pH sensors in E and prM trigger sequential conformational changes at different steps of the viral life cycle to control virus maturation and membrane fusion. There is increasing evidence that incomplete proteolytic cleavage of prM, leading to mosaic particles with patches of envelope proteins in both their immature as well as mature conformations, may be an important factor for certain biological properties of flaviviruses. Dynamic motions of the envelope proteins ("virus breathing") further increase deviations from a picture of static icosahedral structures. The resulting particle heterogeneity causes the presentation of otherwise inaccessible sites for interactions at the surface of infectious virions that can modulate viral attachment to cells and influence the induction of antibodies as well as virus neutralization.

F. X. Heinz (✉) · K. Stiasny
Center for Virology, Medical University of Vienna, Vienna, Austria
e-mail: franz.x.heinz@meduniwien.ac.at; karin.stiasny@meduniwien.ac.at

E. Böttcher-Friebertshäuser et al. (eds.), *Activation of Viruses by Host Proteases*,
https://doi.org/10.1007/978-3-319-75474-1_6

6.1 Introduction

The genus *Flavivirus* in the family *Flaviviridae* comprises about 70 different viruses, most of which are *ar*thropod-*bo*rne (ARBO) and transmitted to their vertebrate hosts by invertebrate vectors such as mosquitos or ticks (Pierson and Diamond 2013). The capacity to replicate efficiently in such phylogenetically divergent hosts is essential for these viruses to be maintained in their different natural cycles. However, two subsets of flaviviruses do not have a dual host requirement for their maintenance in nature, and these are classified as "no-known vector" and "insect-only" flaviviruses (Moureau et al. 2015; Blitvich and Firth 2015). All of the important human-pathogenic flaviviruses are vector-borne, with the most prominent representatives being yellow fever (YF), dengue (Den), Zika, West Nile (WN), Japanese encephalitis (JE), and tick-borne encephalitis (TBE) viruses (Pierson and Diamond 2013; Wilder-Smith et al. 2017). Concerning their ecological cycles, YF, dengue, and Zika viruses are outstanding, because they can be transmitted by highly domesticated mosquitoes (*Aedes aegypti*) and can use humans as their only vertebrate hosts (Wilder-Smith et al. 2017). This combination carries the potential of urban outbreaks in tropical and subtropical countries and can lead to the expansion of regions of endemicity, as most recently exemplified by the explosive spread of Zika virus from Southeast Asia to Pacific Islands and finally to the Americas (Kindhauser et al. 2016).

Structurally, flaviviruses are among the best-studied membrane-containing viruses because of their relatively simple composition and the icosahedral organization of the viral envelope (Pierson and Diamond 2013). High-resolution structures are now available for both immature and mature virions, providing insight into the molecular details of proteolytic activation of noninfectious immature precursor particles to yield infectious viruses that are endowed with a highly efficient membrane fusion machinery (Zhang et al. 2013b; Kostyuchenko et al. 2013, 2016; Sirohi et al. 2016; Prasad et al. 2017). There is increasing evidence that the proteolytic maturation of flaviviruses is incomplete in many instances (Pierson and Diamond 2012), and the formation of partially mature particles may vary among hosts and different tissues in these hosts. Structural heterogeneity is further increased by the dynamics of the viral envelope, often referred to as "virus breathing" (Kuhn et al. 2015). It has been hypothesized that the generation of partially mature virions may expand the capacity of flaviviruses to infect widely divergent hosts and tissues by increasing potential interaction sites with cellular attachment factors (Rey et al. 2017). The heterogeneity of flaviviruses can therefore be a regulator of specific pathogenic properties and is known to have a strong influence on the human antibody response. In this chapter, we review the characteristics of mature and immature flaviviruses, the structural details of proteolytic cleavage activation to generate infectious virions, and how incomplete maturation can impact the biology of flaviviruses.

6.2 Mature Flaviviruses and Virus Entry

Mature flaviviruses contain only three structural proteins, designated C (capsid), E (envelope), and M (membrane) (Fig. 6.1a). Molecular details of protein and particle organizations have been elucidated by X-ray crystallography combined with

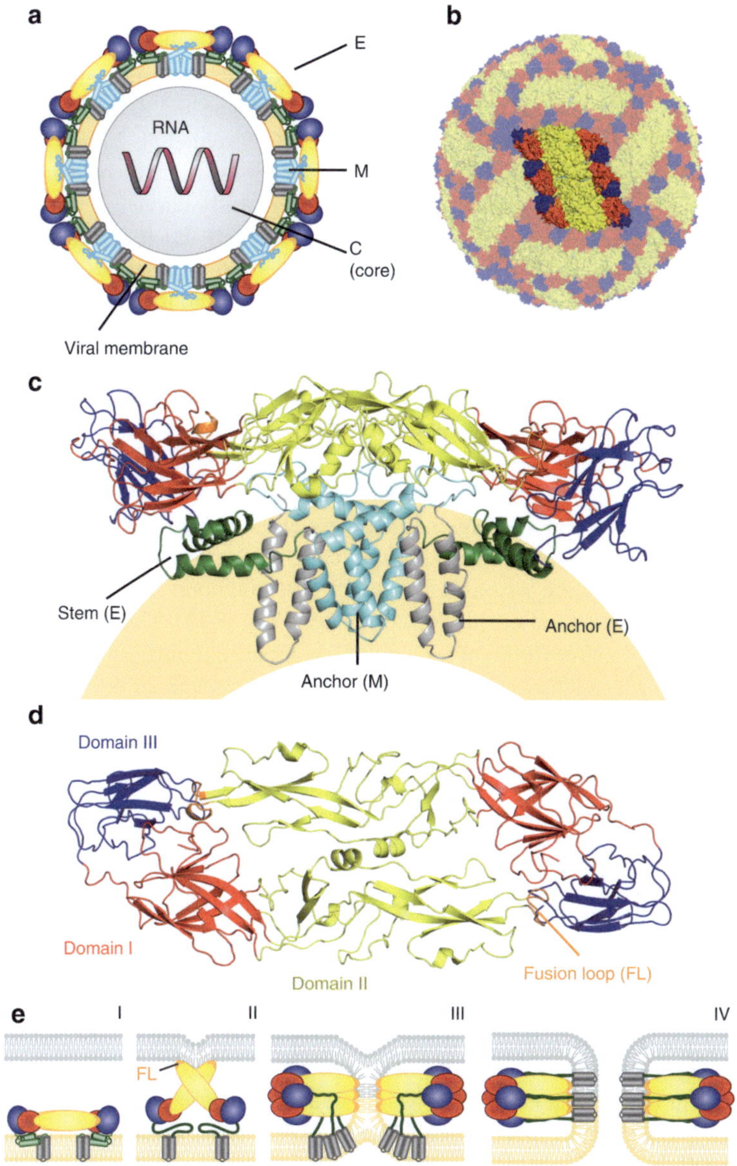

Fig. 6.1 Structural organization of mature flavivirus particles and fusion mechanism. (**a**) Schematic of a mature flavivirus particle composed of a spherical core with the capsid protein C and a viral envelope with two membrane-associated proteins, E (envelope) and M (membrane). (**b**) Surface representation of a mature dengue virus serotype 2 particle, revealing the herringbone-like arrangement of 30 rafts, each consisting of three E homodimers. One of these rafts is highlighted. (**c**) Ribbon diagram of the dengue virus serotype 2 E protein dimer in its side view. The viral membrane is shown in light orange. (**d**) Ribbon diagram of the dengue virus serotype 2 E protein dimer in its top view. (**e**) Schematic representation of the flavivirus fusion mechanism. Viral membrane, light orange. Endosomal target membrane, gray. Color codes of E and M proteins: E: domain I, red; domain II, yellow; domain III, blue; fusion loop, orange; stem, green; transmembrane domains, gray; M: cyan. Panels **b–d** were generated using the structures of dengue virus serotype 2 [PDB: 3J27, (Zhang et al. 2013b)]

electron cryo-microscopy. Little structural information is available for the nucleocapsid core, which is a spherical assembly of the positive stranded RNA genome and an undefined number of C proteins. In contrast, high-resolution structures exist for the envelopes of several flaviviruses [including dengue and Zika viruses, (Zhang et al. 2013b; Kostyuchenko et al. 2013, 2014, 2016; Sirohi et al. 2016)] that can be summarized as follows (Fig. 6.1). The E protein forms a head-to-tail homodimer that is oriented parallel to the viral membrane and gently curved to accommodate the surface curvature of the viral particle (Fig. 6.1c, d). Each of the monomeric subunits contains three domains (DI, DII, DIII) (Fig. 6.1d) that are connected by flexible regions, allowing variations of the hinge angles and the relative orientation of domains to each other in different phases of the life cycle. Importantly, the fusion loop (FL) at the tip of domain II is buried in the E homodimer by interdigitation with a hydrophobic pocket provided by the second monomeric subunit (Fig. 6.1d). The carboxy terminus of DIII is linked to a double-membrane anchor by a sequence element called "stem" (Fig. 6.1c). As revealed by electron cryo-microscopy structures of dengue virus at 3.5 Å resolution, the stem consists of three helices that lie horizontally to the envelope (Fig. 6.1c) (Zhang et al. 2013b). Two of these helices are amphipathic and interact with the outer leaflet of the viral membrane.

The surface of mature virions is formed by a herringbone-like arrangement of E dimers that cover the viral membrane completely (Fig. 6.1b). The E proteins in this icosahedral lattice have different interactions with surrounding subunits, and three E molecules form an icosahedral asymmetric unit (Kuhn et al. 2002). The M protein lies under the E protein shell and interacts with the side of E facing the membrane, primarily through hydrophobic contacts [Fig. 6.1a, c, (Zhang et al. 2013b)]. Similar to E, M has a double-membrane anchor and a horizontally oriented perimembrane helix (Fig. 6.1c).

6.2.1 Viral Attachment and Receptors

A large body of evidence indicates that flaviviruses enter cells by receptor-mediated endocytosis [reviewed in (Acosta et al. 2014; Perera-Lecoin et al. 2014; Cruz-Oliveira et al. 2015)]. Since E is the only component at the surface of fully mature virions (Fig. 6.1b), it was generally believed that interactions of this protein with molecules at the plasma membrane would lead to endocytosis. The search for true entry receptors proved to be difficult and led to the identification of a number of various cellular attachment factors for different flaviviruses, including lectins (such as DC-SIGN), carbohydrates, as well as lipids [reviewed in (Acosta et al. 2014; Perera-Lecoin et al. 2014; Cruz-Oliveira et al. 2015)]. Recent studies, however, have shown that the entry of flaviviruses (including dengue and Zika viruses) can also be mediated through the recognition of the viral lipid membrane by proteins of the TIM (**T**-cell **i**mmunoglobulin **m**ucin domain) and TAM (**T**yro3, **A**xl, and **M**er) receptor families (Perera-Lecoin et al. 2014; Amara and Mercer 2015; Miner and Diamond 2017). The normal function of these plasma membrane proteins is to trigger the phagocytosis of apoptotic cells. Hijacking of this process by viruses was therefore termed "apoptotic mimicry" (Amara and Mercer 2015). As an additional mode of entry, flaviviruses can be internalized by Fcγ receptor-positive cells in the form of virus-antibody complexes (Acosta et al. 2014; Cruz-Oliveira et al. 2015). Such infections can be facilitated not only by antibodies to E but also to the precursor of M (prM), a component of immature virions (see below).

It is an important conclusion of these virus entry studies that the closed E protein shell of fully mature viral particles as depicted in Fig. 6.1b is a structurally idealized version of infectious flaviviruses that cannot explain a number of in vivo data, both with respect to cellular infection and to the interaction with antibodies. It becomes more and more apparent that flaviviruses constitute inhomogeneous populations of virus particles that can display different molecular surfaces, thus probably expanding the capacity of these viruses to enter cells of phylogenetically distant species and different tissues in their hosts (Pierson and Diamond 2012; Rey et al. 2017). The heterogeneity of flavivirus particles is a result of incomplete proteolytic maturation as well as of viral envelope dynamics ("virus breathing"), and these aspects are discussed in more detail below.

6.2.2 The Viral Fusion Machinery

The flavivirus E protein is a class II viral fusion protein that is structurally not only closely related to the fusion proteins of alpha- and bunyaviruses [which, like flaviviruses, form icosahedral symmetric arrays on the mature virion; reviewed in Harrison (2015)] but also to cellular proteins like Eff-1 and HAP2 that mediate cell-cell and gamete fusion, respectively (Pérez-Vargas et al. 2014; Fédry et al. 2017; Doms 2017). Similar to other viruses that enter cells through receptor-mediated endocytosis, the E protein can sense the low pH in endosomes by a set of conserved histidines, which orchestrate structural changes that initiate and drive membrane fusion, leading to the release of the nucleocapsid into the cytoplasm (Figs. 6.1e and 6.3). It has been proposed that E is held in place in its metastable conformation at the surface of mature virions by hydrophobic contacts with the underlying M and that protonation of histidines in both M and E would lead to a repulsion of the two proteins to allow the initiation of the multistep membrane fusion process (Zhang et al. 2013b).

In this mechanism (depicted in Fig. 6.1e), the E dimers dissociate into monomers and rise from the viral membrane, making use of the "stem" that links the ectodomain to the transmembrane (TM) domains (Fig. 6.1c). In the monomers, the fusion loop (FL) at the tip of DII is exposed, allowing its interaction with the endosomal membrane (Fig. 6.1e, panel II). Recent data suggest that trimer formation from a pool of monomers leads to essentially irreversible target membrane engagement, and at least two of the E trimers are required for progression to hemifusion (Chao et al. 2014) (Fig. 6.1e, panel III). Full merger of the two membranes and fusion pore formation are finally driven by a "jackknifing" rearrangement of E that requires the relocation of DIII from its position at the end of the rodlike pre-fusion structure to the side of the E trimer, yielding the post-fusion conformation (Fig. 6.1e, panel IV)

(Harrison 2015; White and Whittaker 2016). The movement of DIII is facilitated by the flexible DI-DIII linker and requires the dissolution of contacts between the two domains in their pre-fusion arrangement. It appears that the release of DIII from its molecular clamp in the E dimer is controlled by the protonation of one or two histidines at the DI-DIII interface (Fritz et al. 2008). Reorganization of the domains into a hairpin-like trimeric and energetically stable post-fusion structure, in which the FL and the TM anchors are juxtaposed, is assumed to provide the energy for membrane fusion (Fig. 6.1e) and is reminiscent of the fusion-related structural reorganizations characteristic of class I and III viral fusion proteins (Harrison 2015; White and Whittaker 2016). Overall, flavivirus membrane fusion is extremely fast and efficient and does not require any interactions with cellular proteins, as revealed by fusion experiments with pure liposomes (Smit et al. 2011; Stiasny and Heinz 2006). There is evidence from studies with dengue viruses (Zaitseva et al. 2010), however, that their fusion may depend on the presence of acidic lipids in the target membrane and therefore is delayed until the virus reaches late endosomes, which are enriched in these lipids.

6.3 Structure of Immature Particles

Flaviviruses are not assembled in their infectious forms but as immature particles with a radically different surface structure compared to mature virions (Fig. 6.2a, b). Assembly of these particles occurs at the endoplasmic reticulum (ER) membrane by a process that involves the formation of a complex between the precursor of M (prM) and E, lateral interactions between prM-E heterodimers, and a poorly defined mechanism that leads to the formation and incorporation of the viral core [Fig. 6.3; (Lindenbach et al. 2013)]. The latter step appears to be inefficient to a certain degree, resulting in the formation of capsid-less subviral particles (SPs) as natural by-products of flavivirus infections (Fig. 6.3) (Apte-Sengupta et al. 2014). At least in certain cells, prM appears to have a chaperone-like function for the correct folding of E (Lorenz et al. 2002).

The structures of several immature flaviviruses were determined by electron cryo-microscopy (Zhang et al. 2003, 2013a; Kostyuchenko et al. 2013; Prasad et al. 2017). The surfaces of these particles display 60 spikes of trimers of prM-E heterodimers that are organized in an icosahedral lattice (Fig. 6.2b). The three heterodimers in a spike, however, are not related by threefold symmetry (Zhang et al. 2003), because the interactions of any one of the prM-E complexes with the two others in the spike are different (Fig. 6.2c). This lack of equivalence apparently stems from the mechanism underlying the assembly process, suggesting a sequential pathway for the generation of immature virions. As revealed by recent studies with Zika virus (Prasad et al. 2017), the spike is held together at its external tips by interactions of the pr part of prM with the FL of E proteins and stabilized at its base by contacts between DIII of E in one spike and DII of E from another spike.

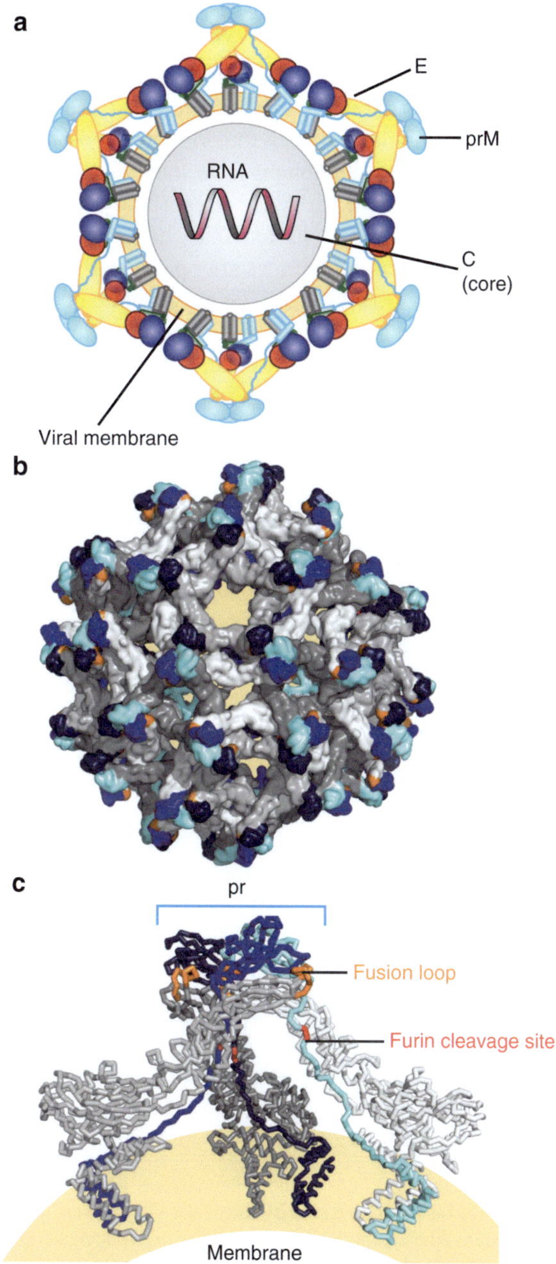

Fig. 6.2 Structural organization of immature flavivirus particles. (**a**) Schematic of an immature flavivirus particle composed of a spherical core with the capsid protein C and a viral envelope with two membrane-associated proteins, E (envelope) and prM (precursor of M). (**b**) Surface representation of an immature dengue virus serotype 1 particle with the underlying viral membrane shown in light orange. The particles are studded with 60 spikes, each consisting of trimers of prM-E heterodimers. The three E proteins in each spike are colored in different shades of gray, the prM proteins in different shades of blue. The fusion loop in E is highlighted in orange. (**c**) Ribbon diagram of a single trimeric prM-E spike. Same color code as in **b**. The first two residues upstream of the furin cleavage site (P1, P2, see also Fig. 6.5) are shown in red. Panels **b** and **c** were generated using the structures of dengue virus serotype 1 [PDB: 4B03, (Kostyuchenko et al. 2013)]

The crystal structure of a dengue prM-E heterodimer was determined to a resolution of 2.2 Å using a recombinant protein in which the transmembrane region of prM was replaced with an 8-amino acid linker (Li et al. 2008). The structure of E in this complex is similar to that in the dimeric pre-fusion form, with some differences of the hinge angles between the domains, consistent with flexibility of these structures to accommodate changes in the oligomeric state of E and other structural changes in the viral life cycle. Importantly, in this heterodimer, the pr peptide is positioned on top of the FL and thus shields the FL from possible premature interactions with cellular membranes after particle assembly and during intracellular transport (see below) (Fig. 6.3). The polypeptide chain of pr extends linearly from the globular head (consisting of seven mostly antiparallel β-strands) along the E protein in the spike [Fig. 6.2c, (Li et al. 2008; Kostyuchenko et al. 2013; Prasad et al. 2017)]. This sequence element of prM contains a furin cleavage site (Fig. 6.2c), and its proteolytic cleavage is essential for generating infectious virions. The trans-Golgi network (TGN) is the key intracellular organelle, where virus maturation takes place during exocytic transport of immature virus particles (Fig. 6.3). In these vesicles, structural rearrangements and proteolytic cleavage of prM occur that finally lead to the release of infectious virions by exocytosis. Flaviviruses are thus members of enveloped viruses that require the cleavage of an accessory protein and not the fusion protein itself to generate an assembly of proteins, primed to undergo triggered membrane fusion during virus entry (White and Whittaker 2016).

6.4 Priming of Fusion by Proteolytic Cleavage and Structural Changes During Maturation

Exposure of immature viruses to slightly acidic pH in the TGN (Fig. 6.3) leads to a dramatic change in the arrangement of prM and E in the envelope of immature particles that precedes proteolytic cleavage activation (Fig. 6.4). A structural comparison of the low-pH-treated immature virus and fully mature viruses (see above) shows that E acquires the herringbone-like organization, typical of mature virions, at this stage in the life cycle [Figs. 6.3 and 6.4; (Yu et al. 2008)]. In order to make such a restructuring happen, E has to be released from its heterodimeric complex with prM and must find a homologous partner to form the final E homodimer. As

Fig. 6.3 Flavivirus life cycle. Flaviviruses enter cells by receptor-mediated endocytosis. Membrane fusion is triggered by the acidic pH in endosomes and leads to the release of the viral genome into the cytoplasm. Viral genomes are replicated at virus-induced intracellular membranes and incorporated into newly formed immature particles by an incompletely understood budding mechanism at the ER membrane. As a by-product of this assembly process, capsid-less subviral particles (SPs) are formed that consist only of a membrane associated with prM and E. The immature particles are transported through the exocytic pathway. The acidic pH in the trans-Golgi network (TGN) induces a structural rearrangement that is required for the cleavage of prM by furin. The cleavage product pr remains associated with the particles at acidic pH but falls off at neutral pH when the particles are released from the cells

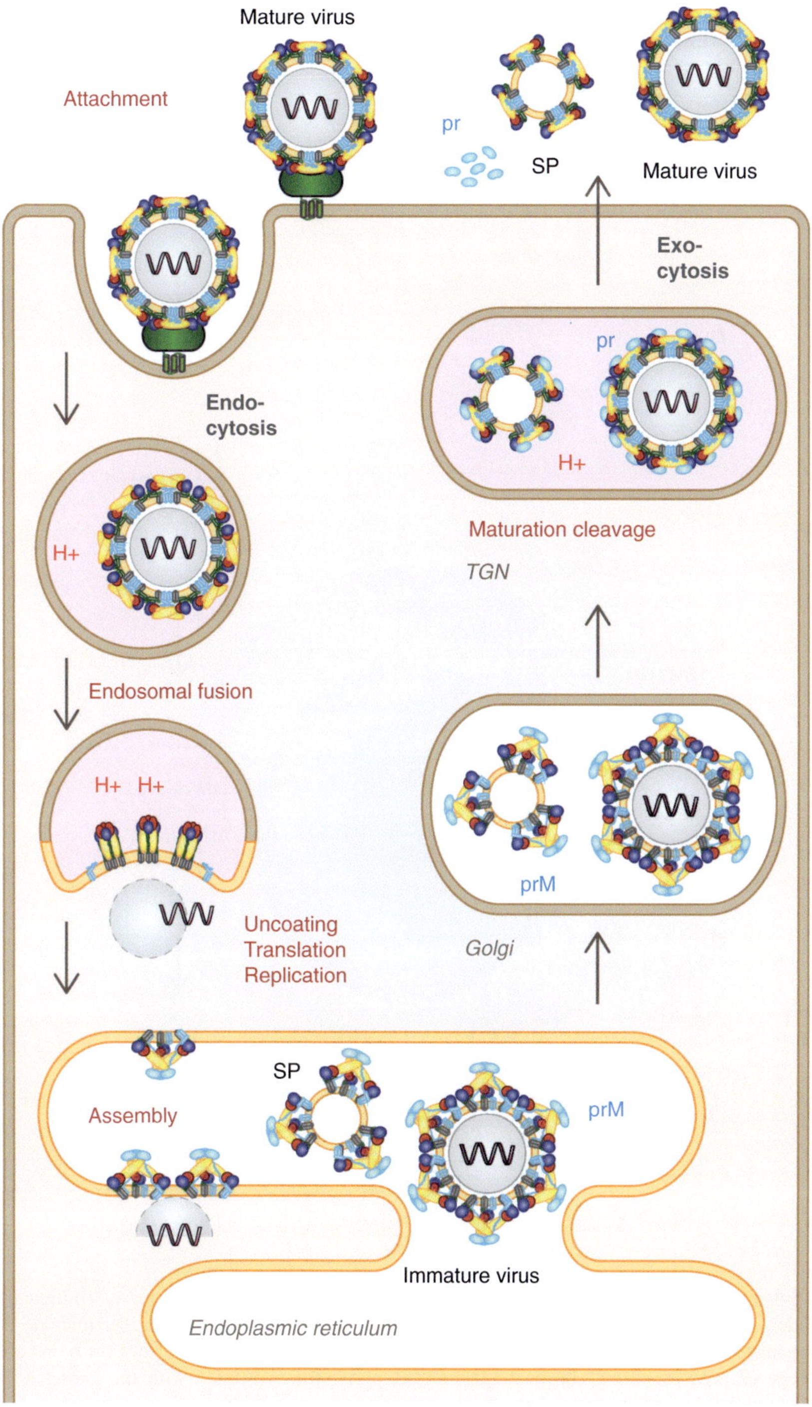

Mature virus
Attachment
pr
SP
Mature virus
Exo-
cytosis
pr
Endo-
cytosis
H+
H+
Maturation cleavage
TGN
Endosomal fusion
H+ H+
prM
Uncoating
Translation
Replication
Golgi
SP
Assembly
prM
Immature virus
Endoplasmic reticulum

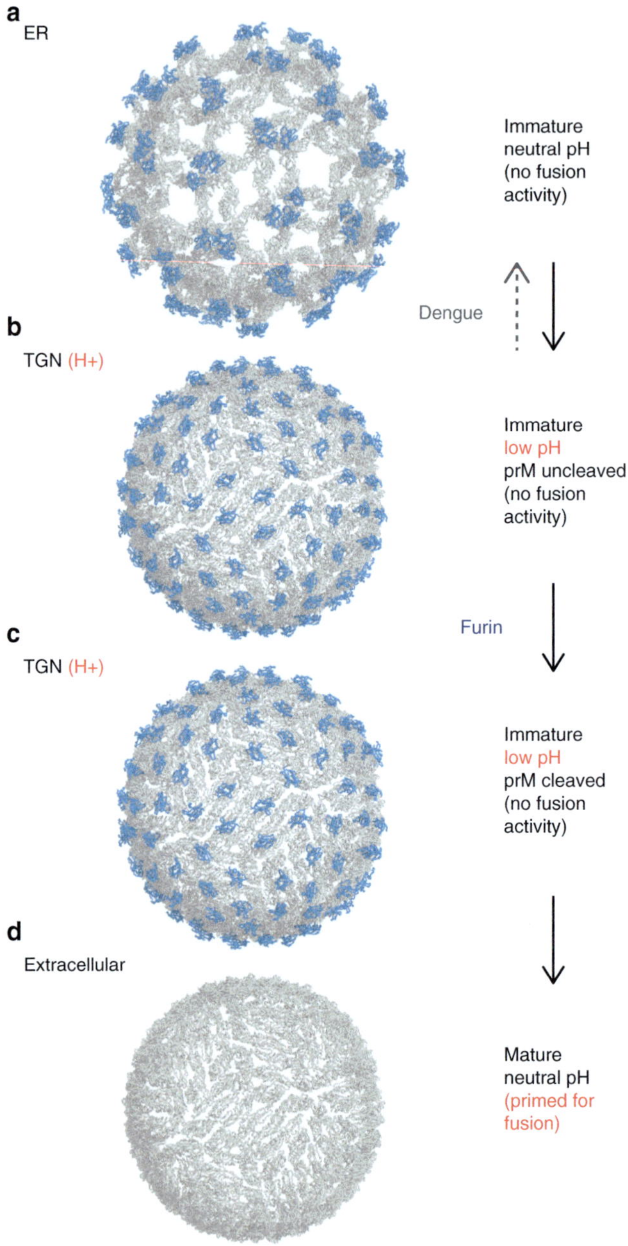

Fig. 6.4 Structural metamorphoses of flavivirus particles from assembly to release. (**a**) Immature particles assembled in the endoplasmic reticulum (ER). (**b**) Formation of the herringbone-like arrangement of E induced by the low pH in the TGN before cleavage of prM. (**c**) Like **b**, but after cleavage of prM by furin. The cleavage product pr is still associated with the particles. (**d**) Extracellular fully mature virus with the E proteins in a metastable conformation, primed for fusion. The cleaved pr has dissociated from the particles. The E proteins are shown in gray, the pr parts of prM in blue. Panels **a**–**d** were generated using the structures of dengue virus serotype 2 [PDB: 3C6D, 3C6R, (Yu et al. 2008)]

revealed by studies with dengue viruses, this pH-dependent switch appears to be controlled by a conserved histidine at position 98 of prM, which upon protonation (possibly in combination with protonation of His 244 in E) produces electrostatic repulsion resulting in the dissociation of prM and E (Zheng et al. 2014). The required rotation of E during the transition from the neutral pH to the low-pH immature form has been proposed to be effected by the linear part of prM, following its globular head (Fig. 6.2c), that acts like a drawstring and tightens upon encountering the low pH in the TGN (Zhang et al. 2013b).

A combination of three-dimensional electron cryo-tomography and two-dimensional image analysis with dengue virus serotype 2 indicated that the maturation-related conformational changes start from a single nucleation center formed by a small area of a few prM and E molecules in the herringbone-like arrangement (Plevka et al. 2011). Afterward, the mature patch spreads around the particle whenever the envelope proteins at the boundary of this region have a suitable conformation and orientation to become integrated in the evolving E herringbone structure (Plevka et al. 2014). Under certain conditions, binding of antibodies to the trimeric spike in immature particles can be strong enough to hold the prM and E proteins together even at acidic pH and thus inhibit the conformational change that would be required for virus maturation (Wang et al. 2013).

Docking studies with the known structure of furin showed that its binding would be hindered in the original spiky form of immature particles (Li et al. 2008) (Fig. 6.2c). The reorganization of the immature envelope is thus necessary to make the furin recognition site accessible and to allow the cleavage of prM into pr and M. Released by this cleavage, the amino-terminal 20 amino acids of M can slip through the holes in the E dimer (Fig. 6.1d) to the same membrane-proximal side of E as the rest of M, which is inserted in the viral membrane by a hairpin of two relatively short helices (Fig. 6.1c) (Zhang et al. 2013b).

6.4.1 The Protective Role of prM During Exocytosis

Since the metastable E herringbone lattice characteristic of mature and fusion-competent flaviviruses is formed already in the presence of prM, one would assume that fusion activity would be activated immediately after the cleavage of prM in the acidic environment of the TGN (Figs. 6.3 and 6.4c). It was therefore unclear why the maturation-cleaved viral particles would not undergo premature fusion already at this stage of their life cycle. The conundrum was resolved by an elegant series of electron cryo-microscopy and biochemical experiments using dengue virus serotype 2 (Yu et al. 2008, 2009), which revealed that the structure of the low-pH immature viruses remained essentially the same after furin cleavage of prM (Fig. 6.4b, c). Thus, the furin cleavage itself does not cause any conformational changes in addition to those induced by low pH. The pr part of prM remains bound at acidic pH to the original position in E, making extensive contacts with DII of one monomer and DI of a neighboring E monomer, burying the FL at the interface between pr and the E dimer. As shown by liposome coflotation experiments, the retention of pr prevents the insertion of the FL into target membranes, as it would occur at acidic pH in the absence of pr. It is therefore concluded that flaviviruses use this intricate mechanism for avoiding premature fusion and inactivation during the process of maturation in the TGN.

Upon release, the virus encounters neutral pH in the extracellular fluid, which leads to the deprotonation of histidines. It has been proposed that the shift in pH disrupts the binding between His 244 in E and Asp 63 of pr (Zhang et al. 2013b) and allows the dissociation of pr from the particles (Fig. 6.4c, d). These are now again poised for a low-pH-induced conformational switch that drives membrane fusion when the virus infects a new cell. The conformational metamorphoses accompanying the maturation of flaviviruses are depicted and summarized in Fig. 6.4.

Overall, acidic pH and a complex network of pH sensors in both prM and E (with histidines playing a major role) orchestrate the maturation as well as the entry of these viruses into their host cells (Fritz et al. 2008; Zheng et al. 2014). Both of these stages can be inhibited by the addition of acidotropic agents or inhibitors of the ATPases required for the acidification of intracellular organelles (Randolph et al. 1990; Guirakhoo et al. 1992; Heinz et al. 1994). In vitro cleavage of immature virions generated in the presence of such agents resulted in a strong increase in infectivity (Stadler et al. 1997; Yu et al. 2008; Junjhon et al. 2008). The proof for an absolute requirement of prM cleavage for acquiring infectivity came from genetic experiments in which the cleavage site in prM was inactivated by a single amino acid deletion in an infectious clone of TBE virus (Elshuber et al. 2003). Changing the furin cleavage site R-T-R-R to R-T-R resulted in the production of immature virions that were completely noninfectious for BHK-21 cells. Since LoVo cells, which lack furin, produce only immature virions, it can be concluded that this cellular protease is indeed responsible for flavivirus maturation (Stadler et al. 1997; Zybert et al. 2008; Zicari et al. 2016).

The fine-tuning of the maturation and entry pathways may vary slightly for different flaviviruses to accommodate specific requirements of the respective life cycles that involve different vectors and vertebrate hosts as well as different tissues in these hosts. Such variations are exemplified by the fact that the low-pH-induced rearrangement of immature viruses and the exposure of the furin cleavage site are reversible in the case of dengue viruses [(Yu et al. 2008), Fig. 6.4a, b], whereas they are irreversible in TBE virus [(Stadler et al. 1997) and unpublished observations]. Furthermore, the patterns of conserved histidines in E vary among phylogenetically distinct groups of flaviviruses, like those transmitted by mosquitos or ticks. Only a subset of the histidine residues is completely conserved among flaviviruses (Nelson et al. 2009) providing further evidence for modulations of the networks of pH sensors that control virus maturation and fusion. Because of their importance for viral infectivity, the structural reorganizations in both the maturation and entry stages of the viral life cycle have been proposed to be potential targets for antiviral agents (Martín et al. 2009; Schmidt et al. 2010; Costin et al. 2010; Ge and Zhou 2014).

6.5 The Furin Cleavage Site and Partially Mature Virions

Consistent with furin cleavage sites in other substrates (Thomas 2002), the flavivirus prM protein contains basic residues at positions 1, 2, and 4 upstream of the cleavage site (P1, P2, P4) (Fig. 6.5). In addition, there are several other charged residues up to residue 13 that are conserved to varying degrees among flavivirus serocomplexes and appear to affect the efficiency of cleavage (Keelapang et al. 2004; Junjhon et al. 2010) (Fig. 6.5). Specifically, dengue viruses and some insect-only flaviviruses have conserved acidic residues at P3, whereas tick-borne viruses lack the otherwise conserved basic residue at P5 (Fig. 6.5). The sequence variations likely reflect differences in the adaptation of these viruses to their hosts, and several experimental data indicate that they can affect the efficiency of prM cleavage (Junjhon et al. 2008, 2010; Keelapang et al. 2004).

Especially dengue viruses appear to produce a high degree of immature or partially mature viruses, both in infected mosquito and mammalian cells [(Junjhon et al. 2008) and references therein], although their relative proportions may vary. The existence of partially mature viral particles was most directly shown by electron cryo-microscopy (Cherrier et al. 2009; Junjhon et al. 2010) as well as immunoprecipitation analyses of dengue viruses from infected C6/36 mosquito cells after concentration and purification by ultracentrifugation (Junjhon et al. 2010). Quantitative analyses indicated that smooth-surfaced mature virions and partially

			P 5 4321			
DEN1	75	VTYGTC-SQTGEHR	RDKR	↓	SVALAP	97
DEN2	75	VTYGTC-TTMGEHR	REKR	↓	SVALVP	97
DEN3	75	VTYGTC-NQAGEHR	RDKR	↓	SVALAP	97
DEN4	75	VMYGTC-TQNGERR	REKR	↓	SVALTP	97
ZIKA	76	VVYGTCHHKKGEAR	RSRR	↓	AVTLPS	99
WN	76	VRYGRC-TKTRHSR	RSRR	↓	SLTVQT	98
YF	73	VAYGKC-DSAGRSR	RSRR	↓	AIDLPT	96
TBE	74	LEYGRC--GKQEGS	RTRR	↓	SVLIPS	96
		N-terminus			C-terminus	

Fig. 6.5 Alignment of prM amino acid sequences from different flaviviruses. The furin cleavage site is indicated by an arrow. The positions upstream of the cleavage site are indicated by P and the respective number. Conserved residues are shown in red; the acidic residues at P3 characteristic for dengue viruses are shown in blue. The furin cleavage motifs are highlighted in yellow. The following sequences were used for the alignment with MAFFT (http://www.ebi.ac.uk/Tools/msa/mafft/): Dengue virus serotype 1 strain SG/07K3640DK1/2008 (DEN1, Genbank accession no: GQ398255), dengue virus serotype 2 strain 16681 (DEN2, U87411), dengue virus serotype 3 strain SG/05K863DK1/2005 (DEN3, EU081190), dengue virus serotype 4 strain SG/06K2270DK1/2005 (DEN4, GQ398256), Zika virus strain H/PF/2013 (ZIKA, KJ776791), West Nile virus strain NY99 (WN, KC407666), yellow fever virus strain Asibi (YF, AY640589), tick-borne encephalitis virus strain Neudoerfl (TBE, U27495)

mature virions were quite abundant, constituting approximately 55% and 42% of the total particles, respectively. In contrast, completely immature particles represented only a minor subpopulation (about 3%).

It can be assumed that the infectivity of partially mature viruses (Mukherjee et al. 2011; Davis et al. 2006) is due to the presence of sufficient E dimers in their unprotected spring-loaded conformation at the mature side of the mosaic particles that can readily mediate membrane fusion, when taken up by cells. As discussed below, partial maturation endows flaviviruses with more options of viral entry than they would have with completely mature particles only.

Dengue viruses and some members from the group of insect-only flaviviruses (Blitvich and Firth 2015) are unique in the conservation of a cleavage-suppressive acidic residue (Asp or Glu) at P3 of the furin cleavage site (Fig. 6.5), which is probably responsible for the abundance of partially mature dengue viruses secreted from infected cells. The negative impact of an acidic amino acid at P3 was revealed by mutational analyses of the sequence upstream of the cleavage site (Junjhon et al. 2008). However, cleavage-augmenting effects were also observed (at P5 Arg and P6 His residues), suggesting an interplay of sequence-specific effects that modulate the efficiency of prM cleavage and thus the proportions of mature and partially mature virions released from infected cells. Interestingly, mutants with reduced cleavage secreted a higher proportion of capsid-less subviral particles, but the mechanisms underlying this effect on virus assembly remained unclear (Junjhon et al. 2008).

It is unknown at present whether exogenous proteases can also contribute to flavivirus maturation at a stage after release from infected cells, similar to what is known to occur in the proteolytic activation of several other enveloped viruses (Böttcher-Friebertshäuser et al. 2013; Smith et al. 2009; Simmons et al. 2013). Using a noninfectious TBE virus mutant with a disabled furin cleavage site (the original cleavage site R-T-R-R was changed to R-T-R), Elshuber et al. (2003) demonstrated that infectivity could be restored by the addition of trypsin, which is able to cleave the mutated site in prM. Interestingly, the lethal mutation could also be overcome by two different resuscitating mutations that occurred during passaging of the mutated virus in vitro (BHK-21 cells) and in vivo in mice (Elshuber and Mandl 2005). In BHK-21 cells, cleavability by furin was reestablished in the disabled mutant by a spontaneous codon duplication that led to a minimal furin cleavage motif (R-R-T-R). Passaging in mouse brains, however, yielded a different pattern of mutations that restored infectivity in mice but not in BHK-21 cells. In this case, the evolving viruses displayed substitutions near the furin cleavage site that changed the original number of six cysteine residues in the wild-type prM either to five or to seven. It was therefore suggested that structural perturbations of prM caused by the cysteine mutations allowed replication under certain conditions (in mouse brains but not in BHK-21 cells) without the need for furin cleavage, probably through the cleavage by extracellular proteases present in mouse brain tissues. Alternatively, structural changes in prM may have affected the balanced network of low-pH-regulated structural interactions between prM and E that control maturation as well as membrane fusion in such a way that infectivity was restored at least at a low level. In a continuation of this work, Fischl et al. were able to adapt the furin-disabled TBE virus mutant to cleavage by chymotrypsin through passaging in the presence of this enzyme (Fischl et al. 2008). Two specific mutations (Ser85Phe and Arg89His) were identified to confer the chymotrypsin-dependent phenotype and were shown to be stable during at least six cell culture passages. These mutants were not neurovirulent, in contrast to the wild-type or trypsin-dependent mutants, presumably because of the lack of chymotrypsin in brain tissues (Fischl et al. 2008).

6.6 Biological Impact of Incomplete prM Cleavage and Particle Heterogeneity

Differences in the extent of prM cleavage can result in the secretion of mosaic particles from infected cells, displaying patches of envelope proteins arranged in their "mature" and "immature" configuration (Fig. 6.6b) (Cherrier et al. 2009; Plevka et al. 2011). Incomplete maturation cleavage can thus result in a substantial

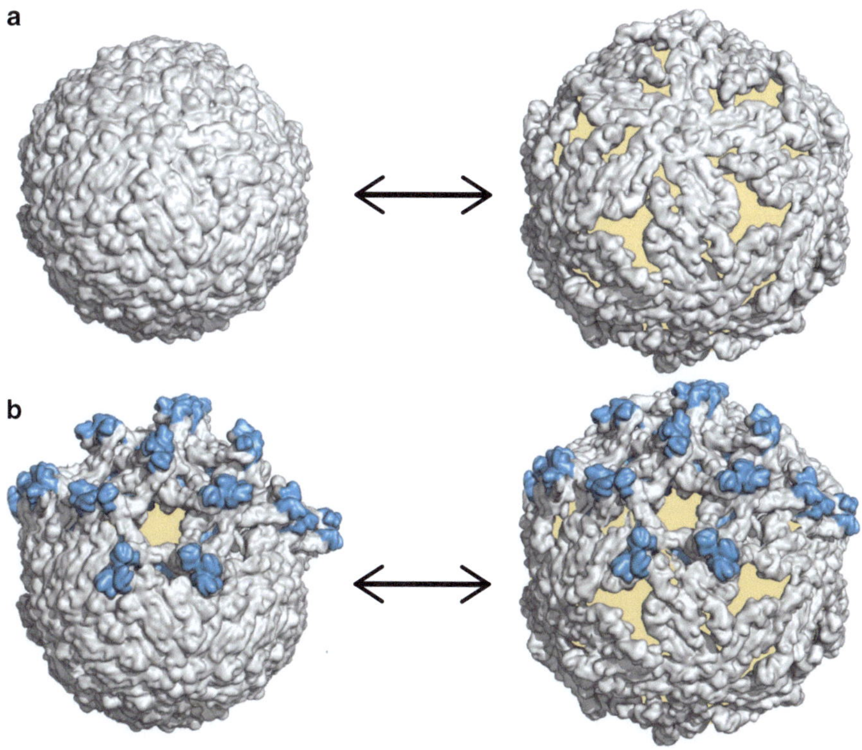

Fig. 6.6 Particle heterogeneities resulting from virus breathing and incomplete prM cleavage. Possible dynamics of the viral envelope (virus breathing) of a fully mature virion (**a**) and a partially mature virion (**b**). The E proteins are shown in gray, the prM proteins in blue, the underlying viral membrane in yellow. The figure was generated using the structures of dengue virus serotypes 1 and 2 [PDB: 4CCT, 4B03, 3ZKO; (Fibriansah et al. 2013; Kostyuchenko et al. 2013)]

heterogeneity of viral particles with more or less prM cleaved, dependent on sequence properties of the virus and the type of cells infected (Pierson and Diamond 2012). Particle heterogeneity is further increased by the dynamics of flavivirus envelopes. This phenomenon, termed "virus breathing," can result in substantial protein movements and the exposure of otherwise cryptic protein surfaces [Fig. 6.6a; (Kuhn et al. 2015)]. Both of these factors, incomplete maturation cleavage (Fig. 6.6b) and virus breathing (Fig. 6.6a, b, right panels), have potential implications for interactions with cellular receptors, the induction of antibodies, the effect of antibody binding on infectivity, pathogenesis, and possibly the maintenance of flaviviruses in their natural ecological cycles (Rey et al. 2017).

A number of experimental data support the concept that particle heterogeneity plays an important role in the biology of flaviviruses. Dengue viruses have played a leading role in these analyses because they appear to be especially prone to incomplete cleavage of prM and breathing of their envelopes (Pierson and Diamond 2012; Kuhn et al. 2015). All of the available structure-related data are derived from viruses produced in cell cultures, and corresponding studies of viruses circulating in natural settings have been elusive to experimental analysis so far. Considering the heterogeneity observed in vitro, it appears however, likely, that this property is also typical for flaviviruses in their natural ecological cycles and may contribute to the characteristics of their host range and pathogenesis. In vitro data have shown that both the efficiency of the maturation cleavage and virus breathing can be affected by single amino acid mutations (Keelapang et al. 2004; Junjhon et al. 2008; Zheng et al. 2014; Dowd et al. 2015; Goo et al. 2017), and particle heterogeneity may therefore even vary between different strains of the same virus.

6.6.1 Effects on Cell Attachment and Entry

Although it has been widely assumed that E mediates receptor binding, because of its prominent position at the surface of mature virions, it becomes increasingly clear that particle heterogeneity offers opportunities to flaviviruses for additional mechanisms of cell attachment. Interaction of the single N-linked glycan attached to the prM protein of partially mature West Nile virus and the lectin DC-SIGNR was shown to be alone sufficient for mediating infection of human cells in vitro (Davis et al. 2006). It has also been proposed that the cluster of positive charges upstream of the furin cleavage site in prM may be involved in binding to heparan sulfate or other glycosaminoglycans at the cell surface (Keelapang et al. 2004). Of special importance are the recently discovered mechanisms of flavivirus entry that are mediated by interactions of the TIM and TAM receptor families with the viral lipid membrane (Amara and Mercer 2015). Such interactions would be impossible with fully mature static virions displaying a closed shell of E proteins at their surface (Fig. 6.1b). However, the viral membrane is accessible in immature virions to a certain degree and can be a target for lipid receptors (Fig. 6.2b).

Immature patches at the surface of partially mature virions (Fig. 6.6b) thus offer an opportunity for TIM- and TAM-mediated infection of cells. Membrane accessibility is probably further increased by irregularities at the interface between mature and immature arrays of envelope proteins as well as by breathing of the fully mature lattice of E proteins (Fig. 6.6a) (Rey et al. 2017). The importance of apoptotic mimicry in flavivirus infections was first shown for dengue viruses (Meertens et al. 2012) but recently also demonstrated to be an important mechanism of Zika virus entry into cells involved in transplacental infections (Richard et al. 2017; Miner and Diamond 2017). In this context, it is important to note that the prM protein was affected by a disproportionately high number of mutations in the evolution of Zika virus from its African-Asian ancestral strains to the currently circulating strains in the Americas (Wang et al. 2016). It can therefore be speculated that some of these mutations may have contributed to the explosive spread and possibly to an altered tissue tropism allowing congenital infections, as observed in the current outbreak (Coyne and Lazear 2016; Miner and Diamond 2017). Studies to assess the effects of these mutations on biological properties of the virus using infectious clones and appropriate animal models (Morrison and Diamond 2017) are ongoing.

6.6.2 Effects on Human Antibody Responses and Antibody-Mediated Neutralization

The heterogeneity of flavivirus particles circulating in infected humans can have a strong influence on the antibody patterns induced by these infections. The formation of prM-specific B cells and antibodies can be regarded as an indirect indicator of circulating particles with incomplete prM cleavage. In dengue virus infections, studies of polyclonal and monoclonal antibody responses have provided evidence that a substantial proportion of antibodies induced is directed against the prM protein (Dejnirattisai et al. 2010; Beltramello et al. 2010; Smith et al. 2012, 2016; Luo et al. 2013). Much less prM-specific antibodies appear to be produced in the course of other flavivirus infections, such as TBE virus infections or YF 17D vaccination (Jarmer et al. 2014; Vratskikh et al. 2013), suggesting that the extent of prM cleavage varies among different flaviviruses under conditions of natural infection or live vaccination. The dominance of antibodies to the FL in E, as shown in dengue and Zika virus infections (Lai et al. 2008; Dejnirattisai et al. 2010; Beltramello et al. 2010; Smith et al. 2012, 2014; de Alwis et al. 2014; Sapparapu et al. 2016), may also be related to particle heterogeneity, since this structural element is more exposed in the trimeric spikes of immature envelopes than in the mature herringbone arrangement (Cherrier et al. 2009), and its accessibility can be subject to further variation by virus- and even strain-specific breathing dynamics (Dowd et al. 2014).

The degree of maturity can also have a strong influence on the neutralizing activity of E protein-specific antibodies (Nelson et al. 2008; Dowd et al. 2014; Mukherjee et al. 2014), because it can modulate the accessibility of epitopes on virus particles, one of the key parameters of virus neutralization (Dowd and Pierson 2011). Structural differences between the immature and mature side of partially mature particles as well as the disordered part between these two sides can be assumed to have divergent effects on antibodies specific for different epitopes. As an example, antibody E53 that recognizes the FL in E bound efficiently to the immature prM-E spike of WN and dengue viruses but not to the mature E dimer (Cherrier et al. 2009; Nelson et al. 2008). Such maturation-dependent variation of antibody binding imposes an additional problem to in vitro virus neutralization assays, because the extent of prM cleavage may differ between different cells (e.g., mammalian or mosquito cells). Thus, the selection of cells for producing the virus to be incubated with the antibodies as well as the cells inoculated with the virus-antibody mixtures can affect the outcome of neutralization assays. Standardization can be improved by the use of furin-overexpressing cells, because they yield homogeneous populations of mature viruses (Mukherjee et al. 2016). It remains to be shown, however, whether the results of these assays will provide the best in vitro correlate of in vivo protection.

6.6.3 Effects on Antibody-Dependent Enhancement of Infection

Antibodies that fail to reach an occupancy sufficient for virus neutralization can even enhance the infection of FcγR-positive cells (Pierson et al. 2007), such as monocytes, macrophages, or dendritic cells (Halstead et al. 2010), by mediating endocytic or phagocytic uptake of infectious virus-antibody complexes. This mechanism can be demonstrated both in in vitro and in vivo using animal models and is implicated in the pathogenesis of severe forms of dengue disease, like hemorrhagic dengue fever and dengue shock syndrome (Halstead 2014; Katzelnick et al. 2017). Antibodies to prM and the FL of E are abundantly produced in dengue virus infections (Lai et al. 2008; Dejnirattisai et al. 2010; Beltramello et al. 2010; Smith et al. 2012, 2014, 2016; de Alwis et al. 2014), especially in the course of secondary infections. Both types of antibodies have no or only low neutralizing activity but can efficiently mediate ADE of infection (Dejnirattisai et al. 2010; Beltramello et al. 2010; Smith et al. 2012, 2014, 2016; de Alwis et al. 2014). Partially mature particles, presenting prM and exposing the FL better than mature particles (Cherrier et al. 2009), may be the predominant viral forms involved in this phenomenon. Because of the possible detrimental effects of prM antibodies, it has been proposed that their induction should be avoided in the context of dengue and possibly other flavivirus vaccines (Dejnirattisai et al. 2010; Smith et al. 2016).

As a special case, completely immature, noninfectious dengue viruses produced in furin-deficient LoVo cells were shown to become highly infectious upon ADE-mediated entry into FcγR-positive cells (Rodenhuis-Zybert et al. 2010). Experiments with specific inhibitors of furin provided convincing evidence that this enzyme, which is not only enriched in the TGN but also cycled to the plasma membrane and endosomes, was responsible for rendering completely immature particles infectious after their endocytic uptake. Such mechanisms can thus potentially contribute to the infection processes of flaviviruses in different hosts and tissues.

Conclusions

Assembly and maturation of flaviviruses are controlled by complex structural interactions of the envelope proteins prM and E. Low-pH-triggered conformational changes in the TGN expose a furin cleavage site in prM and lead to the cleavage of this accessory protein into pr and M. The viral fusion protein (E) itself does not undergo proteolytic cleavage. Priming of the fusion activity of E is also triggered by low pH and caused by structural rearrangements upon entry into cells. This priming requires the loss of the protective and fusion-inhibiting interaction of E with the pr part of prM, which occurs upon virus release from infected cells by exocytosis. Advances in the determination of high-resolution structures of viral particles by electron cryo-microscopy have dramatically increased our understanding of the molecular processes involved in virus assembly, maturation, and entry. One of the key findings gained from these studies, when combined with biochemical and functional analyses, is the recognition of flaviviruses as heterogeneous populations of mature and only partially mature particles, resulting from an incomplete proteolytic maturation cleavage of prM. Many biological properties can be modulated by varying degrees of maturity, and we are only at the beginning of understanding of how flavivirus particle heterogeneity can affect biological properties including pathogenesis. It is quite possible that the capacity of fine-tuning the process of virus maturation and the generation of heterogeneous virus populations is an important asset for flaviviruses to adapt to divergent hosts in their natural cycles. Many aspects in this context are unresolved, such as the composition of virus populations circulating in infected invertebrate and vertebrate hosts as well as their contribution to the maintenance of these viruses in nature, and will likely become topics of intensive future research.

References

Acosta EG, Kumar A, Bartenschlager R. Revisiting dengue virus–host cell interaction: new insights into molecular and cellular virology. In: Karl M, Frederick AM, editors. Advances in virus research, vol. vol 88. San Diego: Academic Press; 2014. p. 1–109. https://doi.org/10.1016/B978-0-12-800098-4.00001-5.

Amara A, Mercer J. Viral apoptotic mimicry. Nat Rev Microbiol. 2015;13(8):461–9. https://doi.org/10.1038/nrmicro3469.

Apte-Sengupta S, Sirohi D, Kuhn RJ. Coupling of replication and assembly in flaviviruses. Curr Opin Virol. 2014;9:134–42. https://doi.org/10.1016/j.coviro.2014.09.020.

Beltramello M, Williams KL, Simmons CP, Macagno A, Simonelli L, Quyen NT, Sukupolvi-Petty S, Navarro-Sanchez E, Young PR, de Silva AM, Rey FA, Varani L, Whitehead SS, Diamond MS, Harris E, Lanzavecchia A, Sallusto F. The human immune response to Dengue virus is dominated by highly cross-reactive antibodies endowed with neutralizing and enhancing activity. Cell Host Microbe. 2010;8(3):271–83. https://doi.org/10.1016/j.chom.2010.08.007.

Blitvich BJ, Firth AE. Insect-specific flaviviruses: a systematic review of their discovery, host range, mode of transmission, superinfection exclusion potential and genomic organization. Virus. 2015;7(4):1927–59. https://doi.org/10.3390/v7041927.

Böttcher-Friebertshäuser E, Klenk HD, Garten W. Activation of influenza viruses by proteases from host cells and bacteria in the human airway epithelium. Pathog Dis. 2013;69(2):87–100. https://doi.org/10.1111/2049-632X.12053.

Chao LH, Klein DE, Schmidt AG, Peña JM, Harrison SC. Sequential conformational rearrangements in flavivirus membrane fusion. eLife. 2014;3:e04389. https://doi.org/10.7554/eLife.04389.

Cherrier MV, Kaufmann B, Nybakken GE, Lok SM, Warren JT, Chen BR, Nelson CA, Kostyuchenko VA, Holdaway HA, Chipman PR, Kuhn RJ, Diamond MS, Rossmann MG, Fremont DH. Structural basis for the preferential recognition of immature flaviviruses by a fusion-loop antibody. EMBO J. 2009;28(20):3269–76. https://doi.org/10.1038/emboj.2009.245.

Costin JM, Jenwitheesuk E, Lok S-M, Hunsperger E, Conrads KA, Fontaine KA, Rees CR, Rossmann MG, Isern S, Samudrala R, Michael SF. Structural optimization and de novo design of dengue virus entry inhibitory peptides. PLoS Negl Trop Dis. 2010;4(6):e721. https://doi.org/10.1371/journal.pntd.0000721.

Coyne CB, Lazear HM. Zika virus—reigniting the TORCH. Nat Rev Microbiol. 2016;14(11):707–15. https://doi.org/10.1038/nrmicro.2016.125.

Cruz-Oliveira C, Freire JM, Conceicao TM, Higa LM, Castanho MA, Da Poian AT. Receptors and routes of dengue virus entry into the host cells. FEMS Microbiol Rev. 2015;39(2):155–70. https://doi.org/10.1093/femsre/fuu004.

Davis CW, Mattei LM, Nguyen H-Y, Ansarah-Sobrinho C, Doms RW, Pierson TC. The location of asparagine-linked glycans on West Nile virions controls their interactions with CD209 (dendritic cell-specific ICAM-3 grabbing nonintegrin). J Biol Chem. 2006;281(48):37183–94. https://doi.org/10.1074/jbc.M605429200.

de Alwis R, Williams KL, Schmid MA, Lai C-Y, Patel B, Smith SA, Crowe JE, Wang W-K, Harris E, de Silva AM. Dengue viruses are enhanced by distinct populations of serotype cross-reactive antibodies in human immune sera. PLoS Pathog. 2014;10(10):e1004386. https://doi.org/10.1371/journal.ppat.1004386.

Dejnirattisai W, Jumnainsong A, Onsirisakul N, Fitton P, Vasanawathana S, Limpitikul W, Puttikhunt C, Edwards C, Duangchinda T, Supasa S, Chawansuntati K, Malasit P, Mongkolsapaya J, Screaton G. Cross-reacting antibodies enhance dengue virus infection in humans. Science. 2010;328(5979):745–8. https://doi.org/10.1126/science.1185181.

Doms RW. What came first—the virus or the egg? Cell. 2017;168(5):755–7. https://doi.org/10.1016/j.cell.2017.02.012.

Dowd KA, DeMaso CR, Pierson TC. Genotypic differences in dengue virus neutralization are explained by a single amino acid mutation that modulates virus breathing. MBio. 2015;6(6):e01559-15. https://doi.org/10.1128/mBio.01559-15.

Dowd KA, Mukherjee S, Kuhn RJ, Pierson TC. Combined effects of the structural heterogeneity and dynamics of flaviviruses on antibody recognition. J Virol. 2014;88(20):11726–37. https://doi.org/10.1128/JVI.01140-14.

Dowd KA, Pierson TC. Antibody-mediated neutralization of flaviviruses: a reductionist view. Virology. 2011;411(2):306–15. https://doi.org/10.1016/j.virol.2010.12.020.

Elshuber S, Allison SL, Heinz FX, Mandl CW. Cleavage of protein prM is necessary for infection of BHK-21 cells by tick-borne encephalitis virus. J Gen Virol. 2003;84(Pt 1):183–91.

Elshuber S, Mandl CW. Resuscitating mutations in a furin cleavage-deficient mutant of the flavivirus tick-borne encephalitis virus. J Virol. 2005;79(18):11813–23. https://doi.org/10.1128/JVI.79.18.11813-11823.2005.

Fédry J, Liu Y, Péhau-Arnaudet G, Pei J, Li W, Tortorici MA, Traincard F, Meola A, Bricogne G, Grishin NV, Snell WJ, Rey FA, Krey T. The ancient gamete fusogen HAP2 is a eukaryotic class II fusion protein. Cell. 2017;168(5):904–915.e910. https://doi.org/10.1016/j.cell.2017.01.024.

Fibriansah G, Ng TS, Kostyuchenko VA, Lee J, Lee S, Wang J, Lok SM. Structural changes in dengue virus when exposed to a temperature of 37°C. J Virol. 2013;87(13):7585–92. https://doi.org/10.1128/JVI.00757-13.

Fischl W, Elshuber S, Schrauf S, Mandl CW. Changing the protease specificity for activation of a flavivirus, tick-borne encephalitis virus. J Virol. 2008;82(17):8272–82. https://doi.org/10.1128/JVI.00587-08.

Fritz R, Stiasny K, Heinz FX. Identification of specific histidines as pH sensors in flavivirus membrane fusion. J Cell Biol. 2008;183(2):353–61.

Ge P, Zhou ZH. Chaperone fusion proteins aid entropy-driven maturation of class II viral fusion proteins. Trends Microbiol. 2014;22(2):100–6. https://doi.org/10.1016/j.tim.2013.11.006.

Goo L, VanBlargan LA, Dowd KA, Diamond MS, Pierson TC. A single mutation in the envelope protein modulates flavivirus antigenicity, stability, and pathogenesis. PLoS Pathog. 2017;13(2):e1006178. https://doi.org/10.1371/journal.ppat.1006178.

Guirakhoo F, Bolin RA, Roehrig JT. The Murray Valley encephalitis virus prM protein confers acid resistance to virus particles and alters the expression of epitopes within the R2 domain of E glycoprotein. Virology. 1992;191(2):921–31. https://doi.org/10.1016/0042-6822(92)90267-S.
Halstead SB. Dengue antibody-dependent enhancement: knowns and unknowns. Microbiol Spectr. 2014;2(6). https://doi.org/10.1128/microbiolspec.AID-0022-2014.
Halstead SB, Mahalingam S, Marovich MA, Ubol S, Mosser DM. Intrinsic antibody-dependent enhancement of microbial infection in macrophages: disease regulation by immune complexes. Lancet Infect Dis. 2010;10(10):712–22. https://doi.org/10.1016/S1473-3099(10)70166-3.
Harrison SC. Viral membrane fusion. Virology. 2015;479–480:498–507. https://doi.org/10.1016/j.virol.2015.03.043.
Heinz FX, Stiasny K, Puschner-Auer G, Holzmann H, Allison SL, Mandl CW, Kunz C. Structural changes and functional control of the tick-borne encephalitis virus glycoprotein E by the heterodimeric association with protein prM. Virology. 1994;198(1):109–17.
Jarmer J, Zlatkovic J, Tsouchnikas G, Vratskikh O, Strauss J, Aberle JH, Chmelik V, Kundi M, Stiasny K, Heinz FX. Variation of the specificity of the human antibody responses after tick-borne encephalitis virus infection and vaccination. J Virol. 2014;88(23):13845–57. https://doi.org/10.1128/JVI.02086-14.
Junjhon J, Edwards TJ, Utaipat U, Bowman VD, Holdaway HA, Zhang W, Keelapang P, Puttikhunt C, Perera R, Chipman PR, Kasinrerk W, Malasit P, Kuhn RJ, Sittisombut N. Influence of pr-M cleavage on the heterogeneity of extracellular dengue virus particles. J Virol. 2010;84(16):8353–8. https://doi.org/10.1128/JVI.00696-10.
Junjhon J, Lausumpao M, Supasa S, Noisakran S, Songjaeng A, Saraithong P, Chaichoun K, Utaipat U, Keelapang P, Kanjanahaluethai A, Puttikhunt C, Kasinrerk W, Malasit P, Sittisombut N. Differential modulation of prM cleavage, extracellular particle distribution, and virus infectivity by conserved residues at nonfurin consensus positions of the dengue virus pr-M junction. J Virol. 2008;82(21):10776–91.
Katzelnick LC, Coloma J, Harris E. Dengue: knowledge gaps, unmet needs, and research priorities. Lancet Infect Dis. 2017;17(3):e88–e100. https://doi.org/10.1016/S1473-3099(16)30473-X.
Keelapang P, Sriburi R, Supasa S, Panyadee N, Songjaeng A, Jairungsri A, Puttikhunt C, Kasinrerk W, Malasit P, Sittisombut N. Alterations of pr-M cleavage and virus export in pr-M junction chimeric dengue viruses. J Virol. 2004;78(5):2367–81. https://doi.org/10.1128/JVI.78.5.2367-2381.2004.
Kindhauser MK, Allen T, Frank V, Santhana R, Dye C. Zika: the origin and spread of a mosquito-borne virus. Bull World Health Organ. 2016;94(9):675–686C. https://doi.org/10.2471/BLT.16.171082.
Kostyuchenko VA, Chew PL, Ng TS, Lok SM. Near-atomic resolution cryo-electron microscopic structure of dengue serotype 4 virus. J Virol. 2014;88(1):477–82. https://doi.org/10.1128/JVI.02641-13.
Kostyuchenko VA, Lim EXY, Zhang S, Fibriansah G, Ng T-S, Ooi JSG, Shi J, Lok S-M. Structure of the thermally stable Zika virus. Nature. 2016;533(7603):425–8. https://doi.org/10.1038/nature17994.
Kostyuchenko VA, Zhang Q, Tan JL, Ng TS, Lok SM. Immature and mature dengue serotype 1 virus structures provide insight into the maturation process. J Virol. 2013;87(13):7700–7. https://doi.org/10.1128/JVI.00197-13.
Kuhn RJ, Dowd KA, Beth Post C, Pierson TC. Shake, rattle, and roll: Impact of the dynamics of flavivirus particles on their interactions with the host. Virology. 2015;479–480C:508–17. https://doi.org/10.1016/j.virol.2015.03.025.
Kuhn RJ, Zhang W, Rossmann MG, Pletnev SV, Corver J, Lenches E, Jones CT, Mukhopadhyay S, Chipman PR, Strauss EG, Baker TS, Strauss JH. Structure of dengue virus: implications for flavivirus organization, maturation, and fusion. Cell. 2002;108(5):717–25.
Lai CY, Tsai WY, Lin SR, Kao CL, Hu HP, King CC, Wu HC, Chang GJ, Wang WK. Antibodies to envelope glycoprotein of dengue virus during the natural course of infection are predominantly

cross-reactive and recognize epitopes containing highly conserved residues at the fusion loop of domain II. J Virol. 2008;82(13):6631–43. https://doi.org/10.1128/JVI.00316-08.
Li L, Lok SM, Yu IM, Zhang Y, Kuhn RJ, Chen J, Rossmann MG. The flavivirus precursor membrane-envelope protein complex: structure and maturation. Science. 2008;319(5871):1830–4.
Lindenbach BD, Murray CL, Thiel HJ, Rice CM. Flaviviridae. In: Knipe DM, Howley PM, Cohen JI, et al., editors. Fields virology. 6th ed. Philadelphia: Lippincott, Williams & Wilkins; 2013. p. 712–46.
Lorenz IC, Allison SL, Heinz FX, Helenius A. Folding and dimerization of tick-borne encephalitis virus envelope proteins prM and E in the endoplasmic reticulum. J Virol. 2002;76(11):5480–91.
Luo Y-Y, Feng J-J, Zhou J-M, Yu Z-Z, Fang D-Y, Yan H-J, Zeng G-C, Jiang L-F. Identification of a novel infection-enhancing epitope on dengue prM using a dengue cross-reacting monoclonal antibody. BMC Microbiol. 2013;13:194. https://doi.org/10.1186/1471-2180-13-194.
Martín CS-S, Liu CY, Kielian M. Dealing with low pH: entry and exit of alphaviruses and flaviviruses. Trends Microbiol. 2009;17(11):514–21. https://doi.org/10.1016/j.tim.2009.08.002.
Meertens L, Carnec X, Lecoin MP, Ramdasi R, Guivel-Benhassine F, Lew E, Lemke G, Schwartz O, Amara A. The TIM and TAM families of phosphatidylserine receptors mediate dengue virus entry. Cell Host Microbe. 2012;12(4):544–57. https://doi.org/10.1016/j.chom.2012.08.009.
Miner JJ, Diamond MS. Zika virus pathogenesis and tissue tropism. Cell Host Microbe. 2017;21(2):134–42. https://doi.org/10.1016/j.chom.2017.01.004.
Morrison TE, Diamond MS. Animal models of Zika virus infection, pathogenesis, and immunity. J Virol. 2017;91(8):pii: e00009-17. https://doi.org/10.1128/jvi.00009-17.
Moureau G, Cook S, Lemey P, Nougairede A, Forrester NL, Khasnatinov M, Charrel RN, Firth AE, Gould EA, de Lamballerie X. New insights into flavivirus evolution, taxonomy and biogeographic history, extended by analysis of canonical and alternative coding sequences. PLoS One. 2015;10(2):e0117849. https://doi.org/10.1371/journal.pone.0117849.
Mukherjee S, Dowd KA, Manhart CJ, Ledgerwood JE, Durbin AP, Whitehead SS, Pierson TC. Mechanism and significance of cell type-dependent neutralization of flaviviruses. J Virol. 2014;88(13):7210–20. https://doi.org/10.1128/JVI.03690-13.
Mukherjee S, Lin TY, Dowd KA, Manhart CJ, Pierson TC. The infectivity of prM-containing partially mature West Nile virus does not require the activity of cellular furin-like proteases. J Virol. 2011;85(22):12067–72. https://doi.org/10.1128/JVI.05559-11.
Mukherjee S, Sirohi D, Dowd KA, Chen Z, Diamond MS, Kuhn RJ, Pierson TC. Enhancing dengue virus maturation using a stable furin over-expressing cell line. Virology. 2016;497:33–40. https://doi.org/10.1016/j.virol.2016.06.022.
Nelson S, Jost CA, Xu Q, Ess J, Martin JE, Oliphant T, Whitehead SS, Durbin AP, Graham BS, Diamond MS, Pierson TC. Maturation of West Nile virus modulates sensitivity to antibody-mediated neutralization. PLoS Pathog. 2008;4(5):e1000060.
Nelson S, Poddar S, Lin T-Y, Pierson TC. Protonation of individual histidine residues is not required for the pH-dependent entry of West Nile virus: evaluation of the "histidine switch" hypothesis. J Virol. 2009;83(23):12631–5. https://doi.org/10.1128/JVI.01072-09.
Perera-Lecoin M, Meertens L, Carnec X, Amara A. Flavivirus entry receptors: an update. Virus. 2014;6(1):69–88. https://doi.org/10.3390/v6010069.
Pérez-Vargas J, Krey T, Valansi C, Avinoam O, Haouz A, Jamin M, Raveh-Barak H, Podbilewicz B, Rey Félix A. Structural basis of eukaryotic cell-cell fusion. Cell. 2014;157(2):407–19. https://doi.org/10.1016/j.cell.2014.02.020.
Pierson TC, Diamond MS. Degrees of maturity: the complex structure and biology of flaviviruses. Curr Opin Virol. 2012;2(2):168–75. https://doi.org/10.1016/j.coviro.2012.02.011.
Pierson TC, Diamond MS. Flaviviruses. In: Knipe DM, Howley PM, Cohen JI, et al., editors. Fields virology. 6th ed. Philadelphia: Lippincott, Williams & Wilkins; 2013. p. 747–94.
Pierson TC, Xu Q, Nelson S, Oliphant T, Nybakken GE, Fremont DH, Diamond MS. The stoichiometry of antibody-mediated neutralization and enhancement of West Nile virus infection. Cell Host Microbe. 2007;1(2):135.

Plevka P, Battisti AJ, Junjhon J, Winkler DC, Holdaway HA, Keelapang P, Sittisombut N, Kuhn RJ, Steven AC, Rossmann MG. Maturation of flaviviruses starts from one or more icosahedrally independent nucleation centres. EMBO Rep. 2011;12(6):602–6. https://doi.org/10.1038/embor.2011.75.

Plevka P, Battisti AJ, Sheng J, Rossmann MG. Mechanism for maturation-related reorganization of flavivirus glycoproteins. J Struct Biol. 2014;185(1):27–31. https://doi.org/10.1016/j.jsb.2013.11.001.

Prasad VM, Miller AS, Klose T, Sirohi D, Buda G, Jiang W, Kuhn RJ, Rossmann MG. Structure of the immature Zika virus at 9 A resolution. Nat Struct Mol Biol. 2017;24(2):184–6. https://doi.org/10.1038/nsmb.3352. Advance online publication

Randolph VB, Winkler G, Stollar V. Acidotropic amines inhibit proteolytic processing of flavivirus prM protein. Virology. 1990;174(2):450–8.

Rey FA, Stiasny K, Heinz FX. Flavivirus structural heterogeneity: implications for cell entry. Curr Opin Virol. 2017;24:132–9.

Richard AS, Shim B-S, Kwon Y-C, Zhang R, Otsuka Y, Schmitt K, Berri F, Diamond MS, Choe H. AXL-dependent infection of human fetal endothelial cells distinguishes Zika virus from other pathogenic flaviviruses. Proc Natl Acad Sci U S A. 2017;114(8):2024–9. https://doi.org/10.1073/pnas.1620558114.

Rodenhuis-Zybert IA, van der Schaar HM, da Silva Voorham JM, van der Ende-Metselaar H, Lei HY, Wilschut J, Smit JM. Immature dengue virus: a veiled pathogen? PLoS Pathog. 2010;6(1):e1000718. https://doi.org/10.1371/journal.ppat.1000718.

Sapparapu G, Fernandez E, Kose N, Bin C, Fox JM, Bombardi RG, Zhao H, Nelson CA, Bryan AL, Barnes T, Davidson E, Mysorekar IU, Fremont DH, Doranz BJ, Diamond MS, Crowe JE. Neutralizing human antibodies prevent Zika virus replication and fetal disease in mice. Nature. 2016;540(7633):443–7. https://doi.org/10.1038/nature20564.

Schmidt AG, Yang PL, Harrison SC. Peptide inhibitors of dengue-virus entry target a late-stage fusion intermediate. PLoS Pathog. 2010;6(4):e1000851. https://doi.org/10.1371/journal.ppat.1000851.

Simmons G, Zmora P, Gierer S, Heurich A, Pöhlmann S. Proteolytic activation of the SARS-coronavirus spike protein: cutting enzymes at the cutting edge of antiviral research. Antivir Res. 2013;100(3):605–14. https://doi.org/10.1016/j.antiviral.2013.09.028.

Sirohi D, Chen Z, Sun L, Klose T, Pierson TC, Rossmann MG, Kuhn RJ. The 3.8 Å resolution cryo-EM structure of Zika virus. Science. 2016;352(6284):467–70.

Smit JM, Moesker B, Rodenhuis-Zybert I, Wilschut J. Flavivirus cell entry and membrane fusion. Virus. 2011;3(2):160–71. https://doi.org/10.3390/v3020160.

Smith EC, Popa A, Chang A, Masante C, Dutch RE. Viral entry mechanisms: the increasing diversity of paramyxovirus entry. FEBS J. 2009;276(24):7217–27. https://doi.org/10.1111/j.1742-4658.2009.07401.x.

Smith SA, de Alwis AR, Kose N, Jadi RS, de Silva AM, Crowe JE Jr. Isolation of dengue virus-specific memory B cells with live virus antigen from human subjects following natural infection reveals the presence of diverse novel functional groups of antibody clones. J Virol. 2014;88(21):12233–41. https://doi.org/10.1128/JVI.00247-14.

Smith SA, Nivarthi UK, de Alwis R, Kose N, Sapparapu G, Bombardi R, Kahle KM, Pfaff JM, Lieberman S, Doranz BJ, de Silva AM, Crowe JE Jr. Dengue virus prM-specific human monoclonal antibodies with virus replication-enhancing properties recognize a single immunodominant antigenic site. J Virol. 2016;90(2):780–9. https://doi.org/10.1128/JVI.01805-15.

Smith SA, Zhou Y, Olivarez NP, Broadwater AH, de Silva AM, Crowe JE Jr. Persistence of circulating memory B cell clones with potential for dengue virus disease enhancement for decades following infection. J Virol. 2012;86(5):2665–75. https://doi.org/10.1128/JVI.06335-11.

Stadler K, Allison SL, Schalich J, Heinz FX. Proteolytic activation of tick-borne encephalitis virus by furin. J Virol. 1997;71(11):8475–81.

Stiasny K, Heinz FX. Flavivirus membrane fusion. J Gen Virol. 2006;87(10):2755–66. https://doi.org/10.1099/vir.0.82210-0.

Thomas G. Furin at the cutting edge: from protein traffic to embryogenesis and disease. Nat Rev Mol Cell Biol. 2002;3(10):753–66.
Vratskikh O, Stiasny K, Zlatkovic J, Tsouchnikas G, Jarmer J, Karrer U, Roggendorf M, Roggendorf H, Allwinn R, Heinz FX. Dissection of antibody specificities induced by yellow fever vaccination. PLoS Pathog. 2013;9(6):e1003458. https://doi.org/10.1371/journal.ppat.1003458.
Wang L, Valderramos SG, Wu A, Ouyang S, Li C, Brasil P, Bonaldo M, Coates T, Nielsen-Saines K, Jiang T, Aliyari R, Cheng G. From mosquitos to humans: genetic evolution of Zika virus. Cell Host Microbe. 2016;19(5):561–5. https://doi.org/10.1016/j.chom.2016.04.006.
Wang Z, Li L, Pennington JG, Sheng J, Yap ML, Plevka P, Meng G, Sun L, Jiang W, Rossmann MG. Obstruction of dengue virus maturation by Fab fragments of the 2H2 antibody. J Virol. 2013;87(16):8909–15. https://doi.org/10.1128/JVI.00472-13.
White JM, Whittaker GR. Fusion of enveloped viruses in endosomes. Traffic. 2016;17(6):593–614. https://doi.org/10.1111/tra.12389.
Wilder-Smith A, Gubler DJ, Weaver SC, Monath TP, Heymann DL, Scott TW. Epidemic arboviral diseases: priorities for research and public health. Lancet Infect Dis. 2017;17(3):e101–6. https://doi.org/10.1016/S1473-3099(16)30518-7.
Yu IM, Holdaway HA, Chipman PR, Kuhn RJ, Rossmann MG, Chen J. Association of the pr peptides with dengue virus at acidic pH blocks membrane fusion. J Virol. 2009;83(23):12101–7. https://doi.org/10.1128/JVI.01637-09.
Yu IM, Zhang W, Holdaway HA, Li L, Kostyuchenko VA, Chipman PR, Kuhn RJ, Rossmann MG, Chen J. Structure of the immature dengue virus at low pH primes proteolytic maturation. Science. 2008;319(5871):1834–7.
Zaitseva E, Yang S-T, Melikov K, Pourmal S, Chernomordik LV. Dengue virus ensures its fusion in late endosomes using compartment-specific lipids. PLoS Pathog. 2010;6(10):e1001131. https://doi.org/10.1371/journal.ppat.1001131.
Zhang W, Kaufmann B, Chipman PR, Kuhn RJ, Rossmann MG. Membrane curvature in flaviviruses. J Struct Biol. 2013a;183(1):86–94. https://doi.org/10.1016/j.jsb.2013.04.005.
Zhang X, Ge P, Yu X, Brannan JM, Bi G, Zhang Q, Schein S, Zhou ZH. Cryo-EM structure of the mature dengue virus at 3.5-A resolution. Nat Struct Mol Biol. 2013b;20(1):105–10. https://doi.org/10.1038/nsmb.2463.
Zhang Y, Corver J, Chipman PR, Zhang W, Pletnev SV, Sedlak D, Baker TS, Strauss JH, Kuhn RJ, Rossmann MG. Structures of immature flavivirus particles. EMBO J. 2003;22(11):2604–13.
Zheng A, Yuan F, Kleinfelter LM, Kielian M. A toggle switch controls the low pH-triggered rearrangement and maturation of the dengue virus envelope proteins. Nat Commun. 2014;5:3877. https://doi.org/10.1038/ncomms4877.
Zicari S, Arakelyan A, Fitzgerald W, Zaitseva E, Chernomordik LV, Margolis L, Grivel J-C. Evaluation of the maturation of individual dengue virions with flow virometry. Virology. 2016;488:20–7. https://doi.org/10.1016/j.virol.2015.10.021.
Zybert IA, van der Ende-Metselaar H, Wilschut J, Smit JM. Functional importance of dengue virus maturation: infectious properties of immature virions. J Gen Virol. 2008;89(Pt 12):3047–51. https://doi.org/10.1099/vir.0.2008/002535-0.

Part II

Activating Proteases

The Role of Secreted Serine Proteases of the Host in Influenza Viral Pathogenesis

7

Hiroshi Kido

Abstract

Influenza A virus (IAV) is one of the most common pathogens causing acute respiratory infections in humans of all age group. IAV infectivity depends on activation of the viral hemagglutinin by proteolytic enzymes of the host, among which secreted trypsin-type serine proteases play a prominent role. Proinflammatory cytokines induced in influenza virus infections upregulate production of the proteases and, thus, enhance virus replication, tissue damage, and metabolic disorders. Application of protease inhibitors counteracting these effects is therefore a promising therapeutic regimen against influenza.

7.1 Introduction

Influenza A virus (IAV), of the *Orthomyxoviridae* family, is the most common infective pathogen in humans, causing significant morbidity and mortality in infants and the elderly (Kim et al. 1979; Lipatov et al. 2004). IAV entry into cells involves membrane fusion mediated by the hemagglutinin (HA), the major glycoprotein of the viral envelope. The fusion activity of HA depends on cleavage by trypsin-type serine proteases that have to be provided by the host, because the viral genome does not encode processing enzymes (Klenk et al. 1975; Klenk and Garten 1994; Kido et al. 2007; Kido 2015). The proteases are not only crucial for IAV entry, but they are also important determinants for infection spread, host tropism, and multiple organ failure (MOF). Once IAV infection ensues, IAV upregulates the HA-processing proteases through induction of proinflammatory cytokines in various organs,

H. Kido
Division of Enzyme Chemistry, Institute for Enzyme Research, Tokushima University, Tokushima, Japan
e-mail: kido@tokushima-u.ac.jp

E. Böttcher-Friebertshäuser et al. (eds.), *Activation of Viruses by Host Proteases*,
https://doi.org/10.1007/978-3-319-75474-1_7

particularly in vascular endothelial cells (Pan et al. 2011; Kido et al. 2012; Kido 2015). Upregulated trypsin-type enzymes potentiate viral multiplication by conversion of the precursor HA_0 into the cleavage products HA_1 and HA_2 and also cause damage through proteinase-activated receptor-2 (PAR-2) in various tissues (Kunzelmann et al. 2002). Therefore, in the initial as well as in the advanced stages of viral infection, trypsin-type HA-processing proteases are important determinants of pathogenesis and play major roles in IAV-induced pneumonia, acute myocarditis, myocardial infarction (Pan et al. 2011), influenza-associated encephalopathy in infants, and MOF (Ichiyama et al. 2007; Wang et al. 2010; Kido 2015). Sendai virus, Newcastle disease virus, and other paramyxoviruses also require trypsin-like processing proteases for entry into cells and multiplication (Tashiro et al. 1992; Klenk and Garten 1994; Kido et al. 1999; Le et al. 2006).

The so-called cytokine storm, marked by increased levels of proinflammatory cytokines, comprising, among others, tumor necrosis factor (TNF)-α, interleukin (IL)-6, and IL-1β, has positive and negative effects on host survival in IAV infection (Wang et al. 2010; Pan et al. 2011; Kido 2015; Indalao et al. 2016). The inflammatory response also affects cell adhesion, vascular permeability, apoptosis, and mitochondrial reactive oxygen that may lead to vascular dysfunction and MOF (Spraque and Khalil 2009; Hiyoshi et al. 2015). For the close interaction of cytokines and cellular ectopic trypsin in IAV infection, we have coined the term "influenza virus–cytokine–trypsin cycle" (Kido et al. 2012; Kido 2015).

This chapter reviews trypsin-type HA-processing secreted serine proteases and their interplay with IAV and cytokines that induce vascular hyperpermeability and MOF in severe influenza. Protease-inhibition-based therapeutic options for IAV-induced MOF are also discussed.

7.2 IAV-Activating Host Proteases

An overview on the secreted trypsin-type proteases responsible for activation of IAV HA is presented in Table 7.1. A fair number of enzymes activating mammalian IAV and low pathogenic avian influenza (LPAI) viruses have been originally characterized as functional entities. They include the secreted trypsin-type serine proteases plasmin (Lazarowitz et al. 1973), factor Xa (Gotoh et al. 1990), urokinase, plasma kallikrein (Scheiblauer et al. 1992), tryptase Clara (Kido et al. 1992), mini-plasmin and micro-plasmin that are truncated derivatives of plasmin (Murakami et al. 2001; Yao et al. 2004), ectopic anionic trypsin (Towatari et al. 2002; Le et al. 2006), porcine lung tryptase (Chen et al. 2000), TC30 (Sato et al. 2003), and kallikrein-related peptidases 5 and 12 (Hamilton and Whittaker 2013). As described in detail in Chap. 8, several type II membrane-bound proteases have been identified more recently at the genetic level, such as human airway trypsin-like protease (HAT) (Yasuoka et al. 1997), transmembrane serine protease (TMPRSS)2 (Böttcher et al. 2006), and TMPRSS4 (Chaipan et al. 2009). All of these endoproteases recognize the carboxyl moiety of a single R residue within the cleavage motif Q/E-X-R (where X is any amino acid except C and basic amino acids) present in HA of seasonal and pandemic human IAV.

Table 7.1 Secreted trypsin-type serine proteases activating influenza and paramyxoviruses

Enzyme	Species	Gene	MW (kDa) SDS-PAGE	Virus activation	Distribution in tissues and cell types	Physiological function	References
Tryptase Clara	Rat	*ND*	30	IAV Sendai virus	Bronchi, Clara (Club) cells	Unknown	Kido et al. (1992), Tashiro et al. (1992)
Mini-plasmin	Rat, mouse	*PLG (rat, mouse)*	28 + 12	IAV Sendai virus	Bronchi, brain, liver, blood, adrenal, kidney	Degradation of fibrin clots (fibrinolysis) and other blood plasma proteins, activation of collagenases and mediators of the complement system	Murakami et al. (2001), Yao et al. (2004)
Ectopic anionic trypsin	Rat	*PRSS2 (human)*	22	IAV Sendai virus	Lung, brain	Unknown	Towatari et al. (2002)
Mast cell tryptase	Pig	*ND*	32–35	IAV Sendai virus	Mast cells	Unknown	Chen et al. (2000)
Tryptase TC30	Pig	*ND*	30	IAV	ND	Unknown	Sato et al. (2003)
Plasmin	Human	*PLG*	90,5	IAV	Liver, blood, adrenal, kidney	Degradation of fibrin clots (fibrinolysis) and other blood plasma proteins, activation of collagenases and mediators of the complement system	Lazarowitz et al. (1973)
Urokinase (uPA)	Human	*PLAU*	54	IAV	Blood, urine, extracellular matrix of many tissues	Activation of plasminogen to plasmin in fibrinolysis	Scheiblauer et al. (1992)
Plasma kallikrein	Human	*KLKB1*	90	IAV	Ubiquitous, blood	Cleaves kininogen to release bradykinin, activation of plasminogen to plasmin	Scheiblauer et al. (1992)

(continued)

Table 7.1 (continued)

Enzyme	Species	Gene	MW (kDa) SDS-PAGE	Virus activation	Distribution in tissues and cell types	Physiological function	References
Kallikrein 5	Human	*KLK5*	32–37	IAV	Skin, breast, testis, ovary, low level in lung	Role in desquamation	Hamilton and Whittaker (2013)
Kallikrein 12	Human	*KLK12*	26–37	IAV	Esophagus, lung, pancreas, prostate, salivary gland, stomach, testis, trachea	Unknown	Hamilton and Whittaker (2013)
Factor Xa	Chicken Human	*ND* *F10*	46	IAV Sendai virus Avirulent NDV	Embryonated chicken eggs, allantois membrane, liver, blood (human)	Conversion of prothrombin to thrombin in blood coagulation (human)	Gotoh et al. (1990)

IAV influenza A virus, *NDV* Newcastle disease virus, *uPA* urokinase plasminogen activator, *PC* proprotein convertase, *MW* molecular weight, *ND* not determined

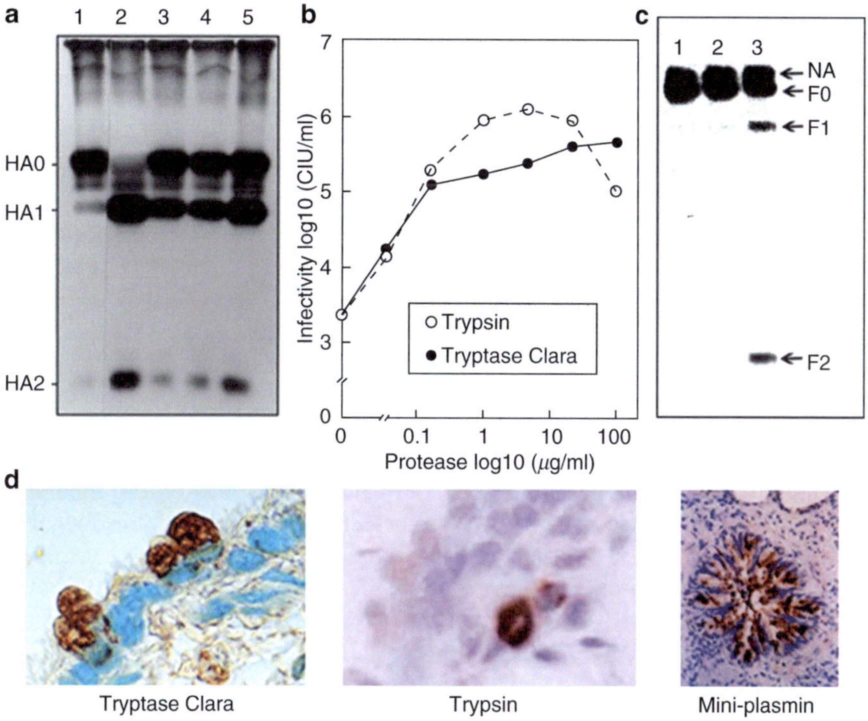

Fig. 7.1 Proteolytic activation of human IAV and Sendai virus and immunohistochemical localization of tryptase Clara, trypsin, and mini-plasmin in rat airway. (**a**) Proteolytic activation of IAV Aichi/2/68 (H3N2) in vitro by tryptase Clara and trypsin followed by SDS-PAGE. [^{3}H] Glucosamine-labeled nonactivated virus isolated from culture media of Madin–Darby canine kidney (MDCK) cells (lane 1) was treated with trypsin (10 μg/ml) for 15 min (lane 2), tryptase Clara (10 μg/ml) for 15 min (lane 3), tryptase Clara (50 μg/ml) for 15 min (lane 4), or tryptase Clara (50 μg/ml) for 30 min (lane 5) at 37 °C. SDS-PAGE was done on a 13% acrylamide gel under reducing conditions. (**b**) Proteolytic activation monitored by viral infectivity. Nonactivated virus propagated in MDCK cells was digested with trypsin or tryptase Clara for 15 min at 37 °C. The infectivity of the activated virus was assayed by the immunofluorescent cell-counting method (Kido et al. 1992). (**c**) [^{3}H]Glucosamine-labeled nonactivated Sendai virus isolated from culture media of LLC-MK_2 cells (lane 1) was treated with purified trypsin from rat brain, 0.15 μg (lane 2) and 0.3 μg (lane 3), for 60 min at 37 °C, and then analyzed by SDS-PAGE. (**d**) Immunohistochemical localization of tryptase Clara, ×410 (Kido et al. 1992); trypsin, ×410 (Towatari et al. 2002); and mini-plasmin, ×140 (Murakami et al. 2001), in rat lungs

Figure 7.1a shows that tryptase Clara purified from rat lung and trypsin from bovine pancreas cleave [^{3}H]glucosamine-labeled influenza virus HA_0 into subunits HA_1 and HA_2 (Kido et al. 1992). Direct amino acid sequencing of the amino-terminus of the HA_2 subunit of IAV Aichi/2/68 (H3N2) revealed residues G-L-F-G-A-I-A-G-, indicating that the cleavage site used by tryptase Clara and trypsin was between R^{325} and G^{326}. Figure 7.1b shows that tryptase Clara and trypsin potentiated IAV multiplication in a dose-dependent manner. Figure 7.1c shows proteolytic processing of [^{3}H]glucosamine-labeled Sendai virus F_0 by ectopic trypsin purified from rat brain in a similar manner as reported for tryptase Clara (Tashiro et al. 1992; Towatari et al. 2002; Le et al. 2006). Although almost all trypsin-type proteases known to date cleave the synthetic short peptide substrates Q/E-X-R at the C-terminal R residue, not all of them are able to activate HA, probably because of a restricted accessibility of the cleavage site. Therefore, [^{3}H]glucosamine-labeled viruses, rather than small peptides, are the best substrates for screening of the proteases responsible for virus activation.

The proteases activating at monobasic cleavage sites show different tissue distribution in rat airway (Fig. 7.1d). Tryptase Clara is located in Clara cells of the bronchioles in the lower respiratory tract (Kido et al. 1992), ectopic lung trypsin is present in stromal cells of peribronchiolar regions (Towatari et al. 2002), and mini-plasmin is found in folded epithelial cells in bronchiolar divisions (Murakami et al. 2001). IAV infection of mice markedly upregulated HA-processing ectopic trypsin in the lung, heart, brain, and vascular endothelial cells (Le et al. 2006; Wang et al. 2010; Pan et al. 2011), and the enzymes played a critical role in the spread of IAV causing vascular hyperpermeability and tissue damage. Mast cell tryptase and tryptase TC30 have also been reported as HA-processing enzymes in porcine lung, but human mast cell tryptase does not activate HA_0 (Chen et al. 2000; Sato et al. 2003). Besides trypsin-type proteases provided by the host, microbial proteases can also activate influenza virus HA in bacterial coinfections of the airways and play important roles in the spread of the virus (Tashiro et al. 1987; Kido et al. 2009).

A large difference in proteolytic potentiation of various virus strains was observed among the HA-processing proteases (Murakami et al. 2001). This is illustrated by a comparative analysis of three strains shown in Fig. 7.2. Trypsin efficiently activated the infectivity of all strains analyzed, as well as mini- and micro-plasmin, though less efficiently than trypsin. In contrast, the proteolytic activity of plasmin was high only with IAV WSN/33 (H1N1), weak with IAV Aichi/2/68 (H3N2), and nonexistent with IAV seal/Massachusetts/1/81 (H7N7). Thus, differences in the tissue distribution of the processing proteases may affect organ tropism and pathogenicity of the viruses.

The concept that proteolytic activation of HA is an important determinant of pathogenicity has been established long ago when it became clear that there are distinct differences at the HA cleavage site of LPAI viruses and highly pathogenic avian influenza (HPAI) viruses. As described in detail in Chap. 1 of this book with HPAI viruses, cleavage of HA_0 occurs intracellularly at the C-terminal R residue of multibasic motifs, mostly R/K-X-K/R-R. The trypsin-type serine proteases activating LPAI viruses and human IAV described above are not responsible for intracellular proteolytic cleavage of HA_0 of HPAI viruses.

The effects of inhibitory compounds of HA-processing proteases have been studied as candidates for therapeutic options against IAV. Among the various trypsin inhibitors, the natural airway inhibitor secretory leukoprotease inhibitor (SLPI), which is secreted from non-ciliated secretory airway Clara and goblet cells (Mooren et al. 1983; Puchelle et al. 1985) and found in bronchoalveolar lavage fluid as well as nasal and salivary secretions, efficiently suppressed proteolytic activation of HA and viral multiplication in cell cultures and experimentally infected rats

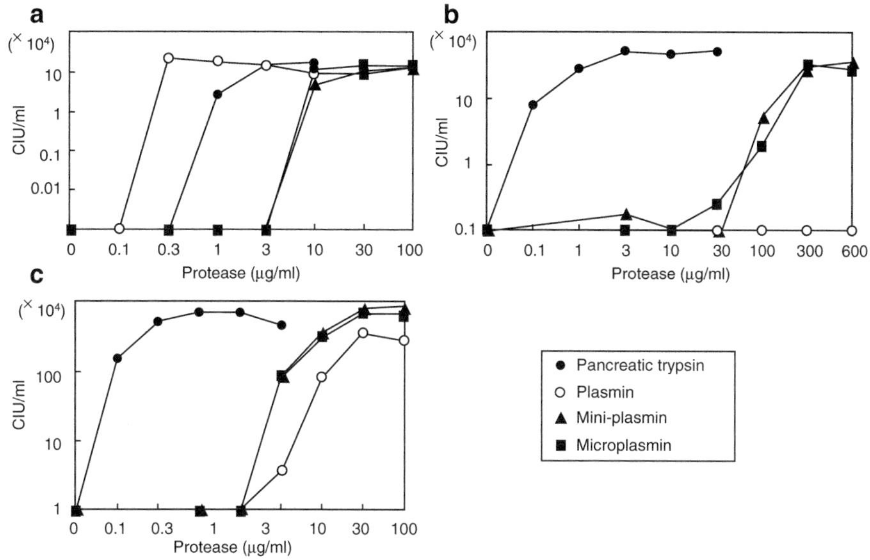

Fig. 7.2 Activation of infectivity by plasmin, mini-plasmin, micro-plasmin, and trypsin of IAV WSN (H1N1) (**a**), IAV seal/Massachusetts/1/81 (H7N7) (**b**), and IAV Aichi/2/68 (H3N2) (**c**). MDCK cell-grown nonactivated IAV strains were treated for 30 min at 37 °C with pancreatic trypsin, plasmin, mini-plasmin, and micro-plasmin at the indicated concentrations. The enzyme reaction was stopped by the addition of 100 μM aprotinin. Infectivity in MDCK cells is shown as cell-infecting units (CIU)

(Beppu et al. 1997; Kido et al. 2007). Aprotinin, a trypsin and plasmin inhibitor purified from bovine lungs, also efficiently inhibited HA cleavage and viral multiplication in embryonated chicken eggs and in mice (Zhirnov et al. 1994, 2011; Beppu et al. 1997; Kido et al. 2007).

7.3 The Influenza Virus–Cytokine–Trypsin Cycle and Its Role in the Pathogenesis of Vascular Hyperpermeability and Multiple Organ Failure

A scheme of the influenza virus–cytokine–trypsin cycle is shown in Fig. 7.3. As mentioned above, nasal IAV infection increases the levels of proinflammatory cytokines, such as TNF-α, IL-6, and IL-1β, in the lung and blood of mice. Immediately after the cytokine storm, there is a marked upregulation of ectopic trypsin in vascular endothelial cells of various organs through activation of nuclear factor kappa B (NF-κB) and activator protein 1 (AP-1) (Wang et al. 2010). Upregulation of trypsin leads, on the one hand, to increased virus titers in the airways and, on the other hand, to activation of matrix metallopeptidase 9 (MMP-9) and protease-activated receptor-2 (PAR-2), resulting in vascular hyperpermeability, edema, and MOF. Some of the observations that provided the basis for this concept are presented in the following.

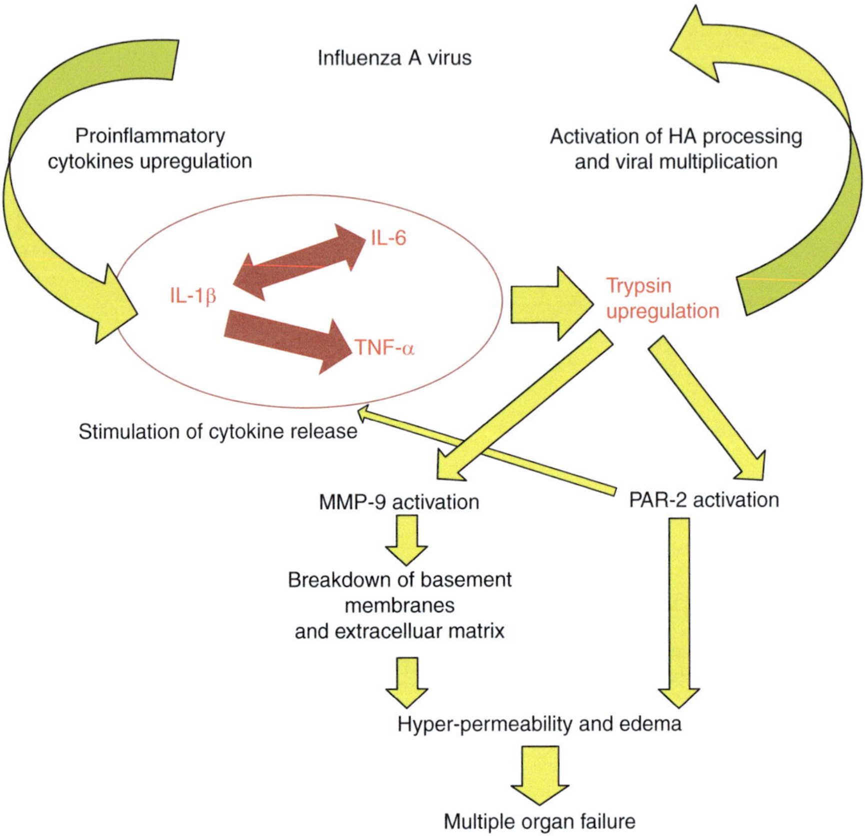

Fig. 7.3 Diagram illustrating the role of the influenza virus–cytokine–trypsin cycle in vascular hyperpermeability, tissue destruction, and MOF in severe influenza. *HA* hemagglutinin, *PAR-2* proteinase-activated receptor-2, *IL-1β* interleukin-1β, *IL-6* interleukin-6, *TNF-α* tumor necrosis factor-α, *MMP* matrix metalloproteinase

Figure 7.4 illustrates the upregulation of cytokines and ectopic trypsin in the lungs of mice after IAV/Puerto Rico/8/34 (H1N1) (PR8) infection. The levels of IL-1β, TNF-α, and IL-6 rapidly increased at day 3 post-infection, sustained until day 6, and then decreased at day 8. Ectopic trypsin also increased markedly at day 3, but the level stayed high until day 8 (Indalao et al. 2016). Since the cytokine responses are associated with activation of transcription factors NF-κB and AP-1 (Mori et al. 2003; Santoro et al. 2003; Nimmerjahn et al. 2004), we treated mice once daily for 4 days with the NF-κB inhibitors pyrrolidine dithiocarbamate (PDTC) and *N*-acetyl-L-cysteine (NAC) and the AP-1 inhibitor nordihydroguaiaretic acid (NDGA). PDTC and NDGA significantly suppressed the upregulation of TNF-α and IL-1β, and NAC suppressed TNF-α and IL-6 at day 4 post-infection

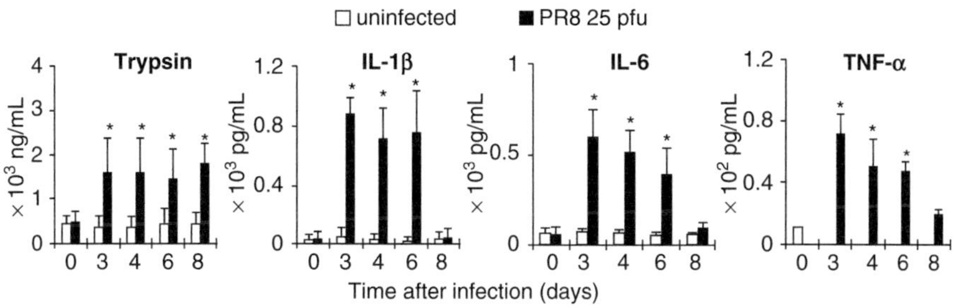

Fig. 7.4 Effects of IAV/PR8/34 (H1N1) infection on the levels of proinflammatory cytokines and trypsin in the lungs of mice. Anesthetized C5B7BL/6 J mice were infected intranasally with 25 pfu of PR8 or instilled saline as a control. Mice were sacrificed at days 0, 3, 4, 6, and 8 after infection. The levels of trypsin, IL-1β, IL-6, and TNF-α in the lungs were analyzed by ELISA. Data are mean ± SD of ten mice per group. $^*P < 0.05$ vs. uninfected group (saline), by the student *t*-test

(Wang et al. 2010). Trypsin induction was inhibited also by treatment with PDTC, NAC, and NDGA, probably via blockade of NF-κB and AP-1 binding in the promoter region of the gene. These results suggest a close interplay between IAV, cytokines, and ectopic trypsin in the pathogenesis of influenza.

Trypsin in mice has three major isoforms ($T_{1–3}$). T_2 is the predominant isoform induced after IAV infection, while the level of T_1 is lower than that of T_2, and T_3 is barely detected (Pan et al. 2011). Since trypsin facilitates IAV entry and replication in target cells, we analyzed the effects of silencing the major trypsin genes T_1 and T_2 on viral replication and cytokine levels in murine H9c2 cardiomyoblasts. Viral replication monitored by NS1 and NP expression was significantly suppressed in T_1- and T_2-knockdown cells; the magnitude of suppression in T_2-knockdown cells was significantly higher than in T_1-knockdown cells (Fig. 7.5). Secretion of proinflammatory cytokines IL-6, IL-1β, and TNF-α into culture media was also suppressed under T_1- and T_2-knockdown conditions (Pan et al. 2011). Based on these results, we proposed the influenza virus–cytokine–trypsin cycle hypothesis as one of the mechanisms underlying MOF in severe influenza (Kido et al. 2012; Kido 2015).

We recently observed that IL-1β is a key cytokine in trypsin upregulation and has a pathological role in MOF (Fig. 7.3) (Indalao et al. 2016). IL-1β treatment upregulated trypsin and increased proinflammatory cytokine secretion in a time-dependent manner in human alveolar A549 cells (Fig. 7.6). The earliest response was IL-1β secretion detected at 2 h in the media, followed by secretion of IL-6, TNF-α, and T_1 at 4, 6, and 8 h, respectively. The levels of these cytokines and T_1 increased continuously in a time-dependent manner during the 8-h experiment. The other major isoforms of trypsin mRNAs, T_2 and T_3, were also upregulated after IL-1β treatment in a similar manner (Indalao et al. 2016). Suppression of the effects of IL-1β by neutralizing monoclonal antibody treatment almost completely suppressed upregulation of trypsin, IL-1β, IL-6, and TNF-α mRNAs in A549 cells (Indalao et al. 2016).

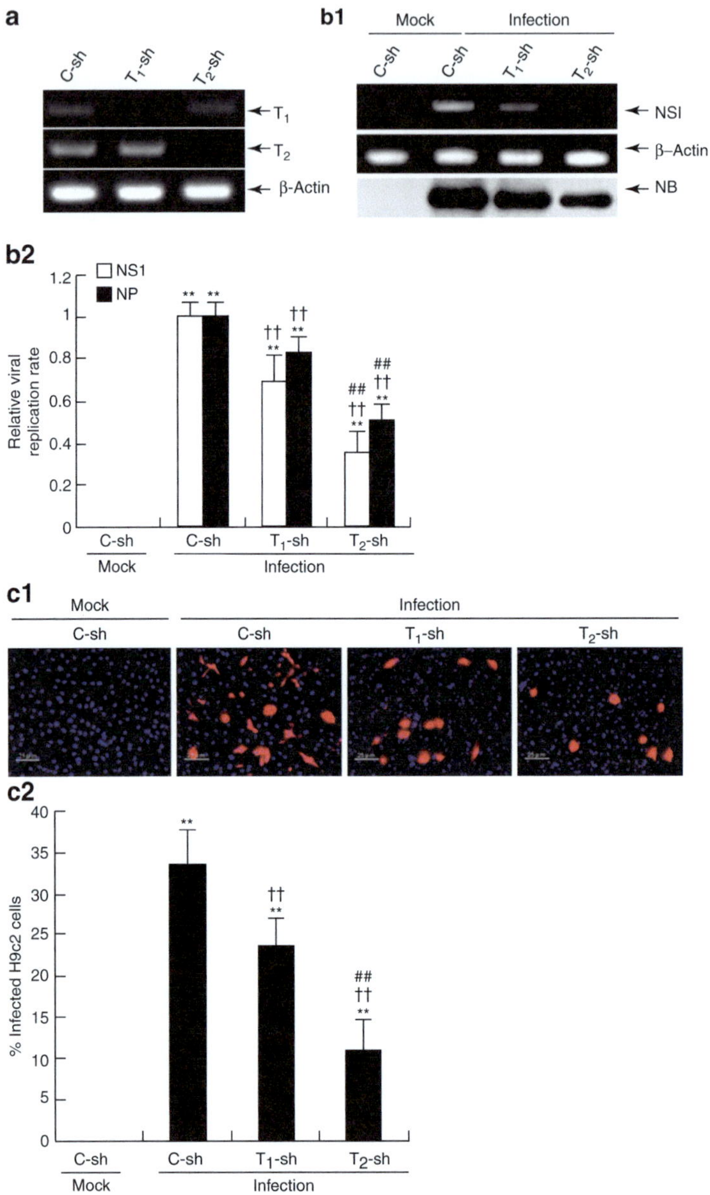

Fig. 7.5 Effects of trypsin knockdown on IAV replication in H9c2 cardiomyoblasts. (**a**) Silencing efficiency of $trypsin_1$ isoform (T_1) and $trypsin_2$ isoform (T_2) genes was determined by RT-PCR in stable transfectant $trypsin_1$-shRNA plasmid (T_1-sh), $trypsin_2$-shRNA plasmid (T_2-sh), and control-shRNA plasmid (C-sh) cell lines. (**b**) Effects of T_1- and T_2-knockdown on viral replication. NS1 gene expression was monitored in cell lysates by RT-PCR, NP was detected in culture media by western immunoblotting (**b1**), and the expression rates in control (C-sh), T_1-sh, and T_2-sh cells were compared (**b2**). (**c1**) Detection of viral antigen by immunofluorescence (red) in H9c2 cells at 24 h after infection. Nuclei were stained by 4,6-diamidino-2-phenylindole (DAPI). Bar = 25 μm. (**c2**) Percentage of infected cells. Data are mean ± SD of three independent experiments. $^{*}P < 0.05$ and $^{**}P < 0.01$ versus mock infection; $^{\dagger\dagger}P < 0.01$ versus infection control; $^{\#\#}P < 0.01$ versus T_1-knockdown

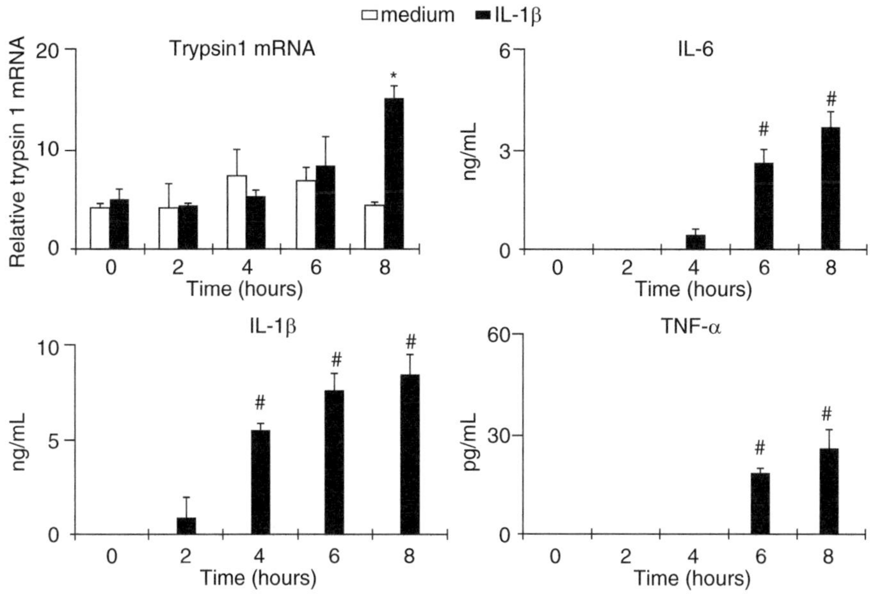

Fig. 7.6 Kinetics of upregulation of trypsin, IL-1β, IL-6, and TNF-α in cultured human A549 cells after treatment with human recombinant IL-1β. A549 cells were primed with 10 ng/ml of recombinant human IL-1β for 1 h, washed with Ca^{2+}- and Mg^{2+}-free phosphate-buffered saline, and then incubated in new serum-free medium for 8 h. Trypsin mRNA levels in the cell lysates and the levels of IL-1β, IL-6, and TNF-α in the culture media were analyzed. Data are mean ± SD ($n = 3$). $^{*}P < 0.05$ vs. untreated group; $^{\#}P < 0.05$ vs. 0 h by the student t-test

7.4 Linkage of the Influenza Virus–Cytokine–Trypsin and Metabolic Disorder–Cytokine Cycles and Options for the Treatment of Multiple Organ Failure

MOF with severe pulmonary edema occurs in the progressive stage of seasonal influenza virus pneumonia, particularly in patients with underlying risk factors (Dolorme and Middleton 1979; Khoufache et al. 2009), and is common in human HPAI virus infection (Fujimoto et al. 1998). In the course of progression of the influenza virus–cytokine–trypsin cycle, increased levels of cytokines induce the metabolic disorder–cytokine cycle in mitochondria mediated by the peroxisome proliferation-activated receptor (PPAR)-γ and pyruvate dehydrogenase kinase 4 (PDK4), as shown in Fig. 7.7 (Yamane et al. 2014; Kido 2015). The cytokine storm suppresses mitochondrial pyruvate dehydrogenase (PDH) activity and glucose oxidation through upregulation of PDK4, resulting in reduced ATP levels in skeletal muscle, liver, lung, and heart, but not brain (Yamane et al. 2014) (Fig. 7.8). The results suggest that PDK4 is a suitable target molecule for the treatment of the metabolic energy disorder induced by severe IAV infection. Among the known PDK inhibitors dichloroacetate (DCA), AZD7545, and radicicol, the pyruvate analog DCA is the classic inhibitor of all PDK isoforms (Bersin and Stacpoole 1997; Kato et al. 2007), although it has side effects, such as peripheral neuropathy. We recently

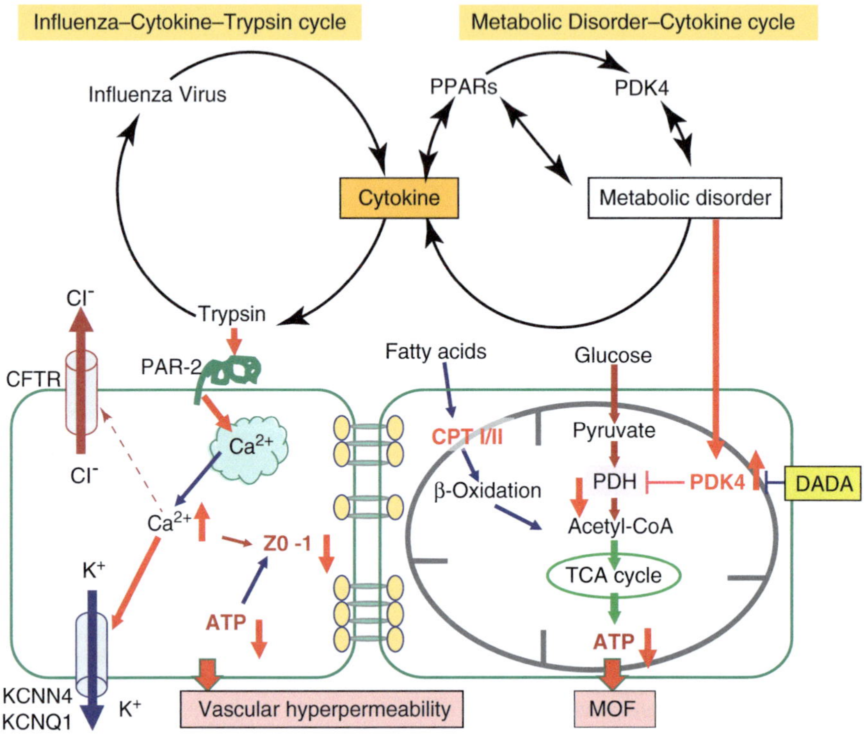

Fig. 7.7 Diagram illustrating the pathogenic role of ATP depletion resulting from the link of the metabolic disorder–cytokine cycle to the influenza virus–cytokine–trypsin cycle. The linked cycles (top) and two overlapping pathways involving cell membrane and cytoplasm (bottom, left) and mitochondria (bottom, right) that lead to vascular hyperpermeability and MOF are shown. *PPARs* peroxisome proliferator-activated receptors, *PDH* pyruvate dehydrogenase, *PDK4* pyruvate dehydrogenase kinase 4, *PAR-2* protease-activated receptor-2, *CPT* carnitine palmitoyltransferase, *CFTR* cystic fibrosis transmembrane conductance regulator, *KCNN4* potassium intermediate/small conductance calcium-activated channel, subfamily N, member 4, *KCNQ1* potassium channel, voltage dependent, KCNQ1(IPR005827), *ZO-1* zonular occludens-1 (for further explanation, see text)

found that diisopropylamine dichloroacetate (DADA) is a selective and safe inhibitor of PDK4. DADA inhibition of PDK4 resulted in significant restoration of PDH activity as well as increased ATP levels in various organs and also improved blood glucose, lactate, and β-hydroxybutyric acid levels (Yamane et al. 2014). Abrogation of PDH suppression in infected mice by DADA led not only to a restoration of the energy metabolism, but it was also associated with a suppression of the cytokine storm. As a consequence, trypsin expression in various organs and IAV replication in the lung were reduced, and all animals recovered from the infection. Figure 7.9 shows a typical example of the effects of DADA on survival rate, body weight, and food and water intake of mice infected with a semilethal dose of IAV PR8 (Yamane et al. 2014). Untreated mice showed progressive avoidance of food and water uptake during days 2–7 post-infection, and the animals started to die after day 7 post-infection. However, infected mice treated with DADA showed no significant decrease in food and water intake as well as no significant reduction in body weight

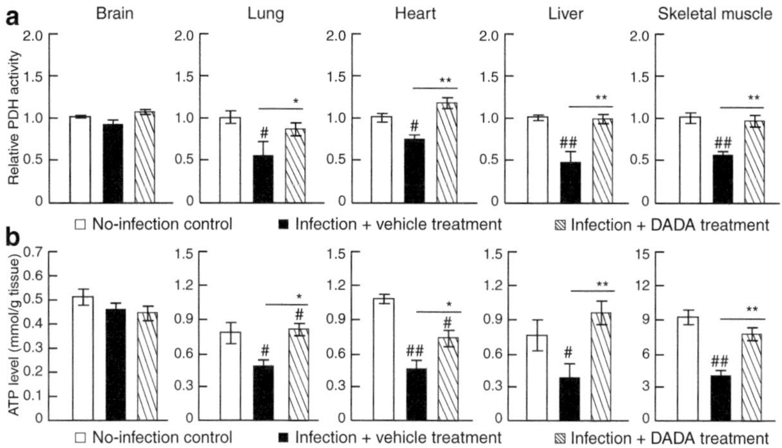

Fig. 7.8 Treatment with DADA restores suppressed PDH activity and ATP levels in skeletal muscle, heart, lung, and liver of IAV PR8-infected mice. Mice infected with PR8 at 120 pfu were treated orally with DADA at 50 mg/kg or 0.5% methylcellulose 400 as a vehicle at 12-h intervals for 14 days, and the levels of PDH activity (**a**) and ATP (**b**) in skeletal muscle, heart, lung, liver, and brain of mice were analyzed at day 7 post-infection. PDH activity levels relative to the values of the control (no infection). Values are mean ± SD of five mice per group. $^{\#}P < 0.05$ and $^{\#\#}P < 0.01$ vs. no infection, $^{*}P < 0.05$ and $^{**}P < 0.01$ vs. infected group treated with vehicle, by one-way analysis of variance (ANOVA) and Tukey's post hoc test

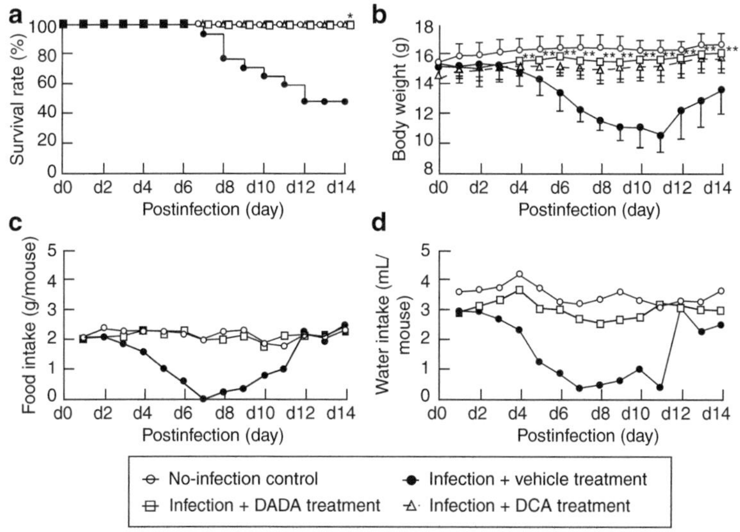

Fig. 7.9 Effects of DADA on survival rate, body weight, and food and water intake. Mice infected with PR8 at an LD_{50} of 60 pfu were treated orally with DADA at 50 mg/kg or 0.5% methylcellulose 400 as a vehicle, or they received dichloroacetate (DCA) intraperitoneally at 28 mg/kg at 12-h intervals for 14 days. Survival rate, body weight, food intake, and water intake of infected mice were monitored. Survival rate (**a**) analyzed by Kaplan–Meier and log-rank tests. Changes in body weight (**b**), food intake (**c**), and water intake (**d**) for each group. Data are mean ± SD of 15 mice per group. $^{*}P < 0.05$ and $^{**}P < 0.01$ vs. infected group treated with vehicle, by two-way ANOVA

during the 14-day experimental period. While untreated mice showed a continuous decrease in the survival, none of the DADA-treated mice died during the experimental period.

Upregulation of trypsin in the course of the influenza virus–cytokine–trypsin cycle stimulates PAR-2. PAR-2 stimulation increases cytokine release (Niu et al. 2008) and leads to lung edema through the activation of chloride and potassium secretion at apical and basolateral membrane channels, respectively, and to increased vascular permeability and relaxation with a rise in intracellular Ca^{2+} concentrations (Palmer et al. 2006; Wang et al. 2010) (Fig. 7.7). Trypsin inhibitors, such as aprotinin, and PAR-2 antagonists suppress these pathological changes induced by IAV (Nakayama et al. 2001; Zhirnov et al. 2011).

Conclusions

It has been known for decades that infectivity, organ tropism, and pathogenicity of IAV and several other viruses, notably Sendai virus and Newcastle disease virus (NDV), are primarily determined by trypsin-type host serine proteases. Although there is a large variety of proteases with trypsin-type substrate specificities, only some of them recognize the tertiary structure of the cleavage site of the viral membrane proteins, thus allowing virus multiplication in the organs and tissues in which they are expressed. As outlined in this review, from the knowledge gained over the past decades on the role of trypsin-type secreted host cellular serine proteases (Table 7.1) in the pathogenesis of IAV infection, the following conclusions can be drawn: (1) Several trypsin-type HA-processing serine proteases in the airway fluids, such as tryptase Clara, ectopic trypsin, and mini-plasmin, are responsible for activation of LPAI viruses and mammalian IAV. In recent years, cell surface-anchored trypsin-like HA-processing proteases have been added to this list. (2) IAV infection is characterized by a marked upregulation of cellular trypsin in various organs and cells induced by proinflammatory cytokines, particularly IL-1β. IAV infection also leads to metabolic disorders with ATP depletion in various cells. Upregulated trypsin in endothelial cells destroys tight junctions through PAR-2 expression on the cell surface. Based on these studies, we propose the influenza virus–cytokine–trypsin cycle hypothesis as one of the main mechanisms underlying MOF. Similar to trypsin knockdown, administration of aprotinin, a trypsin inhibitor, suppresses viral replication and upregulation of trypsin and cytokines as well as ATP depletion, resulting in significant improvement of cellular functions. (3) As viral infection progresses, the influenza virus–cytokine–trypsin cycle interconnects with the metabolic disorder–cytokine cycle. Conjugation of both cycles enhances the severity of IAV infection and MOF. Treatment of disordered glucose oxidation by the PDK4 inhibitor DADA normalizes glucose and lipid metabolism and enhances ATP levels in mitochondria. Normalization of the metabolic disorders further suppresses cytokine production, upregulation of trypsin, and viral replication in the lung and, thus, significantly improves the survival rate of infected mice. (4) These findings support the concept that application of protease inhibitors is a promising therapeutic regimen against influenza virus infection.

These inhibitors will not only interfere with the formation of infectious virus, but, in combination with other compounds, they may also prevent cellular dysfunctions contributing to the development of IAV-induced MOF.

Acknowledgment Our studies were supported in part by grants-in-aid #16H05348 and the Special Coordination Funds for Promoting Science and Technology of the Ministry of Education, Culture, Sports, Science and Technology of Japan and by Health and Labour Sciences Research Grants (grant #12103307) from the Ministry of Health, Labour and Welfare of Japan.

References

Beppu Y, Immamura Y, Tashiro M, Towatari T, Ariga H, Kido H. Human mucus protease inhibitor in airway fluids is a potential defensive compound against infection with influenza A and Sendai viruses. J Biochem. 1997;121:309–16.

Bersin RM, Stacpoole PW. Dichloroacetate as metabolic therapy for myocardial ischemia and failure. Am Heart J. 1997;134:841–55.

Böttcher E, Matrosovich T, Beyerle M, Klenk H-D, Garten W, Matrosovich M. Proteolytic activation of influenza viruses by serine proteases TMPRSS2 and HAT from human airway epithelium. J Virol. 2006;80:9896–8.

Chaipan C, Kobasa D, Bertram S, Glowacka I, Steffen I, Tsegaye TS, Takeda M, Bugge TH, Kim S, Park Y, Marzi A, Pöhlmann S. Proteolytic activation of the 1918 influenza virus hemagglutinin. J Virol. 2009;83:3200–11.

Chen Y, Shiota M, Ohuchi M, Towatari T, Tashiro M, Murakami M, Yano M, Yang B, Kido H. Mast cell tryptase from pig lungs triggers infection by pneumotropic Sendai and influenza viruses. Purification and characterization. Eur J Biochem. 2000;267:3189–97.

Dolorme L, Middleton PJ. Influenza A virus associated with acute encephalopathy. Am J Dis Child. 1979;133:822–4.

Fujimoto S, Kobayashi M, Uemura O, Iwasa M, Ando T, Katoh T, Nakamura C, Maki N, Togari H, Wada Y. PCR on cerebrospinal fluid to show influenza-associated acute encephalopathy or encephalitis. Lancet. 1998;352:873–5.

Gotoh B, Ogasawara T, Toyoda TM, Inocencio N, Hamaguchi M, Nagai Y. An endoprotease homologous to the blood clotting factor X as a determinant of viral tropism in chick embryo. EMBO J. 1990;9:4189–95.

Hamilton BS, Whittaker GR. Cleavage activation of human-adapted influenza virus subtypes by kallikrein-related peptidases 5 and 12. J Biol Chem. 2013;288:17399–407.

Hiyoshi M, Indalao IL, Yano M, Yamane K, Takahashi E, Kido H. Influenza A virus infection of vascular endothelial cells induces GSK-3β-mediated β-catenin degradation in adherens junctions, with a resultant increase in membrane permeability. Arch Virol. 2015;160:225–34.

Ichiyama T, Morishima T, Kajimoto M, Matsushige T, Matsubara T, Furukawa S. Matrix metalloprotease-9 and tissue inhibitors of metalloproteinases I in influenza-associated encephalopathy. Pediatr Infect Dis J. 2007;26:542–4.

Indalao IL, Sawabuchi T, Takahashi E, Kido H. IL-1β is a key cytokine that induces trypsin upregulation in the influenza-cytokine-trypsin cycle. Arch Virol. 2016;162:201–11.

Kato M, Li J, Chuang JL, Chung DT. Distinct structural mechanisms for inhibition of pyruvate dehydrogenase kinase isoforms by AZD7545, dichloroacetate, and radicicol. Structure. 2007;15:992–1004.

Khoufache K, LeBouder F, Morello E, Laurent F, Riffault S, Andrade-Gordon P, Boullier S, Rousset P, Vergnolle N, Riteau B. Protective role for protease-activated receptor-2 against influenza virus pathogenesis via an IFN-γ-dependent pathway. J Immunol. 2009;182:7795–802.

Kido H. Influenza virus pathogenicity regulated by host cellular proteases, cytokines and metabolites, and its therapeutic options. Proc Jpn Acad Ser B. 2015;91:351–68.

Kido H, Murakami M, Oba K, Chen Y, Towatari T. Cellular proteinases trigger the infectivity of influenza A and Sendai viruses. Mol Cells. 1999;9:235–44.

Kido H, Okumura Y, Takahashi E, Pan H, Wang S, Chida J, Le TQ, Yano M. Host envelope glycoprotein processing proteases are indispensable for entry into human cells by seasonal and highly pathogenic avian influenza viruses. J Mol Genet Med. 2009;3:167–75.

Kido H, Okumura Y, Takahashi E, Pan HY, Wang S, Yao D, Yao M, Chida J, Yano M. Role of host cellular proteases in the pathogenesis of influenza and influenza-induced multiple organ failure. Biochim Biophys Acta. 2012;1824:186–94.

Kido H, Okumura Y, Yamada H, Le TQ, Yano M. Proteases essential for human influenza virus entry into cells and their inhibitors as potential therapeutic agents. Curr Pharm Des. 2007;13:405–14.

Kido H, Yokogoshi Y, Sakai K, Tashiro M, Kishino Y, Fukutomi A, Katunuma N. Isolation and characterization of a novel trypsin-like protease found in rat bronchiolar epithelial Clara cells. A possible activator of the viral fusion glycoprotein. J Biol Chem. 1992;267:13573–9.

Kim HM, Brandt CD, Arrobio JO, Murphy B, Chanock RM, Parrott RH. Influenza A and B virus infection in infants and young children during the years 1957-1976. Am J Epidemiol. 1979;109:464–79.

Klenk H-D, Garten W. Host cell proteases controlling virus pathogenicity. Trends Microbiol. 1994;2:39–43.

Klenk H-D, Rott R, Orlich M, Blödom J. Activation of influenza A viruses by trypsin treatment. Virology. 1975;68:426–39.

Kunzelmann K, Schreiber R, König J, Mall M. Ion transport induced by proteinase-activated receptors (PAR2) in colon and airways. Cell Biochem Biophys. 2002;4:31–9.

Lazarowitz SG, Goldberg AR, Choppin PW. Proteolytic cleavage by plasmin of the HA polypeptide of influenza virus: host cell activation of serum plasminogen. Virology. 1973;56:172–80.

Le TQ, Kawachi M, Yamada H, Shiota M, Okumura Y, Kido H. Identification of trypsin I as a candidate for influenza A virus and Sendai virus envelope glycoprotein processing protease in rat brain. Biol Chem. 2006;387:467–75.

Lipatov AS, Govorkova EA, Webby RJ, Ozaki H, Peiris M, Guan Y, Poon L, Webster RG. Influenza: emergence and control. J Virol. 2004;78:8951–9.

Mooren HWD, Kramps JA, Franken C, Meijer CJLM, Dickmen JA. Localization of a low-molecular-weight bronchial protease inhibitor in the peripheral human lung. Thorax. 1983;38(3):180.

Mori I, Goshima F, Koshizuka T, Koide N, Sugiyama T, Yoshida T, Yokochi T, Nishiyama Y, Kimura Y. Differential activation of the c-Jun N-terminal kinase/stress-activated protein kinase and p38 mitogen-activated protein kinase signal transduction pathways in the mouse brain upon infection with neurovirulent influenza A virus. J Gen Virol. 2003;84:2401–8.

Murakami M, Towatari T, Ohuchi M, Shiota M, Akao M, Okumura Y, Parry MA, Kido H. Mini-plasmin found in the epithelial cells of bronchioles triggers infection by broad-spectrum influenza A viruses and Sendai virus. Eur J Biochem. 2001;268:2847–55.

Nakayama T, Hirano K, Nishimura J, Takahashi S, Kanaide H. Mechanism of trypsin-induced endothelium-dependent vasorelaxation in the porcine coronary artery. Br J Pharmacol. 2001;134:815–26.

Nimmerjahn F, Dudziak D, Dirmeier U, Hobom G, Riedel A, Schlee M, Staudt LM, Rosenwald A, Behrends U, Bornkamm GW, Mautner J. Active NF-κB signaling is a prerequisite for influenza virus infection. J Gen Virol. 2004;85:2347–56.

Niu QX, Chen HQ, Chen ZY, Fu YL, Lin JL, He SH. Induction of inflammatory cytokine release from human umbilical vein endothelial cells by agonists of proteinase-activated receptor-2. Clin Exp Pharmacol Physiol. 2008;35:89–96.

Palmer ML, Lee SY, Maniak PJ, Carlson D, Fahrenkrug SC, O'Grady SM. Protease-activated receptor regulation of Cl- secretion in Calu-3 cells requires prostaglandin release and CFTR activation. Am J Physiol Cell Physiol. 2006;290:C1189–C98.

Pan H, Yamada H, Chida J, Wang S, Yano M, Yao M, Zhu J, Kido H. Up-regulation of ectopic trypsins in the myocardium by influenza A virus infection triggers acute myocarditis. Cardiovasc Res. 2011;89:595–603.

Puchelle EJ, Hinnraski J, Tournier JM, Adnet JJ. Ultrastructural localization of bronchial inhibitor in human airways using protein A-gold technique. Biol Cell. 1985;55:151–4.
Santoro MG, Rossi A, Amici C. NF-κB and virus infection: who controls whom. EMBO J. 2003;22:2552–60.
Sato M, Yoshida S, Iida K, Tomozawa T, Kido H, Yamashita M. A novel influenza A virus activating enzyme from porcine lung: purification and characterization. Biol Chem. 2003;384:219–27.
Scheiblauer H, Reinacher M, Tashiro M, Rott R. Interactions between bacteria and influenza A virus in the development of influenza pneumonia. J Infect Dis. 1992;166:783–91.
Spraque AH, Khalil RA. Inflammatory cytokines in vascular dysfunction and vascular disease. Biochem Pharmacol. 2009;78:539–52.
Tashiro M, Ciborowski P, Klenk H-D, Pulverer G, Rott R. Role of staphylococcus protease in the development of influenza pneumonia. Nature. 1987;352:536–7.
Tashiro M, Yokogoshi Y, Tobita K, Seto JT, Rott R, Kido H. Tryptase Clara an activating protease for Sendai virus in rat lungs, is involved in pneumopathogenicity. J Virol. 1992;22:7211–6.
Towatari T, Ide M, Ohba K, Chiba Y, Murakami M, Shiota M, Kawachi M, Yamada H, Kido H. Identification of ectopic anionic trypsin I in rat lungs potentiating pneumotropic virus infectivity and increased enzyme level after virus infection. Eur J Biochem. 2002;269:2613–21.
Wang S, Le TQ, Kurihara N, Chida J, Cisse Y, Yano M, Kido H. Influenza virus-cytokine-protease cycle in the pathogenesis of vascular hyperpermeability in severe influenza. J Infect Dis. 2010;202:991–1001.
Yamane K, Indalao IL, Chida J, Yamamoto Y, Hanawa M, Kido H. Diisopropylamine dichloroacetate, a novel pyruvate dehydrogenase kinase 4 inhibitor, as a potential therapeutic agent for metabolic disorders and multiorgan failure in severe influenza. PLoS One. 2014;9:e98032.
Yao D, Chen Y, Kuwajima M, Shiota M, Kido H. Accumulation of mini-plasmin in the cerebral capillaries causes vascular invasion of the murine brain by a pneumotropic influenza A virus: implications for influenza-associated encephalopathy. Biol Chem. 2004;385:487–92.
Yasuoka S, Ohnishi T, Kawano S, Tsuchihashi S, Ogawara M, Masuda K, Yamaoka K, Takahashi M, Sano T. Purification, characterization, and localization of a novel trypsin-like protease found in the human airway. Am J Respir Cell Mol Biol. 1997;16:300–8.
Zhirnov OP, Golyando PB, Ovcharenko AV. Replication of influenza B virus in chicken embryos is suppressed by exogenous aprotinin. Arch Virol. 1994;135:209–16.
Zhirnov OP, Klenk H-D, Wright PF. Aprotinin and similar protease inhibitors as drugs against influenza. Antivir Res. 2011;92:27–36.

Membrane-Anchored Serine Proteases: Host Cell Factors in Proteolytic Activation of Viral Glycoproteins

8

Eva Böttcher-Friebertshäuser

Abstract

Over one third of all known proteolytic enzymes are serine proteases. Among these, the trypsin-like serine proteases comprise one of the best characterized subfamilies due to their essential roles in blood coagulation, food digestion, fibrinolysis, or immunity. Trypsin-like serine proteases possess primary substrate specificity for basic amino acids. Most of the well-characterized trypsin-like proteases such as trypsin, plasmin, or urokinase are soluble proteases that are secreted into the extracellular environment. At the turn of the millennium, a number of novel trypsin-like serine proteases have been identified that are anchored in the cell membrane, either by a transmembrane domain at the N- or C-terminus or via a glycosylphosphatidylinositol (GPI) linkage. Meanwhile more than 20 membrane-anchored serine proteases (MASPs) have been identified in human and mouse, and some of them have emerged as key regulators of mammalian development and homeostasis. Thus, the MASP corin and TMPRSS6/matriptase-2 have been demonstrated to be the activators of the atrial natriuretic peptide (ANP) and key regulator of hepcidin expression, respectively. Furthermore, MASPs have been recognized as host cell factors activating respiratory viruses including influenza virus as well as severe acute respiratory syndrome (SARS) and Middle East respiratory syndrome (MERS) coronaviruses. In particular, transmembrane protease serine S1 member 2 (TMPRSS2) has been shown to be essential for proteolytic activation and consequently spread and pathogenesis of a number of influenza A viruses in mice and as a factor associated with severe influenza virus infection in humans.

This review gives an overview on the physiological functions of the fascinating and rapidly evolving group of MASPs and a summary of the current knowledge on their role in proteolytic activation of viral fusion proteins.

E. Böttcher-Friebertshäuser
Institute of Virology, Philipps-University Marburg, Marburg, Germany
e-mail: friebertshaeuser@staff.uni-marburg.de

E. Böttcher-Friebertshäuser et al. (eds.), *Activation of Viruses by Host Proteases*,
https://doi.org/10.1007/978-3-319-75474-1_8

8.1 Introduction

The designation trypsin-like serine proteases has originally been used for a large group of soluble proteolytic enzymes, which are involved in digestion, blood coagulation, fibrinolysis, and immunity. Analyses of vertebrate genomes at the turn of the millennium have identified a novel subfamily of trypsin-like serine proteases that are anchored in the cell membrane, either by a carboxy-terminal transmembrane domain (type I), an amino-terminal transmembrane domain (type II), or via a glycosylphosphatidylinositol (GPI) linkage at the carboxy terminus. Over the past two decades, these membrane-anchored serine proteases (MASPs) have emerged as key regulators of mammalian development and homeostasis in processes such as epithelial tight junction formation, skin development, epithelial sodium channel activation, cellular iron homeostasis, blood pressure, inner ear development, placental morphogenesis, neural tube closure, and male fertility. Moreover, dysregulated expression and/or activity of a number of MASPs is observed in many cancer tissues. Within the past decade, MASPs expressed in the human airways have furthermore been identified as host cell factors that may support proteolytic activation and spread of respiratory viruses including influenza virus, human metapneumovirus, severe acute respiratory syndrome (SARS) , and Middle East respiratory syndrome (MERS) coronavirus (CoV).

The largest group of the MASPs is the family of type II transmembrane serine proteases (TTSPs) (Hooper et al. 2001). To date, the TTSP family comprises 18 known members in humans and 20 in mice. All members of the TTSP family share a common domain structure possessing a short N-terminal cytoplasmic domain, a transmembrane domain, a C-terminal serine protease domain, and a variable stem region that may contain 1–11 protein domains of 6 different types (Fig. 8.1). The most prominent member of the TTSP, enteropeptidase, has been identified over a century ago by Pavlov and coworkers due to its essential role in food digestion; however, only the cloning of the enteropeptidase cDNA in 1994 revealed the presence of a membrane anchor and explained its localization to the plasma membrane of duodenal cells. The modular structure of the enteropeptidase protein was found to be similar to that of a previously cloned protease, hepsin (Leytus et al. 1988).

The first of the GPI-anchored serine proteases identified was prostasin (Yu et al. 1994; Chen et al. 2001a, b). A second human GPI-anchored serine protease, testisin, was identified in 1998/1999 (Inoue et al. 1998, Hooper et al. 1999). Tryptase gamma 1 is the only type I transmembrane serine protease identified to date. Prostasin, testisin, and tryptase gamma 1 are composed of a single protease domain linked to a GPI anchor or a transmembrane domain at the C-terminus (Fig. 8.1).

MASPs belong to the chymotrypsin (S1)-like serine protease family (reviewed in Perona and Craik 1995 and Hooper et al. 2001). They are synthesized as inactive single-chain zymogens that are activated by cleavage following an arginine or lysine residue within a highly conserved activation motif preceding the catalytic domain. After activation, the catalytic domain remains linked to the membrane-anchored domains by a disulfide bond but can also be released as a soluble protease (Hooper et al. 2001). All MASPs have a highly conserved S1 serine protease domain that

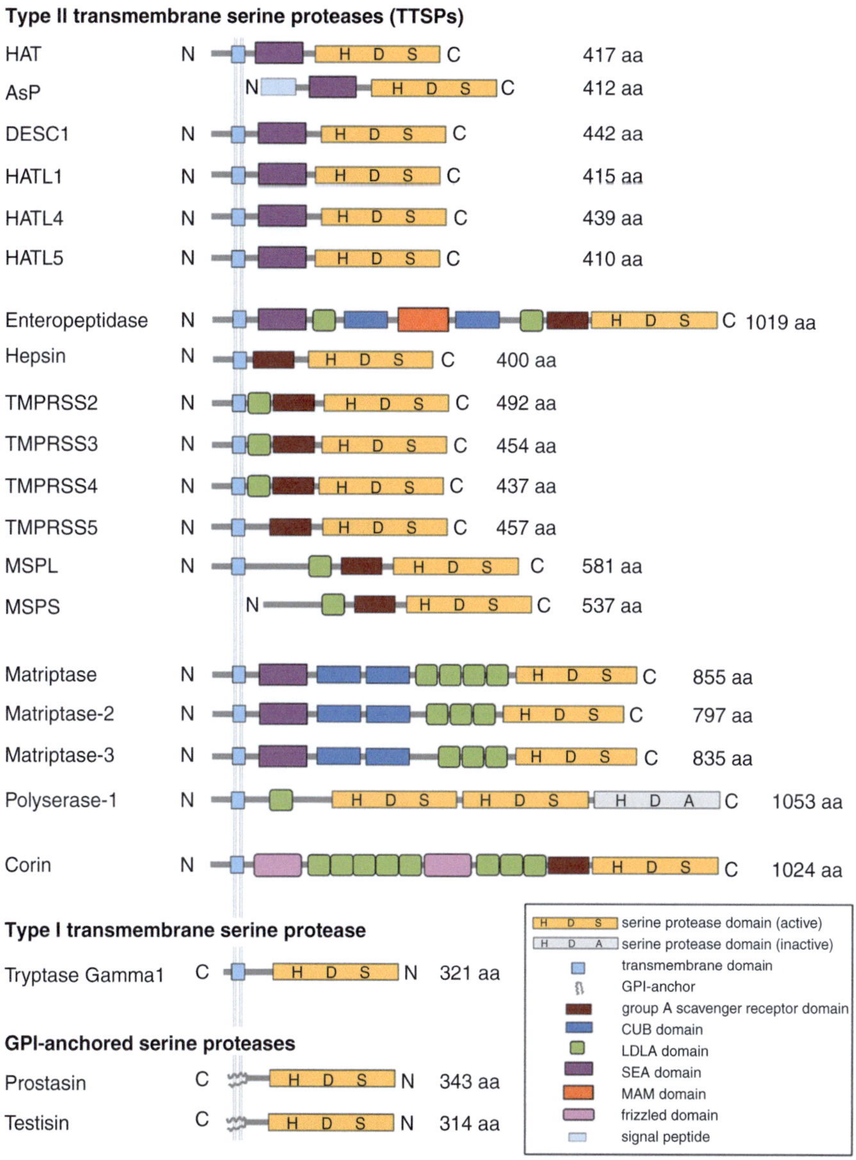

Fig. 8.1 Domain structures of membrane-anchored serine proteases. *HAT* human airway trypsin-like protease; *HATL* HAT-like protease; *AsP* adrenal secretory serine protease; *DESC1*, differentially expressed in squamous cell carcinoma; *TMPRSS*, transmembrane serine protease S1; *MSPL/MSPS* mosaic serine protease large/short-form; *GPI* glycosylphosphatidylinositol; *CUB* Cls/Clr, urchin embryonic growth factor and bone morphogenetic protein-1; *SEA* sea urchin sperm protein, enterokinase, and agrin; *LDLA* low-density lipoprotein receptor A; *MAM* meprin, A5 antigen, and receptor protein phosphatase μ. Amino and carboxy termini are indicated by N and C, respectively. *aa* amino acids

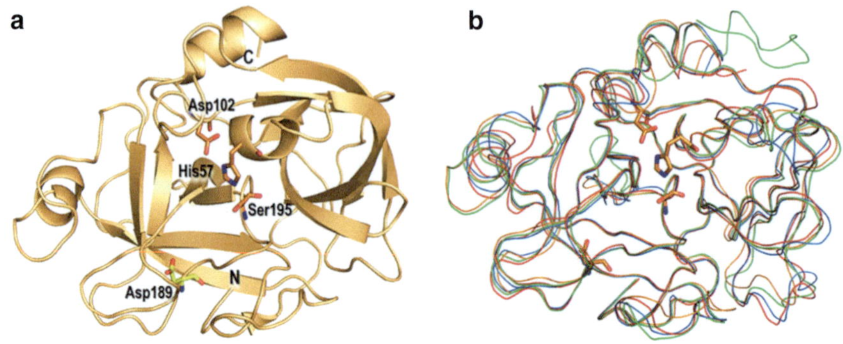

Fig. 8.2 Crystal structure of the catalytic domain of human DESC1. (**a**) DESC1 (PDB: 2OQ5, Kyrieleis et al. 2007) is shown in cartoon style. The residues of the catalytic triad (carbons in orange, nitrogen in blue, and oxygen in red) and of Asp189 (carbons in yellow) are shown as sticks. N-terminus and C-terminus are labeled. (**b**) DESC1 (ribbon style, orange) superimposed with the catalytic domains of human matriptase (blue, PDB: 1EAX, Friedrich et al. 2002), human enteropeptidase (red, PDB: 1EKB, Lu et al. 1999), and human hepsin (green, PDB: 1P57, Somoza et al. 2003). The residues of the active site of DESC1 and Asp189 are shown as sticks (carbons in orange, nitrogen in blue, and oxygen in red)

contains the histidine, aspartate, and serine residues (catalytic triad) necessary for the catalytic activity (Fig. 8.2). In addition, all MASPs show a strong preference for cleavage of substrates after a basic residue (arginine or lysine) due to a negatively charged aspartate located at the bottom of the S1 pocket that is highly conserved in enzymes with trypsin-like activity (reviewed in Perona and Craik 1995 and Hooper et al. 2001). Accordingly, autocatalytic activation of several TTSPs has been observed in vitro, suggesting that some of the TTSPs could function as initiators of proteolytic cascades (Hooper et al. 2001).

This review aims to give an overview on the physiological functions of this fascinating and rapidly evolving group of enzymes and a summary of the current knowledge on their role in proteolytic activation of viral fusion proteins.

8.2 Type II Transmembrane Serine Proteases (TTSPs)

Based on the arrangements of their extracellular protein domain, the phylogenetic analysis of the serine protease domain, and the chromosomal arrangement of the cognate genes, human TTSPs are divided into four subfamilies: the HAT/DESC (human airway trypsin-like protease/differentially expressed in squamous cell carcinoma) family, the hepsin/TMPRSS (transmembrane protease/serine S1) family, the matriptase family, and the corin family (Szabo et al. 2003) (Fig. 8.1). TTSP genes are found in all vertebrate genomes. There also exist two nonmammalian TTSPs in *Drosophila*, stubble-stubbloid (st-sb), and corin, indicating that the TTSP family may have originated from two ancestral genes, one giving rise to the HAT/DESC family and one to the corin family (Appel et al. 1993; Bugge et al. 2009).

8.2.1 HAT/DESC Family

The HAT/DESC family contains five human members: HAT, DESC1, HAT-like 1 protease (HATL1), HATL4, and HATL5. In addition, HATL2 and HATL3 have been described in rodents (Table 8.1). The stem region of all HAT/DESC members is composed of a single SEA (sea urchin sperm protein, enterokinase, and agrin) domain. The genes encoding the HAT/DESC family members are all located in tandem on human chromosome 4 and mouse chromosome 5, respectively (Hobson et al. 2004). To date, the functional significance of the HAT/DESC1 family is largely unknown with HAT being the best studied exemption.

8.2.1.1 HAT: A Trypsin-Like Protease Associated with Airway Diseases

Human airway trypsin-like protease (HAT) was originally isolated from sputum of patients with chronic airway diseases as a soluble active protease with an apparent molecular mass of 27 kDa (Yoshinaga et al. 1998). Subsequent cloning of HAT cDNA from human trachea cDNA revealed that it encodes for a protease of 417 amino acids with a predicted molecular mass of 47 kDa that contains a transmembrane region near the N-terminus (Yamaoka et al. 1998). The HAT zymogen undergoes autocatalytic activation in vitro (Kato et al. 2012). HAT has been shown to be expressed as an active protease on the cell surface of HAT-expressing Madin-Darby canine kidney (MDCK) cells (Böttcher-Friebertshäuser et al. 2010). Surface biotinylation analysis showed that both the zymogen and the mature form of HAT are present on the cell surface, indicating that autoactivation might take place at the cell surface.

HAT is encoded by the TMPRSS11D gene, located on human chromosome 4q13.2. The TMPRSS11D gene is the human ortholog of long splice variants of the airway trypsin-like protease from mouse (MAT1) and rat (RAT1) (Hansen et al. 2004). An alternatively spliced isoform of MAT1 and RAT1 has been identified in rat (RAT2) and mouse (MAT2), respectively. It is also known as adrenal secretory serine protease (AsP) and contains an N-terminal signal peptide instead of a transmembrane domain and a SEA domain (Fig. 8.1). There have been no reports of a short isoform of HAT in humans.

Within human tissues, HAT expression is prominent in the trachea and bronchi and was also detected in the gastrointestinal tract, the skin, and the brain (Sales et al. 2011; Bertram et al. 2012). In the airway epithelium, HAT has been shown to be expressed at the apical membrane of ciliated cells, but not in goblet cells, submucosal glands, and mast cells (Takahashi et al. 2001). A number of studies suggest a potential role for HAT in the pathophysiology of bronchial asthma and chronic bronchitis (Chokki et al. 2004; Matsushima et al. 2006; Yasuoka et al. 1997; Yoshinaga et al. 1998). Among other functions HAT has been shown to cleave fibrinogen, to modulate structure and functions of the urokinase-type plasminogen activator receptor (uPAR, CD87), and to activate the protease-activated receptor 2 (PAR-2). Protease-activated receptors (PARs) are a family of seven transmembrane domain G-protein-coupled receptors that are activated by serine proteases through

Table 8.1 Tissue distribution and physiological functions of MASPs in human and mouse

Protease	Other names	Chromosomal localization human/mouse	Tissue distribution	Physiological functions	Physiological substrate(s)	Phenotype of mutant mice
HAT	TMPRSS11D	4q13.2/5E1	air, bra, bro, es, ey, pro, sk, si, spc, st, tes, to, tr, ub	Unknown	Unknown	Deficiency: no effect on development, postnatal growth, or long-term health
DESC1	TMPRSS11E	4q13.3/5E1	air, epd, es, pro, sk, sg, tes, to, tr	Unknown	Unknown	n.d.
HAT-like 1 (human)	TMPRSS11A, DESC3, ECRG1	4q13.2/E51	bl, eye, es, sk, st, to, tr	Unknown	Unknown	Deficiency: no effect on development, postnatal growth, or long-term health
HAT-like 2 (mouse)	DESC4	-/5E1	air, bl, bra, co, ey, he, li, lu, sg, tg, to	Unknown	Unknown	n.d.
HAT-like 3 (mouse)	TMPRSS11C, neurobin	-/5E1	bra, spc	Unknown	Unknown	n.d.
HAT-like 4	TMPRSS11F	4q13.2/5E1	es, pl, sk, tes, to, tr	Unknown	Unknown	Deficiency: viable and fertile, increased body fluid loss
HAT-like 5	TMPRSS11B	4q13.3/E51	es, eye, tes, to, tr	Unknown	Unknown	n.d.
Enteropeptidase	PRSS7	21q21/16C3.1	psi	Converting trypsinogen to trypsin in the intestine	Trypsinogen	n.d.
Hepsin	TMPRSS1	19q11-q13.2/7A3	hyp, ie, ki, li, lu, pa, pro, thm, thr	Cochlear development and hearing (mice)	Unknown	Deficiency: viable and fertile, deafness, abnormal cochlear development, thyroid hormone deficiency

TMPRSS2	Epitheliasin	21q22.3/16C2	air, bro, co, he, ki, li, lu, pa, pro, sg, si, st, tr	Unknown	Unknown	Deficiency: no effect on development, postnatal growth, or long-term health, no apparent phenotype
TMPRSS3	TADG-12	21q22.3/17A3.3	co, coc, he, ie, ki, li, lu, ova, pa, pl, pro, si, spl, tes, thm	Hearing, survival of cochlear hair cells and spiral ganglion neurons	Unknown	Expression of truncated nonfunctional TMPRSS3 mutant: deafness
TMPRSS4	CAP2	11q23.3/9A5.2	bro, bl, co, es, ki, lu, si, st	Unknown	Unknown	Deficiency: no effect on development, postnatal growth, or long-term health, no apparent phenotype
Spinesin	TMPRSS5	11q23.3/9A5.2	bra, ie, spc	Hearing	Unknown	n.d.
TMPRSS12		12q13.13/15F1	chicken: lu, phx, spl, tr	n.d.	Unknown	n.d.
TMPRSS13	MSPL	11q23.2/9A5.2	lu, pa, pl, pro	Unknown	Unknown	Deficiency: abnormal skin development
Matriptase	MT-SP1, ST14, CAP3, PRSS14, epithin	11q25/9A4	Variety of epithelial tissues	Epidermal and hair follicle development, epithelial barrier functions (tight junction formation) Inhibition of matriptase by HAI-1 is essential for formation of the placental labyrinth in mice	Prostasin	Deficiency: early postnatal mortality, impaired epithelial barrier function leading to severe skin dehydration Hypomorphic (100-fold reduced epidermal matriptase expression): viable and fertile, impaired epidermal prostasin zymogen activation and profilaggrin processing, impaired tight junction formation, ichthyosis, hypotrichosis

(continued)

Table 8.1 (continued)

Protease	Other names	Chromosomal localization human/mouse	Tissue distribution	Physiological functions	Physiological substrate(s)	Phenotype of mutant mice
Matriptase-2	TMPRSS6	22q13.1/15E2	bra, ki, li, lu, spl, tes, ut	Key regulator in iron homeostasis, suppression of hepcidin expression	Hemojuvelin	Deficiency or lack of catalytic activity increases hepcidin levels leading to iron-refractory iron-deficiency anemia (IRIDA)
Matriptase-3	TMPRSS7	3q13.2/16B5	bra, epd, ey, lu, ova, pro, sg, sk, tes, tr, ut	Unknown	Unknown	n.d.
Polyserase-1	TMPRSS9	19p13.3/10C1	bra, he, ki, li, pl, skm	Unknown	Unknown	n.d.
Corin	TMPRSS10	4p13-p12/5C3.2	he, hf, ie, ki, tes, ut	Pro-ANP convertase, processing of peptides critical for regulation of blood pressure, role in specifying the hair shaft pigmentation	Pro-atrial natriuretic peptide (ANP)	Deficiency: fertile, develop normally, increased body weight, salt-sensitive hypertension, reduced sodium excretion, lack of pro-ANP processing, abnormal hair pigmentation
Tryptase gamma 1	PRSS31	16p13.3/17A3.3	mc	Unknown	Unknown	Deficiency: viable, no effect on development

Prostasin	PRSS8, CAP1	16p11.2/7F3	Variety of epithelial tissues	Epidermal and hair follicle development and epithelial barrier functions, regulation of ENaC-mediated alveolar sodium and water transport (mice)	Matriptase, ENaC (gamma subunit)	Deficiency: early postnatal mortality due to severe skin dehydration, impaired profilaggrin processing, impaired tight junction formation Expression of catalytically inactive prostasin: normal prenatal and postnatal survival, no profound defects in epidermal development and barrier function, abnormal hair growth, and delayed skin wound healing
Testisin	PRSS21, ESP-1	16p13.3/17A3.3-B	eos, sperm, tes	Directs epididymal sperm maturation and fertilizing ability	Unknown	Deficiency: subfertility, aberrant sperm morphology

HAT human airway trypsin-like protease, *DESC* differentially expressed in squamous cell carcinoma, *TMPRSS* transmembrane protease serine S1, *ECRG* esophageal carcinoma-related gene, *PRSS* protease serine S1, *CAP* channel-activating protease, *TAGD* tumor-associated differentially expressed gene, *MSPL* mosaic serine protease large-form, *ST* suppression of tumorigenicity, *ESP* eosinophil serine protease. Tissue distribution (human and mouse) experimentally confirmed at mRNA and/or protein level (for references, see text and Bugge et al. 2009). *air* upper airways; *bl* bladder; *bra* brain; *bro* bronchioles; *co* colon; *coc* cochlea; *epd* epididymis; *eos* eosinophils; *es* esophagus; *ey* eye; *he* heart; *hf* hair follicle; *hyp* hypophysis; *ie* inner ear; *ki* kidney; *li* liver; *lu* lung; *mc* mast cells; *ova* ovary; *pa* pancreas; *phx* pharynx; *pl* placenta; *pro* prostate; *psi* proximal small intestine; *sg* salivary gland; *si* small intestine; *sk* skin; *skm* skeletal muscle; *spc* spinal cord; *spl* spleen; *st* stomach; *tes* testis; *tg* tear gland; *thm* thymus; *thr* thyroid; *to* tongue; *tr* trachea; *ub* urinary bladder; *ut* uterus; *n.d.* not determined specific N-terminal proteolytic cleavage and the unmasking of a tethered ligand. Activated PAR-2 plays a pivotal role in cell adhesion and early inflammatory processes and has been reported to mediate allergic airway inflammation of the mouse airways in vivo (Schmidlin et al. 2002). HAT has been described to increase mucin gene expression and to stimulate bronchial fibroblast proliferation in airway epithelial cells through PAR-2-mediated signaling pathways (Yoshinaga et al. 1998; Matsushima et al. 2006).

HAT expression has been shown to be deregulated in skin diseases such as psoriasis vulgaris. Higher expression of HAT was found in psoriatic epidermal lesions (Iwakiri et al. 2004). It has been reported that HAT might promote PAR-2-mediated interleukin 8 (IL-8) production to accumulate inflammatory cells in the epidermal layer of psoriasis vulgaris.

The physiological function of HAT in the airways and in the skin, however, remains unknown. Knockout of HAT/TMPRSS11D expression in mice does not affect development, postnatal growth, or long-term health (Sales et al. 2011), indicating that HAT/TMPRSS11D activity is dispensable or can be compensated by other proteases. A recent study found that HAT expression is lost during the dedifferentiation of epithelial cells in high-grade tumors, a hallmark of squamous cell carcinogenesis (Duhaime et al. 2016). Therefore, HAT might act as an activator and initiator of a proteolytic cascade during terminal differentiation of squamous epithelia.

HAT and TMPRSS2 (see later chapter) were the first human proteases identified to cleave and activate the influenza A virus surface glycoprotein hemagglutinin (HA) with a monobasic cleavage site and to support multicycle replication and spread of the virus in cell culture (Böttcher et al. 2006). The role of HAT and other MASPs in activation of viral glycoproteins will be described in more detail at the end of this review and is summarized in Table 8.2.

Table 8.2 Proteolytic activation of viral glycoproteins by MASPs

Protease	Influenza virus hemagglutinin (HA)	Coronavirus spike protein S	Other viruses	References
HAT/ TMPRSS11D	Avian and mammalian IAV (R/K↓) (subtypes H1, H2, H3, H9, H11, H12) IBV	SARS-CoV MERS-CoV	n.d.	Böttcher et al. (2006), Chaipan et al. (2009), Bertram et al. (2011), Böttcher-Friebertshäuser et al. (2012) and Galloway et al. (2013)
DESC1	IAV (subtypes H1, H2, H3, H17)	SARS-CoV MERS-CoV	n.d.	Zmora et al. (2014) and Hoffmann et al. (2016)
HATL1	n.d.	SARS-CoV	n.d.	Kam et al. (2009)
HATL5	No	No	n.d.	Bertram et al. (2010)
HATL4	No	No	n.d.	Bertram et al. (2010)
Hepsin	No	n.d.	Fusion protein F of subtype B avian metapneumovirus (RKKR↓)	Bertram et al. (2010) and Yun et al. (2016)

Table 8.2 (continued)

Protease	Influenza virus hemagglutinin (HA)	Coronavirus spike protein S	Other viruses	References
TMPRSS2	Avian and mammalian IAV (R/K↓) (subtypes H1–H7, H9–H11, H11–H18) IBV Essential for proteolytic activation of H7N9 and H1N1 IAV in mice	SARS-CoV MERS-CoV Human CoV-229E Porcine epidemic diarrhea virus	Fusion protein F of paramyxoviruses: human metapneumovirus; human parainfluenza viruses 1, 3, 4a, and 4b; and Sendai virus	Böttcher et al. (2006), Shirogane et al. (2008), Matsuyama et al. (2010), Glowacka et al. (2011), Shirato et al. (2013), Shulla et al. (2011), Böttcher-Friebertshäuser et al. (2012), Abe et al. (2013), Bertram et al. (2013), Ferrara et al. (2013), Galloway et al. (2013), Gierer et al. (2013), Hatesuer et al. (2013), Tarnow et al. (2014) Sakai et al. (2014), Hoffmann et al. (2016) and Fan et al. (2017)
TMPRSS3	No	No	n.d.	Bertram et al. (2010) and Zmora et al. (2014)
TMPRSS4	IAV (H1, H3)	No: SARS-CoV	n.d.	Chaipan et al. (2009) and Glowacka et al. (2011)
TMPRSS12	n.d.	n.d.	Fusion protein F of subtype B avian metapneumovirus (RKKR↓), no activation of F with RQSR↓ motif	Yun et al. (2016)
MSPL/ TMPRSS13	Highly pathogenic avian IAV of subtype H5N2 (KKKR↓) IAV (subtypes H1, H2, H3, H17 monobasic motifs)	SARS-CoV MERS-CoV Porcine epidemic diarrhea virus	n.d.	Okumura et al. (2010), Zmora et al. (2014), Hoffmann et al. (2016) and Fan et al. (2017)
Matriptase	IAV (subtype H9 with RSS/RR↓ motif but not VSSR↓ motif, some H1N1 strains (IQSR↓))	n.d.	n.d.	Hamilton et al. (2012), Baron et al. (2013) and Beaulieu et al. (2013)
Matriptase-2	No	n.d.	n.d.	Bertram et al. (2010)

(continued)

Table 8.2 (continued)

Protease	Influenza virus hemagglutinin (HA)	Coronavirus spike protein S	Other viruses	References
Matriptase-3	No	No: SARS-CoV	n.d.	Chaipan et al. (2009) and Glowacka et al. (2011)
Polyserase-1	No	No	n.d.	Zmora et al. (2014)
Corin	No	No	n.d.	Bertram et al. (2010)
Prostasin	No	No	n.d.	Bertram et al. (2010), Böttcher-Friebertshäuser et al. (2010) and Zmora et al. (2014)
Testisin	IAV (subtype H1, H3, H9)	n.d.	n.d.	E. Böttcher-Friebertshäuser, A. Arendt, unpublished data

IAV influenza A virus; *IBV* influenza B virus; *HA* hemagglutinin; *S* spike protein; *F* fusion protein; *SARS* severe acute respiratory syndrome; *MERS* Middle East respiratory syndrome; *CoV* coronavirus; *n.d.* not determined. Amino acids are indicated as single-letter code. The cleavage site is indicated by an arrow

8.2.1.2 DESC1 and TMPRSS11A/HATL1: Tumor Suppressors in Esophagus Cancer

DESC1 (differentially expressed in squamous cell carcinoma, also designated as transmembrane protease serine S1 family member 11E (TMPRSS11E)) is expressed in the human head, neck, skin, prostate, and testis. DESC1 was first identified as a gene downregulated in squamous cell carcinoma of the head and neck (Lang and Schuller 2001). Moreover, induction of normal keratinocyte differentiation by calcium challenge was accompanied by an increase in DESC1 expression in vitro (Sedghizadeh et al. 2006). A recent study found that DESC1 sensitizes cells to apoptosis by downregulating the epidermal growth factor receptor (EGFR)/AKT pathway in esophageal squamous cell carcinoma (Ng et al. 2016). Protease activity was required for this function, suggesting that DESC1 cleaves EGFR, which subsequently leads to the downregulation of the AKT pathway. Moreover, DESC1 has been demonstrated to reduce tumor growth kinetics in an orthotopic nude mouse model for study of esophageal squamous cell carcinoma. Taken together, there is accumulating evidence suggesting the association of DESC1 downregulation with cancer development. However, its tumor suppressive role remains to be characterized in more detail in future studies. The crystal structure of the catalytic domain of DESC1 has been solved (Kyrieleis et al. 2007) and is shown in Fig. 8.2.

TMPRSS11A, also designated as HAT-like 1 protease (HATL1) and esophageal carcinoma-related gene 1 (ECRG1), has been identified as a gene downregulated in esophagus cancers (Li et al. 2006). TMPRSS11A-specific mRNA is present in the eye, testis, glandular stomach, tongue, trachea, bladder, forestomach, and skin of mice (Sales et al. 2011). Knockout of TMPRSS11A expression does not affect development, postnatal growth, or long-term health in mice (Sales et al. 2011). TMPRSS11A/ECRG1 is one among many tumor suppressor genes that may play a role in the initiation and development of esophageal squamous cell carcinoma (Netzel-Arnett et al. 2003). Overexpression of TMPRSS11A/ECRG1 has been shown to inhibit cell growth and to induce G1/S cell cycle arrest through upregulation of p15INK4b expression in esophageal cells in vitro (Zhao et al. 2004).

8.2.1.3 TMPRSS11F/HATL4 and TMPRSS11B/HATL5

There is little known about TMPRSS11F/HATL4 and TMPRSS11B/HATL5. TMPRSS11B/HATL5 expression has been detected in the cervix, esophagus, and oral cavity. It was found to be significantly decreased in squamous cell carcinomas of these tissues as compared to normal and tumor adjacent samples (Miller et al. 2014), and thus it might have a similar physiological function like DESC1 and TMPRSS11A. TMPRSS11F/HATL4-specific mRNA was found to be present in the skin, esophagus, trachea, tongue, eye, bladder, testis, uterus, and stomach (Sales et al. 2011). Recently, TMPRSS11F-knockout mice were shown to be viable and fertile (Zhang et al. 2017). Compared with wild-type controls, TMPRSS11F-deficient newborn mice had greater body fluid loss and higher mortality in a trans-epidermal body fluid loss test, indicating that TMPRSS11F/HATL4 is involved in epidermal barrier function to prevent body fluid loss.

8.2.2 Hepsin/TMPRSS Family

The hepsin/TMPRSS family comprises eight members: hepsin/TMPRSS1, TMPRSS2, TMPRSS3, TMPRSS4, spinesin/TMPRSS5, TMPRSS12, TMPRSS13/MSPL (mosaic serine protease large-form), and enteropeptidase. All members of this family have a group A scavenger receptor domain in their stem region linked to the serine protease domain, preceded by a single LDLA (low-density lipoprotein receptor A) domain in TMPRSS2, TMPRSS3, TMPRSS4, and MSPL. Enteropeptidase is unique for this family, having multiple protein domains between the transmembrane domain and the catalytic domain, including a SEA domain, two CUB (C1s/C1r, urchin epidermal growth factor and bone morphogenetic protein 1) domains, two LDLA domains, a MAM (meprin, A5 protein, tyrosine phosphatase μ) domain, and a group A scavenger receptor domain (Fig. 8.1).

8.2.2.1 Enteropeptidase: Activation of Pancreatic Hydrolases by Converting Trypsinogen to Trypsin

As already mentioned enteropeptidase, originally named enterokinase, has been discovered in 1899 in the laboratory of Ivan Pavlov as an activity of extracts of small intestinal mucosa that was able to activate hydrolytic enzymes in pancreatic fluid. In 1939, Moses Kunitz demonstrated that purified porcine enteropeptidase converts crystalline trypsinogen to trypsin (Kunitz 1939). In the 1970s, purification of enteropeptidase from porcine, bovine, and human intestine revealed that it consists of a heavy chain (82–140 kDa) and a light chain (35–62 kDa) connected by a disulfide bond. Edman degradation of the amino-terminal sequence of the light chain of bovine enteropeptidase revealed its homology to other trypsin-like proteases. However, only cloning of the enteropeptidase cDNA in 1994 revealed the presence of a membrane anchor near the amino-terminus and indicated that the active two-chain form is derived from a single-chain precursor (Kitamoto et al. 1994).

Enteropeptidase activity is almost exclusively localized to the duodenum. The enzyme is localized in the brush border of enterocytes and some goblet cells of the human duodenal mucosa and, at lower levels, in the mucosa of the proximal jejunum (Hermon-Taylor et al. 1977; Yuan et al. 1998). Enteropeptidase specifically cleaves trypsinogen at the activation site DDDDK↓I that is highly conserved among vertebrates (Zheng et al. 2009). Trypsin, in turn, activates a number of pancreatic zymogens such as chymotrypsinogen, proelastase, procarboxypeptidases, and prolipases in the lumen of the gut. Such unique sequence specificity is thought to provide an important mechanism to ensure that trypsinogen is activated only by enteropeptidase, but not by other proteases in pancreas, avoiding damage of the pancreas due to excess protease activities.

The cDNA sequences of enteropeptidase have been determined for several vertebrate species including cattle, swine, mouse, Japanese rice fish (medaka), and man (Zheng et al. 2009). The amino acid sequences of the human and bovine protease are 82% identical. In humans, enteropeptidase is encoded by the PRSS7 gene localized on chromosome 21q21 (Kitamoto et al. 1994, 1995). Enteropeptidase is synthesized as a zymogen of 1019 amino acids, which requires activation by another protease at K784 within the activation site sequence ITPK↓IVGG. Trypsin and the protease duodenase that is secreted by Brunner's glands of the proximal segment of duodenum have been shown to activate enteropeptidase (Zamolodchikova et al. 1997, 2000). However, duodenase is synthesized as a zymogen, too, and requires activation by another protease. Moreover, the measured rate of activation of purified bovine recombinant proenteropeptidase by duodenase was about 70-fold lower than that by trypsin (Zamolodchikova et al. 2000). Thus, the role for duodenase as primary activator of proenteropeptidase remains to be established.

The molecular basis of human enteropeptidase localization to the apical membrane is not yet completely understood. The transmembrane domain anchors enteropeptidase in the brush border of duodenal enterocytes. In addition, mucin-like repeats in the SEA domain and N-linked glycosylation of the catalytic domain have been found necessary for apical delivery in MDCK cells (Zheng and Sadler 2002; Zheng et al. 2009). In addition, enteropeptidase is present as a soluble form in the small intestinal lumen. Shedding may be due to the action of biliary or pancreatic proteases and possibly to local effects of gastrointestinal hormones (Götze et al. 1972). However, it remains unknown whether shedding of enteropeptidase plays a role in regulating its activity in the gut.

The enteropeptidase serine protease domain contains a basic tetrapeptide segment consisting of R/K96-R-R-K99, which is not conserved in other serine proteases (Matsushima et al. 1994; Kitamoto et al. 1994; Yuan et al. 1998). Computer modeling suggested that this basic segment is located on the protein surface where it might bind the acidic P2–P5 residues of trypsinogen activation sites. The crystal structure of the bovine enteropeptidase catalytic domain in complex with the inhibitor V-$(D)_4$-K-chloromethane confirmed this extended substrate binding exosite (Lu et al. 1999). The K99 residue was identified as major determinant for recognition of the P2 and P4 aspartate residues. It is conserved among enteropeptidases of many species. Substitution of the lysine by alanine prevented enteropeptidase from cleaving trypsinogen (Lu et al. 1999). However, two-chain enteropeptidase cleaves trypsinogen about 500-fold more rapidly than does the isolated light chain (Lu et al. 1997), indicating that the heavy chain is necessary for optimal cleavage of trypsinogen.

The physiological importance of enteropeptidase is indicated by severe intestinal malabsorption in congenital deficiency of this enzyme. A number of cases of primary enterokinase deficiency have been reported since it was first described in 1969 (Hadorn et al. 1969). Patients suffer from severe intestinal malabsorption with diarrhea, vomiting, and growth failure in early infancy. Nonsense or frameshift mutations in the *PRSS7* gene have been shown in patients (Holzinger et al. 2002). Congenital enteropeptidase deficiency can be treated successfully by administration of pancreatic extract in early infancy (Zheng et al. 2009).

Because of its unique substrate specificity and protein stability, enteropeptidase is also of biotechnological interest. Introduction of the DDDDK↓I enterokinase recognition sequence is widely used as a tool to specifically cleave and activate recombinant proproteins or fusion proteins. More recently, enteropeptidase from the Japanese rice fish medaka has been cloned and characterized (Ogiwara and Takahashi 2007). The E173A mutant of medaka enteropeptidase showed an even stricter specificity for the DDDDK sequence compared to bovine enteropeptidase and may therefore provide the most appropriate protease to cleave recombinant proteins containing the DDDDK motif.

8.2.2.2 TMPRSS2: Prostate Cancer Progression and Proteolytic Activation of Influenza A Viruses

TMPRSS2 cDNA was originally cloned by exon trapping when the transcription map of human chromosome 21 was developed (Paoloni-Giacobino et al. 1997). The human gene is mapped to 21q22.3 and encodes for a protein of 492 amino acids. Murine TMPRSS2 is also designated as epitheliasin and is encoded on chromosome 16. The modular structure of TMPRSS2 is illustrated in Fig. 8.1.

TMPRSS2 is widely expressed in epithelial cells of the respiratory, gastrointestinal, and urogenital tract with high expression levels in the prostate and colon (Bugge et al. 2009; Bertram et al. 2012). Immunohistochemical studies revealed that TMPRSS2 is also expressed in cardiac myocytes (Bertram et al. 2012). TMPRSS2 is associated with prostate cancer. The protease has been shown to be overexpressed in prostate cancer tissue, and the TMPRSS2 level has been shown to be correlated with prostate cancer progression (Lucas et al. 2008, 2014; Chen et al. 2010). Moreover, fusion of the androgen-regulated TMPRSS2 promoter to E26 transformation-specific (ETS) transcription factor genes, particularly the ETS-regulated gene (ERG), resulting in overexpression of ERG is seen in nearly 50% of patients and is used as a prognostic marker of prostate carcinogenesis (Tomlins et al. 2005; Gasi Tandefelt et al. 2014). The TMPRSS2 gene and the ERG gene are located approximately 3 Mb apart in the same relative orientation on chromosome 21. However, the role of these gene fusions in the development and progression of prostate cancer is not understood in detail. The TMPRSS2 gene is furthermore flanked by the Mx1 (myxovirus resistance 1) gene, encoding a GTPase that is part of the antiviral response induced by type I and type III interferons (Paoloni-Giacobino et al. 1997; Verhelst et al. 2013).

TMPRSS2 promoted prostate cancer cell invasion and metastasis to distant organs in a mouse model of prostate carcinogenesis (Lucas et al. 2014). TMPRSS2 has been shown to activate PAR-2 and hepatocyte growth factor (HGF)/c-Met signaling pathways and to downregulate E-cadherin expression in prostate cancer cells (Wilson et al. 2005; Lucas et al. 2014; Leshem et al. 2011). More recently, TMPRSS2 has been described to promote prostate cancer tumor growth and metastasis, via activation of the TTSP matriptase and degradation of extracellular matrix (ECM) laminin β1 and nidogen-1 in vitro and in a xenograft mouse model of prostate cancer (Ko et al. 2015). TMPRSS2 protein level was shown to correlate with increased levels of active matriptase as well as increased metastases (Ko et al. 2015).

The physiological role of TMPRSS2 is unknown so far. TMPRSS2-deficient mice lack a discernible phenotype (Kim et al. 2006). The protease has been shown to cleave the epithelial sodium channel (ENaC) in *Xenopus* oocytes in vitro and suggested to be involved in regulation of the airway surface liquid (ASL) volume by proteolytic cleavage of ENaC (Donaldson et al. 2002). Interestingly, by using TMPRSS2-deficient mice, three recent studies identified TMPRSS2 as a host cell factor essential for activation and spread of H1N1 and H7N9 influenza A viruses in mice (see below). Intriguingly, knockout of TMPRSS2 expression protected mice from an otherwise lethal infection due to inhibition of activation of progeny virus and, thus, spread along the respiratory tract (Hatesuer et al. 2013; Tarnow et al. 2014; Sakai et al. 2014).

TTSPs are believed to be situated at the cell surface, and shedding of TMPRSS2 has been described from prostate and prostate cancer cells and from human nasal epithelial cells exposed to ozone in vitro (Afar et al. 2001; Kesic et al. 2012). Studies on the subcellular localization of influenza A virus HA cleavage by TMPRSS2 upon co-expression in MDCK cells, however, indicate that TMPRSS2 cleaves HA in the trans-Golgi network (TGN) or during its transport to the plasma membrane, where virus assembly and budding take place (Böttcher et al. 2009; Böttcher-Friebertshäuser et al. 2010, 2013) (see Sect. 8.5). In contrast, TMPRSS2 present on the cell surface and soluble TMPRSS2 shed from MDCK cells showed poor if any enzymatic activity and were not able to cleave HA (Böttcher-Friebertshäuser et al. 2010). The reason for the lack of TMPRSS2 activity on the cell surface and in cell supernatants is unknown and might be related to the expression of protease inhibitors or missing cofactors. However, intracellular activation of HA by TMPRSS2 revealed that TTSPs may not only act as proteolytic enzymes on the cell surface but can process their substrates also (or already) in intracellular compartments.

8.2.2.3 Dysregulation of TMPRSS3, TMPRSS5/Spinesin, and Hepsin Is Associated with Deafness

TMPRSS3 was first identified as a novel multi-domain serine protease overexpressed in ovarian cancers and therefore originally named tumor-associated differentially expressed gene 12 (TADG-12) (Underwood et al. 2000). Independently, mutations in the TMPRSS3 gene were associated with congenital and childhood onset autosomal recessive deafness (Scott et al. 2001). This was the first description of a protease involved in hearing loss. The TMPRSS3 gene maps on chromosome 21 at q22 and encodes for a protein of 454 amino acids with an overall domain structure similar to TMPRSS2 and TMPRSS4. TMPRSS3 is synthesized as a zymogen and undergoes autoactivation (Guipponi et al. 2002). RT-PCR analysis and RNA in situ hybridization experiments revealed expression of TMPRSS3 in the thymus, stomach, testis, ovary, kidney, and eye and in a variety of inner ear tissues, including inner hair cells, stria vascularis, spiral ganglion neurons, modiolus, and organ of Corti (Scott et al. 2001; Guipponi et al. 2002, 2008).

A number of different mutations in the TMPRSS3 gene have been identified in patients with non-syndromic autosomal recessive deafness (DFNB8/10) (Scott et al. 2001; Ben-Yosef et al. 2001; Masmoudi et al. 2001, Guipponi et al. 2008). Mutations occur in all functional domains and have been shown to disrupt the proteolytic activity of TMPRSS3, indicating that TMPRSS3 protease activity is critical during inner ear development (Guipponi et al. 2002; Wattenhofer et al. 2005). It remains to be elucidated how missense mutations in the LDLA and SRCR domains affect the proteolytic activity of TMPRSS3. Studies in a mouse model carrying a protein-truncating nonsense mutation in TMPRSS3, Y260X (X = stop codon), revealed that TMPRSS3 is essential for mouse cochlear hair cell survival at the onset of hearing (Fasquelle et al. 2011). Mice expressing TMPRSS3-Y260X are completely deaf due to rapid and massive degeneration of hair cells. Moreover, loss of spiral ganglion neurons was observed in TMPRSS3-Y260X mice at the age of 4 months (Fasquelle et al. 2011). Proteomic analyses revealed that TMPRSS3 deficiency leads to a decrease in the expression of Kcnma1 potassium channels in inner hair cells (Molina et al. 2013). However, it remains to be investigated in more detail how TMPRSS3 regulates the abundance of functional Kcnma1 channels expression in hair cells. A recent study by Li et al. demonstrated that knockdown of TMPRSS3 inhibited cell viability of spiral ganglion neurons in vitro (Li et al. 2014). Moreover, they observed that microRNA miR-204 suppressed spiral ganglion neuron survival in vitro by targeting TMPRSS3 (Li et al. 2014). TMPRSS3 mRNA was found to have a putative miR-204 binding site within its 3′-UTR that is highly conserved among the vertebrates.

In addition to its role in hearing, TMPRSS3 was found to be overexpressed in pancreatic and ovarian cancer and to promote proliferation, invasion, and migration of ovarian cancer cells via activation of the ERK1/2 pathway in vitro (Wallrapp et al. 2000, Zhang et al. 2016).

Hepsin was identified in cDNA clones obtained from human liver and was the first serine protease characterized to contain a transmembrane domain

(Leytus et al. 1988). Hepsin is abundant in the liver. Other tissues such as kidney, pancreas, lung, stomach, prostate, and thyroid express low levels of hepsin mRNA (Tsuji et al. 1991). The human hepsin gene has been localized to chromosome 19q11-13.2. Hepsin consists of an N-terminal cytoplasmic domain, a transmembrane domain, and an extracellular portion composed of a SRCR domain and a C-terminal protease domain. The crystal structure of a soluble human hepsin that included the SRCR and protease domains has been solved (Somoza et al. 2003).

A number of potential hepsin substrates have been identified in vitro including blood clotting factors VII, IX, and XII, prourokinase, promatriptase, proprostasin, macrophage-stimulating protein (MSP), and laminin-332 (reviewed in Antalis et al. 2011), suggesting that hepsin may play a role in blood coagulation and embryonic development. The evidence for these functions, however, remains inconclusive since hepsin-deficient mice are viable and fertile, and do not exhibit obvious defects in growth and blood coagulation (Wu et al. 1998; Brunati et al. 2015). Unexpectedly, a study by Guipponi et al. found that hepsin-deficient mice exhibited profound hearing loss (Guipponi et al. 2007). Hepsin knockout mice have abnormal cochlea and reduced myelin protein expression in the auditory nerve (Guipponi et al. 2007). Furthermore, low levels of plasma thyroxine, a thyroid secreted hormone important for cochlear development, have been found in hepsin-deficient mice (Hanifa et al. 2010). The molecular mechanisms by which hepsin regulates normal hearing are not understood, and so far it is unknown if hearing loss in hepsin-knockout mice is a result of thyroid hormone deficiency.

Further studies of knockout mouse models of hepsin demonstrated that the protease activates pro-hepatocyte growth factor (pro-HGF) in the liver and is responsible for cleavage and urinary secretion of uromodulin (Tamm-Horsfall protein) (Hsu et al. 2012; Brunati et al. 2015). It was proposed that HGF/c-Met signaling may regulate expression of connexins, gap junction proteins, in hepatocytes in mice. Loss of hepsin was found to increase expression of connexins, resulting in an expansion of hepatocyte size and a concomitant narrowing of sinusoids. Interestingly, systemic delivery of tumor cells by tail-vein injection showed preferential colonization of tumor cells in the liver of hepsin-deficient mice compared to wild-type mice. These data suggest that loss of hepsin enhances the colonization of liver by tumor cells, probably through increased retention of tumor cells because of narrower sinusoids related to enlarged hepatocytes.

Hepsin has been identified as one of the most upregulated genes in prostate cancer. Hepsin increases early in prostate cancer initiation, and its high levels are maintained throughout progression and metastasis and are indicative of poor outcome (Dhanasekaran et al. 2001; Stephan et al. 2004). Overexpression of hepsin has also been shown in many other cancers including breast, ovarian, and endometrial cancer (Murray et al. 2016).

Although a number of studies demonstrated that hepsin is involved in prostate cancer progression, little is known about the basis of its functions. Overexpression of hepsin in a mouse model of non-metastasizing prostate cancer caused disorganization and disruption of basement membrane and promoted primary prostate cancer progression and metastasis to the liver, lung, and bone (Klezovitch et al. 2004).

Overexpression of hepsin also promoted ovarian tumor growth in a mouse model, and proteolytic activity of hepsin was shown to be necessary for promoting tumor progression (Miao et al. 2008). Hepsin and HGF are present in desmosomes. Desmosome dissociation is known to be one of the first steps during HGF-induced epithelial-mesenchymal transition, indicating that increased levels of hepsin and its substrate HGF may play a role in ovarian cancer progression through their interaction with desmosomes (Miao et al. 2008). At present, the mechanism of this interaction and the functional importance of hepsin localization in the desmosomes warrant further investigation.

TMPRSS5/spinesin was cloned from a human spinal cord cDNA library (Yamaguchi et al. 2002). The human TMPRSS5 gene is located on chromosome 11q23 and encodes a simple TTSP of 457 amino acids consisting of a short cytoplasmic domain, a transmembrane domain, a stem region containing a SRCR domain, and a C-terminal serine protease domain (Yamaguchi et al. 2002). Northern blot analyses and immunohistochemical staining revealed that TMPRSS5 is predominantly expressed in the brain and the spinal cord. Guipponi et al. demonstrated that TMPRSS5-specific mRNA is furthermore present in inner ear tissues and the testis of rats (Guipponi et al. 2008). A mutation screen in a cohort of ca. 360 sporadic deafness cases revealed three mutations in the TMPRSS5 gene (A317L, F369L, and Y438X). These TMPRSS5 mutants showed reduced (F369L) or no (A317S, Y438X) proteolytic activity in yeast-based protease assays, suggesting that impaired TMPRSS5 activity might cause hearing impairment. However, TMPRSS5/spinesin has not been characterized in more detail, and it remains to be investigated how it affects hearing and which role it may play in the central nervous system.

8.2.2.4 TMPRSS4: An Emerging Potential Therapeutic Target in Cancer

TMPRSS4 (also known as channel-activating protease 2 (CAP2)) was originally identified as a gene expressed in most pancreatic tumors but not in the healthy pancreas (Wallrapp et al. 2000). Meanwhile, TMPRSS4 has been shown to be highly expressed also in many other cancers including thyroid, colon, gastric, and lung cancers; association with poor prognosis has been consistently described (reviewed in Kim and Lee 2014; Tanabe and List 2017).

The TMPRSS4 gene is located on chromosome 11 at q23.3 and encodes a protein of ~437 amino acids. The domain structure of TMPRSS4 is similar to TMPRSS2 and TMPRSS3. TMPRSS4 undergoes autoactivation in vitro (Antalis et al. 2011). The catalytic domain of TMPRSS4 can be released as an active form in cell culture (Min et al. 2014a). TMPRSS4 mRNA was detected in the bladder, esophagus, stomach, small intestine, colon, kidney, larynx, trachea, bronchi, and lung (Wallrapp et al. 2000; Jung et al. 2008; Böttcher-Friebertshäuser unpublished data). In murine lung TMPRSS4 is expressed in type II pneumocytes (Kühn et al. 2016).

Overexpression of TMPRSS4 has been shown to promote invasion and metastasis of human tumor cells by facilitating epithelial-mesenchymal transition (EMT) (Kim et al. 2010; Jung et al. 2008; Cheng et al. 2009). Knockdown of TMPRSS4 expression in lung and colon cancer cells by using siRNA reduced cell proliferation and invasion (Jung et al. 2008). A number of mechanisms, by which TMPRSS4 may modulate tumor cell proliferation and invasion, have been described in vitro. Overexpression of TMPRSS4 in colon cancer cells promoted EMT through the upregulation of integrin α5, thereby enhancing motility and invasiveness (Kim et al. 2010). In addition, TMPRSS4 was shown to induce invasion through upregulation of both expression and activity of urokinase plasminogen activator (uPA) via activation of the transcription factors AP-1, Sp1, and Sp3 and via processing of pro-uPA precursor into its active form in vitro (Min et al. 2014a, b). The serine protease uPA converts inactive plasminogen to active plasmin, which in turn can degrade most extracellular proteins and activate MMPs. Moreover, uPA and its receptor, uPAR (CD87), interact with integrin coreceptors to activate intracellular signaling pathways for cell migration, invasion, proliferation, and survival (reviewed in Hildenbrand et al. 2008). It has been shown that TMPRSS4 can interact with uPAR (Min et al. 2014b). Furthermore, TMPRSS4 has been shown to induce downregulation of E-cadherin, a well-known hallmark of EMT, via activation of transcriptional repressors Sip1/Zeb2 (Jung et al. 2008). Increased TMPRSS4 expression in cancer could be partially due to epigenetic dysregulation. The TMPRSS4 promoter has been shown to be hypomethylated in hepatocellular carcinoma and non-small cell lung cancer (Stefanska et al. 2011; Villalba et al. 2016). Hypomethylation of TMPRSS4 promoter was associated with worse prognosis in non-small cell lung cancer patients. Taken together, TMPRSS4 may be an important upstream regulator of the EMT and the invasiveness of cancer cells and a useful biomarker for the prognosis of certain types of cancers and could be employed for diagnostics and therapeutics.

Mice deficient in TMPRSS4 are viable, fertile, and do not show any obvious abnormalities (Keppner et al. 2015; Kühn et al. 2016). However, so far no cancer studies have been performed in TMPRSS4-deficient mice. Like TMPRSS2, TMPRSS4 has been shown to be able to activate the epithelial sodium channel (ENaC) when co-expressed in *Xenopus* oocytes (Vuagniaux et al. 2002). TMPRSS4 cleaves the gamma subunit of ENaC at a site distinct from the site that is processed by the GPI-anchored protease prostasin (cf. prostasin chapter). However, a recent study demonstrated that regulation of the ENaC-mediated sodium balance is not affected in TMPRSS4-deficient mice (Keppner et al. 2015), indicating that TMPRSS4 is not crucial for processing of ENaC in vivo.

8.2.2.5 TMPRSS13/MSPL: Preferential Recognition of Paired Basic Residues at the Cleavage Site

TMPRSS13 (transmembrane protease, serine 13, also known as mosaic serine protease large-form (MSPL)) was isolated in 2001 from a human lung cDNA library. The human TMPRSS13 is located on chromosome 11q23.2. Alternatively spliced forms from this gene have been identified, one encoding a type II transmembrane protease (MSPL) and a second form encoding a protease without a transmembrane domain (mosaic serine protease short-form (MSPS) that comprises 581 and 537 amino acids, respectively (Kim et al. 2001). RT-PCR analysis revealed that TMPRSS13-specific mRNA is present in various human tissues including lung, liver, kidney, spleen, pancreas, small intestine, prostate, and placenta (Kido et al. 2008).

The biological functions of TMPRSS13 are not well understood. TMPRSS13-deficient mice display abnormal skin development, leading to a compromised barrier function (Madsen et al. 2014). Interestingly, tight junction formation and profilaggrin processing were not affected in TMPRSS13-deficient mice. Thus, TMPRSS13 supports stratum corneum formation and epidermal barrier formation by a mechanism that is independent of both profilaggrin processing and tight junction formation (see below).

TMPRSS13 shows unique substrate specificity among TTSPs, with preferential recognition of paired basic residues (R/K at P1 and P2 positions) (Kido et al. 2008). TMPRSS13 also efficiently cleaved peptide substrates with K at position P4 that were not cleaved by furin and has been shown to activate the HA of highly pathogenic avian influenza viruses of subtype H5N2 possessing this motif at the cleavage site (Okumura et al. 2010). TMPRSS13-specific mRNA was detected in the blood vessels, lungs, trachea, colon, small intestine, and kidney of chickens.

8.2.3 Matriptase Family

The matriptase family has four members: matriptase, matriptase-2, matriptase-3, and polyserase-1. The matriptases have a SEA domain, two CUB domains, and three or four LDLA domains in their stem region. Polyserase-1 is unique among serine proteases, having one enzymatically inactive and two active serine protease domains (Fig. 8.1).

8.2.3.1 Matriptase: Crucial Roles in Epidermal Differentiation and Tight Junction Formation

Matriptase (also known as membrane-type serine protease 1 (MT-SP1), epithin, suppressor of tumorigenicity 14 (ST14), channel-activating protease 3 (CAP3)) was originally identified in 1993 as a new gelatinolytic activity in conditioned medium from cultured breast cancer cells (Shi et al. 1993). Matriptase is encoded by the *ST14* gene located on human chromosome 11q24–25 and encodes a polypeptide of 855 amino acids with a molecular weight of 95 kDa (Lin et al. 1999). Orthologs of matriptase are present in all vertebrate genomes examined to date (review Miller and List 2013).

Matriptase shows the most ubiquitous expression pattern of TTSPs being expressed in epithelial cells of most embryonic and adult tissues (Miller and List 2013). Matriptase has been shown to be required for postnatal survival in mice and has essential physiological functions in terminal differentiation of the oral and intestinal epithelium and the epidermis (List et al. 2002, 2003; Buzza et al. 2010). Mice deficient in matriptase die within 48 h after birth due to a severe dehydration resulting from an impaired epidermal barrier function. Matriptase is also critical for hair follicle growth and thymic development (List et al. 2002, 2003). Interestingly, it was observed that matriptase-deficient mice and mice deficient in expression of the GPI-anchored serine protease prostasin display identical epidermal phenotypes and that prostasin zymogen activation by matriptase is a key function in epidermal development (List et al. 2002, 2003; Netzel-Arnett et al. 2006; Leyvraz et al. 2005; also see prostasin chapter). These observations led to the hypothesis that matriptase may be part of a matriptase-prostasin proteolytic cascade in the epidermis (Netzel-Arnett et al. 2006). Moreover, this cascade has been shown to regulate different steps in terminal epidermal differentiation: tight junction formation, profilaggrin processing, epidermal lipid synthesis, and induction of desquamation through promoting expression or activation of kallikrein-related peptidases (KLKs) (for review see Szabo and Bugge 2011). It remains to be analyzed in more detail which target substrates of the matriptase-prostasin proteolytic cascade are involved in the different steps of epidermal differentiation. A number of studies during the last years, however, revealed that the matriptase-prostasin cascade is much more complex. While matriptase activates prostasin zymogen in the epidermis, prostasin is crucial for matriptase zymogen activation in the intestine and the placenta (Buzza et al. 2013; Szabo et al. 2016; cf. prostasin chapter).

Activity of matriptase during development is controlled by the transmembrane inhibitors hepatocyte growth factor activator inhibitor 1 (HAI-1) and HAI-2. Loss of HAI-1 or HAI-2 in mice results in embryonic lethality (Szabo et al. 2009; Tanaka et al. 2005). HAI-1 is essential for placental differentiation and overall embryonic and postnatal survival in mice (Nagaike et al. 2008; Szabo et al. 2007; Tanaka et al. 2005). Loss of HAI-2 is associated with defects in neural tube closure in mice (Szabo et al. 2009, 2012). Remarkably, all developmental defects in HAI-1- or HAI-2-deficient mice are rescued in whole or in part by reducing or eliminating the expression of matriptase (Szabo et al. 2009, 2012). These observations indicate that matriptase is not essential for placental development in mice but that its activity in the placenta needs to be regulated.

Activation of the single-chain matriptase zymogen is very complex requiring two sequential proteolytic processing events and the transient interaction with its cognate inhibitors HAI-1 and HAI-2 (Oberst et al. 2003, 2005; Nonboe et al. 2017). The first cleavage occurs after G149 in the SEA domain, releasing the enzyme from its transmembrane anchor. However, matriptase remains membrane-bound due to interactions with the cleaved domain and/or with HAI-1 and HAI-2. The second cleavage, which appears to be autocatalytic, occurs after R614 within the highly conserved R↓VVGG activation motif, converting the single-chain form into the active two-chain form. Co-expression of HAI-1 and HAI-2 has been shown to be necessary for matriptase expression, stability, and intracellular trafficking in vitro (Oberst et al. 2005; Larsen et al. 2013; Nonboe et al. 2017). Matriptase can be shed from the cell surface. Shed matriptase was identified originally in complex with an inhibitor HAI-I in human milk, and additional shed forms have been reported in conditioned media of cultured epithelial cell lines (reviewed in List et al. 2006a).

Apart from its physiological roles, matriptase has been extensively studied in the context of tumor progression. Expression of matriptase is upregulated in a variety of epithelial cancers including breast, prostate, ovarian, cervical, gastric, colon, renal cell, esophageal, and oral squamous cell carcinoma (reviewed in Tanabe and List 2017). List et al. demonstrated that matriptase causes strong proliferation of keratinocytes and formation of squamous cell carcinomas when only modestly overexpressed in the epidermis of transgenic mice (List et al. 2006b). Matriptase may promote tumor growth, invasion, and metastasis by converting pro-HGF and pro-uPA into its active forms, by degrading ECM components, or by activating PAR-2. Activation of matriptase by androgen-induced TMPRSS2 has been shown to promote prostate cancer tumor growth and metastasis in vitro and in a xenograft mouse model of prostate cancer (Ko et al. 2015).

Matriptase has furthermore been linked to other pathological processes. Impaired matriptase proteolytic activity due to mutations in the *ST14* gene was linked to a rare form of skin disease named autosomal recessive ichthyosis syndrome with hypotrichosis (ARIH) (condition of sparse hair) (List et al. 2007; reviewed in Antalis et al. 2011). Hypomorphic mice with 100-fold reduced expression levels of matriptase mRNA have been shown to phenocopy the key features of ARIH (List et al. 2007). Moreover, expression of matriptase is elevated in osteoarthritis, and the protease was identified as initiator of cartilage matrix degradation in osteoarthritis (Milner et al. 2010). In contrast, matriptase expression is significantly downregulated in intestinal tissues of patients with Crohn disease and ulcerative colitis (Netzel-Arnett et al. 2012).

8.2.3.2 TMPRSS6/Matriptase-2: Maintenance of Systemic Iron Homeostasis

TMPRSS6/matriptase-2 is composed of a transmembrane domain, followed by a SEA domain, two CUB domains, three LDLR, and a C-terminal trypsin-like serine protease domain (Wang et al. 2014). Matriptase-2 shares high structural and enzymatic similarities with matriptase, which contains four LDLR repeats instead of three. The matriptase-2 zymogen (90 kDa) undergoes autocatalytic cleavage at R567 within the R↓IVGG activation motif and remains membrane-anchored through a disulfide bond linking the pro- and catalytic domains (Stirnberg et al. 2010). Soluble forms of matriptase-2 were detected in the conditioned medium of transfected cells. Interestingly, shedding of matriptase-2 has been found to be due to cleavage at R404 and/or R437 within the second CUB domain and seems to be required for converting matriptase-2 into the active form via cleavage at R567. Activation of matriptase-2 was prevented in cells expressing matriptase-2 mutant R404E/R437E, which cannot be shed (Stirnberg et al. 2010).

TMPRSS6/matriptase-2 is expressed predominantly in the liver. TMPRSS6/matriptase-2-specific mRNA has been also detected to a lower extent in the kidney, spleen, lung, brain, mammary gland, testis, and uterus (reviewed in Ramsay et al. 2009; Wang et al. 2014).

TMPRSS6/matriptase-2 plays a key role in the maintenance of systemic iron homeostasis. Systemic iron homeostasis is maintained by regulating the iron absorption in the duodenum, by recycling of iron from senescent erythrocytes in macrophages, and by mobilizing stored iron in the liver. Increases in iron levels stimulate the production of the hepatic hormone hepcidin, which blocks iron export into the circulation by binding to and targeting the iron exporter ferroportin on the plasma membrane of duodenal enterocytes, macrophages, and hepatocytes for degradation. Hepcidin production is suppressed in the case of iron deficiency (reviewed in Ganz and Nemeth 2012 and Wang et al. 2014). Therefore, hepcidin is a key regulator of intestinal iron absorption, plasma iron concentrations, and tissue iron distribution. Lack of hepcidin causes juvenile hemochromatosis, a particularly severe form of iron overload disorder. In contrast, inappropriately high levels of hepcidin cause chronic inhibition of iron absorption and consequent anemia (Ganz and Nemeth 2012; Wang et al. 2014).

Mutations in the TMPRSS6 gene were found to cause increased hepcidin expression, which leads to iron-refractory iron-deficiency anemia (IRIDA) (Finberg et al. 2008). Similar phenotypes were also observed in mouse models deficient in TMPRSS6/matriptase-2 expression or with expression of a truncated protease form that lacks the catalytic domain (Du et al. 2008). Thus, results in mice and humans indicated that TMPRSS6/matriptase-2 is required to sense iron deficiency in mammals. Silvestri et al. demonstrated that TMPRSS6/matriptase-2 negatively regulates expression of hepcidin by cleaving the GPI-anchored protein hemojuvelin from the plasma membrane (Silvestri et al. 2008). Hemojuvelin is a coreceptor in the bone morphogenetic protein (BMP)/SMAD signaling pathway that upregulates hepcidin in response to increased iron (Babitt et al. 2006; Finberg et al. 2010). Cleavage of hemojuvelin by TMPRSS6/matriptase-2 downregulates BMP/SMAD signaling and inhibits hepcidin expression.

Studies have shown that TMPRSS6/matriptase-2 expression can be modulated by iron status. The underlying mechanism, however, is not fully understood. Beliveau et al. found that TMPRSS6/matriptase-2 is constitutively internalized from the plasma membrane in cell culture due to motifs within its cytoplasmic tail (Béliveau et al. 2011). Internalized TMPRSS6/matriptase-2 was detected in LAMP-2-labeled vesicles, suggesting that the protease transits to lysosomes, where it is degraded. However, it is still not clear whether this mechanism of protein degradation regulates TMPRSS6/matriptase-2 expression depending on the iron status.

8.2.3.3 Matriptase-3

Matriptase-3 was identified by bioinformatic analysis in 2005 (Szabo et al. 2005). The TMPRSS7 gene encoding matriptase-3 is located on human chromosome 3q13.2 and encodes a N-glycosylated TTSP of ca. 90 kDa expressed on the cell surface in vitro. Orthologs of the matriptase-3 gene are present in all vertebrates analyzed to date, including chimpanzee, dog, rodents, chicken, and fish (Szabo et al. 2005). In human tissues, matriptase-3 mRNA has been detected in the testis, ovary, brain, salivary gland, lung, and trachea (Bugge et al. 2009). The generation of matriptase-3-deficient mice has not been reported so far, and the physiological substrates and function of matriptase-3 are unknown.

8.2.3.4 Polyserase-1

Polyserase-1 (polyserine protease 1, also named TMPRSS9) is a unique TTSP with three tandem serine protease domains, of which two display catalytic activity (Cal et al. 2003). Polyserase-1 was originally cloned from human liver cDNA. The protease is widely expressed in mouse and human tissues. In addition, a shorter splice variant, termed serase-1B, which contains only the first of the three serine protease domains of polyserase-1, has been described in mice and humans, with its highest expression detected in liver, small intestine, pancreas, testes, and peripheral blood CD14+ and CD8+ cells (Okumura et al. 2006). The putative functional advantages derived from the complex structural organization of polyserase-1 and its functional significance remain unknown in both normal and pathological conditions.

8.2.4 Corin Family

8.2.4.1 Corin: The Pro-Atrial Natriuretic Peptide (pro-ANP) Activating Enzyme

Corin is the only member of the corin family and has a complex stem region (Fig. 8.1). At the turn of the millennium, corin was identified as a serine protease in the heart and as the physiological activator of atrial natriuretic peptide (ANP), also called ANF (atrial natriuretic factor) (Yan et al. 1999, 2000). ANP is an important hormone that regulates blood pressure and cardiac function by promoting natriuresis, diuresis, and vasodilation (Li et al. 2017, review). In cardiomyocytes, ANP is synthesized as a precursor protein pro-ANP that is stored in intracellular granules and converted to active ANP by corin upon secretion in response to high blood volume or pressure.

The human corin gene (*TMPRSS10*) located on chromosome 4p12–13 consists of 22 exons and spans ca. 200 kb, making it one of the largest protease genes in the human genome. Human corin consists of 1042 amino acids and includes an N-terminal cytoplasmic tail, a transmembrane domain, and an extracellular region that contains two frizzled domains, eight LDLA repeats, a SRCR domain, and a C-terminal serine protease domain (Li et al. 2017). Corin is the only serine protease containing frizzled-like domains. Corin zymogen is activated by cleavage at a conserved site between Arg801 and Ile802. In 2015, Chen et al. identified the proprotein convertase paired basic amino acid-cleaving enzyme 4 (PACE4), also designated as proprotein convertase subtilisin/kexin 6 (PCSK6), as the long-sought physiological activator of corin (Chen et al. 2015). Co-expression of PACE4 and corin enhanced corin activation in HEK293 cells. Moreover, knockout of PACE4 expression in mice led to impaired corin activation, decreased pro-ANP processing, and development of salt-sensitive hypertension.

Corin is extensively N-glycosylated (Liao et al. 2007; Gladysheva et al. 2008; Wang et al. 2015). Human corin has a predicted molecular mass of 116 kDa; however, native and recombinant corin appears as a protein of 200 kDa. Human and mouse corin have 19 and 16 predicted N-glycosylation sites in its extracellular domains, respectively. It has been shown that N-glycosylation at Asn697 in the SRCR domain and Asn1022 in the protease domain are required for corin cell surface expression and zymogen activation.

Corin is primarily expressed in cardiomyocytes. Furthermore, corin has been detected in the kidney, blood, and urine. The protease is shed from the cell surface of transient corin expressing HEK293 cells and from the surface of cardiomyocytes as three distinct soluble fragments of 180, 160, and 100 kDa, respectively, which represent activated two-chain forms linked by a disulfide bond containing the 40 kDa catalytic domain (Jiang et al. 2011). The metalloproteinase ADAM10 was shown to cleave corin in its juxtamembrane region to release the 180-kDa fragment, corresponding to the nearly entire extracellular region. In contrast, the 160- and 100-kDa fragments were generated by corin autocleavage at Arg164 in frizzled domain 1 and Arg427 in LDLA domain 5, respectively. Further studies revealed that the 180-kDa fragment exhibited the biological activity in processing pro-ANP, whereas the two other fragments had little activity (Li et al. 2017).

The presence of soluble corin in human blood indicates that shed corin can enter the circulation. Remarkably, corin remains active in the presence of human plasma, indicating that circulating protease inhibitors do not block corin activity. To date, no physiological corin inhibitors have been identified.

The importance of corin in regulating blood pressure has been shown in corin-deficient mice, which exhibited reduced sodium excretion and salt-sensitive hypertension due to impaired pro-ANP processing (Chan et al. 2005). Reduced corin expression has been detected in animal models of cardiomyopathies. Furthermore, decreased levels of circulating corin have been reported in patients with hypertension, pre-eclampsia, and cardiovascular diseases including acute myocardial infarction, heart failure, and stroke. Mutations in the corin gene that result in defects in intracellular trafficking of the protease, cell surface expression, and zymogen activation have been found in these patients (reviewed in Li et al. 2017). Moreover, a PCSK6 mutation with impaired corin activation has been identified in a hypertensive patient (Chen et al. 2015). Latest studies show that plasma corin concentrations provide a valuable prognostic marker for risk stratification of patients with acute myocardial infarction (AMI) and low levels of circulating corin are related with poor clinical outcomes (Zhou et al. 2016).

In addition to ANP, the mammalian natriuretic peptide family comprises two additional members: BNP (B-type or brain natriuretic peptide) and CNP (C-type natriuretic peptide). ANP and BNP are primarily expressed in the heart, while CNP is of endothelial origin and present in various tissues. pro-CNP is converted to active CNP by furin (Wu et al. 2003). It has been shown that both corin and furin cleave pro-BNP in vitro (Semenov et al. 2010). However, pro-BNP processing was not abolished in corin-deficient mice, indicating that corin is not essential for pro-BNP cleavage in vivo (Chen et al. 2015).

Corin expression has also been detected in noncardiac tissues including kidney and in human urine (Fang et al. 2013). In rat models of kidney disease, reduced renal corin expression was associated with sodium retention (Polzin et al. 2010). Latest studies show reduced urinary and renal corin levels in patients with chronic kidney disease (Fang et al. 2013). Further studies are necessary to fully understand the role of corin in regulating renal function and sodium homeostasis.

Curiously, corin-deficient mice appear to have a lighter coat color (dirty blond), and this phenotype depends on the agouti gene (Enshell-Seijffers et al. 2008).

In summary, corin is a key enzyme in the natriuretic peptide system, and corin defects may contribute to major diseases such as hypertension, heart failure, preeclampsia, and kidney disease.

8.3 Type I Transmembrane Serine Proteases

8.3.1 Tryptase Gamma 1

Tryptase gamma 1 (also known as transmembrane tryptase (TMT) and protease serine member S31 (PRSS31)) was identified in 1999 as a tryptase present in human and mouse mast cells that differs from all other known tryptases, which are soluble proteases, by containing a C-terminal transmembrane domain (type I transmembrane serine protease) (Wong et al. 1999). The physiological role(s) of tryptase gamma 1 are unknown. PRSS31-deficient mice exhibit no obvious developmental abnormalities but show markedly reduced experimental chronic obstructive pulmonary disease (COPD) and colitis compared to wild-type littermates, indicating a role of tryptase gamma 1 in mast cell-dependent inflammatory diseases (Hansbro et al. 2014).

8.4 GPI-Anchored Serine Proteases

8.4.1 Prostasin: Proteolytic and Non-Proteolytic Functions in Epithelial Development and Tissue Homeostasis

Prostasin was purified and characterized as an active soluble protease from human seminal fluid in 1994 (Yu et al. 1994). The protease is expressed in a variety of epithelial tissues with high expression in the prostate, bronchus, lung, and kidney in mouse and human. Prostasin is also known as channel-activating protease (CAP)-1 and was the first membrane serine protease found to activate the epithelial sodium channel (ENaC) (Vallet et al. 1997).

The *PRSS8* gene encoding prostasin is conserved in all vertebrate species examined. In humans, the *PRSS8* gene is located on chromosome 16p11.2 and encodes a protein of 343 amino acids. Prostasin is GPI-anchored in the cell surface and associates with lipid rafts (Chen et al. 2001b; Verghese et al. 2006). In polarized cells, prostasin is present on the apical membrane. Using substrate libraries, prostasin was shown to have a preference for polybasic substrates with R/K in P4, H/K/R in P3, basic or large hydrophobic amino acids in P2, and R/K in P1 (Shipway et al. 2004). No activity was seen with substrates containing isoleucine in P1′, providing an explanation for prostasin being not capable of undergoing autoactivation (Shipway et al. 2004). Prostasin zymogen conversion in the epidermis requires matriptase (Netzel-Arnett et al. 2006; cf. matriptase chapter). The crystal structure of the catalytic domain of prostasin has been solved (Rickert et al. 2008). Prostasin has been shown to be an essential regulator of the ENaC and thereby regulates the homeostasis of extracellular fluid volume, blood pressure, and intestinal sodium and water absorption (Frateschi et al. 2012; Planès et al. 2010). Prostasin cleaves the ENaC gamma subunit at K186 within the cleavage motif RKRK↓ distal to a furin cleavage site at R144 (RKRR↓) (Bruns et al. 2007). Dual cleavage of the gamma subunit releases a 43-amino acid inhibitory peptide and causes full activation of the channel, resulting in an increased cellular uptake of Na^+. Studies show that prostasin is highly expressed in cystic fibrosis airways and is a strong basal activator of ENaC in cystic fibrosis airway epithelial cells (Donaldson et al. 2002; Tong et al. 2004). Increased levels of soluble prostasin are also found in urine of hypertensive patients (Narikiyo et al. 2002). Prostasin may be released from the cell surface by an endogenous GPI-specific phospholipase D1 or via cleavage in its C-terminal hydrophobic domain (Yu et al. 1994; Verghese et al. 2006). Soluble prostasin purified from human seminal fluid terminates at R323 (Yu et al. 1994). On the other hand, prostasin expression is reduced in a number of cancers including prostate, breast, and colorectal cancers, and prostasin has been shown to inhibit prostate and breast cancer cell invasion in vitro (Chen et al. 2001a; Chen and Chai 2002; Bao et al. 2016).

Prostasin-deficient mice display impaired epidermal barrier function, abnormal hair follicle maturation, impaired profilaggrin processing, defects in tight junction formation, and fatal dehydration (Leyvraz et al. 2005; Netzel-Arnett et al. 2006; Szabo et al. 2016). Moreover, constitutive knockout of prostasin leads to embryonic lethality due to placental insufficiency (Hummler et al. 2013). The identical phenotypes of matriptase- and prostasin-deficient mice suggested that both proteases are components of one proteolytic cascade (reviewed in Netzel-Arnett et al. 2006; cf. matriptase chapter). This hypothesis, however, proved incompatible with studies demonstrating that prostasin acts upstream of matriptase in intestinal epithelial cells and in the placenta through matriptase zymogen activation (Buzza et al. 2013; Szabo et al. 2016). Thus, the prostasin-matriptase cascade turned out to be more complex than previously thought.

Paradoxically, a number of studies revealed that prostasin requires neither zymogen conversion nor catalytic activity to execute its essential functions in the epidermal development (Peters et al. 2014; Friis et al. 2016). Already in 2006/2007, studies by Andreasen et al. and Bruns et al. reported that catalytically inactive prostasin mutant S238A was able to activate ENaC in *Xenopus* oocytes in vitro, indicating that the catalytic activity of prostasin appears to be dispensable for cleavage of the ENaC gamma subunit (Andreasen et al. 2006; Bruns et al. 2007). But cell surface expression of prostasin via the GPI anchor was essential for ENaC activation. More recent studies in mouse models expressing catalytically inactive prostasin mutant S238A or zymogen-locked mutant R44Q revealed that prostasin supports both epidermal development and long-term survival in a non-catalytic manner (Peters et al. 2014; Friis et al. 2016). In contrast, prostasin proteolytic activity was found to be crucial for matriptase zymogen activation in the placenta (Szabo et al. 2016). In addition, mice expressing catalytically inactive or zymogen-locked prostasin displayed impaired hair follicle development and delayed skin wound healing, indicating that some of the physiological functions of prostasin require its proteolytic activity. The V170D (low enzymatic activity) and the G54-P57 deletion mutations of prostasin have been identified in mouse frizzy (fr) and rat hairless (frCR) animals, respectively, and have been proposed to be responsible for their skin phenotypes (Spacek et al. 2010; Szabo et al. 2012).

Taken together, prostasin is unique among trypsin-like proteases in that it has essential functions as an enzymatically active protease as well as an enzymatically inactive zymogen. The specific mechanism(s) by which catalytically inactive prostasin supports epidermal development remain to be established. Matriptase and prostasin form a reciprocal zymogen activation complex *in vitro* that results in the formation of both active matriptase and active prostasin (Friis et al. 2013).

8.4.2 Testisin: Roles in Sperm Maturation and Motility

Testisin (also referred to as PRSS21, testicular protease 5 (TESP5), or eosinophil serine protease 1 (ESP-1)) is aberrantly expressed in male germ cells and sperm and is also expressed in microvascular endothelial cells and in eosinophils (Hooper et al. 1999; Inoue et al. 1998; Aimes et al. 2003). The protease was originally cloned from human eosinophils and from HeLa cells (Inoue et al. 1998; Hooper et al. 1998). Testisin expression is lost in testicular germ cell tumors (Hooper et al. 1999) and is found overexpressed in ovarian tumors (Shigemasa et al. 2000). The testisin gene, *PRSS21*, is located on human chromosome 16p13.3. Several isoforms of human testisin have been identified that are generated by alternative pre-mRNA splicing (Hooper et al. 2000; Inoue et al. 1998).

Testisin is synthesized as a 43-kDa precursor in the testis, and the zymogen is converted into the 42- and 41-kDa active enzymes during sperm transport in the epididymis (Honda et al. 2002). Testisin is anchored to the membrane via a GPI moiety at its carboxy terminus and is included into lipid rafts on the sperm membrane (Honda et al. 2002). Unlike other membrane-anchored serine proteases, testisin has not been found to be naturally shed from the plasma membrane, but the protease can be release from cells in vitro using exogenous bacterial phosphatidylinositol-specific phospholipase C (Honda et al. 2002).

Mammalian fertilization requires sperm to penetrate the cumulus matrix surrounding the oocyte to reach the zona pellucida (ZP), binding and invasion of the ZP, and finally fusion of the sperm and oocyte plasma membranes. The serine protease acrosin has been long believed to participate in limited proteolysis of ZP, thus enabling sperm to penetrate the egg coat. However, acrosin-deficient mice were fully fertile, although they displayed delayed sperm penetration of the ZP at the early stage of fertilization in vitro (Baba et al. 1994), indicating that additional serine protease(s) play important roles in the regulation of male fertility. Mice deficient in testisin expression are fertile, too, but display deformed spermatozoa with an increased tendency toward decapitation and reduced motility (Netzel-Arnett et al. 2009). Testisin was found to direct murine sperm cell maturation and sperm-fertilizing ability during passage of spermatozoa through the epididymis to their site of temporary storage in the cauda (Netzel-Arnett et al. 2009). Combined knockout of acrosin and testisin in mice impairs fertility in vivo and causes complete loss of fertilization ability in vitro, which suggests that sperm trypsin-like activity is indispensable for in vitro fertilization but not particularly for fertilization in vivo in mice. Interestingly, these data suggest that the female reproductive tract partially compensates for the loss of the sperm function, presumably due to the presence of an acrosin-/testisin-like protease (Kawano et al. 2010; reviewed in Szabo and Bugge 2011).

Little is known regarding specific physiological substrates of testisin during epididymal sperm maturation and initiation of sperm motility as well as during testicular and ovarian tumor formation and progression. Recently, testisin was shown to be capable of activating PAR-2 in vitro (Driesbaugh et al. 2015). PAR-2 activation has been associated with the regulation of sperm motility following trypsin activation (Miyata et al. 2012). However, activation of PAR-2 by testisin in vivo and a possible role in sperm motility remain to be demonstrated.

8.5 MASPs in Viral Infections

The majority of viral fusion proteins is synthesized as precursors and requires processing by a host cell protease to trigger fusion of the viral lipid envelope and cellular membranes in order to release the virus genome into the host cell. For a large number of viruses, cleavage of the fusion protein occurs either at a single arginine (R↓) by trypsin-like proteases or at a multibasic motif of the consensus sequence R-X-R/K-R↓ by ubiquitous expressed subtilisin-like proteases such as furin and proprotein convertase 5/6 (PC5/6) (see Chap. 9).

The first report of proteolytic activation of viral fusion proteins was published in 1973 in a study using Sendai virus. It was demonstrated that the viral glycoprotein F is synthesized as an inactive precursor that is converted into its biological active form due to cleavage by a host cell protease and that F cleavage is a prerequisite for infectivity and multicycle virus replication (Homma and Ohuchi 1973; Scheid and Choppin 1974). In the following years, striking differences in glycoprotein activation have been observed with Newcastle disease virus and avian influenza viruses, which proved to be important determinants of the pathogenicity of these viruses (Nagai et al. 1976; Bosch et al. 1981). In 1992, the proprotein convertase furin was identified as protease that activates the hemagglutinin of the highly pathogenic avian influenza virus (HPAIV) strains at multibasic motifs (Stieneke-Gröber et al. 1992; reviewed by W. Garten in another chapter of this book) (Fig. 8.3a, b). Furin which is expressed in all tissues is responsible for the systemic infection typical for these viruses. However, less was known about the identity of the protease(s) that support HA cleavage at a single arginine residue.

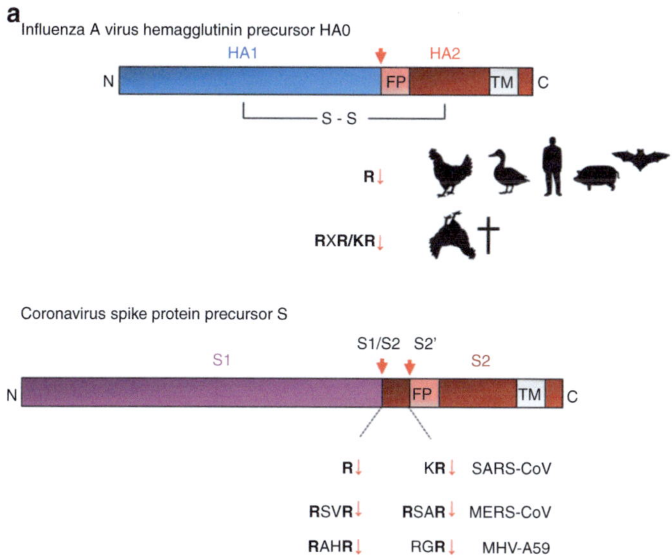

Fig. 8.3 Activation of the envelope glycoprotein hemagglutinin HA of influenza A virus and spike protein S of CoV by host cell proteases. (**a**) Schematic illustration of the precursor proteins HA0 and S of influenza A virus and coronaviruses (CoVs), respectively, and cleavage sites (red arrows). HA0 is cleaved into HA1 and HA2; both subunits remain linked by a disulfide bond. HA0 of highly pathogenic avian influenza A viruses is activated at a multibasic cleavage site, while HA0 of low pathogenic avian influenza A viruses and mammalian influenza A viruses is cleaved at a single basic residue. CoV S protein is cleaved at two distinct sites at monobasic motifs or the minimal furin consensus motif RXXR. The S1 and S2 domains are not held by disulfide bonds but remain associated non-covalently (Millet and Whittaker 2015). FP, fusion peptide. TM, transmembrane domain. Basic amino acids crucial for cleavage by relevant proteases are written in bold letters. (**b**) Compartmentalization of activation of HA and S by host cell proteases. The envelope proteins HA and S mediate virus entry into cells through receptor binding and fusion of the viral envelope with host cell membranes. Fusion delivers the viral genome into the host cell and initiates viral replication and generation of progeny virions. Newly synthesized virions are finally released via budding at the plasma membrane (influenza A virus) or at the ER-Golgi intermediate compartment (ERGIC) and subsequent exocytosis of new virions at the cell surface (CoV). HA and S require cleavage by host cell proteases to gain their fusion capacity. In the case of influenza A virus, newly synthesized HA is cleaved during its transport to the plasma membrane by furin (multibasic cleavage site) or TMPRSS2 (monobasic cleavage site) in the trans-Golgi network (TGN) or at the cell surface by HAT during assembly and budding of progeny virus. Thus, virus containing cleaved HA is released from infected cells. HA can also be cleaved by HAT on the cell surface prior to entry into a new cell. In contrast, TMPRSS2 present on the cell surface does not support HA cleavage. It remains unclear whether HAT is enzymatically active within the cell and thus can cleave HA already during its transit to the plasma membrane. Most CoVs, including SARS- and human 229E-CoVs, are released with non-cleaved S from the infected cells, indicating that S is not activated during its transit through the exocytic pathway. Therefore, the S protein of CoV is activated upon entry into the host cell. Entry can take place via fusion at or close to the plasma membrane or in early endosomes ("early entry") and may be pH-independent or via late fusion in late endosomes ("late entry") in a low pH-dependent manner. Early or late fusion seems to be dependent on the protease(s) that cleave S at the S2' site. Cleavage of S by furin and TTSPs is believed to support an early entry (e.g., MERS-CoV), whereas activation of S by cathepsins supports late entry via fusion in late endosomes (e.g., mouse hepatitis virus (MHV) A59) (Burkard et al. 2014; Millet and Whittaker 2015; Park et al. 2016). Newly synthesized S protein of some CoV, including MERS-CoV, has been reported to be cleaved by furin or TTSPs in the exocytic pathway, but the role of S cleavage in the secretory pathway for virus-cell and cell-cell fusion remains to be investigated in more detail. It also remains unknown whether furin and/or TTSPs such as TMPRSS2 are present as enzymatically active enzymes in the ERGIC and might cleave S in this compartment

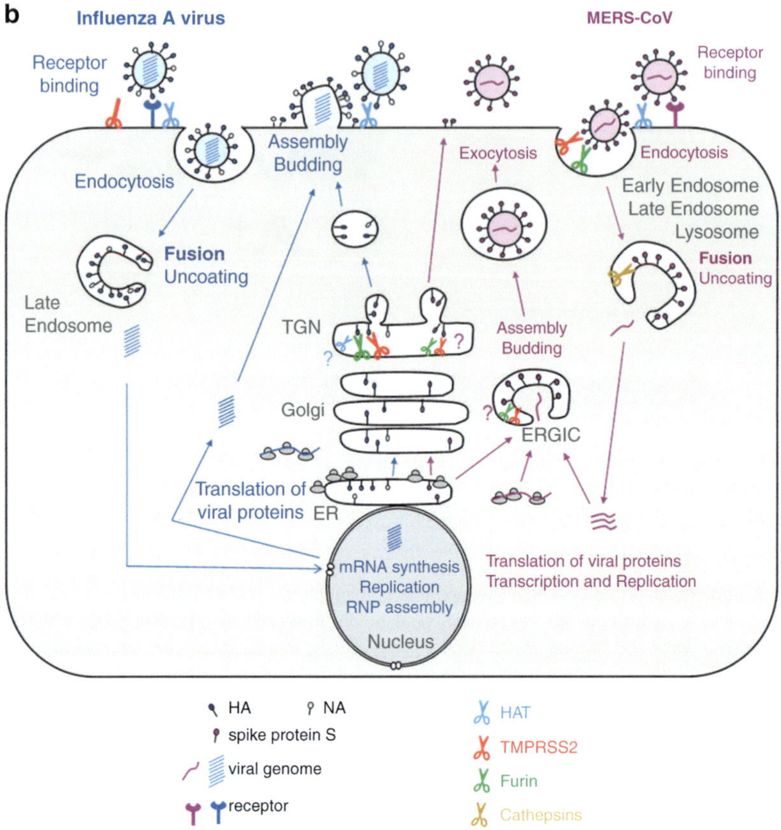

Fig. 8.3 (continued)

8.5.1 The Search for Virus-Activating Proteases in the Airway Epithelium

Human and mammalian influenza A and B viruses and low pathogenic avian influenza A viruses (LPAIV) are activated at monobasic HA cleavage site motifs by trypsin in vitro (Klenk et al. 1975; Lazarowitz and Choppin 1975). A number of soluble trypsin-like proteases isolated from rat and swine lung, such as tryptase Clara, mini-plasmin, or tryptase TC30, as well as a blood clotting factor Xa-homologous protease in embryonated chicken eggs have also been found to activate HA of these viruses as described in a chapter by H. Kido in this book. However, the genetic identity is still unknown for many of these enzymes, and it remains unclear whether they play a role in in vivo infection. Relevant HA-activating proteases in the human airways were unknown for a long time. Cleavage of HA by soluble proteases such as trypsin in cell culture takes place outside the cells during assembly and budding of new virions when HA is present on the plasma membrane or after progeny virus is released from the infected cell. Thus, it was believed for a long time that HA with a monobasic cleavage site is activated extracellularly and, therefore, differs from HA of HPAIV, which is cleaved in the TGN by furin.

In 2006, a number of proteases that possess trypsin-like activity were cloned from primary human tracheobronchial epithelial (HTBE) cells in search for human HA-activating proteases (Böttcher et al. 2006). Among a couple of candidates, the TTSPs HAT and TMPRSS2 were demonstrated to activate HA with monobasic cleavage site in vitro and thus were identified as the first human influenza virus-activating proteases in the respiratory tract (Böttcher et al. 2006) (Table 8.2).

Subsequently, HAT and TMPRSS2 were shown to activate also other respiratory viruses at monobasic cleavage site motifs in vitro, such as human coronaviruses (CoVs) including SARS-CoV and MERS-CoV, the human metapneumovirus, and human parainfluenza viruses (Shirogane et al. 2008; Glowacka et al. 2011; Matsuyama et al. 2010; Shulla et al. 2011; Shirato et al. 2013; Millet and Whittaker 2015; Abe et al. 2013). In addition, further TTSPs were tested for their ability to activate viral fusion proteins, and TMPRSS4, DESC1, TMPRSS11A, MSPL/TMPRSS13, and matriptase have been found to activate influenza virus HA as well as SARS-CoV and MERS-CoV spike protein S in cell cultures (Table 8.2). Other TTSPs, such as hepsin, TMPRSS3, matriptase-2, matriptase-3, or corin, did not support proteolytic activation of influenza viruses or CoV in cell culture, demonstrating that only certain trypsin-like proteases present in the airway epithelium support activation and spread of respiratory viruses. TMPRSS2-homologous proteases have been identified in swine, chicken, and mouse and were shown to be capable of activating HA at a single arginine, suggesting that homologous proteases are involved in HA cleavage in different host species (Bertram et al. 2012; Peitsch et al. 2014; Tarnow et al. 2014).

It has long been known that HAs with monobasic cleavage site differ in their sensitivity to host proteases. For instance, HAs of A/WSN/33 (H1N1), A/PR8/34 (H1N1), A/Asia/1/57 (H2N2), and A/duck/Ukraine/1/63 (H3N8) are activated by trypsin, plasmin, kallikrein, and uPA in vitro, whereas HA of A/chicken/Germany/49 (H10N7) is activated only by trypsin (Lazarowitz et al. 1973; Scheiblauer et al. 1992). More recent studies on cleavage of human influenza A virus HA by kallikrein (KLK) 5 and 12 showed that H1 is cleaved by both KLK5 and KLK12 in vitro, whereas H2 and H3 are only cleaved by KLK12 and KLK5 , respectively (Hamilton and Whittaker 2013). Furthermore, the sensitivity of 16 HA subtypes to cleavage by TMPRSS2, HAT, and pancreatic trypsin was demonstrated to vary significantly among the different subtypes (Galloway et al. 2013; see also Chap. 1). Interestingly, studies in TMPRSS2-deficient mouse models demonstrated that differences in the sensitivity of HA to cleavage by host proteases may affect influenza virus spread and pathogenesis (Hatesuer et al. 2013; Tarnow et al. 2014; Sakai et al. 2014) (see below). The mechanisms underlying protease specificity of HA with monobasic cleavage site, however, are still unknown and may be related to the structure and exposure of the cleavage site loop or steric hindrance by adjacent carbohydrate moieties (Kawaoka et al. 1984; Sakai et al. 2015).

8.5.2 Subcellular Localization and Time Point of Virus Activation by TTSPs

The identification of membrane-bound HA-activating proteases raised the question where in the host cell and at which step of the viral replication cycle the viral glycoproteins are cleaved by TTSPs. The subcellular localization and enzymatic activity of HAT and TMPRSS2 in airway epithelial cells have not been investigated in more detail so far but have been studied in MDCK cells that proved to be a suitable model system (Böttcher et al. 2009). Interestingly, it was demonstrated that activation of influenza virus HA by HAT and TMPRSS2 differs in subcellular localization and can take place at different steps of the viral life cycle (Böttcher et al. 2009; Böttcher-Friebertshäuser et al. 2010, 2013; Fig. 8.3b). Incubation of HAT-expressing MDCK cells with fluorogenic peptides to measure the protease activity on the cell surface revealed that HAT is present as an enzymatically active protease on the cell surface. In addition, it was shown that HAT is capable of cleaving newly synthesized HA0, probably during assembly and budding of new virions on the plasma membrane, as well as HA0 of incoming influenza virus at the stage of entry during attachment to a new host cell. In contrast, only poor if any enzymatic activity of TMPRSS2 was measured on the surface of TMPRSS2-expressing MDCK cells, and the cells were not able to support activation of incoming influenza virions containing HA0, although TMPRSS2 was present as both zymogen and mature form on the surface of MDCK cells (Böttcher-Friebertshäuser et al. 2010). Cleavage of HA was found to take place intracellularly during its transport through the secretory pathway to the plasma membrane (Fig. 8.3b). Immunofluorescence studies of transient TMPRSS2 expression in mammalian cells revealed that the protease accumulates in the TGN, where it co-localizes with furin, suggesting that HA cleavage by TMPRSS2 and furin occurs in the same cellular compartment (Böttcher-Friebertshäuser et al. 2013). These data were in agreement with earlier studies by Zhirnov et al. that showed that cleavage of HA in human respiratory or intestinal epithelial cells occurs intracellularly and is performed by cell-associated proteases (Zhirnov et al. 2002; Zhirnov and Klenk 2003). These studies provided further evidence that cleavage of influenza virus HA can occur at different steps of the viral life cycle: during transport of HA along the secretory pathway to the cell surface, on the plasma membrane during assembly and budding, and late in infection upon entry into new cells (Boycott et al. 1994; Zhirnov et al. 2002; Böttcher-Friebertshäuser et al. 2010). However, shedding of TTSPs may be enhanced under stress, and soluble forms might contribute to virus activation under such conditions. For example, enhanced shedding of TMPRSS2 and HAT from human nasal epithelial cells and hence enhanced influenza virus replication have been observed upon exposure to ozone in vitro (Kesic et al. 2012).

Studies on proteolytic activation of the spike protein S of human CoV including SARS-CoV and MERS-CoV by HAT and TMPRSS2 revealed also that viruses may have different options to be activated by host cell proteases during the viral replication cycle (Fig. 8.3b). CoV S proteins possess two distinct cleavage sites, S1/S2 and S2', and can be cleaved by a number of proteases, including furin, cathepsin L, and trypsin, and trypsin-like proteases such as the TTSP HAT, TMPRSS2, DESC1, and MSPL (reviewed in Millet and Whittaker 2015, and Chap. 4 this book; see also Fig. 8.3). Cleavage of S by cathepsins occurs in late endosomes or lysosomes and is pH dependent, whereas activation by TTSPs may support cathepsin- and low pH-independent CoV entry via fusion in early endosomes in vitro (Burkard et al. 2014; Millet and Whittaker 2015; Park et al. 2016). Both HAT and TMPRSS2 are believed to cleave S at or close to the cell surface (for review see Chap. 4). It remains to be investigated whether TMPRSS2 and/or HAT are present (and active) in endosomes. It further remains to be investigated why TMPRSS2 may cleave CoV S but not influenza virus HA at the cell surface. For some CoV, receptor binding has been shown to be required for proteolytic activation of S. Treatment of SARS-CoV particles with trypsin prior to binding to its receptor angiotensin I-converting enzyme 2 (ACE2) inactivates the virions in vitro, whereas SARS-CoV particles bound to its receptor are activated by trypsin for membrane fusion (cf. Chap. 4). Moreover, TMPRSS2 has been shown to cleave ACE2 and thereby to increase SARS-CoV S uptake into ACE2-expressing cells in vitro (Heurich et al. 2014). Thus, receptor binding of S or interactions of TMPRSS2 and ACE2 might trigger S cleavage by TMPRSS2 at the plasma membrane. In general, CoV are believed to be proteolytically activated upon entry into cells, but newly synthesized S may also be activated prior to virus release from the infected cell (Fig. 8.3b). The subcellular compartmentalization of S cleavage during its transit through the exocytic pathway and/or virus assembly and budding at the ER-Golgi intermediate compartment (ERGIC) (reviewed in Ujike and Taguchi 2015), however, are not understood in detail.

8.5.3 Identification of TMPRSS2 as Host Cell Factor Essential for Influenza A Virus Activation and Spread in Mice

In three recent independent studies, TMPRSS2 was identified as the solely HA-activating protease and as a host cell factor essential for spread and pneumotropism of human H1N1 and H7N9 influenza A viruses in mice (Hatesuer et al. 2013; Tarnow et al. 2014; Sakai et al. 2014). Intriguingly, knockout of TMPRSS2 expression prevented virus activation and consequently spread into the lungs and thereby protected the animals from influenza pathogenesis, whereas wild-type littermates succumbed to severe infection. These studies also revealed that other TTSPs that activate H1N1 virus in vitro, such as HAT/TMPRSS11D, TMPRSS4, DESC1, or MSPL, do not support HA activation in mice in vivo. For HAT/TMPRSS11D, this can be explained by the fact that the protease is expressed in the upper airways, trachea, and bronchi of mice, but is not present in the lungs and, thus, cannot support influenza virus pneumotropism (Sales et al. 2011; Tarnow et al. 2014). The same is probably true for DESC1. However, TMPRSS4 and MSPL/TMPRSS13 are present in lung tissue (Kühn et al. 2016; Kim et al. 2001), and it remains unclear why both proteases do not support activation of H1N1 and H7N9 influenza virus in the airways of TMPRSS2-deficient mice. Noteworthy, replication of human H3N2 influenza A virus and influenza B virus (IBV) was almost independent of TMPRSS2 expression in mice, indicating that

H3 and IBV HA can be activated by additional host proteases in contrast to H1 and H7 (Tarnow et al. 2014; Sakai et al. 2016). In a recent study, knockout of both TMPRSS4 and TMPRSS2 caused reduced body weight loss and mortality upon H3N2 influenza A virus infection in mice in comparison with wild-type animals, indicating that both proteases contribute to activation of H3N2 virus in mice (Kühn et al. 2016). Nonetheless, H3N2 influenza A virus was still proteolytically activated in TMPRSS2-TMPRSS4-knockout mice and caused severe disease with 30% mortality, indicating that another H3-cleaving protease(s) is present in murine airways. MSPL/TMPRSS13 might be a potential candidate.

Taken together, these studies demonstrated for the first time that expression of the appropriate HA cleaving protease along the respiratory tract is essential for influenza virus pneumotropism and pathogenicity in a mammalian host. In addition, TMPRSS2 emerged as a potential drug target for influenza treatment. It will be very interesting to analyze the role of TMPRSS2 and further TTSPs in activation and pathogenicity of other respiratory viruses including SARS-CoV and MERS-CoV and human parainfluenza viruses using suitable mouse models. Interestingly, a single nucleotide polymorphism in the TMPRSS2 gene that results in higher TMPRSS2 expression has been associated with increased susceptibility to H1N1 and H7N9 influenza virus and higher risk of severe infection, suggesting that TMPRSS2 may play a crucial role in influenza virus activation also in humans (Cheng et al. 2015). This view is supported by the observation that knockdown of TMPRSS2 expression in the human airway epithelial cell line Calu-3 strongly suppressed activation and multicycle replication of human H1N1 influenza A viruses (Böttcher-Friebertshäuser et al. 2011). Interestingly, knockdown of TMPRSS2 also strongly suppressed H3N2 virus replication in Calu-3 cells, suggesting that the differences in protease specificity of influenza virus HA observed in mice might be less pronounced in humans. However, further studies are needed to understand the role of TMPRSS2 in influenza virus activation in humans.

8.5.4 Activation of Viral Fusion Proteins with Di-/Multibasic Cleavage Site Motifs by TTSPs

Other TTSPs have been found to activate viral fusion proteins at di- or multibasic amino acid motif. MSPL/TMPRSS13 and the hepsin-related protease TMPRSS12 were able to activate avian H5N2 influenza virus and avian metapneumovirus, respectively, at multibasic cleavage site motifs (Okumura et al. 2010; Yun et al. 2016). This may be relevant particularly for unusual di- or multibasic cleavage site motifs that are not cleaved by furin. H9N2 viruses in Asia and the Middle East have acquired dibasic cleavage site motif R-S-S/R-R that are not activated by furin. A study demonstrated that H9 with R-S-S/R-R at the cleavage site can be activated by matriptase in addition to TMPRSS2 and HAT in vitro (Baron et al. 2013). Matriptase is widely expressed in multiple epithelial tissues and, therefore, may affect H9N2 virus spread, tissue tropism, and pathogenicity. Nephrotropism of H9N2 virus has been described in chickens, and matriptase has been suggested to contribute to H9N2 virus replication in the kidney of chickens (Baron et al. 2013). Some H9N2 isolates have been reported to cause lethal infections in mice with virus replication in the lung and brain (Guo et al. 2000; Li et al. 2012).

Concluding Remarks and Key Research Questions

Described for the first time two decades ago, a number of MASPs have already been established as important regulators in mammalian physiology. However, for many MASPs the physiological functions and/or substrates are not fully characterized or still unknown at all. The TTSP TMPRSS2 was identified as influenza A virus-activating protease in human airway cells and in mice and as host cell factor essential for pneumotropism and pathogenesis of certain influenza A virus strains infecting humans in mice. The role of TMPRSS2 in activation of further respiratory viruses in vivo remains to be investigated, and its important role in influenza virus activation in humans needs to be demonstrated. As mentioned above, a number of TTSPs that have been shown to activate HA in vitro and to be present in the respiratory tract, including TMPRSS4 and TMPRSS13, were not able to compensate for the lack of TMPRSS2 expression in influenza virus activation in mice. Thus, if there is some redundancy among MASPs in processing of physiological substrates, and this is very likely, it is not the case for cleavage of HA. The underlying reasons are unclear so far. Detailed information on the substrate specificity of TMPRSS2 or TMPRSS4 is missing due to the lack of suitable systems for expression and purification of these proteases and available structural information of the protease domains. To date, the crystal structure of the protease domain has been solved for hepsin, DESC1, enteropeptidase, matriptase, and prostasin. Moreover, differences in (1) subcellular compartmentalization and activity, (2) expression levels, and/or (3) distribution in different airway cell types of TMPRSS2 in comparison with other TTSPs might account for the observed differences in activation of influenza A viruses. In particular, the compartmentalization of MASP activity should be investigated in more detail in future studies. MASPs are predicted to act as active enzymes on the cell surface, but studies on activation of influenza A virus HA by TMPRSS2 showed that cleavage takes place intracellularly, probably in the TGN. Thus, some MASPs may process their substrates already (or even exclusively) in intracellular compartments. Expression and enzymatic activity of MASPs in endosomes and lysosomes have not been studied so far but may play a role in processing of both physiological substrates and viral fusion proteins. Furthermore, the role of soluble MASP activity due to shedding of the catalytic domain is poorly understood.

Dysregulated MASP activities are associated with a number of pathophysiological processes, and specific inhibitors may provide promising pharmaceutical tools for the treatment of cancer, iron overload, or respiratory diseases. Potent inhibitors for some MASPs have already been developed (for review see Steinmetzer and Hardes, this book), but no inhibitor of host proteases has been approved for the treatment of virus infections to date. The development of highly selective MASP inhibitors is hampered by the lack of available crystal structures of the catalytic domain as well as knowledge on the role of the different domains of the stem region in substrate specificity, protease conformation, and protein-protein interactions. It remains to be analyzed whether some TTSPs are upregulated during virus infection and thereby may support enhanced virus activation and/or organ tropism. Moreover, it will be interesting to analyze whether dysregulation of MASPs contributes to the susceptibility to virus infection.

The protection of TMPRSS2 knockout mice from influenza A virus pathogenesis strikingly demonstrated the crucial role of virus activation for viral spread in the host. However, we are just beginning to understand in more detail which roles MASPs may play in virus activation, spread, and organ tropism on one hand and whether the physiological functions of these enzymes can be suppressed or compensated during an acute virus infection in order to block virus multiplication by using protease inhibitors on the other hand.

References

Abe M, Tahara M, Sakai K, Yamaguchi H, Kanou K, Shirato K, Kawase M, Noda M, Kimura H, Matsuyama S, Fukuhara H, Mizuta K, Maenaka K, Ami Y, Esumi M, Kato A, Takeda M. TMPRSS2 is an activating protease for respiratory parainfluenza viruses. J Virol. 2013;87:11930–5.

Afar DE, Vivanco I, Hubert RS, Kuo J, Chen E, Saffran DC, Raitano AB, Jakobovits A. Catalytic cleavage of the androgen-regulated TMPRSS2 protease results in its secretion by prostate and prostate cancer epithelia. Cancer Res. 2001;61:1686–92.

Aimes RT, Zijlstra A, Hooper JD, Ogbourne SM, Sit ML, Fuchs S, Gotley DC, Quigley JP, Antalis TM. Endothelial cell serine proteases expressed during vascular morphogenesis and angiogenesis. Thromb Haemost. 2003;89:561–72.

Andreasen D, Vuagniaux G, Fowler-Jaeger N, Hummler E, Rossier BC. Activation of epithelial sodium channels by mouse channel activating proteases (mCAP) expressed in Xenopus oocytes requires catalytic activity of mCAP3 and mCAP2 but not mCAP1. J Am Soc Nephrol. 2006;17:968–76.

Antalis TM, Bugge TH, Wu Q. Membrane-anchored serine proteases in health and disease. Prog Mol Biol Transl Sci. 2011;99:1–50.

Appel LF, Prout M, Abu-Shumays R, Hammonds A, Garbe JC, Fristrom D, Fristrom J. The Drosophila Stubble-stubbloid gene encodes an apparent transmembrane serine protease required for epithelial morphogenesis. Proc Natl Acad Sci U S A. 1993;90:4937–41.

Baba T, Azuma S, Kashiwabara S, Toyoda Y. Sperm from mice carrying a targeted mutation of the acrosin gene can penetrate the oocyte zona pellucida and effect fertilization. J Biol Chem. 1994;269:31845–9.

Babitt JL, Huang FW, Wrighting DM, Xia Y, Sidis Y, Samad TA, Campagna JA, Chung RT, Schneyer AL, Woolf CJ, Andrews NC, Lin HY. Bone morphogenetic protein signaling by hemojuvelin regulates hepcidin expression. Nat Genet. 2006;38:531–9.

Bao Y, Li K, Guo Y, Wang Q, Li Z, Yang Y, Chen Z, Wang J, Zhao W, Zhang H, Chen J, Dong H, Shen K, Diamond AM, Yang W. Tumor suppressor PRSS8 targets Sphk1/S1P/Stat3/Akt signaling in colorectal cancer. Oncotarget. 2016;7:26780–92.

Baron J, Tarnow C, Mayoli-Nüssle D, Schilling E, Meyer D, Hammami M, Schwalm F, Steinmetzer T, Guan Y, Garten W, Klenk HD, Böttcher-Friebertshäuser E. Matriptase, HAT, and TMPRSS2 activate the hemagglutinin of H9N2 influenza A viruses. J Virol. 2013;87:1811–20.

Beaulieu A, Gravel É, Cloutier A, Marois I, Colombo É, Désilets A, Verreault C, Leduc R, Marsault É, Richter MV. Matriptase proteolytically activates influenza virus and promotes multicycle replication in the human airway epithelium. J Virol. 2013;87:4237–51.

Béliveau F, Brulé C, Désilets A, Zimmerman B, Laporte SA, Lavoie CL, Leduc R. Essential role of endocytosis of the type II transmembrane serine protease TMPRSS6 in regulating its functionality. J Biol Chem. 2011;286:29035–43.

Ben-Yosef T, Wattenhofer M, Riazuddin S, Ahmed ZM, Scott HS, Kudoh J, Shibuya K, Antonarakis SE, Bonne-Tamir B, Radhakrishna U, Naz S, Ahmed Z, Riazuddin S, Pandya A, Nance WE, Wilcox ER, Friedman TB, Morell RJ. Novel mutations of TMPRSS3 in four DFNB8/B10 families segregating congenital autosomal recessive deafness. J Med Genet. 2001;38:396–400.

Bertram S, Glowacka I, Blazejewska P, Soilleux E, Allen P, Danisch S, Steffen I, Choi SY, Park Y, Schneider H, Schughart K, Pöhlmann S. TMPRSS2 and TMPRSS4 facilitate trypsin-independent spread of influenza virus in Caco-2 cells. J Virol. 2010;84:10016–25.

Bertram S, Glowacka I, Müller MA, Lavender H, Gnirss K, Nehlmeier I, Niemeyer D, He Y, Simmons G, Drosten C, Soilleux EJ, Jahn O, Steffen I, Pöhlmann S. Cleavage and activation of the severe acute respiratory syndrome coronavirus spike protein by human airway trypsin-like protease. J Virol. 2011;85:13363–72.

Bertram S, Heurich A, Lavender H, Gierer S, Danisch S, Perin P, Lucas JM, Nelson PS, Pöhlmann S, Soilleux EJ. Influenza and SARS-coronavirus activating proteases TMPRSS2 and HAT are expressed at multiple sites in human respiratory and gastrointestinal tracts. PLoS One. 2012;7:e35876.

Bertram S, Dijkman R, Habjan M, Heurich A, Gierer S, Glowacka I, Welsch K, Winkler M, Schneider H, Hofmann-Winkler H, Thiel V, Pöhlmann S. TMPRSS2 activates the human coronavirus 229E for cathepsin-independent host cell entry and is expressed in viral target cells in the respiratory epithelium. J Virol. 2013;87:6150–60.

Bosch FX, Garten W, Klenk HD, Rott R. Proteolytic cleavage of influenza virus hemagglutinins: primary structure of the connecting peptide between HA1 and HA2 determines proteolytic cleavability and pathogenicity of Avian influenza viruses. Virology. 1981;113:725–35.

Böttcher E, Matrosovich T, Beyerle M, Klenk HD, Garten W, Matrosovich M. Proteolytic activation of influenza viruses by serine proteases TMPRSS2 and HAT from human airway epithelium. J Virol. 2006;80:9896–8.

Böttcher E, Freuer C, Steinmetzer T, Klenk HD, Garten W. MDCK cells that express proteases TMPRSS2 and HAT provide a cell system to propagate influenza viruses in the absence of trypsin and to study cleavage of HA and its inhibition. Vaccine. 2009;27:6324–9.

Böttcher-Friebertshäuser E, Freuer C, Sielaff F, Schmidt S, Eickmann M, Uhlendorff J, Steinmetzer T, Klenk HD, Garten W. Cleavage of influenza virus hemagglutinin by airway proteases TMPRSS2 and HAT differs in subcellular localization and susceptibility to protease inhibitors. J Virol. 2010;84:5605–14.

Böttcher-Friebertshäuser E, Stein DA, Klenk HD, Garten W. Inhibition of influenza virus infection in human airway cell cultures by an antisense peptide-conjugated morpholino oligomer targeting the hemagglutinin-activating protease TMPRSS2. J Virol. 2011;85:1554–62.

Böttcher-Friebertshäuser E, Lu Y, Meyer D, Sielaff F, Steinmetzer T, Klenk HD, Garten W. Hemagglutinin activating host cell proteases provide promising drug targets for the treatment of influenza A and B virus infections. Vaccine. 2012;30:7374–80.

Böttcher-Friebertshäuser E, Klenk HD, Garten W. Activation of influenza viruses by proteases from host cells and bacteria in the human airway epithelium. Pathog Dis. 2013;69:87–100.

Boycott R, Klenk HD, Ohuchi M. Cell tropism of influenza virus mediated by hemagglutinin activation at the stage of virus entry. Virology. 1994;203:313–9.

Brunati M, Perucca S, Han L, Cattaneo A, Consolato F, Andolfo A, Schaeffer C, Olinger E, Peng J, Santambrogio S, Perrier R, Li S, Bokhove M, Bachi A, Hummler E, Devuyst O, Wu Q, Jovine L, Rampoldi L. The serine protease hepsin mediates urinary secretion and polymerisation of Zona Pellucida domain protein uromodulin. Elife. 2015;4:e08887.

Bruns JB, Carattino MD, Sheng S, Maarouf AB, Weisz OA, Pilewski JM, Hughey RP, Kleyman TR. Epithelial Na+ channels are fully activated by furin- and prostasin-dependent release of an inhibitory peptide from the gamma-subunit. J Biol Chem. 2007;282:6153–60.

Bugge TH, Antalis TM, Wu Q. Type II transmembrane serine proteases. J Biol Chem. 2009;284:23177–81. Review.
Burkard C, Verheije MH, Wicht O, van Kasteren SI, van Kuppeveld FJ, Haagmans BL, Pelkmans L, Rottier PJ, Bosch BJ, de Haan CA. Coronavirus cell entry occurs through the endo-/lysosomal pathway in a proteolysis-dependent manner. PLoS Pathog. 2014;10:e1004502.
Buzza MS, Netzel-Arnett S, Shea-Donohue T, Zhao A, Lin CY, List K, Szabo R, Fasano A, Bugge TH, Antalis TM. Membrane-anchored serine protease matriptase regulates epithelial barrier formation and permeability in the intestine. Proc Natl Acad Sci U S A. 2010;107:4200–5.
Buzza MS, Martin EW, Driesbaugh KH, Désilets A, Leduc R, Antalis TM. Prostasin is required for matriptase activation in intestinal epithelial cells to regulate closure of the paracellular pathway. J Biol Chem. 2013;288:10328–37.
Cal S, Quesada V, Garabaya C, Lopez-Otin C. Polyserase-I, a human polyprotease with the ability to generate independent serine protease domains from a single translation product. Proc Natl Acad Sci U S A. 2003;100:9185–90.
Chaipan C, Kobasa D, Bertram S, Glowacka I, Steffen I, Tsegaye TS, Takeda M, Bugge TH, Kim S, Park Y, Marzi A, Pöhlmann S. Proteolytic activation of the 1918 influenza virus hemagglutinin. J Virol. 2009;83:3200–11.
Chan JC, Knudson O, Wu F, Morser J, Dole WP, Wu Q. Hypertension in mice lacking the proatrial natriuretic peptide convertase corin. Proc Natl Acad Sci U S A. 2005;102:785–90.
Chen LM, Chai KX. Prostasin serine protease inhibits breast cancer invasiveness and is transcriptionally regulated by promoter DNA methylation. Int J Cancer. 2002;97:323–9.
Chen LM, Hodge GB, Guarda LA, Welch JL, Greenberg NM, Chai KX. Down-regulation of prostasin serine protease: a potential invasion suppressor in prostate cancer. Prostate. 2001a;48:93–103.
Chen LM, Skinner ML, Kauffman SW, Chao J, Chao L, Thaler CD, Chai KX. Prostasin is a glycosylphosphatidylinositol-anchored active serine protease. J Biol Chem. 2001b;276:21434–42. Epub 2001 Mar 26.
Chen YW, Lee MS, Lucht A, Chou FP, Huang W, Havighurst TC, Kim K, Wang JK, Antalis TM, Johnson MD, Lin CY. TMPRSS2, a serine protease expressed in the prostate on the apical surface of luminal epithelial cells and released into semen in prostasomes, is misregulated in prostate cancer cells. Am J Pathol. 2010;176:2986–96.
Chen S, Cao P, Dong N, Peng J, Zhang C, Wang H, Zhou T, Yang J, Zhang Y, Martelli EE, Naga Prasad SV, Miller RE, Malfait AM, Zhou Y, Wu Q. PCSK6-mediated corin activation is essential for normal blood pressure. Nat Med. 2015;21:1048–53.
Cheng H, Fukushima T, Takahashi N, Tanaka H, Kataoka H. Hepatocyte growth factor activator inhibitor type 1 regulates epithelial to mesenchymal transition through membrane-bound serine proteinases. Cancer Res. 2009;69:1828–35.
Cheng Z, Zhou J, To KK, Chu H, Li C, Wang D, Yang D, Zheng S, Hao K, Bossé Y, Obeidat M, Brandsma CA, Song YQ, Chen Y, Zheng BJ, Li L, Yuen KY. Identification of TMPRSS2 as a susceptibility gene for severe 2009 Pandemic A(H1N1) Influenza and A(H7N9) Influenza. J Infect Dis. 2015;212:1214–21.
Chokki M, Yamamura S, Eguchi H, Masegi T, Horiuchi H, Tanabe H, Kamimura T, Yasuoka S. Human airway trypsin-like protease increases mucin gene expression in airway epithelial cells. Am J Respir Cell Mol Biol. 2004;30:470–8.
Dhanasekaran SM, Barrette TR, Ghosh D, Shah R, Varambally S, Kurachi K, Pienta KJ, Rubin MA, Chinnaiyan AM. Delineation of prognostic biomarkers in prostate cancer. Nature. 2001;412:822–6.
Donaldson SH, Hirsh A, Li DC, Holloway G, Chao J, Boucher RC, Gabriel SE. Regulation of the epithelial sodium channel by serine proteases in human airways. J Biol Chem. 2002;277:8338–45.
Driesbaugh KH, Buzza MS, Martin EW, Conway GD, Kao JP, Antalis TM. Proteolytic activation of the protease-activated receptor (PAR)-2 by the glycosylphosphatidylinositol-anchored serine protease testisin. J Biol Chem. 2015;290:3529–41.
Du X, She E, Gelbart T, Truksa J, Lee P, Xia Y, Khovananth K, Mudd S, Mann N, Moresco EM, Beutler E, Beutler B. The serine protease TMPRSS6 is required to sense iron deficiency. Science. 2008;320:1088–92.

Duhaime MJ, Page KO, Varela FA, Murray AS, Silverman ME, Zoratti GL, List K. Cell surface human airway trypsin-like protease is lost during squamous cell carcinogenesis. J Cell Physiol. 2016;231:1476–83.

Enshell-Seijffers D, Lindon C, Morgan BA. The serine protease Corin is a novel modifier of the Agouti pathway. Development. 2008;135:217–25.

Fang C, Shen L, Dong L, Liu M, Shi S, Dong N, Wu Q. Reduced urinary corin levels in patients with chronic kidney disease. Clin Sci (Lond). 2013;124:709–17.

Fasquelle L, Scott HS, Lenoir M, Wang J, Rebillard G, Gaboyard S, Venteo S, François F, Mausset-Bonnefont AL, Antonarakis SE, Neidhart E, Chabbert C, Puel JL, Guipponi M, Delprat B. Tmprss3, a transmembrane serine protease deficient in human DFNB8/10 deafness, is critical for cochlear hair cell survival at the onset of hearing. J Biol Chem. 2011;286:17383–97.

Ferrara F, Molesti E, Böttcher-Friebertshäuser E, Cattoli G, Corti D, Scott S, Temperton N. The human Transmembrane Protease Serine 2 is necessary for the production of Group 2 influenza A virus pseudotypes. J Mol Genet Med. 2013;7:309–14.

Finberg KE, Heeney MM, Campagna DR, Aydinok Y, Pearson HA, Hartman KR, Mayo MM, Samuel SM, Strouse JJ, Markianos K, Andrews NC, Fleming MD. Mutations in TMPRSS6 cause iron-refractory iron deficiency anemia (IRIDA). Nat Genet. 2008;40:569–71.

Finberg KE, Whittlesey RL, Fleming MD, Andrews NC. Down-regulation of Bmp/Smad signaling by Tmprss6 is required for maintenance of systemic iron homeostasis. Blood. 2010;115:3817–26.

Frateschi S, Keppner A, Malsure S, Iwaszkiewicz J, Sergi C, Merillat AM, Fowler-Jaeger N, Randrianarison N, Planès C, Hummler E. Mutations of the serine protease CAP1/Prss8 lead to reduced embryonic viability, skin defects, and decreased ENaC activity. Am J Pathol. 2012;181:605–15.

Friedrich R, Fuentes-Prior P, Ong E, Coombs G, Hunter M, Oehler R, Pierson D, Gonzalez R, Huber R, Bode W, Madison EL. Catalytic domain structures of Mt-Sp1/Matriptase, a matrix-degrading transmembrane serine proteinase. J Biol Chem. 2002;277:2160.

Friis S, Uzzun Sales K, Godiksen S, Peters DE, Lin CY, Vogel LK, Bugge TH. A matriptase-prostasin reciprocal zymogen activation complex with unique features: prostasin as a non-enzymatic co-factor for matriptase activation. J Biol Chem. 2013;288:19028–39.

Friis S, Madsen DH, Bugge TH. Distinct Developmental Functions of Prostasin (CAP1/PRSS8) Zymogen and Activated Prostasin. J Biol Chem. 2016;291:2577–82.

Galloway SE, Reed ML, Russell CJ, Steinhauer DA. Influenza HA subtypes demonstrate divergent phenotypes for cleavage activation and pH of fusion: implications for host range and adaptation. PLoS Pathog. 2013;9:e1003151.

Ganz T, Nemeth E. Hepcidin and iron homeostasis. Biochim Biophys Acta. 2012;1823:1434–43.

Gasi Tandefelt D, Boormans J, Hermans K, Trapman J. ETS fusion genes in prostate cancer. Endocr Relat Cancer. 2014;21:R143–52.

Gierer S, Bertram S, Kaup F, Wrensch F, Heurich A, Krämer-Kühl A, Welsch K, Winkler M, Meyer B, Drosten C, Dittmer U, von Hahn T, Simmons G, Hofmann H, Pöhlmann S. The spike protein of the emerging betacoronavirus EMC uses a novel coronavirus receptor for entry, can be activated by TMPRSS2, and is targeted by neutralizing antibodies. J Virol. 2013;87:5502–11.

Gladysheva IP, King SM, Houng AK. N-glycosylation modulates the cell-surface expression and catalytic activity of corin. Biochem Biophys Res Commun. 2008;373:130–5.

Glowacka I, Bertram S, Müller MA, Allen P, Soilleux E, Pfefferle S, Steffen I, Tsegaye TS, He Y, Gnirss K, Niemeyer D, Schneider H, Drosten C, Pöhlmann S. Evidence that TMPRSS2 activates the severe acute respiratory syndrome coronavirus spike protein for membrane fusion and reduces viral control by the humoral immune response. J Virol. 2011;85:4122–34.

Götze H, Adelson JW, Hadorn HB, Portmann R, Troesch V. Hormone-elicited enzyme release by the small intestinal wall. Gut. 1972;13(6):471.

Guipponi M, Vuagniaux G, Wattenhofer M, Shibuya K, Vazquez M, Dougherty L, Scamuffa N, Guida E, Okui M, Rossier C, Hancock M, Buchet K, Reymond A, Hummler E, Marzella PL, Kudoh J, Shimizu N, Scott HS, Antonarakis SE, Rossier BC. The transmembrane serine protease (TMPRSS3) mutated in deafness DFNB8/10 activates the epithelial sodium channel (ENaC) in vitro. Hum Mol Genet. 2002;11:2829–36.

Guipponi M, Tan J, Cannon PZ, Donley L, Crewther P, Clarke M, Wu Q, Shepherd RK, Scott HS. Mice deficient for the type II transmembrane serine protease, TMPRSS1/hepsin, exhibit profound hearing loss. Am J Pathol. 2007;171:608–16.

Guipponi M, Toh MY, Tan J, Park D, Hanson K, Ballana E, Kwong D, Cannon PZ, Wu Q, Gout A, Delorenzi M, Speed TP, Smith RJ, Dahl HH, Petersen M, Teasdale RD, Estivill X, Park WJ, Scott HS. An integrated genetic and functional analysis of the role of type II transmembrane serine proteases (TMPRSSs) in hearing loss. Hum Mutat. 2008;29:130–41.

Guo YJ, Krauss S, Senne DA, Mo IP, Lo KS, Xiong XP, Norwood M, Shortridge KF, Webster RG, Guan Y. Characterization of the pathogenicity of members of the newly established H9N2 influenza virus lineages in Asia. Virology. 2000;267:279–88.

Hadorn B, Tarlow MJ, Lloyd JK, Wolff OH. Intestinal enterokinase deficiency. Lancet. 1969; 1:812–3.

Hamilton BS, Whittaker GR. Cleavage activation of human-adapted influenza virus subtypes by kallikrein-related peptidases 5 and 12. J Biol Chem. 2013;288:17399–407.

Hamilton BS, Gludish DW, Whittaker GR. Cleavage activation of the human-adapted influenza virus subtypes by matriptase reveals both subtype and strain specificities. J Virol. 2012;86(19):10579–86.

Hanifa S, Scott HS, Crewther P, Guipponi M, Tan J. Thyroxine treatments do not correct inner ear defects in tmprss1 mutant mice. Neuroreport. 2010;21:897–901.

Hansbro PM, Hamilton MJ, Fricker M, Gellatly SL, Jarnicki AG, Zheng D, Frei SM, Wong GW, Hamadi S, Zhou S, Foster PS, Krilis SA, Stevens RL. Importance of mast cell Prss31/transmembrane tryptase/tryptase-γ in lung function and experimental chronic obstructive pulmonary disease and colitis. J Biol Chem. 2014;289:18214–27.

Hansen IA, Fassnacht M, Hahner S, Hammer F, Schammann M, Meyer SR, Bicknell AB, Allolio B. The adrenal secretory serine protease AsP is a short secretory isoform of the transmembrane airway trypsin-like protease. Endocrinology. 2004;145:1898–905.

Hatesuer B, Bertram S, Mehnert N, Bahgat MM, Nelson PS, Pöhlmann S, Schughart K. Tmprss2 is essential for influenza H1N1 virus pathogenesis in mice. PLoS Pathog. 2013;9:e1003774.

Hermon-Taylor J, Perrin J, Grant DA, Appleyard A, Bubel M, Magee AI. Immunofluorescent localisation of enterokinase in human small intestine. Gut. 1977;18:259–65.

Heurich A, Hofmann-Winkler H, Gierer S, Liepold T, Jahn O, Pöhlmann S. TMPRSS2 and ADAM17 cleave ACE2 differentially and only proteolysis by TMPRSS2 augments entry driven by the severe acute respiratory syndrome coronavirus spike protein. J Virol. 2014;88:1293–307.

Hildenbrand R, Gandhari M, Stroebel P, Marx A, Allgayer H, Arens N. The urokinase-system—role of cell proliferation and apoptosis. Histol Histopathol. 2008;23:227–36.

Hobson JP, Netzel-Arnett S, Szabo R, Réhault SM, Church FC, Strickland DK, Lawrence DA, Antalis TM, Bugge TH. Mouse DESC1 is located within a cluster of seven DESC1-like genes and encodes a type II transmembrane serine protease that forms serpin inhibitory complexes. J Biol Chem. 2004;279:46981–94.

Hoffmann M, Krüger N, Zmora P, Wrensch F, Herrler G, Pöhlmann S. The hemagglutinin of Bat-associated influenza viruses is activated by TMPRSS2 for pH-dependent entry into Bat but not human cells. PLoS One. 2016;11:e0152134.

Holzinger A, Maier EM, Bück C, Mayerhofer PU, Kappler M, Haworth JC, Moroz SP, Hadorn HB, Sadler JE, Roscher AA. Mutations in the proenteropeptidase gene are the molecular cause of congenital enteropeptidase deficiency. Am J Hum Genet. 2002;70:20–5.

Homma M, Ohuchi M. Trypsin action on the growth of Sendai virus in tissue culture cells. III. Structural difference of Sendai viruses grown in eggs and tissue culture cells. J Virol. 1973;12:1457–65.

Honda A, Yamagata K, Sugiura S, Watanabe K, Baba T. A mouse serine protease TESP5 is selectively included into lipid rafts of sperm membrane presumably as a glycosylphosphatidylinositol-anchored protein. J Biol Chem. 2002;277:16976–84.

Hooper JD, Nicol DL, Dickinson JL, Eyre HJ, Scarman AL, Normyle JF, Stuttgen MA, Douglas ML, Loveland KA, Sutherland GR, Antalis TM. Testisin, a new human serine proteinase expressed by premeiotic testicular germ cells and lost in testicular germ cell tumors. Cancer Res. 1999;59:3199–205.

Hooper JD, Bowen N, Marshall H, Cullen LM, Sood R, Daniels R, Stuttgen MA, Normyle JF, Higgs DR, Kastner DL, Ogbourne SM, Pera MF, Jazwinska EC, TM. Localization, expression and genomic structure of the gene encoding the human serine protease testisin. Biochim Biophys Acta. 2000;1492:63–71.

Hooper JD, Clements JA, Quigley JP, Antalis TM. Type II transmembrane serine proteases. Insights into an emerging class of cell surface proteolytic enzymes. J Biol Chem. 2001;276:857–60. Review.

Hsu YC, Huang HP, Yu IS, Su KY, Lin SR, Lin WC, Wu HL, Shi GY, Tao MH, Kao CH, Wu YM, Martin PE, Lin SY, Yang PC, Lin SW. Serine protease hepsin regulates hepatocyte size and hemodynamic retention of tumor cells by hepatocyte growth factor signaling in mice. Hepatology. 2012;56:1913–23.

Hummler E, Dousse A, Rieder A, Stehle JC, Rubera I, Osterheld MC, Beermann F, Frateschi S, Charles RP. The channel-activating protease CAP1/Prss8 is required for placental labyrinth maturation. PLoS One. 2013;8:e55796.

Inoue M, Kanbe N, Kurosawa M, Kido H. Cloning and tissue distribution of a novel serine protease esp-1 from human eosinophils. Biochem Biophys Res Commun. 1998;252:307–12.

Iwakiri K, Ghazizadeh M, Jin E, Fujiwara M, Takemura T, Takezaki S, Kawana S, Yasuoka S, Kawanami O. Human airway trypsin-like protease induces PAR-2-mediated IL-8 release in psoriasis vulgaris. J Invest Dermatol. 2004;122:937–44.

Jiang J, Wu S, Wang W, Chen S, Peng J, Zhang X, Wu Q. Ectodomain shedding and autocleavage of the cardiac membrane protease corin. J Biol Chem. 2011;286:10066–72.

Jung H, Lee KP, Park SJ, Park JH, Jang YS, Choi SY, Jung JG, Jo K, Park DY, Yoon JH, Park JH, Lim DS, Hong GR, Choi C, Park YK, Lee JW, Hong HJ, Kim S, Park YW. TMPRSS4 promotes invasion, migration and metastasis of human tumor cells by facilitating an epithelial-mesenchymal transition. Oncogene. 2008;27:2635–47.

Kam YW, Okumura Y, Kido H, Ng LF, Bruzzone R, Altmeyer R. Cleavage of the SARS coronavirus spike glycoprotein by airway proteases enhances virus entry into human bronchial epithelial cells in vitro. PLoS One. 2009;4:e7870.

Kato M, Hashimoto T, Shimomura T, Kataoka H, Ohi H, Kitamura N. Hepatocyte growth factor activator inhibitor type 1 inhibits protease activity and proteolytic activation of human airway trypsin-like protease. J Biochem. 2012;151:179–87.

Kawano N, Kang W, Yamashita M, Koga Y, Yamazaki T, Hata T, Miyado K, Baba T. Mice lacking two sperm serine proteases, ACR and PRSS21, are subfertile, but the mutant sperm are infertile in vitro. Biol Reprod. 2010;83:359–69.

Kawaoka Y, Naeve CW, Webster RG. Is virulence of H5N2 influenza viruses in chickens associated with loss of carbohydrate from the hemagglutinin? Virology. 1984;139:303–16.

Keppner A, Andreasen D, Mérillat AM, Bapst J, Ansermet C, Wang Q, Maillard M, Malsure S, Nobile A, Hummler E. Epithelial sodium channel-mediated sodium transport is not dependent on the membrane-bound serine protease CAP2/Tmprss4. PLoS One. 2015;10:e0135224.

Kesic MJ, Meyer M, Bauer R, Jaspers I. Exposure to ozone modulates human airway protease/anti-protease balance contributing to increased influenza A infection. PLoS One. 2012;7:e35108.

Kido H, Okumura Y, Takahashi E, Pan HY, Wang S, Chida J, Le TQ, Yano M. Host envelope glycoprotein processing proteases are indispensable for entry into human cells by seasonal and highly pathogenic avian influenza viruses. J Mol Genet Med. 2008;3:167–75.

Kim S, Lee JW. Membrane proteins involved in epithelial-mesenchymal transition and tumor invasion: studies on TMPRSS4 and TM4SF5. Genom Inform. 2014;12:12–20.

Kim DR, Sharmin S, Inoue M, Kido H. Cloning and expression of novel mosaic serine proteases with and without a transmembrane domain from human lung. Biochim Biophys Acta. 2001;1518:204–9.

Kim TS, Heinlein C, Hackman RC, Nelson PS. Phenotypic analysis of mice lacking the Tmprss2-encoded protease. Mol Cell Biol. 2006;26:965–75.

Kim S, Kang HY, Nam EH, Choi MS, Zhao XF, Hong CS, Lee JW, Lee JH, Park YK. TMPRSS4 induces invasion and epithelial-mesenchymal transition through upregulation of integrin alpha5 and its signaling pathways. Carcinogenesis. 2010;31:597–606.

Kitamoto Y, Yuan X, Wu Q, McCourt DW, Sadler JE. Enterokinase, the initiator of intestinal digestion, is a mosaic protease composed of a distinctive assortment of domains. Proc Natl Acad Sci U S A. 1994;91:7588–92.

Kitamoto Y, Veile RA, Donis-Keller H, Sadler JE. cDNA sequence and chromosomal localization of human enterokinase, the proteolytic activator of trypsinogen. Biochemistry. 1995;34:4562–8.

Klenk HD, Rott R, Orlich M, Blödorn J. Activation of influenza A viruses by trypsin treatment. Virology. 1975;68:426–39.

Klezovitch O, Chevillet J, Mirosevich J, Roberts RL, Matusik RJ, Vasioukhin V. Hepsin promotes prostate cancer progression and metastasis. Cancer Cell. 2004;6:185–95.

Ko CJ, Huang CC, Lin HY, Juan CP, Lan SW, Shyu HY, Wu SR, Hsiao PW, Huang HP, Shun CT, Lee MS. Androgen-induced TMPRSS2 activates Matriptase and promotes extracellular matrix degradation, prostate cancer cell invasion, tumor growth, and metastasis. Cancer Res. 2015;75:2949–60.

Kühn N, Bergmann S, Kösterke N, Lambertz RL, Keppner A, van den Brand JM, Pöhlmann S, Weiß S, Hummler E, Hatesuer B, Schughart K. The proteolytic activation of (H3N2) Influenza A virus hemagglutinin is facilitated by different type II transmembrane serine proteases. J Virol. 2016;90:4298–307.

Kunitz M. Formation of trypsin from crystalline trypsinogen by means of enterokinase. J Gen Physiol. 1939;22:429–46.

Kyrieleis OJ, Huber R, Ong E, Oehler R, Hunter M, Madison EL, Jacob U. Crystal structure of the catalytic domain of DESC1, a new member of the type II transmembrane serine proteinase family. FEBS J. 2007;274:2148–60.

Lang JC, Schuller DE. Differential expression of a novel serine protease homologue in squamous cell carcinoma of the head and neck. Br J Cancer. 2001;84:237–43.

Larsen BR, Steffensen SD, Nielsen NV, Friis S, Godiksen S, Bornholdt J, Soendergaard C, Nonboe AW, Andersen MN, Poulsen SS, Szabo R, Bugge TH, Lin CY, Skovbjerg H, Jensen JK, Vogel LK. Hepatocyte growth factor activator inhibitor-2 prevents shedding of matriptase. Exp Cell Res. 2013;319:918–29.

Lazarowitz SG, Choppin PW. Enhancement of the infectivity of influenza A and B viruses by proteolytic cleavage of the hemagglutinin polypeptide. Virology. 1975;68:440–54.

Lazarowitz SG, Goldberg AR, Choppin PW. Proteolytic cleavage by plasmin of the HA polypeptide of influenza virus: host cell activation of serum plasminogen. Virology. 1973;56:172–80.

Leshem O, Madar S, Kogan-Sakin I, Kamer I, Goldstein I, Brosh R, Cohen Y, Jacob-Hirsch J, Ehrlich M, Ben-Sasson S, Goldfinger N, Loewenthal R, Gazit E, Rotter V, Berger R. TMPRSS2/ERG promotes epithelial to mesenchymal transition through the ZEB1/ZEB2 axis in a prostate cancer model. PLoS One. 2011;6:e21650.

Leytus SP, Loeb KR, Hagen FS, Kurachi K, Davie EW. A novel trypsin-like serine protease (hepsin) with a putative transmembrane domain expressed by human liver and hepatoma cells. Biochemistry. 1988;27:1067–74.

Leyvraz C, Charles RP, Rubera I, Guitard M, Rotman S, Breiden B, Sandhoff K, Hummler E. The epidermal barrier function is dependent on the serine protease CAP1/Prss8. J Cell Biol. 2005;170:487–96.

Li Y, Zhang X, Huang G, Miao X, Guo L, Lin D, Lu SH. Identification of a novel polymorphism Arg290Gln of esophageal cancer related gene 1 (ECRG1) and its related risk to esophageal squamous cell carcinoma. Carcinogenesis. 2006;27:798–802.

Li X, Qi W, He J, Ning Z, Hu Y, Tian J, Jiao P, Xu C, Chen J, Richt J, Ma W, Liao M. Molecular basis of efficient replication and pathogenicity of H9N2 avian influenza viruses in mice. PLoS One. 2012;7:e40118.

Li Y, Peng A, Ge S, Wang Q, Liu J. miR-204 suppresses cochlear spiral ganglion neuron survival in vitro by targeting TMPRSS3. Hear Res. 2014;314:60–4.

Li H, Zhang Y, Wu Q. Role of corin in the regulation of blood pressure. Curr Opin Nephrol Hypertens. 2017;26:67–73.

Liao X, Wang W, Chen S, Wu Q. Role of glycosylation in corin zymogen activation. J Biol Chem. 2007;282:27728–35.

Lin CY, Anders J, Johnson M, Sang QA, Dickson RB. Molecular cloning of cDNA for matriptase, a matrix-degrading serine protease with trypsin-like activity. J Biol Chem. 1999;274:18231–6.

List K, Haudenschild CC, Szabo R, Chen W, Wahl SM, Swaim W, Engelholm LH, Behrendt N, Bugge TH. Matriptase/MT-SP1 is required for postnatal survival, epidermal barrier function, hair follicle development, and thymic homeostasis. Oncogene. 2002;21:3765–79.

List K, Szabo R, Wertz PW, Segre J, Haudenschild CC, Kim SY, Bugge TH. Loss of proteolytically processed filaggrin caused by epidermal deletion of Matriptase/MT-SP1. J Cell Biol. 2003;163:901–10.

List K, Bugge TH, Szabo R. Matriptase: potent proteolysis on the cell surface. Mol Med. 2006a;12:1–7. Review

List K, Szabo R, Molinolo A, Nielsen BS, Bugge TH. Delineation of matriptase protein expression by enzymatic gene trapping suggests diverging roles in barrier function, hair formation, and squamous cell carcinogenesis. Am J Pathol. 2006b;168:1513–25.

List K, Currie B, Scharschmidt TC, Szabo R, Shireman J, Molinolo A, Cravatt BF, Segre J, Bugge TH. Autosomal ichthyosis with hypotrichosis syndrome displays low matriptase proteolytic activity and is phenocopied in ST14 hypomorphic mice. J Biol Chem. 2007;282:36714–23.

Lu D, Yuan X, Zheng X, Sadler JE. Bovine proenteropeptidase is activated by trypsin, and the specificity of enteropeptidase depends on the heavy chain. J Biol Chem. 1997;272:31293–300.

Lu D, Fütterer K, Korolev S, Zheng X, Tan K, Waksman G, Sadler JE. Crystal structure of enteropeptidase light chain complexed with an analog of the trypsinogen activation peptide. J Mol Biol. 1999;292:361–73.

Lucas JM, True L, Hawley S, Matsumura M, Morrissey C, Vessella R, Nelson PS. The androgen-regulated type II serine protease TMPRSS2 is differentially expressed and mislocalized in prostate adenocarcinoma. J Pathol. 2008;215:118–25.

Lucas JM, Heinlein C, Kim T, Hernandez SA, Malik MS, True LD, Morrissey C, Corey E, Montgomery B, Mostaghel E, Clegg N, Coleman I, Brown CM, Schneider EL, Craik C, Simon JA, Bedalov A, Nelson PS. The androgen-regulated protease TMPRSS2 activates a proteolytic cascade involving components of the tumor microenvironment and promotes prostate cancer metastasis. Cancer Discov. 2014;4:1310–25.

Madsen DH, Szabo R, Molinolo AA, Bugge TH. TMPRSS13 deficiency impairs stratum corneum formation and epidermal barrier acquisition. Biochem J. 2014;461:487–95.

Masmoudi S, Antonarakis SE, Schwede T, Ghorbel AM, Gratri M, Pappasavas MP, Drira M, Elgaied-Boulila A, Wattenhofer M, Rossier C, Scott HS, Ayadi H, Guipponi M. Novel missense mutations of TMPRSS3 in two consanguineous Tunisian families with non-syndromic autosomal recessive deafness. Hum Mutat. 2001;18:101–8.

Matsushima M, Ichinose M, Yahagi N, Kakei N, Tsukada S, Miki K, Kurokawa K, Tashiro K, Shiokawa K, Shinomiya K, et al. Structural characterization of porcine enteropeptidase. J Biol Chem. 1994;269:19976–82.

Matsushima R, Takahashi A, Nakaya Y, Maezawa H, Miki M, Nakamura Y, Ohgushi F, Yasuoka S. Human airway trypsin-like protease stimulates human bronchial fibroblast proliferation in a protease-activated receptor-2-dependent pathway. Am J Physiol Lung Cell Mol Physiol. 2006;290:L385–95.

Matsuyama S, Nagata N, Shirato K, Kawase M, Takeda M, Taguchi F. Efficient activation of the severe acute respiratory syndrome coronavirus spike protein by the transmembrane protease TMPRSS2. J Virol. 2010;84:12658–64.

Miao J, Mu D, Ergel B, Singavarapu R, Duan Z, Powers S, Oliva E, Orsulic S. Hepsin colocalizes with desmosomes and induces progression of ovarian cancer in a mouse model. Int J Cancer. 2008;123:2041–7.

Miller GS, List K. The matriptase-prostasin proteolytic cascade in epithelial development and pathology. Cell Tissue Res. 2013;351:245–53.

Miller GS, Zoratti GL, Murray AS, Bergum C, Tanabe LM, List K. HATL5: a cell surface serine protease differentially expressed in epithelial cancers. PLoS One. 2014;9:e87675.

Millet JK, Whittaker GR. Host cell proteases: critical determinants of coronavirus tropism and pathogenesis. Virus Res. 2015;202:120–34.

Milner JM, Patel A, Davidson RK, Swingler TE, Desilets A, Young DA, Kelso EB, Donell ST, Cawston TE, Clark IM, Ferrell WR, Plevin R, Lockhart JC, Leduc R, Rowan AD. Matriptase is a novel initiator of cartilage matrix degradation in osteoarthritis. Arthritis Rheum. 2010; 62:1955–66.

Min HJ, Lee MK, Lee JW, Kim S. TMPRSS4 induces cancer cell invasion through pro-uPA processing. Biochem Biophys Res Commun. 2014a;446:1–7.

Min HJ, Lee Y, Zhao XF, Park YK, Lee MK, Lee JW, Kim S. TMPRSS4 upregulates uPA gene expression through JNK signaling activation to induce cancer cell invasion. Cell Signal. 2014b;26:398–408.

Miyata H, Thaler CD, Haimo LT, Cardullo RA. Protease activation and the signal transduction pathway regulating motility in sperm from the water strider Aquarius remigis. Cytoskeleton (Hoboken). 2012;69:207–20.

Molina L, Fasquelle L, Nouvian R, Salvetat N, Scott HS, Guipponi M, Molina F, Puel JL, Delprat B. Tmprss3 loss of function impairs cochlear inner hair cell Kcnma1 channel membrane expression. Hum Mol Genet. 2013;22:1289–99.

Murray AS, Varela FA, List K. Type II transmembrane serine proteases as potential targets for cancer therapy. Biol Chem. 2016;397:815–26.

Nagai Y, Klenk HD, Rott R. Proteolytic cleavage of the viral glycoproteins and its significance for the virulence of Newcastle disease virus. Virology. 1976;72:494–508.

Nagaike K, Kawaguchi M, Takeda N, Fukushima T, Sawaguchi A, Kohama K, Setoyama M, Kataoka H. Defect of hepatocyte growth factor activator inhibitor type 1/serine protease inhibitor, Kunitz type 1 (Hai-1/Spint1) leads to ichthyosis-like condition and abnormal hair development in mice. Am J Pathol. 2008;173:1464–75.

Narikiyo T, Kitamura K, Adachi M, Miyoshi T, Iwashita K, Shiraishi N, Nonoguchi H, Chen LM, Chai KX, Chao J, Tomita K. Regulation of prostasin by aldosterone in the kidney. J Clin Invest. 2002;109:401–8.

Netzel-Arnett S, Hooper JD, Szabo R, Madison EL, Quigley JP, Bugge TH, Antalis TM. Membrane anchored serine proteases: a rapidly expanding group of cell surface proteolytic enzymes with potential roles in cancer. Cancer Metastasis Rev. 2003;22:237–58. Review.

Netzel-Arnett S, Currie BM, Szabo R, Lin CY, Chen LM, Chai KX, Antalis TM, Bugge TH, List K. Evidence for a matriptase-prostasin proteolytic cascade regulating terminal epidermal differentiation. J Biol Chem. 2006;281:32941–5.

Netzel-Arnett S, Bugge TH, Hess RA, Carnes K, Stringer BW, Scarman AL, Hooper JD, Tonks ID, Kay GF, Antalis TM. The glycosylphosphatidylinositol-anchored serine protease PRSS21 (testisin) imparts murine epididymal sperm cell maturation and fertilizing ability. Biol Reprod. 2009;81:921–32.

Netzel-Arnett S, Buzza MS, Shea-Donohue T, Désilets A, Leduc R, Fasano A, Bugge TH, Antalis TM. Matriptase protects against experimental colitis and promotes intestinal barrier recovery. Inflamm Bowel Dis. 2012;18:1303–14.

Ng HY, Ko JM, Yu VZ, Ip JC, Dai W, Cal S, Lung ML. DESC1, a novel tumor suppressor, sensitizes cells to apoptosis by downregulating the EGFR/AKT pathway in esophageal squamous cell carcinoma. Int J Cancer. 2016;138:2940–51.

Nonboe AW, Krigslund O, Soendergaard C, Skovbjerg S, Friis S, Andersen MN, Ellis V, Kawaguchi M, Kataoka H, Bugge TH, Vogel LK. HAI-2 stabilizes, inhibits and regulates SEA-cleavage-dependent secretory transport of matriptase. Traffic. 2017;18:378–91.

Oberst MD, Williams CA, Dickson RB, Johnson MD, Lin CY. The activation of matriptase requires its noncatalytic domains, serine protease domain, and its cognate inhibitor. J Biol Chem. 2003;278:26773–9.

Oberst MD, Chen LY, Kiyomiya K, Williams CA, Lee MS, Johnson MD, Dickson RB, Lin CY. HAI-1 regulates activation and expression of matriptase, a membrane-bound serine protease. Am J Physiol Cell Physiol. 2005;289:C462–70.

Ogiwara K, Takahashi T. Specificity of the medaka enteropeptidase serine protease and its usefulness as a biotechnological tool for fusion-protein cleavage. Proc Natl Acad Sci U S A. 2007;104:7021–6.

Okumura Y, Hayama M, Takahashi E, Fujiuchi M, Shimabukuro A, Yano M, Kido H. Serase-1B, a new splice variant of polyserase-1/TMPRSS9, activates urokinase-type plasminogen activator and the proteolytic activation is negatively regulated by glycosaminoglycans. Biochem J. 2006;400:551–61.

Okumura Y, Takahashi E, Yano M, Ohuchi M, Daidoji T, Nakaya T, Böttcher E, Garten W, Klenk HD, Kido H. Novel type II transmembrane serine proteases, MSPL and TMPRSS13, Proteolytically activate membrane fusion activity of the hemagglutinin of highly pathogenic avian influenza viruses and induce their multicycle replication. J Virol. 2010 May;84(10):5089–96.

Paoloni-Giacobino A, Chen H, Peitsch MC, Rossier C, Antonarakis SE. Cloning of the TMPRSS2 gene, which encodes a novel serine protease with transmembrane, LDLRA, and SRCR domains and maps to 21q22.3. Genomics. 1997;44:309–20.

Park JE, Li K, Barlan A, Fehr AR, Perlman S, McCray PBJ, Gallagher T. Proteolytic processing of Middle East respiratory syndrome coronavirus spikes expands virus tropism. Proc Natl Acad Sci U S A. 2016;113(43):12262–7.

Peitsch C, Klenk HD, Garten W, Böttcher-Friebertshäuser E. Activation of influenza A viruses by host proteases from swine airway epithelium. J Virol. 2014;88:282–91.

Perona JJ, Craik CS. Structural basis of substrate specificity in the serine proteases. Protein Sci. 1995;4:337–60. Review.

Peters DE, Szabo R, Friis S, Shylo NA, Uzzun Sales K, Holmbeck K, Bugge TH. The membrane-anchored serine protease prostasin (CAP1/PRSS8) supports epidermal development and postnatal homeostasis independent of its enzymatic activity. J Biol Chem. 2014;289:14740–9.

Planès C, Randrianarison NH, Charles RP, Frateschi S, Cluzeaud F, Vuagniaux G, Soler P, Clerici C, Rossier BC, Hummler E. ENaC-mediated alveolar fluid clearance and lung fluid balance depend on the channel-activating protease 1. EMBO Mol Med. 2010;2:26–37.

Polzin D, Kaminski HJ, Kastner C, Wang W, Krämer S, Gambaryan S, Russwurm M, Peters H, Wu Q, Vandewalle A, Bachmann S, Theilig F. Decreased renal corin expression contributes to sodium retention in proteinuric kidney diseases. Kidney Int. 2010;78:650–9.

Ramsay AJ, Hooper JD, Folgueras AR, Velasco G, López-Otín C. Matriptase-2 (TMPRSS6): a proteolytic regulator of iron homeostasis. Haematologica. 2009;94:840–9.

Rickert KW, Kelley P, Byrne NJ, Diehl RE, Hall DL, Montalvo AM, Reid JC, Shipman JM, Thomas BW, Munshi SK, Darke PL, Su HP. Structure of human prostasin, a target for the regulation of hypertension. J Biol Chem. 2008;283:34864–72.

Sakai K, Ami Y, Tahara M, Kubota T, Anraku M, Abe M, Nakajima N, Sekizuka T, Shirato K, Suzaki Y, Ainai A, Nakatsu Y, Kanou K, Nakamura K, Suzuki T, Komase K, Nobusawa E, Maenaka K, Kuroda M, Hasegawa H, Kawaoka Y, Tashiro M, Takeda M. The host protease TMPRSS2 plays a major role in in vivo replication of emerging H7N9 and seasonal influenza viruses. J Virol. 2014;88:5608–16.

Sakai K, Sekizuka T, Ami Y, Nakajima N, Kitazawa M, Sato Y, Nakajima K, Anraku M, Kubota T, Komase K, Takehara K, Hasegawa H, Odagiri T, Tashiro M, Kuroda M, Takeda M. A mutant H3N2 influenza virus uses an alternative activation mechanism in TMPRSS2 knockout mice by loss of an oligosaccharide in the hemagglutinin stalk region. J Virol. 2015;89:5154–8.

Sakai K, Ami Y, Nakajima N, Nakajima K, Kitazawa M, Anraku M, Takayama I, Sangsriratanakul N, Komura M, Sato Y, Asanuma H, Takashita E, Komase K, Takehara K, Tashiro M, Hasegawa H, Odagiri T, Takeda M. TMPRSS2 independency for haemagglutinin cleavage in vivo differentiates Influenza B virus from Influenza A virus. Sci Rep. 2016;6:29430.

Sales KU, Hobson JP, Wagenaar-Miller R, Szabo R, Rasmussen AL, Bey A, Shah MF, Molinolo AA, Bugge TH. Expression and genetic loss of function analysis of the HAT/DESC cluster proteases TMPRSS11A and HAT. PLoS One. 2011;6:e23261.

Scheiblauer H, Reinacher M, Tashiro M, Rott R. Interactions between bacteria and influenza A virus in the development of influenza pneumonia. J Infect Dis. 1992;166:783–91.

Scheid A, Choppin PW. Identification of biological activities of paramyxovirus glycoproteins. Activation of cell fusion, hemolysis, and infectivity of proteolytic cleavage of an inactive precursor protein of Sendai virus. Virology. 1974;57:475–90.
Schmidlin F, Amadesi S, Dabbagh K, Lewis DE, Knott P, Bunnett NW, Gater PR, Geppetti P, Bertrand C, Stevens ME. Protease-activated receptor 2 mediates eosinophil infiltration and hyperreactivity in allergic inflammation of the airway. J Immunol. 2002;169:5315–21.
Scott HS, Kudoh J, Wattenhofer M, Shibuya K, Berry A, Chrast R, Guipponi M, Wang J, Kawasaki K, Asakawa S, Minoshima S, Younus F, Mehdi SQ, Radhakrishna U, Papasavvas MP, Gehrig C, Rossier C, Korostishevsky M, Gal A, Shimizu N, Bonne-Tamir B, Antonarakis SE. Insertion of beta-satellite repeats identifies a transmembrane protease causing both congenital and childhood onset autosomal recessive deafness. Nat Genet. 2001;27:59–63.
Sedghizadeh PP, Mallery SR, Thompson SJ, Kresty L, Beck FM, Parkinson EK, Biancamano J, Lang JC. Expression of the serine protease DESC1 correlates directly with normal keratinocyte differentiation and inversely with head and neck squamous cell carcinoma progression. Head Neck. 2006;28:432–40.
Semenov AG, Tamm NN, Seferian KR, Postnikov AB, Karpova NS, Serebryanaya DV, Koshkina EV, Krasnoselsky MI, Katrukha AG. Processing of pro-B-type natriuretic peptide: furin and corin as candidate convertases. Clin Chem. 2010;56:1166–76.
Shi YE, Torri J, Yieh L, Wellstein A, Lippman ME, Dickson RB. Identification and characterization of a novel matrix-degrading protease from hormone-dependent human breast cancer cells. Cancer Res. 1993;53:1409–15.
Shi W, Fan W, Bai J, Tang Y, Wang L, Jiang Y, Tang L, Liu M, Cui W, Xu Y, Li Y. TMPRSS2 and MSPL facilitate trypsin-independent porcine epidemic diarrhea virus replication in vero cells. Virus 2017;9.
Shigemasa K, Underwood LJ, Beard J, Tanimoto H, Ohama K, Parmley TH, O'Brien TJ. Overexpression of testisin, a serine protease expressed by testicular germ cells, in epithelial ovarian tumor cells. J Soc Gynecol Investig. 2000;7:358–62.
Shipway A, Danahay H, Williams JA, Tully DC, Backes BJ, Harris JL. Biochemical characterization of prostasin, a channel activating protease. Biochem Biophys Res Commun. 2004;324: 953–63.
Shirato K, Kawase M, Matsuyama S. Middle East respiratory syndrome coronavirus infection mediated by the transmembrane serine protease TMPRSS2. J Virol. 2013;87:12552–61.
Shirogane Y, Takeda M, Iwasaki M, Ishiguro N, Takeuchi H, Nakatsu Y, Tahara M, Kikuta H, Yanagi Y. Efficient multiplication of human metapneumovirus in Vero cells expressing the transmembrane serine protease TMPRSS2. J Virol. 2008;82:8942–6.
Shulla A, Heald-Sargent T, Subramanya G, Zhao J, Perlman S, Gallagher T. A transmembrane serine protease is linked to the severe acute respiratory syndrome coronavirus receptor and activates virus entry. J Virol. 2011;85:873–82.
Silvestri L, Pagani A, Nai A, De Domenico I, Kaplan J, Camaschella C. The serine protease matriptase-2 (TMPRSS6) inhibits hepcidin activation by cleaving membrane hemojuvelin. Cell Metab. 2008;8:502–11.
Somoza JR, Ho JD, Luong C, Ghate M, Sprengeler PA, Mortara K, Shrader WD, Sperandio D, Chan H, McGrath ME, Katz BA. The structure of the extracellular region of human hepsin reveals a serine protease domain and a novel scavenger receptor cysteine-rich (SRCR) domain. Structure. 2003;11:1123–31.
Spacek DV, Perez AF, Ferranti KM, Wu LK, Moy DM, Magnan DR, King TR. The mouse frizzy (fr) and rat 'hairless' (frCR) mutations are natural variants of protease serine S1 family member 8 (Prss8). Exp Dermatol. 2010;19:527–32.
Stefanska B, Huang J, Bhattacharyya B, Suderman M, Hallett M, Han ZG, Szyf M. Definition of the landscape of promoter DNA hypomethylation in liver cancer. Cancer Res. 2011;71:5891–903.
Stephan C, Yousef GM, Scorilas A, Jung K, Jung M, Kristiansen G, Hauptmann S, Kishi T, Nakamura T, Loening SA, Diamandis EP. Hepsin is highly over expressed in and a new candidate for a prognostic indicator in prostate cancer. J Urol. 2004;171:187–91.

Stieneke-Gröber A, Vey M, Angliker H, Shaw E, Thomas G, Roberts C, Klenk HD, Garten W. Influenza virus hemagglutinin with multibasic cleavage site is activated by furin, a subtilisin-like endoprotease. EMBO J. 1992;11:2407–14.

Stirnberg M, Maurer E, Horstmeyer A, Kolp S, Frank S, Bald T, Arenz K, Janzer A, Prager K, Wunderlich P, Walter J, Gütschow M. Proteolytic processing of the serine protease matriptase-2: identification of the cleavage sites required for its autocatalytic release from the cell surface. Biochem J. 2010;430:87–95.

Szabo R, Bugge TH. Membrane-anchored serine proteases in vertebrate cell and developmental biology. Annu Rev Cell Dev Biol. 2011;27:213–35.

Szabo R, Wu Q, Dickson RB, Netzel-Arnett S, Antalis TM, Bugge TH. Type II transmembrane serine proteases. Thromb Haemost. 2003;90:185–93. Review.

Szabo R, Netzel-Arnett S, Hobson JP, Antalis TM, Bugge TH. Matriptase-3 is a novel phylogenetically preserved membrane-anchored serine protease with broad serpin reactivity. Biochem J. 2005;390:231–42.

Szabo R, Molinolo A, List K, Bugge TH. Matriptase inhibition by hepatocyte growth factor activator inhibitor-1 is essential for placental development. Oncogene. 2007;26:1546–56.

Szabo R, Hobson JP, Christoph K, Kosa P, List K, Bugge TH. Regulation of cell surface protease matriptase by HAI2 is essential for placental development, neural tube closure and embryonic survival in mice. Development. 2009;136:2653–63.

Szabo R, Uzzun Sales K, Kosa P, Shylo NA, Godiksen S, Hansen KK, Friis S, Gutkind JS, Vogel LK, Hummler E, Camerer E, Bugge TH. Reduced prostasin (CAP1/PRSS8) activity eliminates HAI-1 and HAI-2 deficiency-associated developmental defects by preventing matriptase activation. PLoS Genet. 2012;8:e10029e37.

Szabo R, Lantsman T, Peters DE, Bugge TH. Delineation of proteolytic and non-proteolytic functions of the membrane-anchored serine protease prostasin. Development. 2016;143: 2818–28.

Takahashi M, Sano T, Yamaoka K, Kamimura T, Umemoto N, Nishitani H, Yasuoka S. Localization of human airway trypsin-like protease in the airway: an immunohistochemical study. Histochem Cell Biol. 2001;115:181–7.

Tanabe LM, List K. The role of type II transmembrane serine protease-mediated signaling in cancer. FEBS J. 2017;284:1421–36.

Tanaka H, Nagaike K, Takeda N, Itoh H, Kohama K, Fukushima T, Miyata S, Uchiyama S, Uchinokura S, Shimomura T, Miyazawa K, Kitamura N, Yamada G, Kataoka H. Hepatocyte growth factor activator inhibitor type 1 (HAI-1) is required for branching morphogenesis in the chorioallantoic placenta. Mol Cell Biol. 2005;25:5687–98.

Tarnow C, Engels G, Arendt A, Schwalm F, Sediri H, Preuss A, Nelson PS, Garten W, Klenk HD, Gabriel G, Böttcher-Friebertshäuser E. TMPRSS2 is a host factor that is essential for pneumotropism and pathogenicity of H7N9 influenza A virus in mice. J Virol. 2014;88:4744–51.

Tomlins SA, Rhodes DR, Perner S, Dhanasekaran SM, Mehra R, Sun XW, Varambally S, Cao X, Tchinda J, Kuefer R, Lee C, Montie JE, Shah RB, Pienta KJ, Rubin MA, Chinnaiyan AM. Recurrent fusion of TMPRSS2 and ETS transcription factor genes in prostate cancer. Science. 2005;310:644–8.

Tong Z, Illek B, Bhagwandin VJ, Verghese GM, Caughey GH. Prostasin, a membrane-anchored serine peptidase, regulates sodium currents in JME/CF15 cells, a cystic fibrosis airway epithelial cell line. Am J Physiol Lung Cell Mol Physiol. 2004;287:L928–35.

Tsuji A, Torres-Rosado A, Arai T, Le Beau MM, Lemons RS, Chou SH, Hepsin KK. a cell membrane-associated protease. Characterization, tissue distribution, and gene localization. J Biol Chem. 1991;266:16948–53.

Ujike M, Taguchi F. Incorporation of spike and membrane glycoproteins into coronavirus virions. Virus. 2015;7:1700–25. Review.

Underwood LJ, Shigemasa K, Tanimoto H, Beard JB, Schneider EN, Wang Y, Parmley TH, O'Brien TJ. Ovarian tumor cells express a novel multi-domain cell surface serine protease. Biochim Biophys Acta. 2000;1502:337–50.

Vallet V, Chraibi A, Gaeggeler HP, Horisberger JD, Rossier BC. An epithelial serine protease activates the amiloride-sensitive sodium channel. Nature. 1997;389:607–10.

Verghese GM, Gutknecht MF, Caughey GH. Prostasin regulates epithelial monolayer function: cell-specific Gpld1-mediated secretion and functional role for GPI anchor. Am J Physiol Cell Physiol. 2006;291:C1258–70.

Verhelst J, Hulpiau P, Saelens X. Mx proteins: antiviral gatekeepers that restrain the uninvited. Microbiol Mol Biol Rev. 2013;77:551–66.

Villalba M, Diaz-Lagares A, Redrado M, de Aberasturi AL, Segura V, Bodegas ME, Pajares MJ, Pio R, Freire J, Gomez-Roman J, Montuenga LM, Esteller M, Sandoval J, Calvo A. Epigenetic alterations leading to TMPRSS4 promoter hypomethylation and protein overexpression predict poor prognosis in squamous lung cancer patients. Oncotarget. 2016;7:22752–69.

Vuagniaux G, Vallet V, Jaeger NF, Hummler E, Rossier BC. Synergistic activation of ENaC by three membrane-bound channel-activating serine proteases (mCAP1, mCAP2, and mCAP3) and serum- and glucocorticoid-regulated kinase (Sgk1) in Xenopus Oocytes. J Gen Physiol. 2002;120:191–201.

Wallrapp C, Hähnel S, Müller-Pillasch F, Burghardt B, Iwamura T, Ruthenbürger M, Lerch MM, Adler G, Gress TMA. novel transmembrane serine protease (TMPRSS3) overexpressed in pancreatic cancer. Cancer Res. 2000;60:2602–6.

Wang CY, Meynard D, Lin HY. The role of TMPRSS6/matriptase-2 in iron regulation and anemia. Front Pharmacol. 2014;5:114.

Wang H, Zhou T, Peng J, Xu P, Dong N, Chen S, Wu Q. Distinct roles of N-glycosylation at different sites of corin in cell membrane targeting and ectodomain shedding. J Biol Chem. 2015;290:1654–63.

Wattenhofer M, Sahin-Calapoglu N, Andreasen D, Kalay E, Caylan R, Braillard B, Fowler-Jaeger N, Reymond A, Rossier BC, Karaguzel A, Antonarakis SE. A novel TMPRSS3 missense mutation in a DFNB8/10 family prevents proteolytic activation of the protein. Hum Genet. 2005;117:528–35.

Wilson S, Greer B, Hooper J, Zijlstra A, Walker B, Quigley J, Hawthorne S. The membrane-anchored serine protease, TMPRSS2, activates PAR-2 in prostate cancer cells. Biochem J. 2005;388:967–72.

Wong GW, Tang Y, Feyfant E, Sali A, Li L, Li Y, Huang C, Friend DS, Krilis SA, Stevens RL. Identification of a new member of the tryptase family of mouse and human mast cell proteases which possesses a novel COOH-terminal hydrophobic extension. J Biol Chem. 1999; 274:30784–93.

Wu Q, Yu D, Post J, Halks-Miller M, Sadler JE, Morser J. Generation and characterization of mice deficient in hepsin, a hepatic transmembrane serine protease. J Clin Invest. 1998;101:321–6.

Wu C, Wu F, Pan J, Morser J, Wu Q. Furin-mediated processing of Pro-C-type natriuretic peptide. J Biol Chem. 2003;278:25847–52.

Yamaguchi N, Okui A, Yamada T, Nakazato H, Mitsui S. Spinesin/TMPRSS5, a novel transmembrane serine protease, cloned from human spinal cord. J Biol Chem. 2002;277:6806–12.

Yamaoka K, Masuda K, Ogawa H, Takagi K, Umemoto N, Yasuoka S. Cloning and characterization of the cDNA for human airway trypsin-like protease. J Biol Chem. 1998;273:11895–901.

Yan W, Sheng N, Seto M, Morser J, Wu Q. Corin, a mosaic transmembrane serine protease encoded by a novel cDNA from human heart. J Biol Chem. 1999;274(21):14926–35.

Yan W, Wu F, Morser J, Wu Q. Corin, a transmembrane cardiac serine protease, acts as a pro-atrial natriuretic peptide-converting enzyme. Proc Natl Acad Sci U S A. 2000;97:8525–9.

Yasuoka S, Ohnishi T, Kawano S, Tsuchihashi S, Ogawara M, Masuda K, Yamaoka K, Takahashi M, Sano T. Purification, characterization, and localization of a novel trypsin-like protease found in the human airway. Am J Respir Cell Mol Biol. 1997;16:300–8.

Yoshinaga S, Nakahori Y, Yasuoka S. Fibrinogenolytic activity of a novel trypsin-like enzyme found in human airway. J Med Invest. 1998;45:77–86.

Yu JX, Chao L, Chao J. Prostasin is a novel human serine proteinase from seminal fluid. Purification, tissue distribution, and localization in prostate gland. J Biol Chem. 1994;269:18843–8.

Yuan X, Zheng X, Lu D, Rubin DC, Pung CY, Sadler JE. Structure of murine enterokinase (enteropeptidase) and expression in small intestine during development. Am J Physiol. 1998;274:G342–9.

Yun B, Zhang Y, Liu Y, Guan X, Wang Y, Qi X, Cui H, Liu C, Zhang Y, Gao H, Gao L, Li K, Gao Y, Wang X. TMPRSS12 is an activating protease for Subtype B Avian Metapneumovirus. J Virol. 2016;90:11231–46.

Zamolodchikova TS, Sokolova EA, Alexandrov SL, Mikhaleva II, Prudchenko IA, Morozov IA, Kononenko NV, Mirgorodskaya OA, Da U, Larionova NI, Pozdnev VF, Ghosh D, Duax WL, Vorotyntseva TI. Subcellular localization, substrate specificity and crystallization of duodenase, a potential activator of enteropeptidase. Eur J Biochem. 1997;249:612–21.

Zamolodchikova TS, Sokolova EA, Lu D, Sadler JE. Activation of recombinant proenteropeptidase by duodenase. FEBS Lett. 2000;466:295–9.

Zhang D, Qiu S, Wang Q, Zheng J. TMPRSS3 modulates ovarian cancer cell proliferation, invasion and metastasis. Oncol Rep. 2016;35:81–8.

Zhang Z, Hu Y, Yan R, Dong L, Jiang Y, Zhou Z, Liu M, Zhou T, Dong N, Wu Q. The transmembrane serine protease HAT-like 4 is important for epidermal barrier function to prevent body fluid loss. Sci Rep. 2017;7:45262.

Zhao N, Wang J, Cui Y, Guo L, Lu SH. Induction of G1 cell cycle arrest and P15INK4b expression by ECRG1 through interaction with Miz-1. J Cell Biochem. 2004;92:65–76.

Zheng X, Sadler JE. Mucin-like domain of enteropeptidase directs apical targeting in Madin-Darby canine kidney cells. J Biol Chem. 2002;277:6858–63.

Zheng XL, Kitamoto Y, Sadler JE. Enteropeptidase, a type II transmembrane serine protease. Front Biosci (Elite Ed). 2009;1:242–9. Review.

Zhirnov O, Klenk HD. Human influenza A viruses are proteolytically activated and do not induce apoptosis in CACO-2 cells. Virology. 2003;313(1):198–212.

Zhirnov OP, Ikizler MR, Wright PF. Cleavage of influenza a virus hemagglutinin in human respiratory epithelium is cell associated and sensitive to exogenous antiproteases. J Virol. 2002;76:8682–9.

Zhou X, Chen J, Zhang Q, Shao J, Du K, Xu X, Kong Y. Prognostic value of plasma soluble corin in patients with acute myocardial infarction. J Am Coll Cardiol. 2016;67:2008–14.

Zmora P, Blazejewska P, Moldenhauer AS, Welsch K, Nehlmeier I, Wu Q, Schneider H, Pöhlmann S, Bertram S. DESC1 and MSPL activate influenza A viruses and emerging coronaviruses for host cell entry. J Virol. 2014;88:12087–97.

Characterization of Proprotein Convertases and Their Involvement in Virus Propagation

9

Wolfgang Garten

Abstract

Proprotein convertases (PCs), also known as eukaryotic subtilases, are a group of serine proteases comprising furin (PACE), PC1 (PC3), PC2, PC4, PACE4, PC5 (PC6), and PC7 (LPC, PC8) that generate bioactive proteins and peptides, such as hormones, receptors, and growth factors by cleaving precursor proteins at multibasic motifs. Two other family members, SKI-1/S1P and PCSK9, cleave regulator proteins involved in cholesterol and fatty acid homeostasis at nonbasic peptide bonds. Furin is ubiquitous in eukaryotic tissues and cells. PACE4, PC5, and PC7 are also widespread, whereas the expression of the other PCs is more restricted. PCs are synthesized as multi-segmented zymogens which are autocatalytically activated. The prodomains have regulatory and inhibitory functions. The catalytic domains are the most conserved domains among the PCs. The architecture of the catalytic active furin domain is known in different binding states. The C-terminal parts of the PCs differ in length and structure and contain encoded peptide signatures guiding the PCs to the subcellular destinations on the secretory pathways: SKI-1/S1P to the cis-Golgi, furin, PC5B, and PC7 to the TGN region but also to the plasma membrane. PACE4, PC5A, and PCSK9 are attached at the cell surface. Truncated, soluble furin and SKI-1/S1P, as well as PC1 and PC2, are released into the extracellular matrix. Many enveloped viruses are activated by furin and furin-like PCs and arenaviruses and a few bunyaviruses by SKI-1/S1P. The PCs cleave the viral fusion glycoprotein to trigger fusion of viral envelopes with cellular membranes to deliver the viral genome into host cells. Cleavage by PCs, occasionally in concert with other endoproteases, enables conformational changes in the viral membrane proteins needed for correct oligomerization of glycoprotein spikes and their effective incorporation into virions. Mutational alterations of PC cleavage sites can reduce the fusion potential of

W. Garten
Institute of Virology, Philipps-University Marburg, Marburg, Germany
e-mail: garten@staff.uni-marburg.de

E. Böttcher-Friebertshäuser et al. (eds.), *Activation of Viruses by Host Proteases*,
https://doi.org/10.1007/978-3-319-75474-1_9

viral surface proteins and thus facilitate the development of secure live attenuated vaccines. Alternatively, agents preventing cleavage of viral surface (glyco) proteins block fusion capacity and multicyclic virus replications. PC inhibitors are suggested as promising antiviral drugs for quite a number of viruses causing severe infections.

9.1 Introduction

Many biologically active proteins are synthesized as larger inactive precursors (proproteins). Posttranslational processing by limited proteolysis of the precursors is a mechanism generating active proteins and peptides that enables cells to regulate many vital processes. In general, processing occurs when the proteins are transferred in the secretory pathway from the endoplasmic reticulum (ER) to other cell compartments and to the cell surface. The precursors are frequently cleaved at amino acid motifs containing single or multiple arginine or lysine residues. For a long time, little was known about the activation proteases. The situation changed when the genomes of a wide variety of organisms were decoded. Roughly 600 different proteases have been identified in each genome allowing expression and functional studies. Differing in mammalian species, roughly 200 are serine proteases that can be divided into about 20 subgroups based on structure, enzymatic properties, and physiological functions. The available evidence indicates that several of these are involved in processing at basic cleavage sites (Puente et al. 2003).

Proinsulin was the first identified precursor polypeptide that contains two dibasic cleavage motifs, arginine-arginine and lysine-arginine (Steiner et al. 1967). Comparison of mature insulin with its precursor suggested that a dibasic-specific endoprotease and a B-type-specific carboxypeptidase are responsible for correct processing, and this concept was further substantiated by the observation that insulin was obtained when the activating enzymes were surrogated by trypsin and carboxypeptidase B in vitro (Kemmler et al. 1971). Subsequently, the number of inactive proproteins with presumed and meanwhile ascertained di- and multibasic cleavage motifs continuously increased. The list started with pro-opiomelanocortin (POMC), proparathyroid hormone, proalbumin, pro-beta-secretase, pro-nerve growth factor, and the proproteins and propeptides of ß-lipotropic hormone, ß-melanocyte-stimulating hormone (β-MSH), γ-lipotropin (γ-LPH), β-lipotropin (β-LPH), and von Willebrand factor and is steadily upgraded (Barr 1991; Seidah 2011). A large number of precursor proteins are cleaved at **KK**↓, **RK**↓, **KR**↓, or **RR**↓, which are frequently combined in the consensus sequences (**K/R**)xn(**K/R**)↓, where a variable number ($n = 0,2,4,6$) of basic and nonbasic amino acids separates the flanking basic amino acids.

The first authentic proprotein converting enzyme identified was the calcium-dependent subtilisin-like protease kexin encoded by the *kex2* gene of *Saccharomyces cerevisiae* that cleaves yeast and mammalian proteins and peptides at dibasic peptide sites (Achstetter and Wolf 1985; Fuller et al. 1989a; Julius et al. 1984; Thomas et al. 1988). The first mammalian orthologue that possesses the capacity for such a proprotein cleavage was furin (Fuller et al. 1989b; Bresnahan et al. 1990; Wise et al. 1990; Hatsuzawa et al. 1990; Misumi et al. 1990b; van de Ven et al. 1990). Furin cleaves many precursor proteins at the C-terminal end of the motif **RX(K/R)R**↓ (Barr 1991; Nakayama 1997). Later additional closely related subtilisin-/kexin-like serine proteases were identified and designated proprotein convertases (PCs) (Seidah 2011). The PC family now contains nine members: furin, PC1/3, PC2, PC4, PACE4, PC5/6 (further on PC5), PC7, SKI-1/S1P, and PCSK9 (Table 9.1). The first seven PCs cleave substrates C-terminally at arginine of multibasic recognition motifs. The last two PCs, SKI-1/S1P and PCSK9, recognize nonbasic scissile peptide bonds. The subtilisin-/kexin-like isoenzyme (SKI-1), also known as site-1 protease (S1P), cleaves proproteins at the motif **R**X(L/I/V)X↓, where X presents any amino acid. The neural apoptosis-regulated convertase 1 (NARC-1), also designated as proprotein convertase subtilisin/kexin type 9 (PCSK9), is autocatalytically cleaved at the amino acid motif VFAQ↓SIP and does not cleave other proproteins in *trans* but has substrate binding and signaling functions (Seidah et al. 2014). The main cleavage site specificities of the individual PCs are shown in Table 9.1. There are several excellent reviews in which the PC field has been described in detail (Artenstein and Opal 2011; Nakayama 1997; Seidah 2011; Seidah et al. 2013; Seidah and Prat 2002, 2012; Steiner 1998; Thomas 2002).

Many cell proproteins, viral envelope glycoproteins, and bacterial toxins exhibit multibasic cleavage sites (reviewed by Klenk and Garten 1994; Gordon and Leppla 1994) (Tables 9.2 and 9.3). Multibasic cleavage was shown first with the hemagglutinin of fowl plague virus (FPV), a highly pathogenic avian influenza virus (HPAIV), and the fusion protein of virulent Newcastle disease virus (NDV) strains (Bosch et al. 1981; Garten et al. 1981, 1982; Toyoda et al. 1987; Nagai 1995). Cleavage of these glycoproteins in practically all cells allows rapid virus spread in the infected host and proved to be a major determinant for the high pathogenicity of these viruses. The first hint for the nature of the activating host proteases came from the observation that activation of FPV was calcium-dependent, a characteristic feature of PCs (Klenk et al. 1984). The final proof was provided by the identification of furin as the enzyme activating the hemagglutinin of HPAIV and the HIV env glycoprotein (Stieneke-Gröber et al. 1992; Hallenberger et al. 1992).

This chapter gives an overview on furin and the other members of the PC family with a focus on those involved in the life cycle of viruses. The structure and function of the proteases and their biosynthesis, subcellular trafficking, and localization in cells will be shown, as well as their occurrence in specific cell types, tissues, and organisms. PCs have a multifunctional role in virus infection. The main function is the proteolytic activation of fusion-competent surface proteins of enveloped viruses required for the delivery of the viral genome into host cells. Beyond that, PCs are responsible for conformation changes of proteins in virus assembly, for receptor recognition, and for release of pathogenicity factors. PCs together with other proteases are involved in complex cleavage patterns of viral surface proteins. The physiological roles of PCs, especially in embryonic development, as revealed by knockdown systems and specific inhibitory agents will also be discussed. Finally, light will be drawn on the use of protease activation mutants for vaccine design and on the use of protease inhibitors for antiviral therapy.

Table 9.1 Proprotein convertases

PCSKs	Alternative names	Initial descriptions	Cleavage site specificity	Distribution in the organism/subcellular compartments/major functions	Phenotypes of PC deficiency PC null mice, *human PC gene defects (in italic)*
PCSK1	PC1/3	Seidah et al. (1990, 1991), Smeekens et al. (1991) Smeekens and Steiner (1990) Nakayama et al. (1992) Mizuno et al. (1989)	**KR**↓ or **RR**↓	Pituitary Secretory granules Processing of prohormone peptides	Viable; pre- and postnatal lethality, dwarfism Hyper-proinsulinemia, chronic diarrhea *Obesity, impaired glucose homeostasis, impaired gastrointestinal function, hypogonatropic hypogonadism*
PCSK2	PC2	Smeekens and Steiner (1990) Seidah et al. (1990)	**KR**↓ or **RR**↓	Pituitary Secretory granules Processing of prohormone peptides	Viable with retarded growth Hyperglycemia, endocrine peptides processing defects
PCSK3	Furin PACE	Roebroek et al. (1986) van de Ven et al. (1990) Hatsuzawa et al. (1990) Bresnahan et al. (1990), Wise et al. (1990), Misumi et al. 1990	**R**X**K/RR**↓ **R**XX**R**↓ RX**R**XXX**R/KR**↓	Ubiquitous TGN, endosome, cell surface, extracellular soluble furin	Lethal at day e10.5–11.5; multiple defects, i.e. Failure of axial rotation, severe ventral closure defects
PCSK4	PC4	Nakayama et al. (1992) Seidah et al. (1992)	KXX**R**↓ **R**X**K/RR**↓	Testes, ovary, placenta Plasma membrane	Viable Reduced fertility
PCSK5	PC5/6A PC5/6B	Lusson et al. (1993) Nakagawa et al. (1993)	**R**X**K/RR**↓	PC5A: ubiquitous; Golgi, granules PC5B: small intestine, kidney, liver, endocrine and non-endocrine cells TGN, endosomes, cell surface	Postnatal death at days 4.5–7.5 Embryo death at days e10.5–11.5
PCSK6	PACE4	Kiefer et al. (1991)	**R**X**K/RR**↓	Ubiquitous TGN	75% mice are viable, 25% are lethal at days e13.5–15.5 Phenotype variable, heart and situs defects, craniofacial malformations

PCSKs	Alternative names	Initial descriptions	Cleavage site specificity	Distribution in the organism/subcellular compartments/major functions	Phenotypes of PC deficiency PC null mice, *human PC gene defects (in italic)*
PCSK7	PC7, PC8 LPC SPC7	Seidah et al. (1996), Bruzzaniti et al. (1996), Meerabux et al. (1996), Constam et al. (1996)	**RXK/RR↓**	Ubiquitous ER, TGN, cell surface	Viable; no change in phenotype
PCSK8	SKI-1/S1P MBTPS1	Seidah et al. (1999) Sakai et al. (1998)	RX(V,L)(K,F,L)↓	Ubiquitous TGN, plasma membrane Lipid metabolism regulation	Die in embryo state Liver specific conditional SKI-1/S1P null mice are viable with reduced synthesis of cholesterol and fatty acids
PCSK9	NARC-1	Abifadel et al. (2003) Seidah et al. (2003)	VFAQ↓SIP	Liver TGN, cell surface A liver-secreted plasma inactive enzyme that drags LDLR to endosomes for degradation	PCSK9-null mice have more LDLR and less insulin in their pancreas than control mice, hypoinsulinemic, hyperglycemic *autosomal dominant hypercholesterolemia (ADH), a risk factor for coronary heart disease*

PCSK proprotein convertase subtilisin/kexin type, *PC* proprotein convertase, *PACE* paired basic amino acid cleaving enzyme, *LPC* lymphoma proprotein convertase, *SKI-1/S1P* subtilisin/kexin-isozyme 1/site-1 protease, *MBTPS1* membrane bound transcription factor peptidase, site 1, *NARC-1* neural apoptosis-regulated convertase-1, *e* embryonal day

Table 9.2 Selected cellular proteins and bacterial toxins cleaved by proprotein convertases

Precursor proteins	Proprotein convertases	Cleavage sites P_7 P_6 P_5 P_4 P_3 P_2 P_1 ↓ P_1' P_2'	References
Growth factors and hormones			
Mouse ß-nerve growth factor	Furin	RTH**RS**K**R**↓SS	Bresnahan et al. (1990)
Human neurotrophin-3	Furin, PC5B	TS**R**RK**R**↓YA	Seidah et al. (1996)
Human insulin-like growth factor-I	PACE4	**K**PA**K**SA**R**↓SV[a]	Duguay et al. (1995) Khatib et al. (2001)
Human parathyroid hormone	Furin	**K**SVK**KR**↓SV[a]	Hendy et al. (1995) Munzer et al. (1997)
Receptors			
Human insulin receptor	PC5A	PS**R**K**RR**↓SL	Robertson et al. (1993)
Human αv integrin	Furin, PC7	PQ**R**R**RR**↓QL	Stawowy et al. (2005)
Plasma proteins			
Human albumin	Furin	**R**GV**RR**↓DA[a]	Oda et al. (1992) Mori et al. (1999)
Human blood factor X	Furin, PC5	LE**R**R**KR**↓SV	Himmelspach et al. (2000)
Human von Willebrand factor	Furin, PACE4, PC7	SH**RS**K**R**↓SL	van de Ven et al. (1990)
Matrix metalloproteinases			
Human stromelysin-3	Furin	ARN**R**Q**KR**↓FV	Pei and Weiss (1995)
Human MT-MMP1	Furin, PACE4	NV**R**R**KR**↓YA	Sato et al. (1996)
Other cellular proteins			
Human furin prodomain[b]	Furin, PACE4	AKR**R**T**KR**↓DV	Anderson et al. (1997)
Mouse 7B2 chaperone	Furin	QR**R**K**RR**↓SV	Paquet et al. (1994)
Bacterial exotoxins			
Anthrax toxin protective antigen	Furin	NS**R**K**KR**↓ST	Molloy et al. (1992)
Diphtheria toxin	Furin	GN**R**V**RR**↓SV	Tsuneoka et al. (1993)
Shiga toxin	Furin, PACE4	AS**R**VA**R**↓MA	Garred et al. (1995)

[a]Unusual cleavage site for furin
[b]Autocatalysis

Table 9.3 Selected viral proteins cleaved by proprotein convertases

Virus	Viral protein	Cleavage motifs[a]	Proprotein convertases	References
Orthomyxoviridae				
A/FPV/Rostock/34 (H7N1)	HA	$_{337}$KK**R**E**KR**↓	Furin	Porter et al. (1979)
A/FPV/Rostock/34 (H7N1)	HA	$_{337}$KK**R**K**KR**↓	Furin, PC5	Garten et al. (1985), Feldmann et al. (2000)
A/turkey/Ireland/1378/85 (H5N8)	HA	$_{338}$**RKRKKR**↓	Furin, PC5	Walker and Kawaoka (1993), Walker et al. (1994), Horimoto et al. (1994)
A/Thailand/1 (KAN-1)/2004 (H5N1)	HA	$_{342}$**RER**RRK**KR**↓	Furin	Puthavathana et al. (2005)
Paramyxoviridae				
Newcastle disease virus (Italian)	F	$_{111}$R**R**Q**RR**↓	Furin, PC5, PACE4	Toyoda et al. (1987), Sakaguchi et al. (1994)
Measles virus	F	$_{108}$R**R**H**KR**↓	Furin	Richardson et al. (1986)
Human parainfluenza virus type 3	F	$_{104}$**P**R**T**KR↓	Furin	Ortmann et al. (1994)
Respiratory syncytial virus (RSV)	F	$_{131}$KK**R**K**RR**↓	Furin	González-Reyes et al. (2001)
Bovine RSV	F	$_{106}$**RAK/RR**↓; $_{131}$KK**R**K**RR**↓	Furin	Zimmer et al. (2001, 2003)
Filoviridae				
Marburg virus	GP	$_{431}$**RKKR**↓	Furin	Volchkov et al. (2000)
Ebola-Zaire virus	GP	$_{497}$R**R**T**RR**↓	Furin	Volchkov et al. (1998)
	sGP	$_{321}$**RVRR**↓	Furin	Volchkova et al. (1999)
Bornaviridae				
Borna virus	GP	$_{245}$K**R**R**RR**↓	Furin	Richt et al. (1998), Eickmann et al. (2005)
Flaviviridae				
Yellow fever virus	pr-M	$_{84}$**RSRR**↓**SRR**↓	Furin	Rice et al. (1985)
Tick-borne encephalitis virus	pr-M	$_{86}$**S**R**T**RR↓	Furin	Stadler et al. (1997)
Dengue virus Types 1, 3	pr-M	$_{86}$R**R**D**KR**↓	Furin	Keelapang et al. (2004)
Dengue virus Types 2, 4	pr-M	$_{86}$R**R**E**KR**↓	Furin	Keelapang et al. (2004)
Zika virus	pr-M	$_{89}$R**R**S**RR**↓	Furin	Kuno and Chang (2007)

(continued)

Table 9.3 (continued)

Virus	Viral protein	Cleavage motifs[a]	Proprotein convertases	References
Togaviridae				
Semliki forest virus	E3/E2	$_{330}$**RHRR**↓	Furin	Jain et al. (1991), Sariola et al. (1995) Zhang et al. (2003)
Sindbis virus	E3/E2	$_{325}$**RSKR**↓	Furin	Strauss et al. (1984)
Chikungunya virus	E3/E2	$_{322}$**RQRR**↓	Furin, PACE4, PC5	Heidner and Johnston (1994), Ozden et al. (2008)
Coronaviridae				
Mouse hepatitis virus	S	$_{713}$R**R**AH**R**↓ (S1/S2)	Furin	Sturman and Holmes (1984), Schmidt et al. (1987), de Haan et al. (2004)
Infectious bronchitis virus (IBV)	S	$_{533}$R**R**F**RR**↓ (S1/S2); $_{687}$**RRKR**↓ (S2′)	Furin Furin	Cavanagh et al. (1986) Bosch et al. (2008)
SARS virus	S	$_{441}$**R**YL**R**↓; $_{758}$**R**NT**R**↓ (S1/S2)	Furin	Follis et al. (2006)
MERS virus	S	$_{626}$**R**QQ**R**↓; $_{691}$**R**ST**R**↓ (S1/S2); $_{884}$**RSAR**↓ (S2′)	Furin Furin	Millet and Whittaker (2014) Millet and Whittaker (2015)
Arenaviridae				
Lassa virus	GP	$_{256}$**RRLL**↓	SKI-1/S1P	Lenz et al. (2001)
LCMV	GP	$_{262}$**RRLA**↓	SKI-1/S1P	Beyer et al. (2003)
Tacaribe virus	GP	$_{246}$**RTLK**↓	SKI-1/S1P	Allison et al. (1991)
Bunyaviridae				
Crimean-Congo hemorrhagic fever virus (*Nairovirus*)	GP38	$_{244}$**RSKR**↓	Furin	Sanchez et al. (2006)
	Pre-GP-N	$_{516}$**RRLL**↓	SKI-1/S1P	Altamura et al. (2007)
	Pre-GP-C	$_{1038}$**RKPL**↓	SKI-1/S1P (?)	Bergeron et al. (2015)

Retrovirus				
HIV-1	env	$_{508}$**REKR**↓	PC7, Furin	McCune et al. (1988), Hallenberger et al. (1997)
	gp120 V3-loop	$_{308}$**RIQR**↓	Furin	Morikawa et al. (1993)
SIV	env	$_{524}$**R**N**KR**↓	Furin	Yamshchikov et al. (1995)
Foamy virus	env leader peptide env leader peptide env	$_{122}$**R**R**RR**↓ $_{122}$**R**TA**R**↓SL**R**↓ $_{572}$**R**K**RR**↓	Furin Furin Furin	Geiselhart et al. (2004) Duda et al. (2004) Duda et al. (2004)
Herpesviridae				
Human cytomegalovirus	gB	$_{354}$**R**TI**R**↓; $_{457}$**R**T**KR**↓	Furin	Brücher et al. (1990), Vey et al. (1995)
Epstein-Barr virus	gB	$_{428}$R**R**R**RR**↓	Furin	Johannsen et al. (2004)
Varicella zoster virus	gB	$_{491}$**R**S**RR**↓; $_{501}$**R**AN**R**↓	Furin	Oliver et al. (2009)
Equine cytomegalovirus	gB	$_{518}$**R**R**RR**↓; $_{495}$**R**SN**R**↓	Furin	Spiesschaert et al. (2016)
Baculovirus				
Culex nigripalpus nucleo-polyhedrovirus (CuniNPV)	F	$_{126}$**R**A**RR**↓	Furin	Long et al. (2006), Wang et al. (2017)

SARS severe acute respiratory syndrome, *MERS* Middle East respiratory syndrome, *LCMV* lymphocytic choriomeningitis virus, *HIV* human immunodeficiency virus, *SIV* simian immunodeficiency virus

[a]Essential amino acid motifs in bold letters

9.2 Structure and Biosynthesis of PCs

All PCs are synthesized as multi-segmented pro-precursors which start with an N-terminal signal peptide; continue with a prodomain (prosegment, propeptide), a catalytic domain, and a P-domain (middle domain); and complete the PC ectodomain with variable C-terminal domains (Fig. 9.1). The convertases PC1/3, PC2, PC4, PC5A, PACE4, and PCSK9 are expressed as soluble- or membrane-attached PCs. Furin, PC5B, PC7, and SKI-1/S1P possess a transmembrane anchor domain and a C-terminal cytoplasmic domain. Furin and SKI-1/S1P can be cleaved at a distinct peptide bond in the ectodomain, and thus both convertases exist also as soluble furin (sfurin) and soluble SKI-1/S1P (sSKI-1).

The prodomain (~ 80 amino acids) of the precursor PCs acts as an intramolecular chaperone that guides the folding and activation of the pro-convertases. Simultaneously, the prodomain and/or their autocatalytically split-off fragments function as PC inhibitors. The catalytic domain is the most conserved domain and covers about 340 amino acids with the reactive amino acids aspartic acid (D), histidine (H), and serine (S) in subtilisin-like arrangement and a specific asparagine (N) residue forming an oxyanion hole (Fig. 9.2). The other domains of the convertases are quite divergent in size, sequence homology, and function. The adjacent P-domain confers structural stability and regulates the enzymatic activity. Furin, PC5, and PACE4 contain a conserved cysteine-rich domain (CRD) (Nakayama 1997). CRD functions as a cell surface anchor and interacts with the tissue inhibitors of metalloproteinases (TIMPs) (Nour et al. 2005). The transmembrane domain of furin, PC5B, PC7, and SKI-1/S1P anchors the cytoplasmic domains in the membrane system of the constitutive exocytic pathway. The cytoplasmic domains possess several intrinsic signals determining the residence for each PCs in specific compartments of the constitutive secretory pathway.

9.2.1 Structure of the Catalytic Domain

Furin is the prototype and by now the best-characterized member of the PC family. Because of difficulties in purifying the enzyme in sufficient amounts and quality, initial studies on the catalytic domain of human furin were based on homology modeling using the crystal structures of the related serine proteinase subtilisin BPN and thermitase of bacterial origin (Siezen et al. 1994). *Than* and colleagues succeeded in establishing the first crystal structure of a truncated form of furin which was expressed in sufficient quantities and purified from mammalian cells as enzymatically active molecule by *Lindberg* and coworkers (Henrich et al. 2003; Fig. 9.2). The catalytic pocket carried the covalently bound inhibitor decanoyl-Arg-Val-Lys-Arg-chloromethylketone (Dec-RVKR-CMK), which belongs to the first small peptidyl inhibitors designed for furin (Garten et al. 1989, 1994). This approach revealed an arrangement of highly negatively charged amino acids, i.e., aspartic and glutamic acids, around the catalytic pocket which explains the binding of substrates with the preferential basic amino acids in distinct positions.

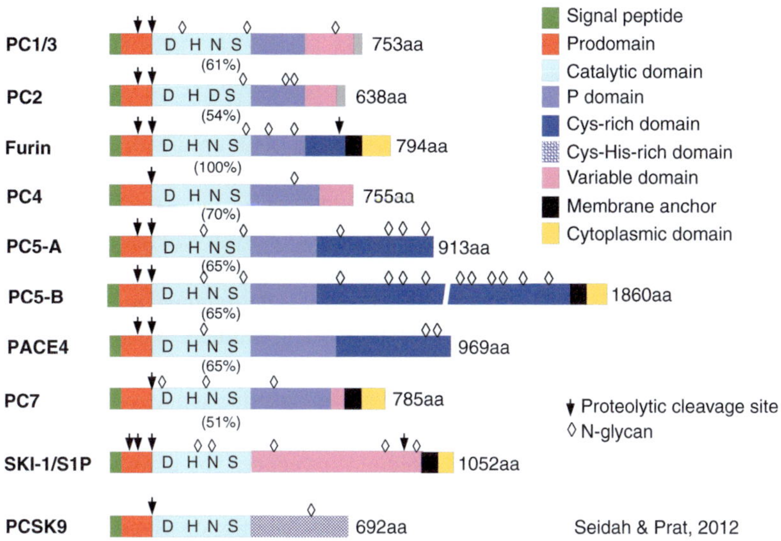

Fig. 9.1 Domain structure of proprotein convertases. Individual domains are illustrated by colored bars, and the total lengths of the human PCs are indicated by the number of amino acids. D, H, N, and S are catalytic active amino acids of the catalytic triad; N contributes to the oxyanion hole. Autocatalytic cleavage sites between and within pro- and catalytic domains of all PCs and the cleavage sites generating soluble forms of furin and SKI-1/S1P are indicated by arrows. C-terminal gray boxes indicate peptide sequences required for cell membrane attachment. The percentage numbers indicate amino acid identity of the catalytic domains relative to furin (Seidah and Prat 2012; Nakayama 1997)

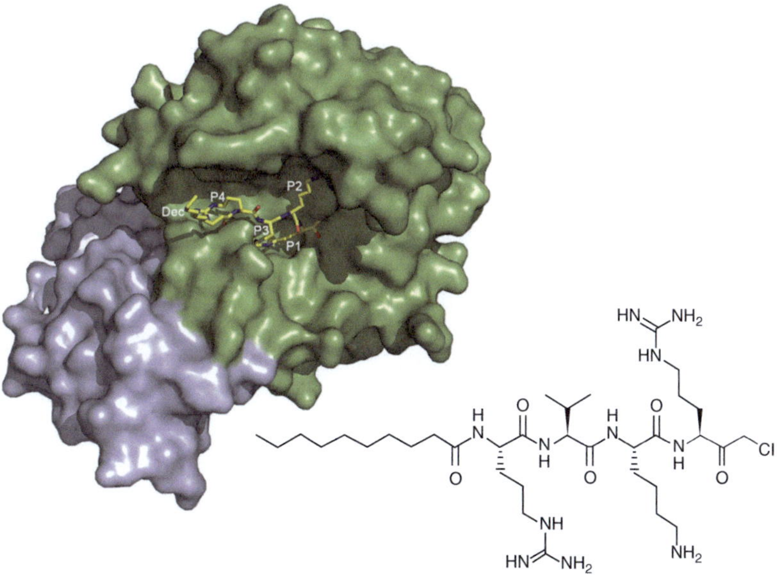

Fig. 9.2 Crystal structure of the inhibitor Dec-Arg-Val-Lys-Arg-CMK in complex with mouse furin. The inhibitor is shown in sticks with carbon atoms in yellow, nitrogen in blue, and oxygen in red; the P1–P4 residues and the N-terminal decanoyl group are labeled. The catalytic domain of furin is presented with its surface in green and the P-domain in light blue (Henrich et al. 2003)

As found with all other members of the PC family, the P-domain of furin stabilizes the catalytic domain (Zhou et al. 1998; Than et al. 2005). Detailed knowledge on enzyme-substrate binding properties was achieved by crystallographic analysis of furin loaded with inhibitors or bound to an inhibitory antibody (Dahms et al. 2014, 2016a, b). Very recently, the structure of an unliganded form of furin was determined (Dahms et al. 2016b), where furin exists in a so-called off state, which is incompatible with substrate binding. Moreover, two structures of ethylenediaminetetraacetic acid (EDTA)-treated forms of furin have been determined, one unliganded and the second in complex with a substrate analogue inhibitor. These studies revealed different affinities of three calcium ions, because only two distinct calcium ions have been removed by EDTA treatment. The transition from the off to the on state is triggered by ligand binding and appears to be the precondition for the preferential recognition of the four-residue sequence motif of furin substrates. Moreover, the comparison with EDTA-treated furin structures revealed that the ligation by the presence of calcium influences the active-site geometry and thus modulates furin activity (Dahms et al. 2016b). Recently, furin was complexed with a small non-substrate-like non-peptidic inhibitor which induces structural distortions of the active site of the enzyme (Dahms et al. 2017). Based on these observations taken together, the high substrate specificity of furin can be explained by conformational changes triggered by binding of substrate and calcium. The detailed structural studies also allowed the rational design of novel inhibitors of furin (see Chap. 11).

The only x-ray structure of another PC is that of human PCSK9 which provides detailed insight into its exceptional biochemical characteristics and biological function (Cunningham et al. 2007; Piper et al. 2007). The full-length PCSK9 precursor (amino acids 31–692) was used for crystallization, and the architecture of a very high portion of the polypeptide chain (amino acids 61–683) was determined at 2.3 Å resolution (Fig. 9.3). The structure model comprises nearly the complete prodomain with only two smaller disorders in both terminal peptide regions remaining unsolved. There is a strong interaction between the prodomain and the catalytic domain which explains the inaccessibility of any extern substrate to the catalytic cleft of PCSK9. The C-terminal domain (termed V domain) of this enzyme has a unique structure that allows binding to the low-density lipoprotein receptor (LDLR), which is a target of PCSK9.

Three-dimensional structures of other human PCs are currently not available, but their catalytic domains have been modeled based on the crystal structure of furin. The catalytic domains of PC4, PACE4, PC5/6, and furin resemble each other, whereas those of PC1/3, PC2, and PC7 are less similar (Henrich et al. 2005).

9.2.2 Structure and Function of the Prodomain

The three-dimensional prodomain structures were solved with PCSK9 and PC1/3. The prodomain of PCSK9 was determined by x-ray crystallography, that of PC1/3 by NMR spectroscopy. The prodomain shows a well-ordered core consisting of a four-stranded antiparallel β-sheet with two α-helices packed against one side of this sheet (Tangrea et al. 2002; Cunningham et al. 2007). Both prodomain structures are

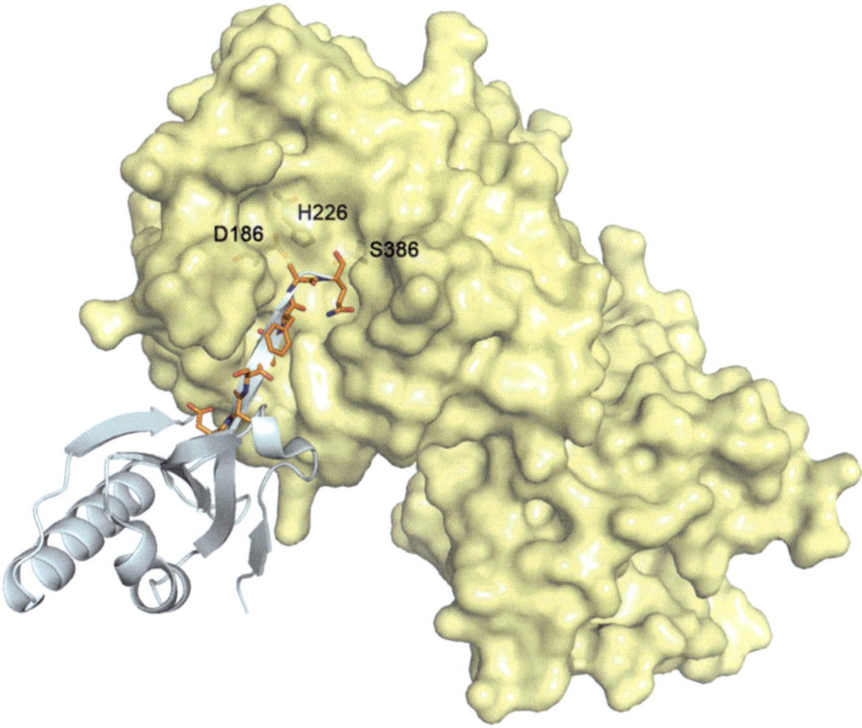

Fig. 9.3 Structure of PCSK9 in its autocatalytically cleaved form. The prodomain is shown in cartoon style (cyan) with its C-terminal heptapeptide segment $_{146}$DSSVFAQ$_{152}$ as sticks with carbon atoms in orange, which is bound and inhibits the active site. The catalytic domain is shown with a transparent surface in yellow; the residues of the catalytic triad (S_{386}, H_{226}, and D_{186}) are labeled and provided as sticks (PDB: 2P4E; Cunningham et al. 2007)

similar assuming that all other PC prodomains are structurally closely related. The prodomain is an independent separate domain which possesses three functions: (1) it masks the catalytic domain, (2) it blocks the enzyme activity a priori, and (3) it regulates the dissociation of autocatalytic cleavage fragments at distinct stages on the secretory pathway in a pH-dependent manner. The mechanism how the PCs are differentially processed in space and time has been elucidated in recent years. The concept of pH sensors is based on the histidine content present in the prodomains of the PCs traveling along the pH gradient of the secretory pathway. Histidine-69 (human furin) is conserved in the prodomains of all PCs and functions as the master histidine-encoded pH sensor which regulates enzyme activation at distinct pH values depending on the numbers of the residual histidine residues present in the prodomain. The prodomain of furin is activated at pH ~6.5 within the trans-Golgi network (TGN), whereas the PC1/3 prodomain is activated at pH ~5.5 within the dense-core secretory granules, whereby the prodomain of furin has twice the content of histidine residues in comparison with the prodomain of PC1/3. The correlation of pH-dependent activation and histidine content was demonstrated by swapping the prodomains between both closely related PC family members (Dillon et al. 2012; Shinde and Thomas 2011; Williamson et al. 2013).

9.2.3 Biosynthesis, Maturation, and Subcellular Localization of PCs

Furin, PC7, PC5B, and SKI-1 are synthesized as class I membrane proteins and the other PCs as soluble secretory proteins (Figs. 9.1 and 9.4). The preproproteins start with the signal peptides which translocate the nascent polypeptide chains from the cytoplasm into the lumen of the ER where the signal peptides are cotranslationally removed from the adjacent prosegment (prodomain). As mentioned above, the newly synthesized prodomains facilitate the folding of the polypeptide chains and block the proteolytic activity of the proPCs until the prodomain fragments are split off and removed at distinct stages of the secretory pathway in a pH-dependent manner. PC1/3 and PC2 become active in secretory granules, furin, PACE4, and PC5B in the TGN region, and PC7 is accumulated in an active form in a separate vesicular

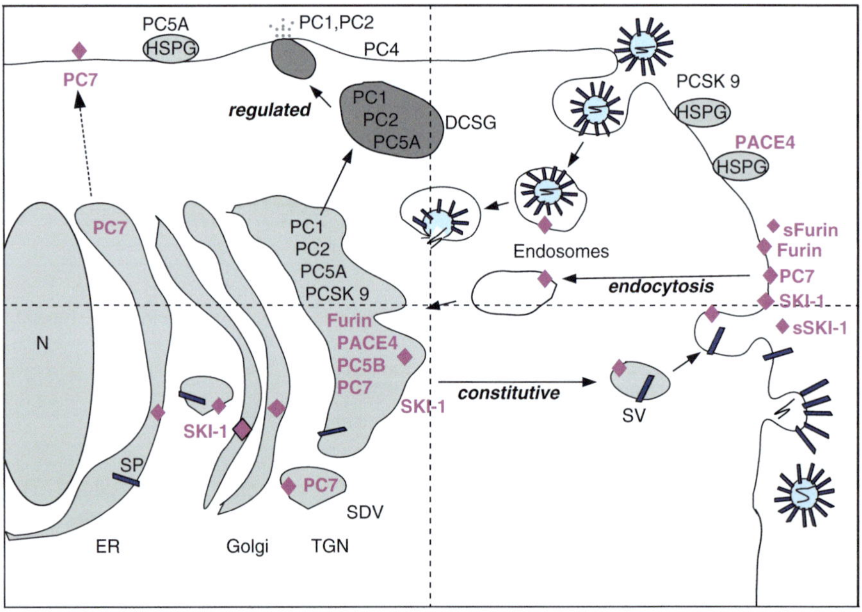

Fig. 9.4 Subcellular trafficking and residences of PCs in a virus-infected cell. The scheme shows the initial infection steps of an enveloped virus with glycoprotein spikes (blue) and the biosynthetic pathway of virus spikes from the endoplasmic reticulum (ER) to the plasma membrane where assembly and release of the virus take place. PCs relevant for virus maturation (magenta) are shown on the exocytic pathway from ER via the Golgi apparatus, the trans-Golgi network (TGN), and the secretory vesicles (SV) to the plasma membrane. PCs recycle via endosomes to their residences: furin and PC5B and PACE4 accumulate in the TGN. PC7 is enriched and enzymatically active in sialyltransferase-deficient vesicles (SDV) and in minor quantities directly exported to the plasma membrane circumventing the Golgi/TGN complex. SKI-1/S1P is mainly active in the cis-Golgi region. The furin ectodomain is cut off by an unknown protease, and SKI1/S1P is probably released by autocatalysis. Soluble enzymatically active furin (sfurin) and soluble sSKI-1 are released into the extracellular matrix; PACE4, PCSK9, and PC5A interact with tissue inhibitors of metalloproteases (TIMPs) and form tertiary complexes with heparan sulfate proteoglycans (HSPGs) ready for cleavage of extracellular substrates. PC1/3, PC2, PC5A, and PCSK9 (black) are most probably irrelevant for processing of viral proteins

structure close to the TGN. SKI-1/S1P is functionally active in the endoplasmic reticulum/Golgi (ER/Golgi) and in the post-Golgi/TGN area (Brown and Goldstein 1999; Lenz et al. 2001; Beyer et al. 2003). PC2 requires the neuroendocrine bifunctional chaperone 7B2 for egress of the ER; its N-terminal domain facilitates the maturation of proPC2, and its C-terminal peptide simultaneously functions as a potent inhibitor (Braks and Martens 1994; Zhu and Lindberg 1995). PC4, PC7, and PCSK9 are processed at a single cleavage site at the boundary between the prodomain and catalytic domain. The other convertases have a second cleavage site, except for SKI-1/S1P which possesses a third cleavage site (Burri et al. 2012a; da Palma et al. 2014; Fig. 9.1). SKI-1/S1P becomes enzymatically active in *trans* before all prodomain fragments are dissociated from the enzyme (Elagoz et al. 2002; da Palma et al. 2016).

The furin-like PCs furin, PACE4, PC5B, and PC7 are trafficking along the constitutive secretory pathway from the ER to the plasma membrane. They show the highest activity in the TGN region sharing the common route with their substrates (Fig. 9.4). PCs recycle to the TGN and to the cis-Golgi. A minor portion of PC7 is routed form the ER directly to the plasma membrane. PC5 exists in two forms, soluble PC5A and membrane-anchored PC5B, due to different gene splicing. Both forms diverge from the TGN in different routes; PC5A migrates via the regulated pathway passing the secretory granules to the plasma membrane where the enzymatically active convertase is tethered to heparan sulfate proteoglycans like PACE4 (Nour et al. 2005). PC7 is arrested as an active integral protease in the TGN-derived sialyltransferase-deficient vesicles (SDV) (Wouters et al. 1998; Declercq et al. 2017). SKI-1/S1P resides preferentially in the late ER and cis-Golgi region where it cleaves the sterol regulatory element-binding protein (SREBP) among other substrates (Brown and Goldstein 1999). PCSK9 is transported via the secretory pathway and secreted in the medium outside from the cell. PC1/3 and PC2 are transported along the regulatory pathway to the dense-core secretory granules, from where they are released by neuronal or hormonal stimuli into the cellular environment (Thomas 2002; Seidah and Prat 2002; Lee and Lindberg 2008).

Subcellular localization of the individual PCs is determined by intrinsic sorting signals (Fig. 9.5). Furin was one of the first molecules shown to accumulate in the TGN under steady-state conditions governed by a molecular address in the cytoplasmic domain (Schäfer et al. 1995; Bosshart et al. 1994; Takahashi et al. 1995). The address is necessary and sufficient for TGN localization and for recycling by the clathrin endocytosis pathway through endosomes. It consists of several destination-determining signals in the form of short peptide sections, which together are necessary for an efficient accumulation of furin in the TGN. They include (1) the acidic signal $CPSDSEEDEG_{783}$ containing two casein kinase II (CKII) phosphorylation sites, (2) the internalization signal $YKGL_{765}$, (3) a leucine-isoleucine signal LI_{760}, and (4) the signal F_{790} (Vey et al. 1994; Schäfer et al. 1995; Molloy et al. 1999; Teuchert et al. 1999a, b; Stroh et al. 1999; Voorhees et al. 1995; Thomas 2002).

The cytoplasmic domain of PC5B contains signal elements homologous to furin: the motif YXXL/I, acidic peptide stretches containing serine residues as potential casein kinase II phosphorylation sites, and dileucine motives (Fig. 9.5). Therefore, the transport pathways of PC5B and furin are similar (De Bie et al. 1996).

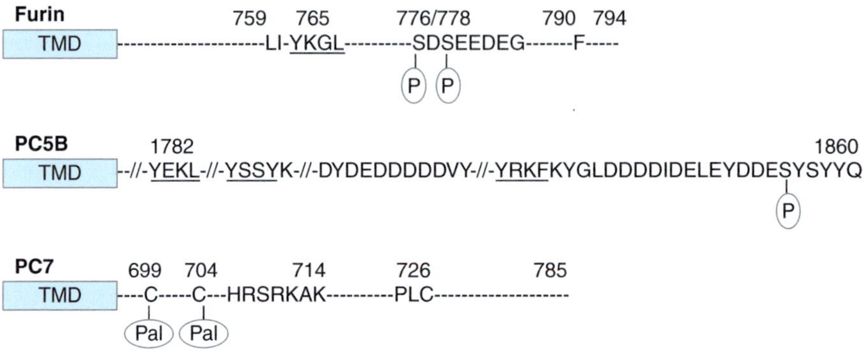

Fig. 9.5 Sorting signals of membrane-anchored PCs. Cytoplasmic domains of furin, PC5B, and PC7 are shown. Furin accumulates in the TGN. Furin endocytosis signals are a di-leucine signal (LI), a YXXL signal (YKGL), and a single phenylalanine signal (F). In concert with an acidic peptide containing two phosphorylation sites at serine, they are responsible for TGN localization. Similar peptide elements are found in the cytoplasmic domain of PC5B indicating the TGN destination. PC7 accumulates in TGN-derived sialyltransferase-deficient vesicles (SDV) using a localization signal composed of the peptide segment PLC, the basic peptide HRSRKAK, and two palmitoylated cysteine residues

PC7 is also transported on the constitutive secretory pathway via the TGN to the plasma membrane, from where it recycles via late endosomes to the TGN region, and concentrates SDV, a post-Golgi compartment distinguishable from the TGN (Wouters et al. 1998). PC7 shuttling between the plasma membrane and the TGN region depends on sequences in the cytoplasmic domain. The sorting signal for SDV localization consists of the following motifs: (1) peptide PLC_{726}, (2) the basic amino acid sequence $HRSRKAK_{708}$, and (3) two cysteines, C_{558} and C_{563}, which are iteratively palmitoylated during the shuttle between TGN and plasma membrane (van de Loo et al. 2000; Declercq et al. 2012, 2017) (Fig. 9.5). Interestingly, a small fraction of PC7 reaches the cell surface through a brefeldin A and coat protein complex II (COPII)-independent unconventional secretory pathway. This may explain the rapid (<10 min) transit of PC7 from the ER to the cell surface (Rousselet et al. 2011), whereas the cleavage of the propeptide of PC7 is a slow process which takes hours rather than minutes (Creemers et al. 2000).

PC1/3 targeting to dense-core secretory granules (DCSG) resides in signals of the carboxy terminal ectodomain (617–753) which contains two α-helices, helix 1 (722–728) and helix 2 (738–750), of which the last one is sufficient for targeting a constitutively secreted protein to dense-core secretory granules (Dikeakos et al. 2009).

SKI-1/S1P cleaves cellular substrates in the ER/cis-Golgi area. There are differences, however, with arenaviral glycoproteins. The Lassa virus glycoprotein is cleaved by SKI-1/S1P before reaching the cis-Golgi, whereas cleavage of the LMCV glycoprotein occurs in the late or post-Golgi compartment (Lenz et al. 2001; Beyer et al. 2003). The substrate recognition of cleavage site variants is dependent on the auto-processing of SKI-1/S1P, suggesting differences in the processing of cellular and viral substrates (Burri et al. 2012a).

9.3 Proprotein Convertases Activating Viruses

Furin and the furin-like PCs, cleaving peptide bonds after basic residues, as well as SKI-1/S1P which cleaves peptide bonds after nonbasic residues, are the PCs which cleave viral glycoproteins (Table 9.3). Furin and furin-like PCs (PC5/6, PACE4, and PC7) are widely or ubiquitously expressed and are responsible for most of the processing events occurring in the constitutive secretory pathway or in endosomes. This leads to the activation/inactivation of receptors, ligands, enzymes, viral glycoproteins, or growth factors. Although these PCs exhibit a certain degree of functional redundancy when overexpressed in cell lines, their inactivation in mice or human beings results in specific phenotypes revealing that, in vivo, each PC primarily fulfills unique processing events and/or functions (Seidah et al. 2013). Involvement of other proteases with similar specificity cannot be excluded, but has not been demonstrated so far.

9.3.1 Furin

Furin, also named PACE or PCSK3, is the prototype of subtilisin-/kexin-like proprotein convertases (PCSKs). Furin is encoded by a transcription unit in the upstream region of the c-fes/fps proto-oncogene (Roebroek et al. 1986). Furin is expressed in all cells and tissues of eukaryotic organisms. It is synthesized as pro-furin with a molecular mass of 100 kDa, which is autocatalytically cleaved into the mature form with a molecular mass of 85 kDa. The first cleavage occurs between the prodomain and the catalytic domain at the C-terminus of the motif $_{101}$AKR**RAKR**↓ and the second one within the prodomain at the amino acid motif $_{70}$**R**GVT**KR**↓ (Leduc et al. 1992; Anderson et al. 1997). Endoproteolytic cleavage and removal of the propeptide fragments are prerequisite for efficient transport out of the endoplasmic reticulum into the TGN where furin acquires full enzymatic activity (Creemers et al. 1995). Endogenous furin was partially purified from Madin-Darby bovine kidney (MDBK) cells and identified by reaction with a furin-specific antiserum (Stieneke-Gröber et al. 1992; Vey et al. 1994). Furin is partially cleaved at arginine (R683) present in the ectodomain by an unknown endoprotease residing at the plasma membrane. The truncated soluble furin is catalytically active outside of cells (Plaimauer et al. 2001).

Furin is the central proprotein convertase that processes most diverse proproteins at multibasic structures on the constitutive secretory pathway. An analysis of the human proteome revealed an estimated number of about 500 potential proprotein candidates susceptible to furin cleavage (Remacle et al. 2008; Shiryaev et al. 2013). The high number of potential substrates together with the ubiquitous expression implicates that furin activates a wide variety of membrane-anchored proteins and membrane-secreted proteins, including precursors of growth factors, cell receptors, adhesion molecules, matrix metalloproteinases, blood plasma proteins, and factors for embryonal development which play important roles in the regulation of many life processes (Table 9.2). Furin also generates MHC class I antigens (Gil-Torregrosa et al. 2000). The majority of the proproteins are cleaved at the multibasic motif

RXK/RR and less frequently at the minimal basic motif **RXXR**. A few precursor proteins possess exceptional motifs, e.g., consensus sequences without arginine or lysine at position P4 as observed with the prodomain of furin (**R**GVT**KR**), proalbumin (**R**GVF**RR**), proprotein C (**R**SHL**KR**), and proparathyroid hormone (**K**SVK**KR**) (Canaff et al. 1999; Mori et al. 1999; Essalmani et al. 2017). Single amino acid positions of the furin motif were extensively studied (Rockwell et al. 2002). Proteins with lysine at position P4 are poor furin substrates but are readily cleaved by TMPRSS13 belonging to the family of transmembrane serine proteases (cf. Chap. 8).

The fact that the hemagglutinin precursor (HA0) of FPV containing a multibasic cleavage site was correctly cleaved and is biologically active after expression in insect cells and in insect larvae indicated that endogenous furin exists in insects (Kuroda et al. 1986, 1989). Furin of *Spodoptera frugiperda* showed the same cleavage properties as furin of vertebrates (Cieplik et al. 1998). In *Drosophila melanogaster*, two genes homologous to human furin, called Dfur1 and Dfur2, have been identified. The Dfur1 gene undergoes differential splicing to generate several type I membrane-bound isoenzymes differing in their C-terminal sequences. They are released as soluble dfurin forms which show cleavage specificity like furin of mammalian species (De Bie et al. 1995; Roebroek et al. 1991).

Furin is expressed in the mouse embryo at embryonic day e7.5. Inactivation of the *fur* locus by homologous recombination in the mouse causes embryonic death shortly after e10.5 due to hemodynamic insufficiency and failure of ventral closure and axial rotation in embryos. The furin-deficient mouse embryos failed to develop large vessels despite the presence of endothelial cell precursors (Roebroek et al. 1998). Among numerous proteins which play crucial roles during embryonic development are transforming growth factor β1 (TGFβ1), bone morphogenetic proteins 5 and 7 (BMP5, BMP7), vascular cell adhesion molecule (VCAM-1), and α-integrins (Scamuffa et al. 2006). To overcome the lethality of furin knockout mice, conditional knockout mutants were constructed. When furin expression was switched off in an interferon-inducible Mx-Cre/loxP knockout mouse model, the animals showed no obvious adverse effects. Histological analysis of the liver did not reveal any overt deviations from normal morphology. Variable degrees of redundancy were observed for the processing of numerous substrates, but none of the tested substrates displayed a complete block of processing. The absence of a severe phenotype raises the possibility of using furin as a local therapeutic target in the treatment of pathologies like cancer and viral infections, although the observed redundancy may require combination therapy or the development of a more broad-spectrum convertase inhibitor (Roebroek et al. 2004; Creemers and Khatib 2008).

Furin plays an important role in virus activation. As mentioned above, the first viral protein found to be processed by furin was the hemagglutinin (subtype H7) of fowl plague virus (FPV) (Stieneke-Gröber et al. 1992). Studies on the hemagglutinin of FPV and viruses of subtype H5 had indicated the presence of a multibasic cleavage site (Bosch et al. 1981; Kawaoka et al. 1987; Kawaoka and Webster 1988), and mutational analyses of the H7 cleavage site clearly defined the characteristic

RXK/RR motif. Cleavage occurs only when this motif is presented in the correct sequence position in loop formation. A shift of the motif by only one amino acid can abrogate cleavage (Garten et al. 1991; Vey et al. 1992). The importance of conserved amino acids of the hemagglutinin of HPAIV was corroborated, especially an arginine at P1 position proved to be essential (Walker and Kawaoka 1993; Walker et al. 1994). The hemagglutinin cleavability of HPAIV is influenced by the amino acid immediately downstream of the cleavage site (Horimoto and Kawaoka 1995). Similar results were obtained for the cleavage motifs of the highly pathogenic Newcastle disease virus (NDV) strains (Pritzer et al. 1990; Gotoh et al. 1992).

Over nearly three decades of research, an increasing number of fusion-competent glycoproteins have been identified which are activated by furin or furin-like proteases. Most glycoproteins belong to the enveloped RNA viruses, but also some enveloped DNA viruses use furin cleavage for maturation (Table 9.3). Accessory proteins forming a complex with fusion proteins, such as the prM protein of flaviviruses, are also activated by furin (cf. Chap. 6). However, furin can be replaced by another furin-like protease in certain cell types and tissues. Such examples were observed with HPAIV and with HIV-1 (Feldmann et al. 2000; Horimoto et al. 1994; Hallenberger et al. 1992; Anderson et al. 1993; Gu et al. 1995; Ohnishi et al. 1994). The glycoproteins of influenza viruses and paramyxoviruses are activated by furin next to the fusion peptide. Other viral glycoproteins are cleaved by furin at more than one site, as is the case with respiratory syncytial virus (RSV) (*see below*), or they are cleaved by furin in concert with other endoproteases as found with coronaviruses (cf. Chap. 4).

Detailed lists of more furin-cleavable viral membrane proteins are given in Table 9.3 and in a previous review (Klenk and Garten 1994). All of these proteins have multibasic cleavage sites, but experimental evidence for furin cleavage has not been obtained in all cases.

Various inhibitors blocking the catalytic activity have been used to substantiate the role of furin as processing enzyme. The first inhibitory agents were acylated basic tetrapeptidyl chloromethyl ketones, such as Dec-RVKR-CMK, polyarginines (nona-D-arginine amide), and serpin inhibitors, such as furin-adapted α-1-antitrypsin Portland (α-1-PDX) (Garten et al. 1989, 1994; Misumi et al. 1990a; Molloy et al. 1992; Jean et al. 1998; Cameron et al. 2000; Kacprzak et al. 2004; Hardes et al. 2015, 2017) (cf. Chap. 11). Acylated peptidyl chloromethyl ketones containing the **RXK/RR** motif bind covalently to the catalytic site of furin and prevent cleavage activation of HPAIV hemagglutinin and a wide array of other fusion-competent viral glycoproteins (Garten et al. 1994). In early years, when furin was still the only PC known substrate, homologous inhibitors were thought to be useful for the identification of furin as the activating protease of a virus. However, this perspective changed when other PCs were discovered. Because many of these enzymes have similar substrate specificities, the inhibitors were of limited use to discriminate them from furin. More convincing approaches for discrimination of PCs with closely related substrate specificity are investigations which exploit cells or animals with selective deficiency of a single PC or approaches which knock down the mRNA of a distinct PC.

9.3.2 PACE4

Paired basic amino acid-cleaving enzyme 4 (PACE4), also known as subtilisin-/kexin-like proprotein convertase (PCSK6), exists as alternatively spliced transcript variants encoding various isoforms which are differently present in tissues and cells. PACE4 is mainly expressed in liver, spleen, gut, brain, and neuroendocrine cells (Seidah et al. 2008). PACE4 migrates on the constitutive pathway and is secreted into the extracellular matrix, where it is attached to heparan sulfate proteoglycans (HSPGs) (Tsuji et al. 2003). PACE4 shares many substrates with furin, such as TGFβ-related proteins, proalbumin, pro-von Willebrand factor, zymogens of the a disintegrin and metalloprotease with thrombospondin type I motif (ADAMT) family, and the precursor of the low-density lipoprotein receptor (pro-LDL receptor). However, it has a more stringent substrate specificity and more limited operating parameters than furin (Longpré and Leduc 2004; Seidah et al. 2013; Wong et al. 2015) (Table 9.2). PACE4 processes diphtheria toxin and anthrax toxin protective antigen, but not Pseudomonas exotoxin A which is processed by furin (Moehring et al. 1993; Sucic et al. 1999). PACE4 cleaves at **RXK/RR**, **R**XX**R** and to a much lesser extent at **RR** and **KR** motifs (Gordon et al. 1997). In contrast to furin, PACE4 cannot process pro-factor IX and is not inhibited by the alpha1-antitrypsin Portland variant (Rehemtulla et al. 1993; Mains et al. 1997). In the absence of PACE4, mouse embryos developed specific deficiencies and survived to 75% compared with 100% lethality of furin knockout mice (Table 9.1). This suggests that, although PACE4 and furin share the ability to process similar substrates, they may also process one or more different substrates during the processes of embryonic development (Scamuffa et al. 2006). There are only few studies demonstrating cleavage of viral glycoproteins by PACE4 (Table 9.3).

9.3.3 PC5

PC5 is identical with PC6 and often designated PC5/6 or PCSK5. The proprotein convertase 5 gene is transcribed into two mRNAs; consequently two different but related enzymes are formed: PC5A and PC5B comprising 913 and 1860 amino acids, respectively. The A and B isoforms have identical pro-, catalytic, and P-domains. The catalytic domain (amino acids 148–439) contains D171, H212, N313, and S386 in the active-site pocket. Both isoforms differ in the cysteine-rich domain (CRD) (Fig. 9.1). PC5A has a CRD domain in comparable length with other PCs and terminates in the ectodomain as a soluble enzyme. PC5A migrates on the regulated secretory route and is packaged into dense-core granules. In contrast, PC5B, containing besides an unusually long CRD domain a transmembrane and a cytoplasmic domain, is transported on the constitutive branch of the secretory pathway. PC5B can be shed from the membrane, and the soluble PC5A can be attached to the cell surface by interaction of its CRD with proteoglycans. A PC5B transcript was found mainly in the intestine and kidney, while PC5A transcripts were detected in various tissues indicating different locations and roles for PC5A and PC5B (Nakagawa et al. 1993). PC5A is the major isoform in most tissues analyzed, except in the liver, where the transcripts are expressed in equivalent amounts. The complete knockout of PC5 in mice causes death at birth, with the embryos exhibiting multiple morphogenic defects (Table 9.1). Conditional knockout mice revealed that growth differentiation factor 11 (Gdf11), also known as bone morphogenetic protein 11 (BMP-11) and BMP-2, is a favorite substrate of PC5; other substrate precursor proteins are vascular endothelial growth factors (Essalmani et al. 2008; Lee et al. 2015). PC5 plays also a dominant role in pregnancy establishment by proteolytic activation of several important factors such as BMP2, caldesmon 1, calmodulin- and actin-binding protein (CALD1), and α-integrins.

The deduced cDNA structures of mouse PC5 and rat PC5 showed that the closest homologue is PACE4. Furthermore, like furin, *Drosophila melanogaster* dfurin2, and PACE4, PC5 shows the presence of a C-terminal cysteine-rich domain containing either five (PC5 and PACE4) or ten (dfurin2) repeats of the consensus motif Cys-Xaa2-Cys-Xaa3-Cys-Xaa (5-7)-Cys-Xaa2-Cys-Xaa (8-15)-Cys-Xaa3-Cys-Xaa (9-16). The richest sources of rat PC5 mRNA (3.8 kb) are the adrenal gland and gut, but it can also be detected in many other endocrine and nonendocrine tissues (Lusson et al. 1993).

PC5 activates the hemagglutinin of HPAIV like furin (Horimoto et al. 1994; Feldmann et al. 2000) (Table 9.3).

The decapeptide $_{107}$QQVV<u>K</u>K**R**T**KR**$_{116}$ mimicking a part of the prodomain of proPC5 is a nanomolar inhibitor of furin, PACE4, and PC5. A mutation at position P6 (<u>K</u>111<u>H</u>) makes the inhibitor more selective for PC5 than for furin indicating that a modification around the basic motif may influence the selectivity of PCs (Nour et al. 2003).

9.3.4 PC7

PC7 is identical with PC8, PCSK7, and lymphoma proprotein convertase (LPC) as it was originally discovered in a high-grade lymphoma carrying a translocation (Meerabux et al. 1996). PC7 is the most ancient and conserved member of the PC family. It is synthesized as proenzyme (101 kDa) and autocatalytically processed into mature PC7 (89 kDa) at R**RAKR**$_{141}$↓. There is no truncated soluble form of PC7 (Declercq et al. 2017). PC7 is invariably expressed as a membrane-anchored enzyme in spleen, thymus, prostate, testis, ovary, small intestine, colon, and peripheral blood leukocytes. High levels of PC7 mRNA are found in cells of the immune system, particularly in CD8+ cells, but also in CD4+, NK cells, and bone marrow cells. This may indicate a role in immune functions. The first described PC7-specific processing reaction was the activation of epidermal growth factor receptor (EGFR) at the cell surface (Rousselet et al. 2011). PC7 cleaves the unusual peptide **K**SVK**KR**↓SVSEIQL derived from proparathyroid hormone. Coexpression of PC7 and human transferrin receptor 1 (hTfR1) indicated that PC7 is the only convertase that sheds this receptor from cells into the medium, whereby the cleavage occurs at the site $_{95}$**K**TECE**R**↓ LA resembling the cleavage site of parathyroid hormone (Guillemot et al. 2013). The cleavage specificity of PC7 is largely similar to that of furin with the motifs (**K/R**)**R**↓ or (**R/K**)X$_n$R↓ ($n = 2, 4$, or 6 amino acids). A 24-mer peptide fragment of the PC7 prosegment (residues 81–104) is a strong inhibitor, K(i) = 7 nM of PC7, comparable to that of the full-length (104 residue) prosegment (Bhattacharjya et al. 2000). Unlike other PC-deficient mice, PC7-null mouse embryos did not show an apparent abnormal phenotype supporting the view that PC7 expression extensively overlaps with that of furin (Seidah 2011). However, PC7 is essential for zebrafish development and bioavailability of TGFβ1a. When the PCSK7 function in developing larvae was inhibited, defects in various organs including the brain and eye were observed, and the larvae died within 7 days post-fertilization (Turpeinen et al. 2013). Expression of PC7 revealed an increased ADAM10 maturation resulting in enhanced α-secretase-mediated processing of amyloid precursor protein (Anders et al. 2001; Lopez-Perez et al. 2001). PC7 cleaves specifically and in a cell-type-specific manner gp160 of HIV into gp120g/p41, suggesting that both furin and PC7 are the major convertases in T4 lymphocytes (Decroly et al. 1997; Hallenberger et al. 1997).

In summary, the furin-like PCs PACE4, PC5, and PC7 may selectively compensate the activation of viral glycoproteins in furin-deficient cells and tissues with different efficiencies. Furin compensation has been demonstrated (1) for virulent NDV F protein activated by PC5, but not by PACE4 (Fujii et al. 1999); (2) for processing of E3/E2 from Chikungunya virus by PC5A, PC5B, and PACE4, but not by PC7 (Ozden et al. 2008); (3) for cleavage of peptides homologous to SARS coronavirus S glycoprotein by PC5, but not by PC7 (Basak et al. 2007); and (4) for activating the env of HIV by PC7 (Decroly et al. 1997; Hallenberger et al. 1997).

9.3.5 SKI-1/S1P

This enzyme is known under the names subtilisin-/kexin-isozyme-1 (SKI-1), site 1 protease (S1P) membrane-bound transcription factor peptidase site 1, sterol regulatory element-binding protein site 1 (SREBP S1) protease, membrane-bound transcription factor protease site 1 (MBTPS1), and PCSK8. The protease plays an important role in lipid metabolism. The lipid composition of animal cells is controlled by SREBPs, transcription factors released from membranes by sterol-regulated proteolysis. By comparing cDNA of protease-competent and protease-deficient cells, the group of *Goldstein* and *Brown* identified the enzyme, which they called S1P, as an intraluminal membrane-bound subtilisin-like protease, 1052 amino acid long, which cleaves SRBPs at the motif **RSVL**↓ in the ER luminal loop between two membrane-spanning regions (Sakai et al. 1998). Using reverse transcriptase (RT)-PCR and degenerated oligonucleotides derived from the active-site residues of subtilisin-/kexin-like serine proteinases, *Seidah* and colleagues independently identified a highly conserved and phylogenetically ancestral human, rat, and mouse type I membrane-bound proteinase which they called subtilisin-/kexin-isozyme-1 (SKI-1) (Seidah et al. 1999). The tissue distribution of SKI-1/S1-P mRNA is ubiquitous. SKI-1/S1P accumulates in the perinuclear region, predominantly in the cis-Golgi, and small punctual SKI-1/S1P containing material is seen in the endosomal/lysosomal compartments by immunochemical staining. Studies with brefeldin A indicated that substrates are cleaved in the early Golgi. proSKI-1/S1P is processed into two membrane-bound forms of SKI-1 (120 and 106 kDa) differing by the nature of their N-glycosylation (da Palma et al. 2014, 2016). At late stages of the secretory pathway, part of the membrane-bound enzyme is shed into the medium in a 98-kDa form. SKI-1/S1P exhibits a wide pH optimum for cleavage (Seidah et al. 1999). Recombinant SKI-1/S1P was expressed, purified, and characterized (Bodvard et al. 2007).

A physiological SKI-1/S1P substrate different from SREBPs is ATF6, a membrane-bound transcription factor that activates genes in ER stress, such as cholesterol deprivation. When unfolded proteins accumulate in response to ER stress, ATF6 is cleaved at **RHLL**↓ to release its cytoplasmic domain, which enters the nucleus (Ye et al. 2000). Other substrates are (pro)renin receptor and brain-derived neurotrophic factor (BDNF) (Nakagawa et al. 2017). BDNF is a member of the neurotrophin family of growth factors found in the brain and the periphery which is cleaved at **RGLT**↓. SKI-1/S1P is required for the transcription of many bone matrix and mineralization-related genes, such as fibronectin and fibrillin in bone osteoblasts and osteocytes. The irreversible inhibitor Dec-Arg-Arg-Leu-Leu-CMK blocks transcription of the corresponding genes and inhibits mineralization (Gorski et al. 2011). These results demonstrated that the differentiated phenotype of osteoblastic cells and possibly osteocytes depends upon SKI-1/S1P. Knockdown of SKI-1/S1P in zebrafish leads to the zebrafish *gonzo* mutant showing a defect in chondrocyte morphogenesis (Schlombs et al. 2003).

Since SKI-1/S1P plays an essential role in cell physiology, it is evident that gene deletion is lethal at an early state of the embryonal development (Table 9.1). Conditional SKI-1/S1P knockout mice are viable (Yang et al. 2001; Seidah 2011).

The SKI-1/S1P processing motifs contain basic and hydrophobic residues at P4 and P2, respectively, with a relatively relaxed acceptance of amino acids at P1 and P3, i.e., R_4-X_3-X_2-X_1 ↓, where $X_{1,3}$ are any amino acids and X_2 is often leucine, isoleucine, and valine. A favorable motif is **RRLL**↓ found in the glycoprotein of Lassa virus (Maisa et al. 2009). The first viral proteins recognized to be cleaved by SKI-1/S1P were the glycoproteins of Lassa virus and lymphocytic choriomeningitis virus (LCMV) (Lenz et al. 2000, 2001; Beyer et al. 2003; Kunz et al. 2003). A high variability of cleavage motifs is found with glycoproteins of arenaviruses and Crimean-Congo hemorrhagic fever virus (CCHFV) (Burri et al. 2012b, 2013; Altamura et al. 2007; Sanchez et al. 2002, 2006; Vincent et al. 2003) (cf. Chap. 3). SKI-1/S1P inhibition effectively blocks hepatitis C virus (HCV) from establishing infection in hepatoma cells (Olmstead et al. 2012). However, the function of SKI-1/S1P in HCV replication is not known.

9.4 PCs Not Known to Activate Viruses

9.4.1 PC1/3

PC1/3, also designated neuroendocrine convertase 1 (NEC1), prohormone convertase 3 (PC3), proprotein convertase 1 (PC1), or PCSK1, occurs in many eukaryotic organisms. The human gene of PC1/3 located on chromosome 5 is transcribed from 13 exons and translated into the preproprotein of PC1/3 comprising 753 amino acids. The catalytic domain (143–441 aa) contains the triad D167, H208, and S382. N309 contributes to the oxyanion hole and stabilizes the transition state of the enzymatic reaction. The catalytic domain has 61% sequence homology with human furin. The prodomain contains 83 amino acids with 30 to 40% sequence identity among the eukaryotic PC1/3s. PC1/3 is present in dense-core vesicles of the regulated secretory pathway in neuroendocrine tissues. It cleaves prohormones and other precursor proteins C-terminally at arginine-arginine or lysine-arginine motifs (Table 9.1). Typical substrates are POMC, proinsulin, proglucagon, and other precursors of neuro-sensing and regulating hormone peptides, such as the hormone-like endopeptidase renin, enkephalin, dynorphin, somatostatin, ghrelin, and agouti-related protein (AGRP) (Creemers et al. 2006). Interestingly, PC1/3 deficiency has quite different phenotypic effects in mice and man, a phenomenon not observed with other PCs. PC1/3 knockout mice are viable but exhibit growth retardation and multiple defects in hormone precursor processing. In humans, PC1/3 deficiency causes obesity, hypogonadism, reactive hypoglycemia, hypoadrenalism, and small-intestinal absorptive dysfunction due to impaired processing of prohormones (Taylor et al. 2003; Farooqi et al. 2007; Seidah 2011) (Table 9.1).

9.4.2 PC2

PC2 is also known as neuroendocrine convertase 2, Kex-like endoprotease 2, or PCSK2. Maturation of PC2 is unusual since it depends on support by the neuroendocrine chaperone 7B2 that prevents auto-aggregation of PC2 (Ramos-Molina and Lindberg 2015). After binding of the chaperone to nascent proPC2, the proPC2/7B2 complex is transported from the ER to the TGN. On the route from the TGN to dense secretory granules, the chaperone dissociates from the complex in the acidic environment, and the prodomain is removed after autocatalytic cleavage (Seidah 2011). Embedding of the C-terminal domain of PC2, and likewise of PC5A, in glycosphingolipid- and cholesterol-rich microdomains appears to be necessary for sorting into secretory granules (Creemers et al. 1996; De Bie et al. 1996). Evidence has also been obtained that an amphipathic α-helix at the C-terminus serves to fix PC5A at the cell membrane (Assadi et al. 2004). Like PC1/3, PC2 cleaves at dibasic amino acid motifs of neuroendocrine peptide and protein precursors, such as pro-opiomelanocortin (POMC), proinsulin, and proglucagon, and the precursors of chromogranin A, neurotensin, and pro-enkephalin (Pan et al. 2006).

PC2$^{-/-}$ mice appear normal at birth but show retarded growth with chronic fasting, hypoglycemia, and reduced glucagon levels (Seidah 2011).

9.4.3 PC4

PC4, also known as proprotein convertase subtilisin/kexin type 4 (PCSK4), is expressed from 15 exons which code for the mRNA of a polypeptide containing 655 amino acids. PC4 is found exclusively in germ cells, suggesting a possible reproductive function of this enzyme (Basak et al. 1999; Nakayama et al. 1992; Seidah et al. 1992). The catalytic domain with its enzymatic triad D158, H189, S373 and the oxyanion hole-forming N300 processes various proproteins by cleavage at paired basic amino acids of the general motif (**R**,**K**)X(R,K,X)**R**. PC4 is closely related to furin. It is primarily found in testicular germ cells and in sperm but also in ovary macrophages. It controls testicular and ovarian physiology.

PC4 is expressed in the human placenta and cleaves the precursor of the insulin-like growth factor II, an important regulator of fetoplacental growth (Qiu et al. 2005). Another specific substrate of PC4 in the testis is pituitary adenylate cyclase-activating polypeptide (PACAP) which is solely processed by PC4 (Seidah 2011). The fertility of *PC4*$^{-/-}$ mice is significantly reduced (Scamuffa et al. 2006), and there are also defects in embryonic development (Mbikay et al. 1997).

9.4.4 PCSK9

PCSK9, also called neural apoptosis-regulated convertase-1 (NARC-1), is a new member of the pyrolysin and proteinase K subfamilies of subtilases (Abifadel et al. 2003). It is highly expressed in the liver as a 74 kDa protein that is autocatalytically cleaved. The prodomain remains tightly bound to the enzyme which is catalytically inactive but functions as a binding protein that interacts with the low-density lipoprotein receptor (LDLR) and plays therefore a crucial role in plasma cholesterol homeostasis. PSCK9 is not involved in proteolytic protein processing.

Loss-of-function PCSK9 mutations were also identified. Two nonsense mutations found in ~2% of black Africans were associated with a ~40% decrease in LDL cholesterol and a ~88% risk reduction of cardiovascular disease (CVD), suggesting that PCSK9 inhibition may be a promising approach to treat hypercholesterolemia and prevent CVD. Therapeutic monoclonal antibodies have been developed to reduce low-density lipoprotein (LDL) cholesterol levels and the risk of coronary artery disease (Weider et al. 2016; Le et al. 2015; Seidah et al. 2017).

9.5 Functions of PCs in Virus Replication

9.5.1 Cleavage Activation of Viral Fusion Proteins

Enveloped viruses induce fusion of viral and cellular membranes to deliver their genomes into the cytoplasm. All viral fusion proteins are C-terminally anchored in the viral membrane and possess a hydrophobic "fusion peptide" (class I fusion proteins) or "fusion loop" (class II and class III fusion proteins) which interacts with the target membrane. Exposure of the hydrophobic domains depends on a conformational change of the fusion protein triggered by low pH in endosomes or by interaction with the receptor-binding protein at the cell surface (Harrison 2008; White and Whittaker 2016; Jardetzky and Lamb 2014). In most cases, the conformational change can only be triggered if the proteins are primed by proteolytic cleavage.

For many viruses, like influenza viruses, paramyxoviruses, retroviruses, and arenaviruses, the cleavage site is a distinct peptide bond next to the "fusion peptide" which, after cleavage, leads to a new exposed N-terminus of the membrane-anchored fusogenic subunit of the viral glycoproteins. The fusion peptide is characterized by a stretch of non-charged hydrophobic amino acids which are strongly conserved within a virus family. In contrast to the conserved fusion peptides, the preceding cleavage sites and the activating proteases vary even when the viruses are closely related. Multibasic cleavage sites recognized by furin and furin-like proprotein convertases PACE4, PC5/6, and PC7 are present in most enveloped viruses, whereby furin is the master key for cleavage of viral glycoproteins.

Furin is also involved in a different fusion mechanism observed with flaviviruses including tick-borne encephalitis virus (TBEV), yellow fever virus, West Nile virus, dengue virus, and Zika virus. Envelope glycoprotein E of these viruses, a class II fusion protein, is not processed by proteolysis. Fusion activity depends, however, on furin cleavage of the tightly associated accessory protein prM. During intracellular virion assembly, prM prevents premature exposure of the fusion loop located on E. After cleavage of prM and release of the prepeptide, mature virions invade cells by endocytosis and exposure of the fusion loop at low pH (cf. Chap. 6). Blockage of prM cleavage by furin inhibitors prevents multiple replication cycles of TBEV and dengue virus in cell cultures (Stadler et al. 1997; Elshuber et al. 2003; Kouretova et al. 2017).

Enveloped DNA viruses utilize the fusion mechanism for host cell invasion, too. Membrane fusion during herpesvirus entry into host cells is a complex process. All herpesviruses express the gB and gHgL complex as well as various non-conserved glycoproteins of individual herpesviruses which interact with selected receptor proteins determining cell tropism (Eisenberg et al. 2012). The gHgL complex acts as an "activator" of entry, and the glycoprotein gB acts as the membrane "fusogen." It has been shown that gB of several herpes viruses is cleaved by furin (Table 9.3). gB is a class III fusion protein with a "fusion loop" and has structural domain similarities with VSV G protein and the baculovirus fusion protein. The cleavage of Epstein-Barr virus gB is required for cell-cell fusion (Sorem and Longnecker 2009), whereas bovine herpesvirus 1 gB and pseudorabies virus gB are not necessarily cleaved (Kopp et al. 1994). Baculovirus nucleopolyhedrovirus (NPV) group II members have a fusion protein, which must be cleaved by furin to mediate fusion (Long et al. 2006; Wang et al. 2017).

9.5.2 Cooperation of Proprotein Convertases with Other Endoproteases

Virus surface glycoproteins of several viruses are activated by various endoproteases at different sites. The spike protein S of coronaviruses which possesses both receptor-binding and fusion functions, is cleaved at several sites (Table 9.3). For example, the severe acute respiratory syndrome (SARS) coronavirus glycoprotein S is cleaved at two sites, $_{441}$**RYLR**↓ and $_{758}$**RNTR**↓ (S1/S2 site), by furin/furin-like proteases, and at the S2′cleavage site $_{796}$**KR**↓ by cathepsin L or transmembrane serine protease 2 (TMPRSS2). The proteolytic activation of protein S at amino acid position 797 is adjacent to the "fusion peptide" of the S protein, which proved to be crucial for fusogenicity. Similarly, the Middle East respiratory syndrome (MERS) coronavirus possesses a glycoprotein S that is cleavable at three distinct sites by furin (Table 9.3). Modulations of the spike cleavage S2′ next to the fusion peptide have profound effects on tropism and pathogenicity (Belouzard et al. 2009; Millet and Whittaker 2015; cf. Chap. 4).

Glycoprotein GP of Ebola virus mediating receptor binding and fusion is first processed to GP1/2 at a conserved furin cleavage site $_{497}$**R**R**T**R**R**↓ which is remote from the fusion loop located on GP2 (amino acids 524 to 540) (Volchkov et al. 1998, 2000). Furin processing is followed by cathepsin cleavage at about amino acid 200 resulting in the removal of a large C-terminal fragment of GP1 and the exposure of a receptor-binding site (Chandran et al. 2005). Both processing steps are therefore essential for the function of GP in virus entry. There is evidence that furin and cathepsin can be replaced in GP processing by other proteases. Furin cleavage of a nonstructural glycoprotein (sGP) of Ebola virus has also been observed (cf. Chap. 5).

The genome segment M of Crimean-Congo hemorrhagic fever virus (CCHFV) polyprotein encodes a polyprotein with four transmembrane anchors separating three luminal/extracellular domains from two cytoplasmic domains. Furin and SKI-1/S1P are involved in the activation mechanism of this polyprotein (Sanchez et al. 2002, 2006; Altamura et al. 2007). Cleavages by SKI-1/S1P and by furin are necessary for producing the nonstructural glycoprotein GP38 and the structural glycoprotein Gn, an important step for gaining fusion capacity (Bergeron et al. 2015).

The envelope glycoprotein (env) of foamy virus, a spuma retrovirus, shows an unusual biosynthesis undergoing cleavage by furin at two different positions. The precursor protein has a type III membrane topology with both the N and C termini located in the cytoplasm. The processing of env into the particle-associated env leader protein (elp) and into the surface (SU) and transmembrane (TM) subunits occurs posttranslationally during transport to the cell surface. Furin or a furin-like protease is responsible for both, the late signal peptidase-like processing and the maturation cleavage needed for fusion (Duda et al. 2004; Geiselhart et al. 2004).

9.5.3 Generation of Biologically Active Peptides

The fusion protein of the human and bovine respiratory syncytial viruses (RSV) is synthesized as an inactive precursor F0 that is proteolytically processed at two multibasic sequences, $_{131}$KK**R**K**RR**↓ and $_{106}$**R**A**RR**↓ (bovine RSV). Both furin consensus sequences must be cleaved to activate the fusion protein (Zimmer et al. 2001; González-Reyes et al. 2001). Cleavage of the bovine RSV fusion protein results in the release of a small peptide that is converted into the biologically active virokinin by additional posttranslational C-terminal modifications, namely, truncation by carboxypeptidase of B type and stabilization by enzymatic amidation. The amino acids 106–139 of the viral fusion protein align with amino acids 45–78 of the human tachykinin precursor type 1 that functions as a tissue hormone and produces rapid contraction of smooth muscles. Virokinin is secreted by virus-infected cells and was found to desensitize tachykinin receptors in the mammalian respiratory tract with potent effects on local inflammatory and immune processes (Zimmer et al. 2003).

9.5.4 Glycoprotein Trimerization and Incorporation into Virus Particles

Cleavage of the Lassa virus glycoprotein by SKI-1/S1P is necessary not only for fusion of the virus envelope with endosomal membranes at cell entry but also for efficient incorporation of the viral glycoprotein into virus particles (Lenz et al. 2001). Virions contain almost exclusively homotrimeric spikes of the cleaved glycoprotein form, whereas at the cell surface of infected cells, monomers and oligomers of the uncleaved form prevail (Schlie et al. 2010a). Glycosylation mutations on the ectodomains of Lassa virus glycoprotein showed that 6 of 11 N-glycans are necessary for glycoprotein cleavage indicating that N-glycans are needed for correct conformation of the precursor glycoprotein to be cleaved by SKI-1/S1P. Interestingly, the glycoprotein precursor is transported to the cell surface in a completely endo H-sensitive form suggesting that cleavage is a prerequisite for completion of complex N-glycosylation (Eichler et al. 2006). Moreover, a mutation in the cytoplasmic domain of Lassa virus glycoprotein abolished the maturation cleavage by SKI-1/S1P within the ectodomain indicating conformational changes across the membrane (Schlie et al. 2010b). These data suggest that, with some viruses, cleavage by PCs may be important for correct folding, oligomerization, and N-glycosylation, which is assumed to be a prerequisite for an effective incorporation of functional virus spikes into virions. Extensive structural rearrangements in the Lassa surface glycoprotein were observed at different pH's and in the presence of the functional secondary intracellular receptor, human LAMP-1, by using high-resolution electron cryo-microscopy and tomography techniques (Li et al. 2016). Such a substantial rearrangement of subdomains of the Lassa viral glycoproteins would not be possible in the absence of SKI-1/S1P cleavage.

The glycoproteins of Borna disease virus (BDV) and simian immunodeficiency virus (SIV) also need to be cleaved by furin for correct trimerization and effective insertion into virus particles (Eickmann et al. 2005; Yamshchikov et al. 1995).

9.5.5 PC Cleavage of Non-Envelope Proteins

Viral protein R (Vpr) is a HIV-1 protein 96 amino acid long that plays an important role in regulating nuclear import of the HIV-1 pre-integration complex and is required for virus replication in nondividing cells such as macrophages. It has been detected in soluble form in the sera and cerebrospinal fluids of HIV-1-infected patients and contributes to HIV pathogenesis. Vpr was found to undergo proteolytic processing at the PC cleavage site, $_{85}$**RQRR**↓ located within the functionally important C-terminal arginine-rich domain. Interestingly, Vpr processing occurred extracellularly upon close contact to cells and most likely involved cell surface-associated furin or furin-like PC (Xiao et al. 2008).

Hepatitis B virus e-antigen (HBeAg) is a secreted version of hepatitis B virus (HBV) core protein raised from the core gene. HBeAg has a cleavable signal peptide and four furin cleavage sites in the C-terminal part. HBeAg promotes immune tolerance and is essential for the development of chronic hepatitis B virus infection. Furin plays a key role in processing the HBeAg precursor into mature HBeAg (Ito et al. 2009; Pang et al. 2013). The furin inhibitor decanoyl-RVKR-chloromethylketone (CMK) combined with the nucleoside analogue entecavir reduced HBV replication and HBeAg secretion in Hep cells and was suggested as a therapeutic regimen for treatment of chronic hepatitis B (Yang et al. 2014).

Papillomaviruses (PV) are non-enveloped viruses that enter cells by endocytosis. Entry involves removal of protein L2 from the surface of the virus particles by furin (Day et al. 2008; Richards et al. 2006).

9.5.6 Protease Activation Mutants for Vaccine Design

A large number of the viruses activated by furin which are shown in Table 9.3 cause important diseases. Highly pathogenic avian H5 and H7 influenza viruses (HPAIV) are not only a problem for the poultry industry but are also a threat to human health. In 1997, an H5N1 virus caused human infections in Hong Kong during a poultry outbreak. Six of eighteen patients succumbed to this virus. Since 2003 the H5N1 virus spread rapidly killing millions of birds and continued to be transmitted to humans. Thus, there is a need for the development of human and animal vaccines against these viruses. For safety reasons the viruses used for the production of such vaccines should contain an HA that is modified at the furin cleavage site to diminish their pathogenic potential. This approach is recommended to meet short-termed demand of H5 or H7 vaccines (Neumann et al. 2008; Webby et al. 2004), and a vaccine with a monobasic HA cleavage site was licensed for humans in responsiveness to the pandemic alert in 2003 (Webby et al. 2004).

When an HA with a monobasic cleavage site is present in live vaccines, virus revertants with a furin cleavage site may emerge by spontaneous mutation. Therefore, safer vaccines are demanded that are produced from attenuated viruses containing a nonbasic cleavage site (Böttcher-Friebertshäuser et al. 2014). This concept is based on insights from early investigations in which attenuated mutants of Sendai and influenza viruses activated by elastase or chymotrypsin have been analyzed (Scheid and Choppin 1976; Orlich et al. 1995). More recently, this strategy was pursued to exchange the monobasic HA cleavage site of influenza strain A/WSN/33 (Stech et al. 2005) and the multibasic HA cleavage motif of an H7N7 influenza virus (Gabriel et al. 2008) for amino acids susceptible to elastase cleavage. The mutants were strictly elastase-dependent, grew equally well as the wild type in cell culture, and were attenuated in mice unlike the lethal wild type. Immunization with an H7N7 mutant at 10^6 pfu dosage protected mice against disease and induced sterile immunity; vaccination with homosubtypic or heterosubtypic reassortants led to cross protection. These observations demonstrate that a mutated HA requiring elastase cleavage can serve as an attenuating component of a safe live vaccine against HPAIV and other influenza viruses.

Flaviviruses with an altered furin cleavage site of prM have been shown to produce single-round infectious particles (Elshuber et al. 2003). Thus, it appears that protease activation mutants may also be a useful approach for the development of safe vaccines against other viruses.

9.5.7 Inhibitors of PCs for Host-Directed Antiviral Therapy

The high dependence of many viruses on PCs makes PCs promising targets for antiviral therapy. Furin inhibitors comprising substrate homologous peptidomimetics, furin- or SKI-1/SP1-adapted serpines, and blocking antibodies as well as peptide-conjugated phosphorodiamidate morpholino oligonucleotides (PPMOs) suppressing the expression of individual PCs are described in detail in Chap. 11. Combinatorial treatment with protease inhibitors and virus-targeting agents such as ribavirin and favipiravir allows a drastic reduction of the drug dosages while maintaining full inhibitory efficacy. Moreover, the application of PC inhibitors prevents the development of resistance to virus-targeting drugs (Lu et al. 2015; Garten et al. 2015).

References

Abifadel M, Varret M, Rabès JP, Allard D, Ouguerram K, Devillers M, Cruaud C, Benjannet S, Wickham L, Erlich D, Derré A, Villéger L, Farnier M, Beucler I, Bruckert E, Chambaz J, Chanu B, Lecerf JM, Luc G, Moulin P, Weissenbach J, Prat A, Krempf M, Junien C, Seidah NG, Boileau C. Mutations in PCSK9 cause autosomal dominant hypercholesterolemia. Nat Genet. 2003;34(2):154–6.

Seidah NG, Abifadel M, Prost S, Boileau C, Prat A. The proprotein convertases in hypercholesterolemia and cardiovascular diseases: emphasis on Proprotein Convertase Subtilisin/Kexin 9. Pharmacol Rev. 2017;69(1):33–52. Review.

Achstetter T, Wolf DH. Hormone processing and membrane-bound proteinases in yeast. EMBO J. 1985;4(1):173–1.

Allison LMC, Salter MWAP, Kiguwa S, Howard CR. Analysis of the glycoprotein gene of Tacaribe virus and neutralization-resistant variants. J Gen Virol. 1991;72:2025–9.

Altamura LA, Bertolotti-Ciarlet A, Teigler J, Paragas J, Schmaljohn CS, Doms RW. Identification of a novel C-terminal cleavage of Crimean-Congo hemorrhagic fever virus PreGN that leads to generation of an NSM protein. J Virol. 2007;81(12):6632–42.

Anders A, Gilbert S, Garten W, Postina R, Fahrenholz F. Regulation of the alpha-secretase ADAM10 by its prodomain and proprotein convertases. FASEB J. 2001;15(10):1837–9.

Anderson ED, Thomas L, Hayflick JS, Thomas G. Inhibition of HIV-1 gp160-dependent membrane fusion by a furin-directed alpha 1-antitrypsin variant. J Biol Chem. 1993;268(33):24887–91.

Anderson ED, VanSlyke JK, Thulin CD, Jean F, Thomas G. Activation of the furin endoprotease is a multiple-step process: requirements for acidification and internal propeptide cleavage. EMBO J. 1997;16(7):1508–18.

Artenstein AW, Opal SM. Proprotein convertases in health and disease. N Engl J Med. 2011;365(26):2507–18. Review.

Assadi M, Sharpe JC, Snell C, Loh YP. The C-terminus of prohormone convertase 2 is sufficient and necessary for Raft association and sorting to the regulated secretory pathway. Biochemistry. 2004;43(24):7798–807.

Barr PJ. Mammalian subtilisins: the long-sought dibasic processing endoproteases. Cell. 1991;66(1):1–3. Review.

Basak A, Touré BB, Lazure C, Mbikay M, Chrétien M, Seidah NG. Enzymic characterization in vitro of recombinant proprotein convertase PC4. Biochem J. 1999;343(Pt 1):29–37.

Basak A, Mitra A, Basak S, Pasko C, Chrétien M, Seaton P. A fluorogenic peptide containing the processing site of human SARS corona virus S-protein: kinetic evaluation and NMR structure elucidation. ChemBioChem. 2007;8(9):1029–37.

Belouzard S, Chu VC, Whittaker GR. Activation of the SARS coronavirus spike protein via sequential proteolytic cleavage at two distinct sites. Proc Natl Acad Sci U S A. 2009;106(14):5871–6. https://doi.org/10.1073/pnas.0809524106. Epub 2009 Mar 24.

Bergeron É, Zivcec M, Chakrabarti AK, Nichol ST, Albariño CG, Spiropoulou CF. Recovery of recombinant Crimean Congo hemorrhagic fever virus reveals a function for non-structural glycoproteins cleavage by Furin. PLoS Pathog. 2015;11(5):e1004879.

Beyer WR, Pöpplau D, Garten W, von Laer D, Lenz O. Endoproteolytic processing of the lymphocytic choriomeningitis virus glycoprotein by the subtilase SKI-1/S1P. Virology. 2003;77(5):2866–72.

Bhattacharjya S, Xu P, Zhong M, Chrétien M, Seidah NG, Ni F. Inhibitory activity and structural characterization of a C-terminal peptide fragment derived from the prosegment of the proprotein convertase PC7. Biochemistry. 2000;39(11):2868–77.

Bodvard K, Mohlin J, Knecht W. Recombinant expression, purification, and kinetic and inhibitor characterisation of human site-1-protease. Protein Expr Purif. 2007;51(2):308–19.

Bosch FX, Garten W, Klenk HD, Rott R. Proteolytic cleavage of influenza virus hemagglutinins: primary structure of the connecting peptide between HA1 and HA2 determines proteolytic cleavability and pathogenicity of Avian influenza viruses. Virology. 1981;113(2):725–35.

Bosch BJ, Bartelink W, Rottier PJ. Cathepsin L functionally cleaves the severe acute respiratory syndrome coronavirus class I fusion protein upstream of rather than adjacent to the fusion peptide. J Virol. 2008;82(17):8887–90.

Bosshart H, Humphrey J, Deignan E, Davidson J, Drazba J, Yuan L, Oorschot V, Peters PJ, Bonifacino JS. The cytoplasmic domain mediates localization of furin to the trans-Golgi network en route to the endosomal/lysosomal system. J Cell Biol. 1994;126(5):1157–72.

Böttcher-Friebertshäuser E, Garten W, Matrosovich M, Klenk HD. The hemagglutinin: a determinant of pathogenicity. Curr Top Microbiol Immunol. 2014;385:3–34. Review.

Braks JA, Martens GJ. 7B2 is a neuroendocrine chaperone that transiently interacts with prohormone convertase PC2 in the secretory pathway. Cell. 1994;78(2):263–73.

Bresnahan PA, Leduc R, Thomas L, Thorner J, Gibson HL, Brake AJ, Barr PJ, Thomas G. Human fur gene encodes a yeast KEX2-like endoprotease that cleaves pro-beta-NGF in vivo. J Cell Biol. 1990;111(6 Pt 2):2851–9.

Brown MS, Goldstein JL. A proteolytic pathway that controls the cholesterol content of membranes, cells, and blood. Proc Natl Acad Sci U S A. 1999;96(20):11041–8. Review.

Brücher KH, Garten W, Klenk HD, Shaw E, Radsak K. Inhibition of endoproteolytic cleavage of cytomegalovirus (HCMV) glycoprotein B by palmitoyl-peptidyl-chloromethyl ketone. Virology. 1990;178(2):617–20.

Bruzzaniti A, Goodge K, Jay P, Taviaux SA, Lam MH, Berta P, Martin TJ, Moseley JM, Gillespie MT. PC8 [corrected], a new member of the convertase family. Biochem J. 1996;314(Pt 3):727–31.

Burri DJ, Pasqual G, Rochat C, Seidah NG, Pasquato A, Kunz S. Molecular characterization of the processing of arenavirus envelope glycoprotein precursors by subtilisin kexin isozyme-1/site-1 protease. J Virol. 2012;86(9):4935–46.

Burri DJ, da Palma JR, Kunz S, Pasquato A. Envelope glycoprotein of arenaviruses. Virus. 2012;4(10):2162–81.

Burri DJ, da Palma JR, Seidah NG, Zanotti G, Cendron L, Pasquato A, Kunz S. Differential recognition of Old World and New World arenavirus envelope glycoproteins by subtilisin kexin isozyme 1 (SKI-1)/site 1 protease (S1P). J Virol. 2013;87(11):6406–14.

Cameron A, Appel J, Houghten RA, Lindberg I. Polyarginines are potent furin inhibitors. Biol Chem. 2000;275(47):36741–9.

Canaff L, Bennett HP, Hou Y, Seidah NG, Hendy GN. Proparathyroid hormone processing by the proprotein convertase-7: comparison with furin and assessment of modulation of parathyroid convertase messenger ribonucleic acid levels by calcium and 1,25-dihydroxyvitamin D3. Endocrinology. 1999;140(8):3633–42.

Cavanagh D, Davis PJ, Pappin DJ, Binns MM, Boursnell ME, Brown TD. Coronavirus IBV: partial amino terminal sequencing of spike polypeptide S2 identifies the sequence Arg-Arg-Phe-Arg-Arg at the cleavage site of the spike precursor propolypeptide of IBV strains Beaudette and M41. Virus Res. 1986;4(2):133–43.

Chandran K, Sullivan NJ, Felbor U, Whelan SP, Cunningham JM. Endosomal proteolysis of the Ebola virus glycoprotein is necessary for infection. Science. 2005;308(5728):1643–5.

Cieplik M, Klenk HD, Garten W. Identification and characterization of spodoptera frugiperda furin: a thermostable subtilisin-like endopeptidase. Biol Chem. 1998;379(12):1433–40.

Constam DB, Calfon M, Robertson EJ. SPC4, SPC6, and the novel protease SPC7 are coexpressed with bone morphogenetic proteins at distinct sites during embryogenesis. J Cell Biol. 1996;134(1):181–91.

Creemers JW, Khatib AM. Knock-out mouse models of proprotein convertases: unique functions or redundancy? Front Biosci. 2008;13:4960–71. Review.

Creemers JW, Vey M, Schäfer W, Ayoubi TA, Roebroek AJ, Klenk HD, Garten W, Van de Ven WJ. Endoproteolytic cleavage of its propeptide is a prerequisite for efficient transport of furin out of the endoplasmic reticulum. J Biol Chem. 1995;270(6):2695–70.

Creemers JWM, Usac EF, Bright NA, Van de Loo JW, Jansen E, Van de Ven WJM, Hutton JC. Identification of a transferable sorting domain for the regulated pathway in the prohormone convertase PC2. J Biol Chem. 1996;271:25284–91.

Creemers JW, van de Loo JW, Plets E, Hendershot LM, Van De Ven WJ. Binding of BiP to the processing enzyme lymphoma proprotein convertase prevents aggregation, but slows down maturation. J Biol Chem. 2000;275(49):38842–7.

Creemers JW, Pritchard LE, Gyte A, Le Rouzic P, Meulemans S, Wardlaw SL, Zhu X, Steiner DF, Davies N, Armstrong D, Lawrence CB, Luckman SM, Schmitz CA, Davies RA, Brennand JC, White A. Agouti-related protein is posttranslationally cleaved by proprotein convertase 1 to generate agouti-related protein (AGRP)83-132: interaction between AGRP83-132 and melanocortin receptors cannot be influenced by syndecan-3. Endocrinology. 2006;147(4):1621–31.

Cunningham D, Danley DE, Geoghegan KF, Griffor MC, Hawkins JL, Subashi TA, Varghese AH, Ammirati MJ, Culp JS, Hoth LR, et al. Structural and biophysical studies of PCSK9 and its mutants linked to familial hypercholesterolemia. Nat Struct Mol Biol. 2007;14:413–9.

da Palma JR, Burri DJ, Oppliger J, Salamina M, Cendron L, de Laureto PP, Seidah NG, Kunz S, Pasquato A. Zymogen activation and subcellular activity of subtilisin kexin isozyme 1/site 1 protease. J Biol Chem. 2014;289:35743–56.

da Palma JR, Cendron L, Seidah NG, Pasquato A, Kunz S. Mechanism of folding and activation of Subtilisin Kexin Isozyme-1 (SKI-1)/Site-1 Protease (S1P). J Biol Chem. 2016;291(5):2055–66.

Dahms SO, Hardes K, Becker GL, Steinmetzer T, Brandstetter H, Than ME. X-ray structures of human furin in complex with competitive inhibitors. ACS Chem Biol. 2014;9(5):1113–8.

Dahms SO, Creemers JW, Schaub Y, Bourenkov GP, Zögg T, Brandstetter H, Than ME. The structure of a furin-antibody complex explains non-competitive inhibition by steric exclusion of substrate conformers. Sci Rep. 2016;6:34303.

Dahms SO, Arciniega M, Steinmetzer T, Huber R, Than ME. Structure of the unliganded form of the proprotein convertase furin suggests activation by a substrate-induced mechanism. Proc Natl Acad Sci U S A. 2016;113(40):11196–201.

Dahms SO, Jiao GS, Than ME. Structural studies revealed active site distortions of human furin by a small molecule inhibitor. ACS Chem Biol. 2017.

Day PM, Lowy DR, Schiller JT. Heparan sulfate-independent cell binding and infection with furin-precleaved papillomavirus capsids. J Virol. 2008;82(24):12565–8.

De Bie I, Savaria D, Roebroek AJ, Day R, Lazure C, Van de Ven WJ, Seidah NG. Processing specificity and biosynthesis of the Drosophila melanogaster convertases dfurin1, dfurin1-CRR, dfurin1-X, and dfurin2. J Biol Chem. 1995;270(3):1020–8.

De Bie I, Marcinkiewicz M, Malide D, Lazure C, Nakyama K, Bendayan M, Seidah NG. The isoforms of proprotein convertase PC5 are sorted to different subcellular compartments. sorting signals YxxL, LL und acidic serine-casein kinase phosphorylation, PC5, PC7, kexin, furin. J Cell Biol. 1996;135:1261–75.
de Haan CA, Stadler K, Godeke GJ, Bosch BJ, Rottier PJ. Cleavage inhibition of the murine coronavirus spike protein by a furin-like enzyme affects cell-cell but not virus-cell fusion. J Virol. 2004;78(11):6048–54.
Declercq J, Meulemans S, Plets E, Creemers JW. I nternalization of proprotein convertase PC7 from plasma membrane is mediated by a novel motif. J Biol Chem. 2012;287(12): 9052–60.
Declercq J, Ramos-Molina B, Sannerud R, Brouwers B, Pruniau VPEG, Meulemans S, Plets E, Annaert W, Creemers JWM. Endosome to trans-Golgi network transport of Proprotein Convertase 7 is mediated by a cluster of basic amino acids and palmitoylated cysteines. Eur J Cell Biol. 2017. pii: S0171-9335(16)30198-4.
Decroly E, Benjannet S, Savaria D, Seidah NG. Comparative functional role of PC7 and furin in the processing of the HIV envelope glycoprotein gp160. FEBS Lett. 1997;405(1):68–72.
Dikeakos JD, Di Lello P, Lacombe MJ, Ghirlando R, Legault P, Reudelhuber TL, Omichinski JG. Functional and structural characterization of a dense core secretory granule sorting domain from the PC1/3 protease. Proc Natl Acad Sci U S A. 2009;106(18):7408–13.
Dillon SL, Williamson DM, Elferich J, Radler D, Joshi R, Thomas G, Shinde U. Propeptides are sufficient to regulate organelle-specific pH-dependent activation of furin and proprotein convertase 1/3. J Mol Biol. 2012;423(1):47–62.
Duda A, Stange A, Lüftenegger D, Stanke N, Westphal D, Pietschmann T, Eastman SW, Linial ML, Rethwilm A, Lindemann D. Prototype foamy virus envelope glycoprotein leader peptide processing is mediated by a furin-like cellular protease, but cleavage is not essential for viral infectivity. J Virol. 2004;78(24):13865–70.
Duguay SJ, Lai-Zhang J, Steiner DF. Mutational analysis of the insulin-like growth factor I prohormone processing site. J Biol Chem. 1995;270(29):17566–74.
Eichler R, Lenz O, Garten W, Strecker T. The role of single N-glycans in proteolytic processing and cell surface transport of the Lassa virus glycoprotein GP-C. Virol J. 2006;3:41.
Eickmann M, Kiermayer S, Kraus I, Gössl M, Richt JA, Garten W. Maturation of Borna disease virus glycoprotein. FEBS Lett. 2005;579(21):4751–6.
Eisenberg RJ, Atanasiu D, Cairns TM, Gallagher JR, Krummenacher C, Cohen GH. Herpes virus fusion and entry: a story with many characters. Virus. 2012;4(5):800–32. Review.
Elagoz A, Benjannet S, Mammarbassi A, Wickham L, Seidah NG. Biosynthesis and cellular trafficking of the convertase SKI-1/S1P: ectodomain shedding requires SKI-1 activity. J Biol Chem. 2002;277(13):11265–75.
Elshuber S, Allison SL, Heinz FX, Mandl CW. Cleavage of protein prM is necessary for infection of BHK-21 cells by tick-borne encephalitis virus. J Gen Virol. 2003;84(Pt 1):183–91.
Essalmani R, Zaid A, Marcinkiewicz J, Chamberland A, Pasquato A, Seidah NG, Prat A. In vivo functions of the proprotein convertase PC5/6 during mouse development: Gdf11 is a likely substrate. Proc Natl Acad Sci U S A. 2008;105(15):5750–5.
Essalmani R, Susan-Resiga D, Guillemot J, Kim W, Sachan V, Awan Z, Chamberland A, Asselin MC, Ly K, Desjardins R, Day R, Prat A, Seidah NG. Thrombin activation of protein C requires prior processing by a liver proprotein convertase. J Biol Chem. 2017;292(25):10564–73.
Farooqi IS, Volders K, Stanhope R, Heuschkel R, White A, Lank E, Keogh J, O'Rahilly S, Creemers JW. Hyperphagia and early-onset obesity due to a novel homozygous missense mutation in prohormone convertase 1/3. J Clin Endocrinol Metab. 2007;92(9):3369–73.
Feldmann A, Schäfer MK, Garten W, Klenk HD. Targeted infection of endothelial cells by avian influenza virus A/FPV/Rostock/34 (H7N1) in chicken embryos. J Virol. 2000;74(17):8018–27.
Follis KE, York J, Nunberg JH. Furin cleavage of the SARS coronavirus spike glycoprotein enhances cell-cell fusion but does not affect virion entry. Virology. 2006;350(2):358–69.
Fujii Y, Sakaguchi T, Kiyotani K, Yoshida T. Comparison of substrate specificities against the fusion glycoprotein of virulent Newcastle disease virus between a chick embryo fibro-

blast processing protease and mammalian subtilisin-like proteases. Microbiol Immunol. 1999;43(2):133–40.
Fuller RS, Brake A, Thorner J. Yeast prohormone processing enzyme (KEX2 gene product) is a Ca2+-dependent serine protease. Proc Natl Acad Sci U S A. 1989a;86(5):1434–8.
Fuller RS, Brake AJ, Thorner J. Intracellular targeting and structural conservation of a prohormone-processing endoprotease. Science. 1989b;246(4929):482–6.
Gabriel G, Garn H, Wegmann M, Renz H, Herwig A, Klenk HD, Stech J. The potential of a protease activation mutant of a highly pathogenic avian influenza virus for a pandemic live vaccine. Vaccine. 2008;26(7):956–65.
Garred O, van Deurs B, Sandvig K. Furin-induced cleavage and activation of Shiga toxin. J Biol Chem. 1995;270(18):10817–21.
Garten W, Bosch FX, Linder D, Rott R, Klenk HD. Proteolytic activation of the influenza virus hemagglutinin: The structure of the cleavage site and the enzymes involved in cleavage. Virology. 1981;115(2):361–74.
Garten W, Linder D, Rott R, Klenk HD. The cleavage site of the hemagglutinin of fowl plague virus. Virology. 1982;122(1):186–90.
Garten W, Kuroda K, Schuy W, Naruse H, Scholtissek C, Klenk HD. Haemagglutinin transport mutants. Vaccine. 1985;3(3 Suppl):227–9.
Garten W, Stieneke A, Shaw E, Wikstrom P, Klenk HD. Inhibition of proteolytic activation of influenza virus hemagglutinin by specific peptidyl chloroalkyl ketones. Virology. 1989;172(1):25–31.
Garten W, Vey M, Ohuchi R, Ohuchi M, Klenk HD. Modification of the cleavage activation of the influenza virus hemagglutinin by site-specific mutagenesis. Behring Inst Mitt. 1991;89:12–22.
Garten W, Hallenberger S, Ortmann D, Schäfer W, Vey M, Angliker H, Shaw E, Klenk HD. Processing of viral glycoproteins by the subtilisin-like endoprotease furin and its inhibition by specific peptidyl chloroalkyl ketones. Biochimie. 1994;76(3-4):217–25. Review
Garten W, Braden C, Arendt A, Peitsch C, Baron J, Lu Y, Pawletko K, Hardes K, Steinmetzer T, Böttcher-Friebertshäuser E. Influenza virus activating host proteases: identification, localization and inhibitors as potential therapeutics. Eur J Cell Biol. 2015;94(7-9):375–83.
Geiselhart V, Bastone P, Kempf T, Schnölzer M, Löchelt M. Furin-mediated cleavage of the feline foamy virus. J Virol. 2004;78(24):13573–81.
Gil-Torregrosa BC, Castano AR, Lopez D, Del Val M. Generation of MHC class I peptide antigens by protein processing in the secretory route by furin. Traffic. 2000;1:641–51.
González-Reyes L, Ruiz-Argüello MB, García-Barreno B, Calder L, López JA, Albar JP, Skehel JJ, Wiley DC, Melero JA. Cleavage of the human respiratory syncytial virus fusion protein at two distinct sites is required for activation of membrane fusion. Proc Natl Acad Sci U S A. 2001;98(17):9859–64.
Gordon VM, Leppla SH. Proteolytic activation of bacterial toxins: role of bacterial and host cell proteases. Infect Immun. 1994;62(2):333–40. Review.
Gordon VM, Rehemtulla A, Leppla SH. A role for PACE4 in the proteolytic activation of anthrax toxin protective antigen. Infect Immun. 1997;65(8):3370–5.
Gorski JP, Huffman NT, Chittur S, Midura RJ, Black C, Oxford J, Seidah NG. Inhibition of proprotein convertase SKI-1 blocks transcription of key extracellular matrix genes regulating osteoblastic mineralization. J Biol Chem. 2011;286(3):1836–49.
Gotoh B, Ohnishi Y, Inocencio NM, Esaki E, Nakayama K, Barr PJ, Thomas G, Nagai Y. Mammalian subtilisin-related proteinases in cleavage activation of the paramyxovirus fusion glycoprotein: superiority of furin/PACE to PC2 or PC1/PC3. J Virol. 1992;66(11):6391–7.
Gu M, Rappaport J, Leppla SH. Furin is important but not essential for the proteolytic maturation of gp160 of HIV-1. FEBS Lett. 1995;365(1):95–7.
Guillemot J, Canuel M, Essalmani R, Prat A, Seidah NG. Implication of the proprotein convertases in iron homeostasis: proprotein convertase 7 sheds human transferrin receptor 1 and furin activates hepcidin. Hepatology. 2013;57(6):2514–24.
Hallenberger S, Bosch V, Angliker H, Shaw E, Klenk HD, Garten W. Inhibition of furin-mediated cleavage activation of HIV-1 glycoprotein gp160. Nature. 1992;360(6402):358–61.

Hallenberger S, Moulard M, Sordel M, Klenk HD, Garten W. The role of eukaryotic subtilisin-like endoproteases for the activation of human immunodeficiency virus glycoproteins in natural host cells. J Virol. 1997;71(2):1036–45.

Hardes K, Becker GL, Lu Y, Dahms SO, Köhler S, Beyer W, Sandvig K, Yamamoto H, Lindberg I, Walz L, von Messling V, Than ME, Garten W, Steinmetzer T. Novel Furin inhibitors with potent anti-infectious activity. ChemMedChem. 2015;10(7):1218–31.

Hardes K, Ivanova T, Thaa B, McInermey GM, Klokk TI, Sandvig K, Kunzel S, Lindberg I, Steinmetzer T. Elongated and shortened peptidomimetic inhibitors of the proprotein convertase furin. ChemMedChem. 2017;12:613–20.

Harrison SC. Viral membrane fusion. Nat Struct Mol Biol. 2008;15(7):690–8.

Hatsuzawa K, Hosaka M, Nakagawa T, Nagase M, Shoda A, Murakami K, Nakayama K. Structure and expression of mouse furin, a yeast Kex2-related protease. Lack of processing of coexpressed prorenin in GH4C1 cells. J Biol Chem. 1990;265(36):22075–8.

Heidner HW, Johnston RE. The amino-terminal residue of Sindbis virus glycoprotein E2 influences virus maturation, specific infectivity for BHK cells, and virulence in mice. J Virol. 1994;68(12):8064–70.

Hendy GN, Bennett HP, Gibbs BF, Lazure C, Day R, Seidah NG. Proparathyroid hormone is preferentially cleaved to parathyroid hormone by the prohormone convertase furin. A mass spectrometric study. J Biol Chem. 1995;270(16):9517–25.

Henrich S, Cameron A, Bourenkov GP, Kiefersauer R, Huber R, Lindberg I, Bode W, Than ME. The crystal structure of the proprotein processing proteinase furin explains its stringent specificity. Nat Struct Biol. 2003;10(7):520–6. PDB: 1PJ8. Erratum in: Nat Struct Biol. 2003; 10(8):669.

Henrich S, Lindberg I, Bode W, Than ME. Proprotein convertase models based on the crystal structures of furin and kexin: explanation of their specificity. J Mol Biol. 2005;345(2):211–27.

Himmelspach M, Pfleiderer M, Fischer BE, Plaimauer B, Antoine G, Falkner FG, Dorner F, Schlokat U. Recombinant human factor X: high yield expression and the role of furin in proteolytic maturation in vivo and in vitro. Thromb Res. 2000;97(2):51–67.

Horimoto T, Kawaoka Y. The hemagglutinin cleavability of a virulent avian influenza virus by subtilisin-like endoproteases is influenced by the amino acid immediately downstream of the cleavage site. Virology. 1995;210(2):466–70.

Horimoto T, Nakayama K, Smeekens SP, Kawaoka Y. Proprotein-processing endoproteases PC6 and furin both activate hemagglutinin of virulent avian influenza viruses. J Virol. 1994;68(9):6074–8.

Ito K, Kim KH, Lok AS, Tong S. Characterization of genotype-specific carboxyl-terminal cleavage sites of hepatitis B virus e antigen precursor and identification of furin as the candidate enzyme. J Virol. 2009;83(8):3507–17.

Jain SK, De Candido S, Kielian M. Processing of the p62 envelope precursor protein of Semliki Forest virus. J Biol Chem. 1991;266(9):5756–61.

Jardetzky TS, Lamb RA. Activation of paramyxovirus membrane fusion and virus entry. Curr Opin Virol. 2014;5:24–33. Review.

Jean F, Stella K, Thomas L, Liu G, Xiang Y, Reason AJ, Thomas G. alpha1-Antitrypsin Portland, a bioengineered serpin highly selective for furin: application as an antipathogenic agent. Proc Natl Acad Sci U S A. 1998;95(13):7293–8.

Johannsen E, Luftig M, Chase MR, Weicksel S, Cahir-McFarland E, Illanes D, Sarracino D, Kieff E. Proteins of purified Epstein-Barr virus. Proc Natl Acad Sci U S A. 2004;101(46):16286–91.

Julius D, Brake A, Blair L, Kunisawa R, Thorner J. Isolation of the putative structural gene for the lysine-arginine cleaving endopeptidase required for processing of yeast prepro-alpha-factor. Cell. 1984;37:1075–89.

Kacprzak MM, Peinado JR, Than ME, Appel J, Henrich S, Lipkind G, Houghten RA, Bode W, Lindberg I. Inhibition of furin by polyarginine-containing peptides: nanomolar inhibition by nona-D-arginine. J Biol Chem. 2004;279(35):36788–94.

Kawaoka Y, Webster RG. Sequence requirements for cleavage activation of influenza virus hemagglutinin expressed in mammalian cells. Proc Natl Acad Sci U S A. 1988;85(2):324–8.

Kawaoka Y, Nestorowicz A, Alexander DJ, Webster RG. Molecular analyses of the hemagglutinin genes of H5 influenza viruses: origin of a virulent turkey strain. Virology. 1987;158(1):218–27.

Keelapang P, Sriburi R, Supasa S, Panyadee N, Songjaeng A, Jairungsri A, Puttikhunt C, Kasinrerk W, Malasit P, Sittisombut N. Alterations of pr-M cleavage and virus export in pr-M junction chimeric dengue viruses. J Virol. 2004;78(5):2367–81.

Kemmler W, Peterson JD, Steiner DF. Studies on the conversion of proinsulin to insulin. I. Conversion in vitro with trypsin and carboxypeptidase B. J Biol Chem. 1971;246(22): 6786–91.

Khatib AM, Siegfried G, Prat A, Luis J, Chrétien M, Metrakos P, Seidah NG. Inhibition of proprotein convertases is associated with loss of growth and tumorigenicity of HT-29 human colon carcinoma cells: importance of insulin-like growth factor-1 (IGF-1) receptor processing in IGF-1-mediated functions. J Biol Chem. 2001;276(33):30686–93.

Kiefer MC, Tucker JE, Joh R, Landsberg KE, Saltman D, Barr PJ. Identification of a second human subtilisin-like protease gene in the fes/fps region of chromosome 15. DNA Cell Biol. 1991;10(10):757–69.

Klenk HD, Garten W. Activation cleavage of viral spike proteins by host proteases. In: Wimmer E, editor. Cellular receptors for animal viruses. Cold Spring Harbor: Cold Spring Harbor Laboratory Press; 1994. p. 241–80.

Klenk HD, Garten W, Rott R. Inhibition of proteolytic cleavage of the hemagglutinin of influenza virus by the calcium-specific ionophore A23187. EMBO J. 1984;3(12):2911–5.

Kopp A, Blewett E, Misra V, Mettenleiter TC. Proteolytic cleavage of bovine herpesvirus 1 (BHV-1) glycoprotein gB is not necessary for its function in BHV-1 or pseudorabies virus. J Virol. 1994;68(3):1667–74.

Kouretova J, Hammamy MZ, Epp A, Hardes K, Kallis S, Zhang L, Hilgenfeld R, Bartenschlager R, Steinmetzer T. Effects of NS2B-NS3 protease and furin inhibition on West Nile and Dengue virus replication. J Enzyme Inhib Med Chem. 2017;32(1):712–21.

Kuno G, Chang GJ. Full-length sequencing and genomic characterization of Bagaza, Kedougou, and Zika viruses. Arch Virol. 2007;152(4):687–96.

Kunz S, Edelmann KH, de la Torre JC, Gorney R, Oldstone MB. Mechanisms for lymphocytic choriomeningitis virus glycoprotein cleavage, transport, and incorporation into virions. Virology. 2003;314(1):168–78.

Kuroda K, Hauser C, Rott R, Klenk HD, Doerfler W. Expression of the influenza virus haemagglutinin in insect cells by a baculovirus vector. EMBO J. 1986;5(6):1359–65.

Kuroda K, Gröner A, Frese K, Drenckhahn D, Hauser C, Rott R, Doerfler W, Klenk HD. Synthesis of biologically active influenza virus hemagglutinin in insect larvae. J Virol. 1989;63(4):1677–85.

Le QT, Blanchet M, Seidah NG, Labonté P. Plasma Membrane Tetraspanin CD81 complexes with Proprotein Convertase Subtilisin/Kexin Type 9 (PCSK9) and low density lipoprotein receptor (LDLR), and its levels are reduced by PCSK9. J Biol Chem. 2015;290(38):23385–400.

Leduc R, Molloy SS, Thorne BA, Thomas G. Activation of human furin precursor processing endoprotease occurs by an intramolecular autoproteolytic cleavage. J Biol Chem. 1992;267(20):14304–8.

Lee SN, Lindberg I. 7B2 prevents unfolding and aggregation of prohormone convertase. Endocrinology. 2008;149(8):4116–27.

Lee SN, Lee DH, Lee MG, Yoon JH. Proprotein convertase 5/6a is associated with bone morphogenetic protein-2-induced squamous cell differentiation. Am J Respir Cell Mol Biol. 2015;52(6):749–61.

Lenz O, ter Meulen J, Feldmann H, Klenk HD, Garten W. Identification of a novel consensus sequence at the cleavage site of the Lassa virus glycoprotein. J Virol. 2000;74(23):11418–21.

Lenz O, ter Meulen J, Klenk HD, Seidah NG, Garten W. The Lassa virus glycoprotein precursor GP-C is proteolytically processed by subtilase SKI-1/S1P. Proc Natl Acad Sci U S A. 2001;98(22):12701–5. Epub 2001 Oct 16.

Li S, Sun Z, Pryce R, Parsy ML, Fehling SK, Schlie K, Siebert CA, Garten W, Bowden TA, Strecker T, Huiskonen JT. Acidic pH-induced conformations and LAMP1 binding of the Lassa Virus glycoprotein spike. PLoS Pathog. 2016;12(2):e1005418.

Long G, Pan X, Westenberg M, Vlak JM. Functional role of the cytoplasmic tail domain of the major envelope fusion protein of group II baculoviruses. J Virol. 2006;80(22):11226–34.
Longpré JM, Leduc R. Identification of prodomain determinants involved in ADAMTS-1 biosynthesis. J Biol Chem. 2004;279(32):33237–45.
Lopez-Perez E, Zhang Y, Frank SJ, Creemers J, Seidah N, Checler F. Constitutive alpha-secretase cleavage of the beta-amyloid precursor protein in the furin-deficient LoVo cell line: involvement of the pro-hormone convertase 7 and the disintegrin metalloprotease ADAM10. J Neurochem. 2001;76(5):1532–9.
Lu Y, Hardes K, Dahms SO, Böttcher-Friebertshäuser E, Steinmetzer T, Than ME, Klenk HD, Garten W. Peptidomimetic furin inhibitor MI-701 in combination with oseltamivir and ribavirin efficiently blocks propagation of highly pathogenic avian influenza viruses and delays high level oseltamivir resistance in MDCK cells. Antiviral Res. 2015;120: 89–100.
Lusson J, Vieau D, Hamelin J, Day R, Chrétien M, Seidah NG. cDNA structure of the mouse and rat subtilisin/kexin-like PC5: a candidate proprotein convertase expressed in endocrine and nonendocrine cells. Proc Natl Acad Sci U S A. 1993;90(14):6691–5.
Mains RE, Berard CA, Denault JB, Zhou A, Johnson RC, Leduc R. PACE4: a subtilisin-like endoprotease with unique properties. Biochem J. 1997;321(Pt 3):587–93.
Maisa A, Ströher U, Klenk HD, Garten W, Strecker T. Inhibition of Lassa virus glycoprotein cleavage and multicycle replication by site 1 protease-adapted alpha(1)-antitrypsin variants. PLoS Negl Trop Dis. 2009;3(6):e446.
Mbikay M, Tadros H, Ishida N, Lerner CP, De Lamirande E, Chen A, El-Alfy M, Clermont Y, Seidah NG, Chrétien M, Gagnon C, Simpson EM. Impaired fertility in mice deficient for the testicular germ-cell protease PC4. Proc Natl Acad Sci U S A. 1997;94(13):6842–6.
McCune JM, Rabin LB, Feinberg MB, Lieberman M, Kosek JC, Reyes GR, Weissman IL. Endoproteolytic cleavage of gp160 is required for the activation of human immunodeficiency virus. Cell. 1988;53(1):55–67.
Meerabux J, Yaspo ML, Roebroek AJ, Van de Ven WJ, Lister TA, Young BD. A new member of the proprotein convertase gene family (LPC) is located at a chromosome translocation breakpoint in lymphomas. Cancer Res. 1996;56(3):448–51.
Millet JK, Whittaker GR. Host cell entry of Middle East respiratory syndrome coronavirus after two-step, furin-mediated activation of the spike protein. Proc Natl Acad Sci U S A. 2014;111:15214–9.
Millet JK, Whittaker GR. Host cell proteases: critical determinants of coronavirus tropism and pathogenesis. Virus Res. 2015;202:120–34.
Misumi Y, Ohkubo K, Sohda M, Takami N, Oda K, Ikehara Y. Intracellular processing of complement pro-C3 and proalbumin is inhibited by rat alpha 1-protease inhibitor variant (Met352DOUBLEHYPHENDOUBLEHYPHENArg) in transfected cells. Biochem Biophys Res Commun. 1990;171(1):236–42.
Misumi Y, Sohda M, Ikehara Y. Sequence of the cDNA encoding rat furin, a possible propeptide-processing endoprotease. Nucleic Acids Res. 1990;18(22):6719.
Mizuno K, Nakamura T, Matsuo H. A unique membrane-bound, calcium-dependent endopeptidase with specificity toward paired basic residues in rat liver Golgi fractions. Biochem Biophys Res Commun. 1989;164(2):780–7.
Moehring JM, Inocencio NM, Robertson BJ, Moehring TJ. Expression of mouse furin in a Chinese hamster cell resistant to Pseudomonas exotoxin A and viruses complements the genetic lesion. J Biol Chem. 1993;268(4):2590–4.
Molloy SS, Bresnahan PA, Leppla SH, Klimpel KR, Thomas G. Human furin is a calcium-dependent serine endoprotease that recognizes the sequence Arg-X-X-Arg and efficiently cleaves anthrax toxin protective antigen. J Biol Chem. 1992;267(23):16396–402.
Molloy SS, Anderson ED, Jean F, Thomas G. Bi-cycling the furin pathway: from TGN localization to pathogen activation and embryogenesis. Trends Cell Biol. 1999;9(1):28–35. Review.
Mori K, Imamaki A, Nagata K, Yonetomi Y, Kiyokage-Yoshimoto R, Martin TJ, Gillespie MT, Nagahama M, Tsuji A, Matsuda Y. Subtilisin-like proprotein convertases, PACE4 and

PC8, as well as furin, are endogenous proalbumin convertases in HepG2 cells. J Biochem. 1999;125(3):627–33.
Morikawa Y, Barsov E, Jones I. Legitimate and illegitimate cleavage of human immunodeficiency virus glycoproteins by furin. J Virol. 1993;67(6):3601–4.
Munzer JS, Basak A, Zhong M, Mamarbachi A, Hamelin J, Savaria D, Lazure C, Hendy GN, Benjannet S, Chrétien M, Seidah NG. In vitro characterization of the novel proprotein convertase PC7. J Biol Chem. 1997;272(32):19672–81.
Nagai Y. Virus activation by host proteinases. A pivotal role in the spread of infection, tissue tropism and pathogenicity. Microbiol Immunol. 1995;39(1):1–9. Review.
Nakagawa T, Hosaka M, Torii S, Watanabe T, Murakami K, Nakayama K. Identification and functional expression of a new member of the mammalian Kex2-like processing endoprotease family: its striking structural similarity to PACE4. J Biochem. 1993;113(2):132–5.
Nakagawa T, Suzuki-Nakagawa C, Watanabe A, Asami E, Matsumoto M, Nakano M, Ebihara A, Uddin MN, Suzuki F. Site-1 protease is required for the generation of soluble (pro)renin receptor. J Biochem. 2017;161(4):369–79.
Nakayama K. Furin: a mammalian subtilisin/Kex2p-like endoprotease involved in processing of a wide variety of precursor proteins. Biochem J. 1997;327(Pt 3):625–35.
Nakayama K, Kim WS, Torii S, Hosaka M, Nakagawa T, Ikemizu J, Baba T, Murakami K. Identification of the fourth member of the mammalian endoprotease family homologous to the yeast Kex2 protease. Its testis-specific expression. J Biol Chem. 1992;267(9):5897–900.
Neumann G, Horimoto T, Kawaoka Y. Reverse genetics of influenza viruses–applications in research and vaccine design. In: Klenk H-D, Matrosovich MN, Stech J, editors. Avian Influenza. Monogr Virol, vol. 27. Basel: Karger; 2008. p. 118–33.
Nour N, Basak A, Chrétien M, Seidah NG. Structure-function analysis of the prosegment of the proprotein convertase PC5A. J Biol Chem. 2003;278(5):2886–95.
Nour N, Mayer G, Mort JS, Salvas A, Mbikay M, Morrison CJ, Overall CM, Seidah NG. The cysteine-rich domain of the secreted proprotein convertases PC5A and PACE4 functions as a cell surface anchor and interacts with tissue inhibitors of metalloproteinases. Mol Biol Cell. 2005;16(11):5215–26.
Oda K, Misumi Y, Ikehara Y, Brennan SO, Hatsuzawa K, Nakayama K. Proteolytic cleavages of proalbumin and complement Pro-C3 in vitro by a truncated soluble form of furin, a mammalian homologue of the yeast Kex2 protease. Biochem Biophys Res Commun. 1992;189(3):1353–61.
Ohnishi Y, Shioda T, Nakayama K, Iwata S, Gotoh B, Hamaguchi M, Nagai Y. A furin-defective cell line is able to process correctly the gp160 of human immunodeficiency virus type 1. J Virol. 1994;68(6):4075–9.
Oliver SL, Sommer M, Zerboni L, Rajamani J, Grose C, Arvin AM. Mutagenesis of varicella-zoster virus glycoprotein B: putative fusion loop residues are essential for viral replication, and the furin cleavage motif contributes to pathogenesis in skin tissue in vivo. J Virol. 2009;83(15):7495–506.
Olmstead AD, Knecht W, Lazarov I, Dixit SB, Jean F. Human subtilase SKI-1/S1P is a master regulator of the HCV Lifecycle and a potential host cell target for developing indirect-acting antiviral agents. PLoS Pathog. 2012;8(1):e1002468.
Orlich M, Linder D, Rott R. Trypsin-resistant protease activation mutants of an influenza virus. J Gen Virol. 1995;76(Pt 3):625–33.
Ortmann D, Ohuchi M, Angliker H, Shaw E, Garten W, Klenk HD. Proteolytic cleavage of wild type and mutants of the F protein of human parainfluenza virus type 3 by two subtilisin-like endoproteases, furin and Kex2. J Virol. 1994;68(4):2772–6.
Ozden S, Lucas-Hourani M, Ceccaldi PE, Basak A, Valentine M, Benjannet S, Hamelin J, Jacob Y, Mamchaoui K, Mouly V, Desprès P, Gessain A, Butler-Browne G, Chrétien M, Tangy F, Vidalain PO, Seidah NG. Inhibition of Chikungunya virus infection in cultured human muscle cells by furin inhibitors: impairment of the maturation of the E2 surface glycoprotein. J Biol Chem. 2008;283(32):21899–908.

Pan H, Che FY, Peng B, Steiner DF, Pintar JE, Fricker LD. The role of prohormone convertase-2 in hypothalamic neuropeptide processing: a quantitative neuropeptidomic study. J Neurochem. 2006;98(6):1763–77.
Pang YJ, Tan XJ, Li DM, Zheng ZH, Lei RX, Peng XM. Therapeutic potential of furin inhibitors for the chronic infection of hepatitis B virus. Liver Int. 2013;33(8):1230–8.
Paquet L, Bergeron F, Boudreault A, Seidah NG, Chrétien M, Mbikay M, Lazure C. The neuroendocrine precursor 7B2 is a sulfated protein proteolytically processed by a ubiquitous furin-like convertase. J Biol Chem. 1994;269(30):19279–85.
Pei D, Weiss SJ. Furin-dependent intracellular activation of the human stromelysin-3 zymogen. Nature. 1995;375(6528):244–7.
Piper DE, Jackson S, Liu Q, Romanow WG, Shetterly S, Thibault ST, Shan B, Walker NP. The crystal structure of PCSK9: a regulator of plasma LDL-cholesterol. Structure. 2007;15(5):545–52.
Plaimauer B, Mohr G, Wernhart W, Himmelspach M, Dorner F, Schlokat U. 'Shed' furin: mapping of the cleavage determinants and identification of its C-terminus. Biochem J. 2001;354(Pt 3):689–95.
Porter AG, Barber C, Carey NH, Hallewell RA, Threlfall G, Emtage JS. Complete nucleotide sequence of an influenza virus haemagglutinin gene from cloned DNA. Nature. 1979;282(5738):471–7.
Pritzer E, Kuroda K, Garten W, Nagai Y, Klenk HD. A host range mutant of Newcastle disease virus with an altered cleavage site for proteolytic activation of the F protein. Virus Res. 1990;15(3):237–42.
Puente XS, Sánchez LM, Overall CM, López-Otín C. Human and mouse proteases: a comparative genomic approach. Nat Rev Genet. 2003;4(7):544–58. Review.
Puthavathana P, Auewarakul P, Charoenying PC, Sangsiriwut K, Pooruk P, Boonnak K, Khanyok R, Thawachsupa P, Kijphati R, Sawanpanyalert P. Molecular characterization of the complete genome of human influenza H5N1 virus isolates from Thailand. J Gen Virol. 2005;86(Pt 2):423–33.
Qiu Q, Basak A, Mbikay M, Tsang BK, Gruslin A. Role of pro-IGF-II processing by proprotein convertase 4 in human placental development. Proc Natl Acad Sci U S A. 2005;102(31):11047–52.
Ramos-Molina B, Lindberg I. Phosphorylation and alternative splicing of 7B2 reduce Prohormone Convertase 2 activation. Mol Endocrinol. 2015;29(5):756–64.
Rehemtulla A, Barr PJ, Rhodes CJ, Kaufman RJ. PACE4 is a member of the mammalian propeptidase family that has overlapping but not identical substrate specificity to PACE. Biochemistry. 1993;32(43):11586–90.
Remacle AG, Shiryaev SA, Oh ES, Cieplak P, Srinivasan A, Wei G, Liddington RC, Ratnikov BI, Parent A, Desjardins R, Day R, Smith JW, Lebl M, Strongin AY. Substrate cleavage analysis of furin and related proprotein convertases. A comparative study. J Biol Chem. 2008;283(30):20897–906.
Rice CM, Lenches EM, Eddy SR, Shin SJ, Sheets RL, Strauss JH. Nucleotide sequence of yellow fever virus: implications for flavivirus gene expression and evolution. Science. 1985;229(4715):726–33.
Richards RM, Lowy DR, Schiller JT, Day PM. Cleavage of the papillomavirus minor capsid protein, L2, at a furin consensus site is necessary for infection. Proc Natl Acad Sci U S A. 2006;103:1522–7.
Richardson C, Hull D, Greer P, Hasel K, Berkovich A, Englund G, Bellini W, Rima B, Lazzarini R. The nucleotide sequence of the mRNA encoding the fusion protein of measles virus (Edmonston strain): a comparison of fusion proteins from several different paramyxoviruses. Virology. 1986;155(2):508–23.
Richt JA, Fürbringer T, Koch A, Pfeuffer I, Herden C, Bause-Niedrig I, Garten W. Processing of the Borna disease virus glycoprotein gp94 by the subtilisin-like endoprotease furin. J Virol. 1998;72(5):4528–33.
Robertson BJ, Moehring JM, Moehring TJ. Defective processing of the insulin receptor in an endoprotease-deficient Chinese hamster cell strain is corrected by expression of mouse furin. J Biol Chem. 1993;268(32):24274–7.

Rockwell NC, Krysan DJ, Komiyama T, Fuller RS. Precursor processing by kex2/furin proteases. Chem Rev. 2002;102(12):4525–48.

Roebroek AJ, Schalken JA, Bussemakers MJ, van Heerikhuizen H, Onnekink C, Debruyne FM, Bloemers HP, Van de Ven WJ. Characterization of human c-fes/fps reveals a new transcription unit (fur) in the immediately upstream region of the proto-oncogene. Mol Biol Rep. 1986;11(2):117–25.

Roebroek AJ, Pauli IG, Zhang Y, van de Ven WJ. cDNA sequence of a Drosophila melanogaster gene, Dfur1, encoding a protein structurally related to the subtilisin-like proprotein processing enzyme furin. FEBS Lett. 1991;289(2):133–7.

Roebroek AJ, Umans L, Pauli IG, Robertson EJ, van Leuven F, Van de Ven WJ, Constam DB. Failure of ventral closure and axial rotation in embryos lacking the proprotein convertase Furin. Development. 1998;125(24):4863–76.

Roebroek AJ, Taylor NA, Louagie E, Pauli I, Smeijers L, Snellinx A, Lauwers A, Van de Ven WJ, Hartmann D, Creemers JW. Limited redundancy of the proprotein convertase furin in mouse liver. J Biol Chem. 2004;279(51):53442–50.

Rousselet E, Benjannet S, Hamelin J, Canuel M, Seidah NG. The proprotein convertase PC7: unique zymogen activation and trafficking pathways. J Biol Chem. 2011;286(4):2728–38.

Sakaguchi T, Fujii Y, Kiyotani K, Yoshida T. Correlation of proteolytic cleavage of F protein precursors in paramyxoviruses with expression of the fur, PACE4 and PC6 genes in mammalian cells. J Gen Virol. 1994;75(Pt 10):2821–7.

Sakai J, Rawson RB, Espenshade PJ, Cheng D, Seegmiller AC, Goldstein JL, Brown MS. Molecular identification of the sterol-regulated luminal protease that cleaves SREBPs and controls lipid composition of animal cells. Mol Cell. 1998;2(4):505–14.

Sanchez AJ, Vincent MJ, Nichol ST. Characterization of the glycoproteins of Crimean-Congo hemorrhagic fever virus. J Virol. 2002;76:7263–75.

Sanchez AJ, Vincent MJ, Erickson BR, Nichol ST. Crimean-congo hemorrhagic fever virus glycoprotein precursor is cleaved by Furin-like and SKI-1 proteases to generate a novel 38-kilodalton glycoprotein. J Virol. 2006;80(1):514–25.

Sariola M, Saraste J, Kuismanen E. Communication of post-Golgi elements with early endocytic pathway: regulation of endoproteolytic cleavage of Semliki Forest virus p62 precursor. J Cell Sci. 1995;108(Pt 6):2465–75.

Sato H, Kinoshita T, Takino T, Nakayama K, Seiki M. Activation of a recombinant membrane type 1-matrix metalloproteinase (MT1-MMP) by furin and its interaction with tissue inhibitor of metalloproteinases (TIMP)-2. FEBS Lett. 1996;393(1):101–4.

Scamuffa N, Calvo F, Chrétien M, Seidah NG, Khatib AM. Proprotein convertases: lessons from knockouts. FASEB J. 2006;20(12):1954–63. Review.

Schäfer W, Stroh A, Berghöfer S, Seiler J, Vey M, Kruse ML, Kern HF, Klenk HD, Garten W. Two independent targeting signals in the cytoplasmic domain determine trans-Golgi network localization and endosomal trafficking of the proprotein convertase furin. EMBO J. 1995;14(11):2424–35.

Scheid A, Choppin PW. Protease activation mutants of sendai virus. Activation of biological properties by specific proteases. Virology. 1976;69(1):265–77.

Schlie K, Maisa A, Lennartz F, Ströher U, Garten W, Strecker T. Characterization of Lassa virus glycoprotein oligomerization and influence of cholesterol on virus replication. J Virol. 2010;84(2):983–92.

Schlie K, Strecker T, Garten W. Maturation cleavage within the ectodomain of Lassa virus glycoprotein relies on stabilization by the cytoplasmic tail. FEBS Lett. 2010;584(21):4379–82.

Schlombs K, Wagner T, Scheel J. Site-1 protease is required for cartilage development in zebrafish. Proc Natl Acad Sci U S A. 2003;100(24):14024–9.

Schmidt I, Skinner M, Siddell S. Nucleotide sequence of the gene encoding the surface projection glycoprotein of coronavirus MHV-JHM. J Gen Virol. 1987;68(Pt 1):47–56.

Seidah NG. What lies ahead for the proprotein convertases? Ann N Y Acad Sci. 2011;1220:149–61. Review.

Seidah NG, Prat A. Precursor convertases in the secretory pathway, cytosol and extracellular milieu. Essays Biochem. 2002;38:79–94. Review.
Seidah NG, Prat A. The biology and therapeutic targeting of the proprotein convertases. Nat Rev Drug Discov. 2012;11(5):367–83. Review.
Seidah NG, Gaspar L, Mion P, Marcinkiewicz M, Mbikay M, Chrétien M. cDNA sequence of two distinct pituitary proteins homologous to Kex2 and furin gene products: tissue-specific mRNAs encoding candidates for pro-hormone processing proteinases. DNA Cell Biol. 1990;9(6):415–24. Erratum in: DNA Cell Biol. 1990;9(10):789.
Seidah NG, Day R, Marcinkiewicz M, Benjannet S, Chrétien M. Mammalian neural and endocrine pro-protein and pro-hormone convertases belonging to the subtilisin family of serine proteinases. Enzyme. 1991;45(5-6):271–84. Review.
Seidah NG, Day R, Hamelin J, Gaspar A, Collard MW, Chrétien M. Testicular expression of PC4 in the rat: molecular diversity of a novel germ cell-specific Kex2/subtilisin-like proprotein convertase. Mol Endocrinol. 1992;6(10):1559–70.
Seidah NG, Benjannet S, Pareek S, Chrétien M, Murphy RA. Cellular processing of the neurotrophin precursors of NT3 and BDNF by the mammalian proprotein convertases. FEBS Lett. 1996;379(3):247–50.
Seidah NG, Mowla SJ, Hamelin J, Mamarbachi AM, Benjannet S, Touré BB, Basak A, Munzer JS, Marcinkiewicz J, Zhong M, Barale JC, Lazure C, Murphy RA, Chrétien M, Marcinkiewicz M. Mammalian subtilisin/kexin isozyme SKI-1: a widely expressed proprotein convertase with a unique cleavage specificity and cellular localization. Proc Natl Acad Sci U S A. 1999;96(4):1321–6.
Seidah NG, Benjannet S, Wickham L, Marcinkiewicz J, Jasmin SB, Stifani S, Basak A, Prat A, Chretien M. The secretory proprotein convertase neural apoptosis-regulated convertase 1 (NARC-1): liver regeneration and neuronal differentiation. Proc Natl Acad Sci U S A. 2003;100(3):928–33.
Seidah NG, Mayer G, Zaid A, Rousselet E, Nassoury N, Poirier S, Essalmani R, Prat A. The activation and physiological functions of the proprotein convertases. Int J Biochem Cell Biol. 2008;40(6-7):1111–25. Review.
Seidah NG, Sadr MS, Chrétien M, Mbikay M. The multifaceted proprotein convertases: their unique, redundant, complementary, and opposite functions. J Biol Chem. 2013;288(30):21473–81.
Seidah NG, Awan Z, Chrétien M, Mbikay M. PCSK9: a key modulator of cardiovascular health. Circ Res. 2014;114(6):1022–36. Review.
Shinde U, Thomas G. Insights from bacterial subtilases into the mechanisms of intramolecular chaperone-mediated activation of furin. Methods Mol Biol. 2011;768:59–106. Review.
Shiryaev SA, Chernov AV, Golubkov VS, Thomsen E, Chudin E, Chee M, Kozlov I, Strongin AY, Cieplak P. High-resolution analysis and functional mapping of cleavage sites and substrate proteins of furin in the human proteome. PLoS One. 2013;8:e54290.
Siezen RJ, Creemers JW, Van de Ven WJ. Homology modelling of the catalytic domain of human furin. A model for the eukaryotic subtilisin-like proprotein convertases. Eur J Biochem. 1994;222(2):255–66.
Smeekens SP, Steiner DF. Identification of a human insulinoma cDNA encoding a novel mammalian protein structurally related to the yeast dibasic processing protease Kex2. J Biol Chem. 1990;265(6):2997–3000.
Smeekens SP, Avruch AS, LaMendola J, Chan SJ, Steiner DF. Identification of a cDNA encoding a second putative prohormone convertase related to PC2 in AtT20 cells and islets of Langerhans. Proc Natl Acad Sci U S A. 1991;88(2):340–4.
Sorem J, Longnecker R. Cleavage of Epstein-Barr virus glycoprotein B is required for full function in cell-cell fusion with both epithelial and B cells. J Gen Virol. 2009;90(Pt 3):591–5.
Spiesschaert B, Stephanowitz H, Krause E, Osterrieder N, Azab W. Glycoprotein B of equine herpesvirus type 1 has two recognition sites for subtilisin-like proteases that are cleaved by furin. J Gen Virol. 2016;97(5):1218–28.
Stadler K, Allison SL, Schalich J, Heinz FX. Proteolytic activation of tick-borne encephalitis virus by furin. J Virol. 1997;71:8475–81.

Stawowy P, Kallisch H, Borges Pereira Stawowy N, Stibenz D, Veinot JP, Gräfe M, Seidah NG, Chrétien M, Fleck E, Graf K. Immunohistochemical localization of subtilisin/kexin-like proprotein convertases in human atherosclerosis. Virchows Arch. 2005;446(4):351–9.

Stech J, Garn H, Wegmann M, Wagner R, Klenk HD. A new approach to an influenza live vaccine: modification of the cleavage site of hemagglutinin. Nat Med. 2005;11(6):683–9.

Steiner DF. The proprotein convertases. Curr Opin Chem Biol. 1998;2:31–9.

Steiner DF, Cunningham D, Spigelman L, Aten B. Insulin biosynthesis: evidence for a precursor. Science. 1967;157(3789):697–700.

Stieneke-Gröber A, Vey M, Angliker H, Shaw E, Thomas G, Roberts C, Klenk HD, Garten W. Influenza virus hemagglutinin with multibasic cleavage site is activated by furin, a subtilisin-like endoprotease. EMBO J. 1992;11(7):2407–14.

Strauss EG, Rice CM, Strauss JH. Complete nucleotide sequence of the genomic RNA of Sindbis virus. Virology. 1984;133(1):92–110.

Stroh A, Schäfer W, Berghöfer S, Eickmann M, Teuchert M, Bürger I, Klenk HD, Garten W. A mono phenylalanine-based motif (F790) and a leucine-dependent motif (LI760) mediate internalization of furin. Eur J Cell Biol. 1999;78(3):151–60.

Sturman LS, Holmes KV. Proteolytic cleavage of peplomeric glycoprotein E2 of MHV yields two 90K subunits and activates cell fusion. Adv Exp Med Biol. 1984;173:25–35.

Sucic JF, Moehring JM, Inocencio NM, Luchini JW, Moehring TJ. Endoprotease PACE4 is Ca2+-dependent and temperature-sensitive and can partly rescue the phenotype of a furin-deficient cell strain. Biochem J. 1999;339(Pt 3):639–47.

Takahashi S, Nakagawa T, Banno T, Watanabe T, Murakami K, Nakayama K. Localization of furin to the trans-Golgi network and recycling from the cell surface involves Ser and Tyr residues within the cytoplasmic domain. J Biol Chem. 1995;270(47):28397–401.

Tangrea MA, Bryan PN, Sari N, Orban J. Solution structure of the pro-hormone convertase 1 prodomain from Mus musculus. J Mol Biol. 2002;320(4):801–12.

Taylor NA, Van De Ven WJ, Creemers JW. Curbing activation: proprotein convertases in homeostasis and pathology. FASEB J. 2003;17(10):1215–27. Review.

Teuchert M, Schäfer W, Berghöfer S, Hoflack B, Klenk HD, Garten W. Sorting of furin at the trans-Golgi network. Interaction of the cytoplasmic tail sorting signals with AP-1 Golgi-specific assembly proteins. J Biol Chem. 1999;274(12):8199–207.

Teuchert M, Berghöfer S, Klenk HD, Garten W. Recycling of furin from the plasma membrane. Functional importance of the cytoplasmic tail sorting signals and interaction with the AP-2 adaptor medium chain subunit. J Biol Chem. 1999;274(51):36781–9.

Than ME, Henrich S, Bourenkov GP, Bartunik HD, Huber R, Bode W. The endoproteinase furin contains two essential Ca2+ ions stabilizing its N-terminus and the unique S1 specificity pocket. Acta Crystallogr D Biol Crystallogr. 2005;61(Pt 5):505–12.

Thomas G. Furin at the cutting edge: from protein traffic to embryogenesis and disease. Nat Rev Mol Cell Biol. 2002;3(10):753–66. Review.

Thomas G, Thorne BA, Thomas L, Allen RG, Hruby DE, Fuller R, Thorner J. Yeast KEX2 endopeptidase correctly cleaves a neuroendocrine prohormone in mammalian cells. Science. 1988;241(4862):226–30.

Toyoda T, Sakaguchi T, Imai K, Inocencio NM, Gotoh B, Hamaguchi M, Nagai Y. Structural comparison of the cleavage-activation site of the fusion glycoprotein between virulent and avirulent strains of Newcastle disease virus. Virology. 1987;158(1):242–7.

Tsuji A, Sakurai K, Kiyokage E, Yamazaki T, Koide S, Toida K, Ishimura K, Matsuda Y. Secretory proprotein convertases PACE4 and PC6A are heparin-binding proteins which are localized in the extracellular matrix. Potential role of PACE4 in the activation of proproteins in the extracellular matrix. Biochim Biophys Acta. 2003;1645(1):95–104.

Tsuneoka M, Nakayama K, Hatsuzawa K, Komada M, Kitamura N, Mekada E. Evidence for involvement of furin in cleavage and activation of diphtheria toxin. J Biol Chem. 1993;268(35):26461–5.

Turpeinen H, Oksanen A, Kivinen V, Kukkurainen S, Uusimäki A, Rämet M, Parikka M, Hytönen VP, Nykter M, Pesu M. Proprotein convertase subtilisin/kexin type 7 (PCSK7) is essential for the zebrafish development and bioavailability of transforming growth factor β1a (TGFβ1a). J Biol Chem. 2013;288(51):36610–23. https://doi.org/10.1074/jbc.M113.453183.

van de Loo JW, Teuchert M, Pauli I, Plets E, Van de Ven WJ, Creemers JW. Dynamic palmitoylation of lymphoma proprotein convertase prolongs its half-life, but is not essential for trans-Golgi network localization. Biochem J. 2000;352(Pt 3):827–33.

van de Ven WJ, Voorberg J, Fontijn R, Pannekoek H, van den Ouweland AM, van Duijnhoven HL, Roebroek AJ, Siezen RJ. Furin is a subtilisin-like proprotein processing enzyme in higher eukaryotes. Mol Biol Rep. 1990;14(4):265–75.

Vey M, Orlich M, Adler S, Klenk HD, Rott R, Garten W. Hemagglutinin activation of pathogenic avian influenza viruses of serotype H7 requires the protease recognition motif R-X-K/R-R. Virology. 1992;188(1):408–13.

Vey M, Schäfer W, Berghöfer S, Klenk HD, Garten W. Maturation of the trans-Golgi network protease furin: compartmentalization of propeptide removal, substrate cleavage, and COOH-terminal truncation. J Cell Biol. 1994;127(6 Pt 2):1829–42.

Vey M, Schäfer W, Reis B, Ohuchi R, Britt W, Garten W, Klenk HD, Radsak K. Proteolytic processing of human cytomegalovirus glycoprotein B (gpUL55) is mediated by the human endoprotease furin. Virology. 1995;206(1):746–9.

Vincent MJ, Sanchez AJ, Erickson BR, Basak A, Chretien M, Seidah NG, Nichol ST. Crimean-Congo hemorrhagic fever virus glycoprotein proteolytic processing by subtilase SKI-1. J Virol. 2003;77(16):8640–9.

Volchkov VE, Feldmann H, Volchkova VA, Klenk HD. Processing of the Ebola virus glycoprotein by the proprotein convertase furin. Proc Natl Acad Sci U S A. 1998;95(10):5762–7.

Volchkov VE, Volchkova VA, Ströher U, Becker S, Dolnik O, Cieplik M, Garten W, Klenk HD, Feldmann H. Proteolytic processing of Marburg virus glycoprotein. Virology. 2000;268(1):1–6.

Volchkova VA, Klenk HD, Volchkov VE. Delta-peptide is the carboxy-terminal cleavage fragment of the nonstructural small glycoprotein sGP of Ebola virus. Virology. 1999;265(1):164–71.

Voorhees P, Deignan E, van Donselaar E, Humphrey J, Marks MS, Peters PJ, Bonifacino JS. An acidic sequence within the cytoplasmic domain of furin functions as a determinant of trans-Golgi network localization and internalization from the cell surface. EMBO J. 1995;14(20):4961–75.

Walker JA, Kawaoka Y. Importance of conserved amino acids at the cleavage site of the haemagglutinin of a virulent avian influenza A virus. J Gen Virol. 1993;74(Pt 2):311–4.

Walker JA, Molloy SS, Thomas G, Sakaguchi T, Yoshida T, Chambers TM, Kawaoka Y. Sequence specificity of furin, a proprotein-processing endoprotease, for the hemagglutinin of a virulent avian influenza virus. J Virol. 1994;68(2):1213–8.

Wang M, Shen S, Wang H, Hu Z, Becnel J, Vlak JM. Deltabaculoviruses encode a functional type I budded virus envelope fusion protein. J Gen Virol. 2017;98(4):847–52.

Webby RJ, Perez DR, Coleman JS, Guan Y, Knight JH, Govorkova EA, McClain-Moss LR, Peiris JS, Rehg JE, Tuomanen EI, Webster RG. Responsiveness to a pandemic alert: use of reverse genetics for rapid development of influenza vaccines. Lancet. 2004;363(9415):1099–103.

Weider E, Susan-Resiga D, Essalmani R, Hamelin J, Asselin MC, Nimesh S, Ashraf Y, Wycoff KL, Zhang J, Prat A, Seidah NG. Proprotein Convertase Subtilisin/Kexin Type 9 (PCSK9) single domain antibodies are potent inhibitors of low density lipoprotein receptor degradation. J Biol Chem. 2016;291(32):16659–71.

White JM, Whittaker GR. Fusion of enveloped viruses in endosomes. Traffic. 2016;17(6):593–614.

Williamson DM, Elferich J, Ramakrishnan P, Thomas G, Shinde U. The mechanism by which a propeptide-encoded pH sensor regulates spatiotemporal activation of furin. J Biol Chem. 2013;288(26):19154–65. <prodomain, His69 function as a pH-sensor>.

Wise RJ, Barr PJ, Wong PA, Kiefer MC, Brake AJ, Kaufman RJ. Expression of a human proprotein processing enzyme: correct cleavage of the von Willebrand factor precursor at a paired basic amino acid site. Proc Natl Acad Sci U S A. 1990;87(23):9378–82.

Wong E, Maretzky T, Peleg Y, Blobel CP, Sagi I. The functional maturation of A Disintegrin and Metalloproteinase (ADAM) 9, 10, and 17 requires processing at a newly identified Proprotein Convertase (PC) cleavage site. J Biol Chem. 2015;290(19):12135–46.

Wouters S, Leruth M, Decroly E, Vandenbranden M, Creemers JW, van de Loo JW, Ruysschaert JM, Courtoy PJ. Furin and proprotein convertase 7 (PC7)/lymphoma PC endogenously expressed in rat liver can be resolved into distinct post-Golgi compartments. Biochem J. 1998;336:311–6.

Xiao Y, Chen G, Richard J, Rougeau N, Li H, Seidah NG, Cohen EA. Cell-surface processing of extracellular human immunodeficiency virus type 1 Vpr by proprotein convertases. Virology. 2008;372:384–97.

Yamshchikov GV, Ritter GD, Vey M, Compans RW. Assembly of SIV virus-like particles containing envelope proteins using a baculovirus expression system. Virology. 1995;214(1):50–8.

Yang J, Goldstein JL, Hammer RE, Moon YA, Brown MS, Horton JD. Decreased lipid synthesis in livers of mice with disrupted Site-1 protease gene. Proc Natl Acad Sci U S A. 2001;98(24):13607–12.

Yang HY, Zheng NQ, Li DM, Gu L, Peng XM. Entecavir combined with furin inhibitor simultaneously reduces hepatitis B virus replication and e antigen secretion. Virol J. 2014;11:165.

Ye J, Rawson RB, Komuro R, Chen X, Dave UP, Prywes R, Brown MS, Goldstein JL. ER stress induces cleavage of membrane-bound ATF6 by the same proteases that process SREBPs. Mol Cell. 2000;6:1355–64.

Zhang X, Fugère M, Day R, Kielian M. Furin processing and proteolytic activation of Semliki Forest virus. J Virol. 2003;77(5):2981–9.

Zhou A, Martin S, Lipkind G, LaMendola J, Steiner DF. Regulatory roles of the P domain of the subtilisin-like prohormone convertases. J Biol Chem. 1998;273(18):11107–14.

Zhu X, Lindberg I. 7B2 facilitates the maturation of proPC2 in neuroendocrine cells and is required for the expression of enzymatic activity. J Cell Biol. 1995;129(6):1641–50.

Zimmer G, Budz L, Herrler G. Proteolytic activation of respiratory syncytial virus fusion protein. Cleavage at two furin consensus sequences. J Biol Chem. 2001;276(34):31642–50.

Zimmer G, Rohn M, McGregor GP, Schemann M, Conzelmann KK, Herrler G. Virokinin, a bioactive peptide of the tachykinin family, is released from the fusion protein of bovine respiratory syncytial virus. J Biol Chem. 2003;278(47):46854–61.

Host Cell Proteases: Cathepsins

10

Klaudia Brix

Abstract

Cathepsins are proteolytic enzymes with a broad spectrum of substrates. They are known to reside within endo-lysosomes where they acquire optimal conditions for proteolytic activity and substrate cleavage. However, cathepsins have been detected in locations other than the canonical compartments of the endocytotic pathway. They are often secreted from cells in either proteolytically inactive proform or as mature and active enzyme; this may happen in both physiological and pathological conditions. Moreover, cytosolic and nuclear forms of cathepsins have been described and are currently an emerging field of research aiming at understanding their functions in such unexpected cellular locations. This chapter summarizes the canonical pathways of biosynthesis and transport of cathepsins in healthy cells. We further describe how cathepsins can reach unexpected locations such as the extracellular space or the cytosol and the nuclear matrix. No matter where viruses and cathepsins encounter, several outcomes can be perceived. Thus, scenarios are discussed on how cathepsins may support virus entry into host cells, involve in viral fusion factor and polyprotein processing in different host cell compartments, or help in packaging of viral particles during maturation. It is of note to mention that this review is not meant to comprehensively cover the present literature on viruses encountering cathepsins but rather illustrates, on some representative examples, the possible roles of cathepsins in replication of viruses and in the course of disease.

K. Brix
Department of Life Sciences and Chemistry,
Jacobs University Bremen,
Bremen, Germany
e-mail: k.brix@jacobs-university.de

E. Böttcher-Friebertshäuser et al. (eds.), *Activation of Viruses by Host Proteases*,
https://doi.org/10.1007/978-3-319-75474-1_10

10.1 Biosynthesis and Default Trafficking Pathway

Cathepsins are synthesized as preproenzymes, whereby the signal peptide targets for entry of the nascent chains into the lumen of the rough endoplasmic reticulum (rER). The signal peptide (pre) is co-translationally cleaved off by signal peptidase upon entry into the ER lumen. The propeptides (pro) keep procathepsins in the zymogen form, that is, proteolytically inactive during their transport from the rER via the Golgi apparatus and the *trans*-Golgi network (TGN) to late endosomes (Fig. 10.1) (Brix et al. 2008, 2015; Erickson et al. 2013). Procathepsins need to pass the quality control of the rER before they become posttranslationally modified while traveling to the TGN. Interestingly, proper folding of cathepsins can be conveyed by their propeptides as some of these bear chaperone functions. The most important and best-studied posttranslational modification of cathepsins is mannose 6-phosphorylation. It is believed to occur in the cis-Golgi and to be required for recognition by cation-dependent mannose 6-phosphate receptors (CD-MPR) of the TGN, which sort the zymogens into clathrin-coated transport vesicles (Fig. 10.1) (De Duve and Wattiaux 1966; Kornfeld and Mellman 1989; von Figura 1991; Kornfeld 1992; Erickson et al. 2013; Brix et al. 2015). Upon arrival in late endosomes, the pH drops such that procathepsins dissociate from the sorting CD-MPR, which is recycled back to the TGN via retromer-coated vesicles. In late endosomes, procathepsins are proteolytically processed to acquire the mature and proteolytically active state (Fig. 10.1). Therefore, either asparaginyl endopeptidase (AEP), which is better known as legumain, or cathepsins themselves cleave procathepsins in *trans* such that the propeptides are removed and the mature forms are delivered for further functioning as soluble enzymes in the compartments of the endocytic pathway. Thus, proteolytic processing for maturation and activation of procathepsins takes place in the late endosome.

As indicated above, the molecular architecture of cathepsins features N-terminal signal peptides that are typically followed by inhibitory propeptides and the peptidase domains (Tables 10.1, 10.2 and 10.3). However, the mature, single-chain form of cathepsin B, in particular, can be processed further, thereby yielding a two-chain form that consists of a light and a heavy chain, which remain bound to each other by disulfide bonds (Mort and Buttle 1997). Both, single- and two-chain forms of cathepsin B are proteolytically active as hydrolases. The molecular architecture of procathepsin-activating AEP/legumain differs from that of cathepsins, in that the peptidase domain directly follows the signal peptide and the pro-domain is found at the C-terminus (Table 10.2). Legumain is further exceptional, because it acts not only as a peptidase but features also peptide ligase activity depending on the conditions it is exposed to (Dall and Brandstetter 2016).

10.2 Proteolytic Activity and Substrate Cleavage Preferences of Aspartic, Cysteine, and Serine Cathepsins

Cathepsins belong to either of three classes of proteolytic enzymes, namely, aspartic, cysteine, or serine proteases (Rawlings 2013; Rawlings et al. 2016). Hence, cathepsins are classified according to the amino acids of their active sites that are

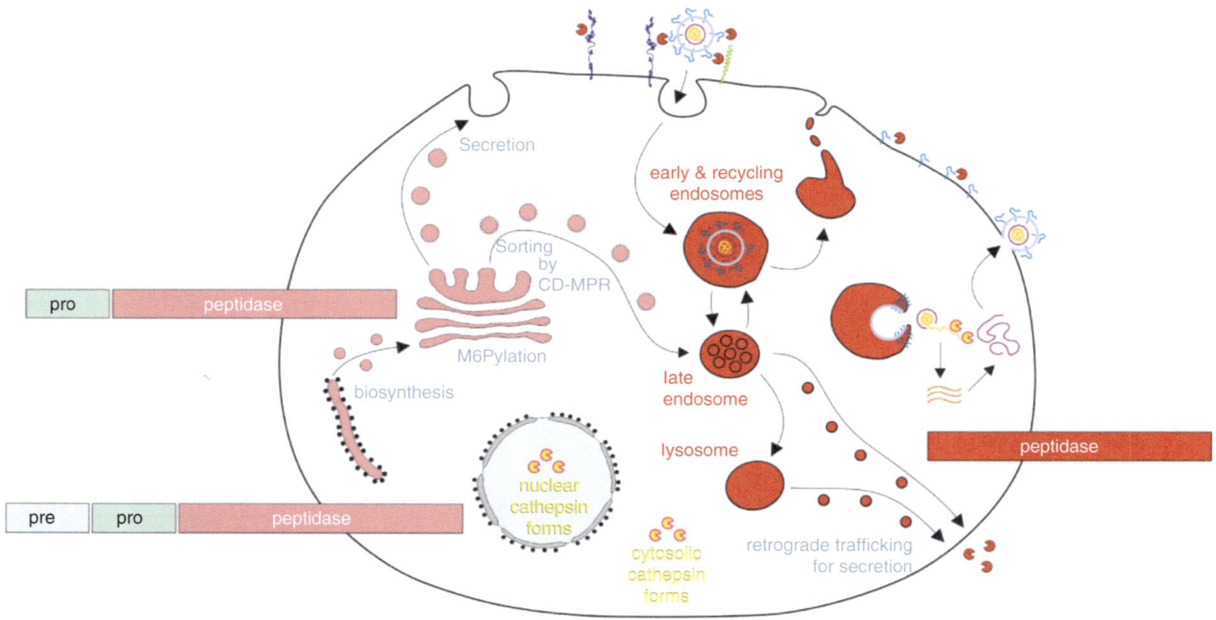

Fig. 10.1 Canonical and noncanonical transport pathway of cathepsins explaining possible interactions with viruses. Biosynthesis as preprocathepsins (light red) is at the rough endoplasmic reticulum from where the zymogen forms are transported via the Golgi apparatus and the *trans*-Golgi network to late endosomes for proteolytic processing yielding mature cathepsins (red). Procathepsins can also follow the secretory pathway when skipping recognition by CD-MPR at the TGN (left). Upon secretion, the procathepsins can be recaptured with the help of CI-MPR (green) at the plasma membrane (top, center). Retrograde transport of vesicles containing cathepsins recruited out of late endosomes or lysosomes explains the secretion of mature cathepsins into the extracellular space (bottom, right). LDL receptor-related proteins (LRPs; blue) serve as alternative cell surface receptors binding pro- and mature cathepsins for localized pericellular action (top, center) or re-internalization and delivery to endo-lysosomes. Specific forms of cathepsins are also present in the cytosol and in the nuclear matrix (pink-yellow), where they are involved in processing of transcription factors, core, and linker histones, in part by being bound to DNA. Secreted, extracellular cathepsins can interact, in principle, with viruses (cyan-violet-dark yellow) before endocytic entry (top, center), especially when bound to cell surface receptors like the CI-MPR (green) or an LRP (blue). In addition, cathepsins in the pericellular space may interfere by proteolytically processing viral proteins or by involving in shedding of cellular constituents, such as the glycocalyx, thereby enhancing the likelihood of viral entry. Moreover, it is possible, in principle, that cytosolic cathepsins (right, middle) mediate viral polyprotein processing, a proposal that remains speculative at this point

Table 10.1 Characteristics of aspartic cathepsins

Cathepsin	Gene symbol	Catalytic type	Clan family subfamily	Uniprot accession number	Locus	Domain architecture (peptidase unit from aa to aa) length in aa	N-Glycosylation, predicted	Disulfide bonds, predicted	Endogenous inhibitor/s	Pathways KEGG
Cathepsin D Conner (2004)	*CTSD*	Asp	AA A1.009	P07339	11p15.5	Pre-pro-mature (66–410) 345 aa	2	4	Alpha-2-macroglobulin, pepstatin	**Lysosome** Tuberculosis
Cathepsin E Kay and Tatnell (2004)	*CTSE*	Asp	AA A1.010	P14091	1q31	Pre-pro-mature (61–396) 336 aa	2	3	Pepstatin	**Lysosome**

Associations with human diseases: Cathepsin D: mutation in congenital neuronal ceroid lipofuscinosis (mutation in *CTSD*) (Siintola et al. 2006); genetic association with Alzheimer's disease (Davidson et al. 2006)
Data compilation based on MEROPS—the peptidase database—accessible at merops.sanger.ac.uk (Rawlings et al. 2016)

Table 10.2 Characteristics of AEP/legumain and cysteine cathepsins

Cathepsin or legumain	Gene symbol	Catalytic type	Clan family subfamily	Uniprot accession number	Locus	Domain architecture (peptidase unit from aa to aa) length in aa	N-Glycosylation, predicted	Disulfide bonds, predicted	Endogenous inhibitor/s	Pathways KEGG
Asparaginyl endopeptidase (AEP)/legumain (animal type (Chen et al. 1997)	LGMN	Cys	CD C13.004	Q99538	14q32.1	Pre-mature-pro (25–286) 262 aa	4	0	Cystatin C, cystatin E/M, cystatin F	Antigen processing and presentation **Lysosome**
Cathepsin B1 (cathepsin B, cathepsin II) (Mort 2004)	*CTSB*	Cys	CA C01.060	P07858	8p22	Pre-pro-mature-pro and Pre-pro-LC-pro-HC-pro (80–331) 252 aa	1	7	Serpin SPI2, cystatin A, cystatin B, cystatin C, cystatin D, cystatin E/M, cystatin SN, alpha-2-macro-globulin, kininogen	Antigen processing and presentation **Lysosome**
Cathepsin C (dipeptidyl peptidase I) (Turk et al. 2004)	*CTSC*	Cys	CA C1.070	P80067	11q14.1-q14.3	Pre-pro-mature (229–459) 231 aa	4	0	Cystatin A, cystatin B, cystatin C, cystatin SN	**Lysosome**

(continued)

Table 10.2 (continued)

Cathepsin F (Bromme 2004a)	*CTSF*	Cys	CA C01.018	Q9UBX1	11q13.1-q13.3	Pre-pro-mature (270–484) 215 aa	5	3		**Lysosome**
Cathepsin H (Kirschke 2004a)	*CTSH*	Cys	CA C01.040	P09668	15q24-q25	Pre-pro-mature (115–334) 220 aa	1	4		**Lysosome**
Cathepsin K (cathepsin O, cathepsin O2) (Bromme 2004b)	*CTSK*	Cys	SA C01.036	P43235	1q21	Pre-pro-mature (115–329) 215 aa	1	3	SSCA1 (squamous cell carcinoma antigen 1), serpin B13, propeptides of cathepsins L, S, K, V	**Lysosome** Osteoclast differentiation Rheumatoid arthritis Toll-like receptor signaling pathway

Cathepsin L (Kirschke 2004b)	*CTSL1*	Cys	CA C01.032	P07711	9q21-q22	Pre-pro-mature (113–333) 221 aa	2	3	Protein C inhibitor, SSCA1, serpin B13, colligin 2, cystatin A, cystatin B, cystatin C, cystatin D, cystatin E/M, cystatin SN, histidine-rich glycoprotein inhibitor unit 1 (HRG), propeptides of cathepsins K, L, S, V, MHC II invariant chain p41 form, testican-1, alpha-2-macroglobulin, kininogen	Antigen processing and presentation **Lysosome** Phagosome Proteoglycans in cancer Rheumatoid arthritis

(continued)

Table 10.2 (continued)

Cathepsin V (cathepsin L2, cathepsin U) (Brömme et al. 1999)	*CTSL2*	Cys	CA C01.009	O60911	9q22.2	Pre-pro-mature (113–334) 222 aa	2	3	cystatin E/M, propeptide of cathepsin V	**Lysosome**
Cathepsin O (Velasco and Lopez-Otin 2004)	*CTSO*	Cys	CA C01.035	P43234	4q31-q32	Pre-pro-mature (103–321) 219 aa	2	3		**Lysosome**
Cathepsin S (Kirschke 2004c)	*CTSS*	Cys	CA C01.034	P25774	1q21	Pre-pro-mature (114–331) 218 aa	1	4	SCCA1, cystatin A, cystatin B, cystatin C, cystatin D, propeptides of cathepsins K, L, S, V	Antigen processing and presentation **Lysosome** Phagosome Tuberculosis
Cathepsin W (Dalton and Brindley 2004)	*CTSW*	Cys	CA C01.037	P56202	11q13.1	Pre-pro-mature (127–364) 238 aa	2	3		**Lysosome**

Cathepsin X (carboxypeptidase LB, cathepsin B2, cathepsin IV, cathepsin Y, cysteine-type carboxypeptidase) (Menard and Sulea 2004)	*CTSZ*	Cys	CA C01.013	Q9UBR2	20q13	Pre-pro-mature (57–302) 246 aa	2	5		**Lysosome**

Associations with human diseases: Cathepsin C: functional absence in Papillon-Lefevre syndrome (Hart et al. 1999, 2000; Toomes et al. 1999), Haim-Munk syndrome (mutation in CTSC) (Cury et al. 2005)
Cathepsin K: functional absence in pycnodysostosis (Gelb et al. 1996; Johnson et al. 1996; Hou et al. 1999)
Cathepsin V: involvement in *Myasthenia gravis* suggested (Tolosa et al. 2003)
Cathepsin S: involvement in immune responses (Driessen et al. 1999)
Cathepsin X/Z: involvement in immune response to *Helicobacter pylori* (Krueger et al. 2005; Obermajer et al. 2009)
Data compilation based on MEROPS—the peptidase database—accessible at merops.sanger.ac.uk (Rawlings et al. 2016)

Table 10.3 Characteristics of serine cathepsins

Cathepsin	Gene symbol	Catalytic type	Clan family subfamily	Uniprot accession number	Locus	Domain architecture (peptidase unit from aa to aa) length in aa	N-Glycosylation, predicted	Disulfide bonds, predicted	Endogenous inhibitor/s	Pathways KEGG
Cathepsin A (serine carboxypeptidase A, cathepsin I) (Pshezhetsky 2004)	*CTSA*	Ser	SC S10.002	P10619	20q13.1	Pre-pro-mature (39–480) 442 aa	2	4		**Lysosome** Renin-angiotensin system
Cathepsin G (Salvesen 2004)	*CTSG*	Ser	PA S01.133	P08311	14q11.2	Pre-pro-mature (21–245) 225 aa	1	3	Tissue factor pathway inhibitor 1 (TFPI), alpha-1-peptidase inhibitor (serpin A1), alpha-1-antichymotrypsin (serpin A3), serpin B1, SCCA1, SCCA2, serpin B6, SLPI (secretory leukocyte peptidase inhibitor), alpha-2-macroglobulin, SPINK5 g.p.	Amoebiasis **Lysosome** Neuroactive ligand-receptor interaction Renin-angiotensin system Systemic lupus erythematosus

Associations with human diseases: Cathepsin A: functional absence in galactosialidosis (Zhou et al. 1995)
Data compilation based on MEROPS—the peptidase database—accessible at merops.sanger.ac.uk (Rawlings et al. 2016)

responsible for catalytic hydrolysis of peptide bonds. Aspartic acid or cysteine residues are part of the catalytic dyad of the aspartic and cysteine cathepsins, respectively, while serine is part of a catalytic triad in which an additional residue is required to stabilize the oxyanion hole in the acyl intermediate upon interaction between the enzyme and its substrate. Aspartic cathepsins are acting as water nucleophiles, whereas cysteine and serine cathepsins act as protein nucleophiles (Baici et al. 2013). Hence, the nucleophile is provided by a side chain of an amino acid in the active site, namely, the sulfhydryl or the hydroxyl group of either cysteine or serine, respectively, in cysteine and serine cathepsins. For example, serine bears the nucleophilic hydroxyl group of serine cathepsins, while histidine acts as the general base. In addition, aspartate helps to orient the imidazolium ring of histidine such that it activates serine to perform the nucleophilic attack on the peptide bond of the substrate, whereby a temporary complex between the enzyme and its substrate forms, i.e., the acyl intermediate. This breaks down rapidly, resulting in protonation of the general base histidine. Subsequent hydrolysis of the scissile bond occurs when a water molecule enters. The reaction mechanism of peptide bond hydrolysis catalyzed by cathepsins therefore involves two substrates, the protein or peptide substrate and a water molecule, and two products are generated, namely, an N- and a C-terminal peptide product.

Most of the cathepsins act as endopeptidases (Fig. 10.2). However, the cysteine cathepsins B and X, the serine cathepsin A, and legumain are also acting as carboxypeptidases, whereas cysteine cathepsin H acts as an aminopeptidase and cysteine cathepsin C forms dimers acting as dipeptidyl peptidase (Fig. 10.2). Thus, cathepsins are mostly acting on their peptide or protein substrates as monomers, but some can dimerize or even multimerize, thereby eventually altering substrate specificity (see also below).

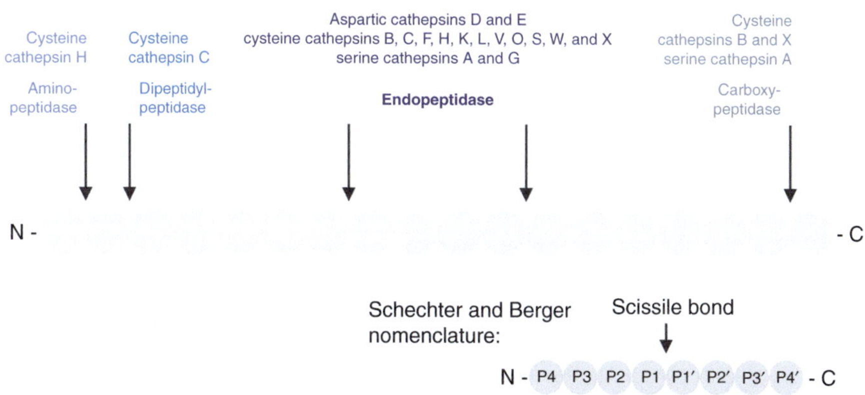

Fig. 10.2 Schematic drawing depicting distinct cathepsins acting as endo- and exopeptidases on a hypothetical peptide substrate. Amino acids are represented as beads on a string from N- to C-terminus in the top panel. Cathepsins are listed according to their mechanism of substrate cleavage. Bottom panel denotes naming of amino acids on both sides of the scissile bond according to the Schechter and Berger nomenclature

Proteolytic activity and substrate specificity are typically determined with recombinant enzyme preparations in well-defined in vitro assays. According to the Schechter and Berger nomenclature (Schechter and Berger 1968), the amino acids next to the scissile bond toward the N-terminus of the substrate are named P1, P2, P3, and P4, while the C-terminal amino acid next to the scissile bond is named P1′, and the following amino acids of the substrate are referred to as P2′, P3′, and P4′, respectively. The substrate-binding pockets in the cathepsins are named accordingly as S1–S4 and S1′–S4′, respectively, as for every other peptidase. Standard assays to determine the activity of cathepsins are available, as are synthetic substrates, which are preferentially or specifically cleaved by the enzymes. Standard cathepsin activity assays mainly consider peptide cleavage at optimal conditions, whereby a reducing environment of acidic to slightly acidic pH—mimicking the conditions in the endo-lysosomal compartments—is considered as the main determinant. The reader is referred to the *Handbook of Proteolytic Enzymes* (Barrett 2004) and to the "MEROPS database" (www.merops.sanger.ac.uk) (Rawlings et al. 2016), which provides a comprehensive and excellent compendium of proteases, their substrates, and inhibitors, for further up-to-date information.

It should be noted that cathepsins vary in their substrate specificities, but many exhibit overlapping substrate cleavage preferences, making it sometimes difficult to distinguish cathepsin activities. This ambiguity in substrate specificity often complicates their specific inhibition. In addition, cathepsins are redundantly expressed (Brix et al. 2008; Reiser et al. 2010; Gansz et al. 2013; Sloane et al. 2013). Thus, a cathepsin can be upregulated to take over the function of a related enzyme, when the latter is eliminated by gene knockout or inhibited by pharmacological interventions. This phenomenon of redundant regulation of cathepsin expression is observed especially when covalent and irreversible instead of transient reversible inhibition is applied.

10.3 Endogenous Inhibitors of Cathepsins

Cathepsin activities are counterbalanced by endogenous inhibitors (Tables 10.1, 10.2 and 10.3). The aspartic cathepsins D and E, which are mostly detected in lysosomes, are inhibited by pepstatin (Dean 1979) or by a natural product isolated from the sea anemone *Actinia equina*, equistatin. This latter inhibitor interestingly also interacts with cysteine peptidases (Lenarcic et al. 1997; Lenarcic and Turk 1999) including the cysteine cathepsins that are found preferentially in endo-lysosomes but also frequently in extra- and pericellular locations. Furthermore, aberrant or alternative forms of cysteine cathepsins are detected in the cytosol, as well as the nuclear and mitochondrial matrices (Brix et al. 2015). Cytosolic cystatins A and B, also known as stefins A and B (Machleidt et al. 1983; Brzin et al. 1983), as well as secreted cystatins C, D, E/M, F, and SN (Barrett 1986; Sloane et al. 1990; Turk and Bode 1991; Alvarez-Fernandez et al. 1999; Abrahamson et al. 2003; Turk et al. 2008; Zeeuwen et al. 2009) serve as endogenous inhibitors of the cysteine cathepsins. In addition, alpha-2-macroglobulin has been identified as an inhibitor of cysteine cathepsins (Fritz 1979; Travis 1988). Cysteine cathepsins, in principle, can also be inhibited by serpins, which are cross-class inhibitors of serine proteases (Silverman et al. 2001). Like the cystatins, serpins are present intra- and extracellularly. Serpins have a reactive center loop that serves as a bait for the attacking serine protease and, upon cleavage, remains bound by covalent linkage to the enzyme (Huntington 2006; Whisstock et al. 2010).

10.4 Tools to Study Expression, Trafficking, and Activities of Cathepsins

Numerous tools are available to study the cathepsins. Specific antibodies are used in immunostaining experiments for subsequent inspection by light and electron microscopy. Such strategies in using antibodies for the detection of cathepsins often bear specificity problems (Weber et al. 2015), because cathepsins are evolutionary conserved and show extensive sequence homologies.

Synthetic substrates, like chromogenic or fluorogenic peptides, are available for in vitro activity assays (see, MEROPS database for further details). Likewise, cyto- or histochemical approaches have been described, which employ synthetic cathepsin substrates that are converted into products less diffusible and readily detected by means of microscopy (Spiess et al. 1994; Brix et al. 1996). Enzymography in polyacrylamide gels is another means of determining multiple cathepsin activities at a time in cell or tissue lysates (Chen and Platt 2011; Platt et al. 2016).

A number of chimeric proteins consisting of cathepsins and fluorescent proteins have been generated which can be expressed in mammalian cells (for reviews, see Brix et al. 2008; Arampatzidou et al. 2011). Tagged cathepsins follow the regular transport pathways like the endogenous enzymes (Naganawa et al. 1999; Linke et al. 2002a, b; Arampatzidou et al. 2011; Frizler et al. 2013; Tamhane et al. 2015; Tamhane et al. 2016). Trafficking studies in living cells have been combined with co-localization experiments in which compartment-specific markers have been used to spot full-length or N-terminally truncated specific forms of cathepsins (see below) while they are on the move through the cell (Linke et al. 2002a; Zwicky et al. 2003; Müntener et al. 2004; Mayer et al. 2008). Intervention by either permanently blocking or transiently interfering with specific transport routes has been applied in order to investigate distinct enzyme trafficking in a variety of eukaryotic cells.

The most elegant way to visualize cathepsin activities "on the spot" and in a quantitative manner is by using so-called activity-based probes (ABPs) (Greenbaum et al. 2002; Jessani et al. 2004; Blum et al. 2005; Blum 2008; Edgington et al. 2011; Blais et al. 2012; Zou et al. 2012; Grzywa and Sieńczyk 2013; Sanman and Bogyo 2014). ABPs consist of a reactive functional group (so-called warhead) that binds to the target protease, a linker region that resembles a peptide substrate, and additional tagging groups. The tagging groups are typically fluorogenic, but biotin or iodinatable moieties have also been chemically engineered onto ABPs. Hence, a family or subfamilies of cathepsins are functionally addressed and become covalently tagged upon cleavage of the ABP's linker. ABPs are available as broad-spectrum probes used in approaching the proteolytic activities of aspartic, cysteine, and serine cathepsins. In addition, specific ABPs have been synthesized that can be cleaved by one or only few cathepsins. Moreover, ABPs with propeptide-mimicking features in the peptide backbone have been used (Schaschke et al. 1998; Diederich et al. 2012). Such highly specific ABPs are, in principle, able to distinguish even closely related cathepsins.

The one-to-one binding principle of ABPs renders them powerful tools to quantify cathepsin activities. This is to say that most ABPs act as inhibitors that are attacked and cleaved in a substrate-like fashion, but remain bound to the targeted enzyme, thus forming an irreversible, covalent complex with the target protease. The concept of ABP labeling of cathepsins can be compared with the suicidal inhibition mechanism of the serpins on their target proteases (see above). The principle of specific and irreversible binding bears the advantage of rendering ABPs into quantitative probes that allow to detect only mature, proteolytically active cathepsins. However, the disadvantage of many ABPs is equally intrinsic, namely, they act as protease inhibitors and may eventually lead to pharmacological knockdown of proteolytic activities when applied to living cells.

The above sketched labeling options to visualize, localize, and quantify cathepsins can be and have been combined in all possible variations (Baruch et al. 2004; Blum et al. 2005, 2007; Brix and Jordans 2005; Blum 2008; Sadaghiani et al. 2007; Brix et al. 2008; Arampatzidou et al. 2011; Salpeter and Blum 2013). Such experiments have helped to uncover the transport pathways that are followed by pro- and mature cathepsins in different eukaryotic cells. Importantly, it is now state of the art to distinguish proteolytically active from inactive cathepsins. Moreover, it became clear that each and every cell type may use cathepsins strategically in its own, distinct way in support of the specific cell's function (Brix et al. 2008, 2015; Mohamed and Sloane 2006; Sloane et al. 2013; Weiss-Sadan et al. 2017). Hence, subcellular locations of cathepsin activities have been detected that encompass the expected canonical endo-lysosomal compartments. However, quite often locations of cathepsin activities were detected which were, by all means, unexpected. Thus, cathepsins exhibit an astonishing variety of transport pathways to reach numerous intra- and extracellular locations where the proteases may function in a spatially confined and temporally regulated manner (Mohamed and Sloane 2006; Gocheva and Joyce 2007; Brix et al. 2013, 2015; Akkari et al. 2016).

10.5 Unexpected Locations Reached by Cathepsins Are Explained by Noncanonical Trafficking

10.5.1 Alternative Pathways of Activation upon Secretion of Procathepsins

Procathepsins may skip recognition by the CD-MPR at the TGN and become further sorted and transported along the secretory pathway with destination to the cell surface for subsequent secretion into the extracellular space (Fig. 10.1, left transport route). Pericellularly, the procathepsins may become activated by soluble endopeptidases or through the action of plasma membrane-bound ectoenzymes. Yet another pathway is used in particular in cells of the innate immune system, like macrophages, that express the cation-independent mannose 6-phosphate receptor (CI-MPR, also referred to as IGFII/M6PR) (Mason et al. 1987; Pohlmann et al. 1995; Collette et al. 2004). This cell surface receptor acts as a scavenger receptor and functions in rescuing faulty or excessively secreted procathepsins for subsequent internalization by endocytosis and sorting along the endocytic pathway. Hence, during their transport from the extracellular space via early endosomes to late endosomes (Fig. 10.1, right transport route), such secreted forms of cathepsins eventually are activated in the same compartment as those following the default transport pathway but delayed in time.

10.5.2 Secretory Lysosomes

Distinct cell types feature so-called secretory lysosomes (Andrews 2000; Brix et al. 2008). These are endocytic compartments, which resemble late endosomes and lysosomes in their characteristic biochemical composition. Hence, secretory lysosomes or cathepsins recruited from within late endosomes and lysosomes are transported in a retrograde fashion, such that the vesicles fuse with the plasma membrane, typically upon signaling. This alternative route of cathepsin transport (Fig. 10.1, bottom right transport route) is observed in a number of specific cell types and allows for secretion of proteolytically active cathepsins into the extracellular space (Linke et al. 2002a, b; Büth et al. 2004).

10.5.3 Cathepsins in Extracellular Locations and Bound to the Cell Surface

Cathepsins are present in the extracellular space as mature and active enzymes or as inactive proforms. While procathepsins that underwent mannose 6-phosphorylation bind to the CI-MPR (see above) (von Figura 1991; Erickson et al. 2013; Brix et al. 2015), mature cathepsins and procathepsins can also become bound by alpha-2-macroglobulin (Arkona and Wiederanders 1996; Peloille et al. 1997). Moreover, this protein serves as a regulator of a number of proteases, namely, it is an inhibitor of matrix metalloproteinases and cysteine cathepsins alike. Other receptors known to interact with secreted cathepsins are those of the diverse low-density lipoprotein (LDL) receptor-like protein (LRP) family of transmembrane proteins present at the cell surface of almost all cell types (Poller et al. 1995; Willnow et al. 1996; Herz and Strickland 2001). Megalin/gp330 is one of the LRPs that is supposed to mediate internalization and endo-lysosomal delivery of cathepsins, if previously secreted by mistake (Nielsen et al. 2007). Thus, LRPs function in a way similar to the CI-MPR but interact also with pro- and mature cathepsins lacking mannose 6-phosphorylation (Fig. 10.1, right transport route).

10.5.4 Regulation of Cathepsin Activities in the Extracellular Space

Pericellular and extracellular localization of procathepsins and mature cathepsins has been observed under both physiological and pathological conditions. For instance, cathepsin-mediated extracellular proteolysis has been shown to promote cancer cell invasion and tumor progression through extracellular matrix (ECM) degradation (Rochefort et al. 2000; Joyce and Hanahan 2004; Liaudet-Coopman et al. 2006; Mohamed and Sloane 2006; Gocheva and Joyce 2007; Sloane et al. 2013). It remains to be seen, however, if the enhanced cathepsin secretion responsible for excessive ECM degradation is triggered by the acidic environment in tumor tissue or by other mechanisms (Cavallo-Medved and Sloane 2003).

Depending on the extracellular conditions, cathepsins may acquire altered or additional substrate preferences. For example, the cysteine cathepsin K is secreted under physiological conditions as an active enzyme and acts in the acidic resorption lacuna formed by osteoclasts during bone turnover (Gelb et al. 1996; Saftig et al. 1998; Rachner et al. 2011; Fonović and Turk 2014; Brömme et al. 2016). However, cathepsin K is also responsible for degradation of thyroglobulin in the extracellular thyroid follicle lumen, namely, at neutral pH and in oxidizing conditions (Tepel et al. 2000; Friedrichs et al. 2003; Jordans et al. 2009). Moreover, secreted cathepsin K monomers can interact with each other, thereby forming polymeric ringlike structures, depending on the molecular composition of the extracellular space (Li et al. 2002). It has been shown that ECM components like glycosaminoglycans and other constituents are essential as scaffolding factors supporting extracellular interactions of cathepsin K molecules (Li et al. 2004). Moreover, cathepsin K can be stabilized extracellularly by binding to cell surface molecules like clusterin (Novinec et al. 2012), which also interacts with protease receptors of the LRP family, e.g., megalin/gp330 (Lemansky et al. 1999). Importantly, the activity of this particular enzyme—cysteine cathepsin K—differs in monomeric and dimeric as well as in soluble and ECM- or membrane-bound form (Aguda et al. 2014). Hence, different substrates can be cleaved by the same cathepsin, and a given substrate can be processed differently by that enzyme, depending on whether the cathepsin is secreted as an active monomer or dimer, scaffolded by ECM constituents, or bound by cell surface receptors complexed with allosteric cofactors. It is important to note that cysteine cathepsin K is well-studied, in particular, because of its significance as an anti-osteoporotic drug target.

These examples highlight that more research is required to understand how the enzymatic activities of cathepsins are regulated, in particular, in unexpected locations. Thus, the well-accepted concept that cathepsins optimally cleave protein and peptide substrates at acidic pH in reducing environments, as found in endo-lysosomes, must be broadened. This is all the more important when cathepsin activities are investigated under conditions of cellular stress as is the case with viral infection.

10.5.5 Cytosolic and Nuclear Cathepsins

As detailed above, cathepsins belong to the obligate constituencies of the compartments of the endocytic pathway where they exert their functions most optimally. However, the phenomenon of leaky lysosomes has also been known for long. Cellular stress as is the case with cancer cells under prolonged drug treatment or UV irradiation can cause leakiness of endo-lysosomal membranes. Mature cathepsins may then be released into the cytosol, where proteolytic activity is controlled by different means, such as endogenous inhibitors that specifically bind to the enzymes or other biochemical properties of the cytosol that are not optimal for cathepsin-mediated cleavage. However, when escaping these safeguarding measures by, for instance, inhibitor downregulation, proteolytically active cathepsins may be present in the cytosol for prolonged time intervals and induce apoptotic, necrotic, or pyroptotic cascades resulting in cell death (Turk et al. 2000; Luke et al. 2007; Turk and Turk 2009; Aits and Jäättelä 2013; Flütsch and Grütter 2013). Furthermore, cytosolic and nuclear cathepsins may modulate cell cycle progression (Goulet and Nepveu 2004; Brix et al. 2015; Tamhane et al. 2016).

10.5.6 Alternative Cathepsin Forms

N-terminally truncated forms of the canonical preprocathepsins translated from, e.g., alternative transcripts are believed to lack the signal peptide and parts of the propeptide (Mehtani et al. 1998; Zwicky et al. 2003; Müntener et al. 2004; Baici et al. 2006; Schilling et al. 2009; Tholen et al. 2014; Brix et al. 2015). They are therefore not targeted for entry into the ER lumen and will not follow the secretory pathway. Instead, N-terminally truncated cathepsins are retained in the cytosol and can even fold properly as they acquire a proteolytically active state in this unexpected location (Goulet and Nepveu 2004; Luke et al. 2007; Reiser et al. 2010; Tedelind et al. 2010). Some of these alternative and aberrant forms of the cathepsins occur even in the nuclear matrix (Fig. 10.1, center at bottom). The mechanism by which they are transported through the nuclear pore complexes is not known, because most endo-lysosomal enzymes (except AEP/legumain) lack a nuclear localization sequence (NLS). It is reasonable to assume, however, that such nuclear cathepsins, as well as the cytosolic forms, are involved in processing of transcription factors, core, and/or linker histones. In vitro experiments further revealed an important role of DNA as a potential scaffolding factor that interferes with serpin-mediated control of cathepsin activities in environments that mimic unexpected cellular locations like the nuclear matrix (Ong et al. 2007).

10.6 Cathepsins Meeting Viruses, Viruses Meeting Cathepsins

10.6.1 Transient Encounters when Traveling Along the Endocytic Pathway

Cathepsins are well known to process the spike proteins of SARS and MERS coronaviruses, thereby activating viral fusogens and enabling host cell entry from within endosomes (Millet and Whittaker 2015; Simmons et al. 2013; Heald-Sargent and Gallagher 2012). Thus, treatment options of preventing host cell infection with

SARS-CoV have been proposed that involve cysteine cathepsin L inhibitors (Tong 2006).

A very complex mechanism of proteolytic activation of the fusion proteins of henipaviruses (HNV) has been described as an essential prerequisite for infectivity and pathogenicity of these highly pathogenic paramyxoviruses (Weis and Maisner 2015). The non-fusogenic F0 protein of HNV is translated in host cells and is subsequently transported along the secretory pathway in its inactive form to the plasma membrane. Upon re-internalization and processing in recycling endosomes by cathepsins B and L (Pager and Dutch 2005; Meulendyke et al. 2005; Vogt et al. 2005; Pager et al. 2006; Diederich et al. 2005, 2012), activated F1/F2 complex travels back to the plasma membrane, where it is incorporated into budding virus particles, or mediates fusion of an infected cell with a neighboring cell. Hence, endosomal cysteine cathepsins B and L play an essential role in promoting spread of infection and formation of syncytia.

Thus, there are significant differences in cathepsin-mediated activation of viral fusion proteins: with HNV it occurs at a late stage of replication, whereas SARS and MERS coronaviruses are activated upon virus entry into host cells. Endosomal cysteine cathepsins B and L are also involved in the processing of the envelope glycoproteins of Marburg and Ebola viruses. Cathepsin cleavage enables the glycoprotein to interact with the Niemann-Pick disease type C1 (NDC-1) protein of the host cell, which is an essential step in filovirus entry (Hunt et al. 2012). Likewise, endosomal cathepsins are utilized by non-enveloped reoviruses for host cell entry. After removal of the outer capsid protein σ3 by cathepsins, the viral protein μ1 is exposed, which is a fusion protein, promoting endosomal membrane rupture (Danthi et al. 2010).

10.6.2 Altered Cathepsin Expression in Virus-Infected Cells

Virus infection may affect transcriptional regulation of cathepsin genes leading to disbalanced cell functions. In addition to the N-terminally truncated forms of cathepsins and those reaching the cytosol as full-length enzymes due to release from endo-lysosomes, it is conceivable that alternate cathepsin forms may derive from altered genes. These may result from gene mutations or chromosomal aberrations, as occurring in cancer cells, or due to upregulated translation of alternative transcripts. Viral oncogenes may affect amplification of cathepsin genes (Mohamed and Sloane 2006), and it was proposed that endogenous retroviruses or elements thereof may activate placenta-specific genes encoding cysteine cathepsins of mice (Rawn and Cross 2008).

Moreover, HIV-infected macrophages have been reported to upregulate both cytosolic cysteine protease inhibitor cystatin B (stefin B) and cysteine cathepsin B, believed to trigger neuronal cell death in HIV-1-associated neurocognitive disorder (HAND) (Rivera et al. 2014). Similarly, virus transformation of cultured cells in vitro is known to cause upregulation and secretion of the so-called major excreted protein (MEP), which was found in the secretion media of transformed fibroblasts and identified as proteolytically active cathepsin L (Mason et al. 1987; Rubin 2005).

Thus, viruses may not only affect transcriptional regulation of cathepsin genes but may thereby also cause mis-trafficking of the proteases. In hepatocellular carcinoma, caused by infection with hepatitis B and C viruses, altered cathepsin trafficking also results in their secretion. In this particular case of virus-induced liver cancer, the over-secretion of cathepsins is further complicated by the defective functioning of the IGFII/M6P receptor (CI-MPR) pathway, normally acting as a recapture mechanism for internalization of faulty and excessively secreted cathepsins (see above) (Scharf and Braulke 2003). Thus, in the absence of re-internalization cues, the cathepsins may be present in enhanced amounts at the cell surface of hepatocellular carcinoma cells.

The findings summarized above show that another, so far only rarely considered scenario may be likely. Namely, viruses may interact with cathepsins present in the extracellular space of, e.g., cancer cells, even before entering the host cell's endocytic compartments. In acknowledging that cathepsins may well act as proteolytic enzymes already in the pericellular space (see above), it becomes clear at this point that some viruses and their constituents can be processed, in principle, by extracellular cathepsins, that is, before actually entering the host cell by endocytosis (Fig. 10.1, top, center).

It has also to be mentioned in this context that extracellular cathepsin B-mediated shedding of constituents of the glycocalyx of endothelial cells was proposed as a process in support of infection with viruses causing hemorrhagic fevers (Becker et al. 2015). Likewise, cysteine cathepsins B-, L-, and S-mediated shedding of E-cadherin, an important cell-cell adhesion molecule, was suggested to cause epithelial cell damage, thereby promoting disease progression in patients with viral infections (Grabowska and Day 2012).

Taken together, these observations illustrate different mechanisms by which viruses may upregulate expression and stimulate secretion of cathepsins. They also suggest that cathepsins activate viruses not only in endocytotic compartments but also at the cell surface. Finally, these findings support the concept that cathepsins contribute to pathogenesis not only by activating the fusion capacity of viruses but also by other mechanisms promoting cell and tissue damage.

10.6.3 Endo-Lysosomal Cathepsins and the Immune Response to Viral Infections

Viruses entering host cells by endocytosis are known to trigger a Toll-like receptor (TLR)-mediated immune response which eventually leads to interferon-alpha production (Sun et al. 2010). Hence, viral nucleic acids are recognized as PAMPs (pathogen-associated molecular patterns) by proteolytically processed transmembrane pattern recognition receptors like TLR9. Proteolytic activation of TLR9 is catalyzed by endosomal AEP/legumain and cathepsins (Bauer 2013). Thus, cathepsins are also involved in the immune response to viral infection.

Cathepsin-mediated processing of viral proteins—typically protein fusogens—happens in early and recycling endosomes (see above), which are connected with other compartments of the endocytic pathway, namely, also with late endosomes or multivesicular bodies (MVBs). MVBs are believed to serve as a place for generation of exosomes (Cocucci and Meldolesi 2015; Hurley 2015). In this regard, it is interesting to note that infection with the filoviruses EBOV and MARV involves the molecular machinery of the ESCRT pathway, which is required for MVB formation and which is important for virus replication, nucleocapsid formation, and maturation in a compartment positive for late endosomal markers (Dolnik et al. 2015).

Another vicious cycle is working in cross-presentation, which follows from initial phagocytosis of portions or entire virus-infected cells by dendritic cells (Rock and Shen 2005). These professional antigen-presenting cells depend on cysteine cathepsin S-mediated endosomal processing of internalized proteins, which then results in antigen presentation in the context of MHC class II. In the unfortunate case of endosomal processing of viral proteins, the resulting viral antigens may therefore be presented on the surface of dendritic cells via MHC class II, instead of MHC class I. When antigen presentation happens in the context of MHC class I, alerted cytotoxic T cells eliminate the virus-infected cells. Upon cross-presentation, however, tolerance may be a nonproductive outcome.

Concluding Remarks

Finally, a chapter on cathepsins in a book on viruses cannot end without placing a special note in the context of therapeutic approaches aiming at eliminating viruses from the host. In particular, the retroviral HIV protease is important for maturation of the virus particles and, hence, targeted therapeutically (Moyle and Gazzard 1996; Cooper 2002). HIV proteinase structurally resembles the ubiquitously expressed aspartic cathepsin D, denoting the similarities between host cell's aspartic cathepsins and retroviral aspartic proteases, which are—in the case of HIV—successfully inhibited when approached by transition-state inhibitors. Therefore, and in conclusion, basic science researchers, virologists, and clinicians have learned a lot from the structural similarities of host cell cathepsins and viral proteases. We deduce that interactions between cell biologists and virologists bear more interesting facts to be gathered in the future. Moreover, many more potential therapeutic answers are to be developed from the encounters of cathepsins and the various viruses utilizing the proteases in their own favor. Hence, some of the comments in this chapter are meant in support of stimulating future discussions in the spirit of thinking "out of the box," and, like the cathepsins, following paths beyond the canonical pathways. In our opinion, this is an endeavor worth to be undertaken and continued in the future.

References

Abrahamson M, Alvarez-Fernandez M, Nathanson CM. Cystatins. Biochem Soc Symp. 2003;70:179–99.

Aguda AH, Panwar P, Du X, Nguyen NT, Brayer GD, Brömme D. Structural basis of collagen fiber degradation by cathepsin K. Proc Natl Acad Sci U S A. 2014;111(49):17474–9.

Aits S, Jäättelä M. Lysosomal cell death at a glance. J Cell Sci. 2013;126(Pt 9):1905–12.

Akkari L, Gocheva V, Quick ML, Kester JC, Spencer AK, Garfall AL, Bowman RL, Joyce JA. Combined deletion of cathepsin protease family members reveals compensatory mechanisms in cancer. Genes Dev. 2016;30(2):220–32.

Alvarez-Fernandez M, Barrett AJ, Gerhartz B, Dando PM, Ni J, Abrahamson M. Inhibition of mammalian legumain by some cystatins is due to a novel second reactive site. J Biol Chem. 1999;274(27):19195–203.

Andrews NW. Regulated secretion of conventional lysosomes. Trends Cell Biol. 2000;10(8):316–21.

Arampatzidou M, Rehders M, Dauth S, Yu DMT, Tedelind S, Brix K. Imaging of protease functions—current guide to spotting cysteine cathepsins in classical and novel scenes of action in mammalian epithelial cells and tissues. Ital J Anat Embryol. 2011;116(1):1–19.

Arkona C, Wiederanders B. Expression, subcellular distribution and plasma membrane binding of cathepsin B and gelatinases in bone metastatic tissue. Biol Chem. 1996;377(11):695–702.

Baici A, Müntener K, Willimann A, Zwicky R. Regulation of human cathepsin B by alternative mRNA splicing: homeostasis, fatal errors and cell death. Biol Chem. 2006;387(8):1017–21.

Baici A, Novinec M, Lenarčič B. Kinetics of the interaction of peptidases with substrates and modifiers. In: Brix K, Stöcker W, editors. Proteases: structure and function. Berlin: Life Sciences, springer.com; 2013. p. 37–84.

Barrett AJ. The cystatins: a diverse superfamily of cysteine peptidase inhibitors. Biomed Biochim Acta. 1986;45(11–12):1363–74.

Barrett AJ. Bioinformatics of proteases in the MEROPS database. Curr Opin Drug Discov Dev. 2004;7(3):334–41.

Baruch A, Jeffery DA, Bogyo M. Enzyme activity—it's all about image. Trends Cell Biol. 2004;14(1):29–35.

Bauer S. Toll-like receptor 9 processing: the key event in Toll-like receptor 9 activation? Immunol Lett. 2013;149(1–2):85–7.

Becker BF, Jacob M, Leipert S, Salmon AH, Chappell D. Degradation of the endothelial glycocalyx in clinical settings: searching for the sheddases. Br J Clin Pharmacol. 2015;80(3):389–402.

Blais DR, Nasheri N, McKay CS, Legault MC, Pezacki JP. Activity-based protein profiling of host-virus interactions. Trends Biotechnol. 2012;30(2):89–99.

Blum G. Use of fluorescent imaging to investigate pathological protease activity. Curr Opin Drug Discov Dev. 2008;11(5):708–16.

Blum G, Mullins SR, Keren K, Fonovic M, Jedeszko C, Rice MJ, Sloane BF, Bogyo M. Dynamic imaging of protease activity with fluorescently quenched activity-based probes. Nat Chem Biol. 2005;1(4):203–9.

Blum G, von Degenfeld G, Merchant MJ, Blau HM, Bogyo M. Noninvasive optical imaging of cysteine protease activity using fluorescently quenched activity-based probes. Nat Chem Biol. 2007;3(10):668–77.

Brix K, Jordans S. Watching proteases in action. Nat Chem Biol. 2005;1(4):186–7.

Brix K, Lemansky P, Herzog V. Evidence for extracellularly acting cathepsins mediating thyroid hormone liberation in thyroid epithelial cells. Endocrinology. 1996;137(5):1963–74.

Brix K, Dunkhorst A, Mayer K, Jordans S. Cysteine cathepsins: cellular roadmap to different functions. Biochimie. 2008;90(2):194–207.

Brix K, Scott CJ, Heck MMS. Compartmentalisation of proteolysis. In: Brix K, Stöcker W, editors. Proteases: structure and function. Berlin: Life Sciences, springer.com; 2013. p. 85–125.

Brix K, McInnes J, Al-Hashimi A, Rehders M, Tamhane T, Haugen MH. Proteolysis mediated by cysteine cathepsins and legumain-recent advances and cell biological challenges. Protoplasma. 2015;252(3):755–74.

Bromme D. Cathepsin F. In: Barrett AJ, Rawlings ND, Woessner JF, editors. Handbook of proteolytic enzymes. 2nd ed. London: Elsevier; 2004a. p. 1087–8.

Bromme D. Cathepsin K. In: Barrett AJ, Rawlings ND, Woessner JF, editors. Handbook of proteolytic enzymes. 2nd ed. London: Elsevier; 2004b. p. 1092–7.

Brömme D, Li Z, Barnes M, Mehler E. Human cathepsin V functional expression, tissue distribution, electrostatic surface potential, enzymatic characterization, and chromosomal localization. Biochemistry. 1999;38(8):2377–85.

Brömme D, Panwar P, Turan S. Cathepsin K osteoporosis trials, pycnodysostosis and mouse deficiency models: commonalities and differences. Expert Opin Drug Discovery. 2016;11(5):457–72.

Brzin J, Kopitar M, Turk V, Machleidt W. Protein inhibitors of cysteine proteinases. I. Isolation and characterization of stefin, a cytosolic protein inhibitor of cysteine proteinases from human polymorphonuclear granulocytes. Hoppe Seylers Z Physiol Chem. 1983;364(11):1475–80.

Büth H, Wolters B, Hartwig B, Meier-Bornheim R, Veith H, Hansen M, Sommerhoff CP, Schaschke N, Machleidt W, Fusenig NE, Boukamp P, Brix K. HaCaT keratinocytes secrete lysosomal cysteine proteinases during migration. Eur J Cell Biol. 2004;83(11–12):781–95.

Cavallo-Medved D, Sloane BF. Cell-surface cathepsin B: understanding its functional significance. Curr Top Dev Biol. 2003;54:313–41.

Chen B, Platt MO. Multiplex zymography captures stage-specific activity profiles of cathepsins K, L, and S in human breast, lung, and cervical cancer. J Transl Med. 2011;9:109.

Chen JM, Dando PM, Rawlings ND, Brown MA, Young NE, Stevens RA, Hewitt E, Watts C, Barrett AJ. Cloning, isolation, and characterization of mammalian legumain, an asparaginyl endopeptidase. J Biol Chem. 1997;272:8090–8.

Cocucci E, Meldolesi J. Ectosomes and exosomes: shedding the confusion between extracellular vesicles. Trends Cell Biol. 2015;25(6):364–72.

Collette J, Bocock JP, Ahn K, Chapman RL, Godbold G, Yeyeodu S, Erickson AH. Biosynthesis and alternate targeting of the lysosomal cysteine protease cathepsin L. Int Rev Cytol. 2004;241:1–51.

Conner GE. Cathepsin D. In: Barrett AJ, Rawlings ND, Woessner JF, editors. Handbook of proteolytic enzymes. 2nd ed. London: Elsevier; 2004. p. 43–52.

Cooper JB. Aspartic proteinases in disease: a structural perspective. Curr Drug Targets. 2002;3(2):155–73.

Cury VF, Gomez RS, Costa JE, Friedman E, Boson W, De Marco L. A homozygous cathepsin C mutation associated with Haim-Munk syndrome. Br J Dermatol. 2005;152:353–6.

Dall E, Brandstetter H. Structure and function of legumain in health and disease. Biochimie. 2016;122:126–50.

Dalton JP, Brindley PJ. Cathepsin W. In: Barrett AJ, Rawlings ND, Woessner JF, editors. Handbook of proteolytic enzymes. 2nd ed. 2004. p. 1109–12.

Danthi P, Guglielmi KM, Kirchner E, Mainou B, Stehle T, Dermody TS. From touchdown to transcription: the reovirus cell entry pathway. Curr Top Microbiol Immunol. 2010;343:91–119.

Davidson Y, Gibbons L, Pritchard A, Hardicre J, Wren J, Tian J, Shi J, Stopford C, Julien C, Thompson J, Payton A, Thaker U, Hayes AJ, Iwatsubo T, Pickering-Brown SM, Pendleton N, Horan MA, Burns A, Purandare N, Lendon CL, Neary D, Snowden JS, Mann DM. Genetic associations between cathepsin D exon 2 C-->T polymorphism and Alzheimer's disease, and pathological correlations with genotype. J Neurol Neurosurg Psychiatry. 2006;77(4):515–7.

De Duve C, Wattiaux R. Functions of lysosomes. Annu Rev Physiol. 1966;28:435–92.

Dean RT. Lysosomes and protein degradation. Ciba Found Symp. 1979;75:139–49.

Diederich S, Moll M, Klenk HD, Maisner A. The Nipah virus fusion protein is cleaved within the endosomal compartment. J Biol Chem. 2005;280(33):29899–903.

Diederich S, Sauerhering L, Weis M, Altmeppen H, Schaschke N, Reinheckel T, Erbar S, Maisner A. Activation of the Nipah virus fusion protein in MDCK cells is mediated by cathepsin B within the endosome-recycling compartment. J Virol. 2012;86(7):3736–45.

Dolnik O, Stevermann L, Kolesnikova L, Becker S. Marburg virus inclusions: a virus-induced microcompartment and interface to multivesicular bodies and the late endosomal compartment. Eur J Cell Biol. 2015;94(7–9):323–31.

Driessen C, Bryant RA, Lennon-Dumenil AM, Villadangos JA, Bryant PW, Shi GP, Chapman HA, Ploegh HL. Cathepsin S controls the trafficking and maturation of MHC class II molecules in dendritic cells. J Cell Biol. 1999;147:775–90.

Edgington LE, Verdoes M, Bogyo M. Functional imaging of proteases: recent advances in the design and application of substrate-based and activity-based probes. Curr Opin Chem Biol. 2011;15(6):798–805.

Erickson AH, Isidoro C, Mach L, Mort JS. Cathepsins: getting in shape for lysosomal proteolysis. In: Brix K, Stöcker W, editors. Proteases: structure and function. Berlin: Life Sciences, springer.com; 2013. p. 127–73.
Flütsch A, Grütter MG. Proteases in death pathways. In: Brix K, Stöcker W, editors. Proteases: structure and function. Berlin: Life Sciences, springer.com; 2013. p. 265–318.
Fonović M, Turk B. Cysteine cathepsins and extracellular matrix degradation. Biochim Biophys Acta. 2014;1840(8):2560–70.
Friedrichs B, Tepel C, Reinheckel T, Deussing J, von Figura K, Herzog V, Peters C, Saftig P, Brix K. Thyroid functions of mouse cathepsins B, K, and L. J Clin Invest. 2003;111(11):1733–45.
Fritz H. Proteinase inhibitors in severe inflammatory processes (septic shock and experimental endotoxaemia): biochemical, pathophysiological and therapeutic aspects. Ciba Found Symp. 1979;75:351–79.
Frizler M, Yampolsky IV, Baranov MS, Stirnberg M, Gütschow M. Chemical introduction of the green fluorescence: imaging of cysteine cathepsins by an irreversibly locked GFP fluorophore. Org Biomol Chem. 2013;11(35):5913–21.
Gansz M, Kern U, Peters C, Reinheckel T. Exploring systemic functions of lysosomal cysteine proteases: the perspective of genetically modified mouse models. In: Brix K, Stöcker W, editors. Proteases: structure and function. Berlin: Life Sciences, springer.com; 2013. p. 217–33.
Gelb BD, Shi GP, Chapman HA, Desnick RJ. Pycnodysostosis, a lysosomal disease caused by cathepsin K deficiency. Science. 1996;273(5279):1236–8.
Gocheva V, Joyce JA. Cysteine cathepsins and the cutting edge of cancer invasion. Cell Cycle. 2007;6(1):60–4.
Goulet B, Nepveu A. Complete and limited proteolysis in cell cycle progression. Cell Cycle. 2004;3(8):986–9.
Grabowska MM, Day ML. Soluble E-cadherin: more than a symptom of disease. Front Biosci (Landmark Ed). 2012;17:1948–64.
Greenbaum D, Baruch A, Hayrapetian L, Darula Z, Burlingame A, Medzihradszky KF, Bogyo M. Chemical approaches for functionally probing the proteome. Mol Cell Proteomics. 2002;1(1):60–8.
Grzywa R, Sieńczyk M. Phosphonic esters and their application of protease control. Curr Pharm Des. 2013;19(6):1154–78.
Hart TC, Hart PS, Bowden DW, Michalec MD, Callison SA, Walker SJ, Zhang Y, Firatli E. Mutations of the cathepsin C gene are responsible for Papillon-Lefevre syndrome. J Med Genet. 1999;36:881–7.
Hart PS, Zhang Y, Firatli E, Uygur C, Lotfazar M, Michalec MD, Marks JJ, Lu X, Coates BJ, Seow WK, Marshall R, Williams D, Reed JB, Wright JT, Hart TC. Identification of cathepsin C mutations in ethnically diverse Papillon-Lefevre syndrome patients. J Med Genet. 2000;37:927–32.
Heald-Sargent T, Gallagher T. Ready, set, fuse! The coronavirus spike protein and acquisition of fusion competence. Virus. 2012;4(4):557–80.
Herz J, Strickland DK. LRP: a multifunctional scavenger and signaling receptor. J Clin Invest. 2001;108(6):779–84.
Hou WS, Bromme D, Zhao Y, Mehler E, Dushey C, Weinstein H, Miranda CS, Fraga C, Greig F, Carey J, Rimoin DL, Desnick RJ, Gelb BD. Characterization of novel cathepsin K mutations in the pro and mature polypeptide regions causing pycnodysostosis. J Clin Invest. 1999;103:731–8.
Hunt CL, Lennemann NJ, Maury W. Filovirus entry: a novelty in the viral fusion world. Virus. 2012;4(2):258–75.
Huntington JA. Shape-shifting serpins--advantages of a mobile mechanism. Trends Biochem Sci. 2006;31(8):427–35.
Hurley JH. ESCRTs are everywhere. EMBO J. 2015;34(19):2398–23407.
Jessani N, Humphrey M, McDonald WH, Niessen S, Masuda K, Gangadharan B, Yates JR 3rd, Mueller BM, Cravatt BF. Carcinoma and stromal enzyme activity profiles associated with breast tumor growth in vivo. Proc Natl Acad Sci U S A. 2004;101(38):13756–61.

Johnson MR, Polymeropoulos MH, Vos HL, Ortiz de Luna RI, Francomano CA. A nonsense mutation in the cathepsin K gene observed in a family with pycnodysostosis. Genome Res. 1996;6:1050–5.

Jordans S, Jenko-Kokalj S, Kuhl NM, Tedelind S, Sendt W, Bromme D, Turk D, Brix K. Monitoring compartment-specific substrate cleavage by cathepsins B, K, L, and S at physiological pH and redox conditions. BMC Biochem. 2009;10:23.

Joyce JA, Hanahan D. Multiple roles for cysteine cathepsins in cancer. Cell Cycle. 2004;3(12):1516–619.

Kay J, Tatnell PJ. Cathepsin E. In: Barrett AJ, Rawlings ND, Woessner JF, editors. Handbook of proteolytic enzymes. 2nd ed. London: Elsevier; 2004. p. 33–8.

Kirschke H. Cathepsin H. In: Barrett AJ, Rawlings ND, Woessner JF, editors. Handbook of proteolytic enzymes. 2nd ed. London: Elsevier; 2004a. p. 1089–92.

Kirschke H. Cathepsin L. In: Barrett AJ, Rawlings ND, Woessner JF, editors. Handbook of proteolytic enzymes. 2nd ed. London: Elsevier; 2004b. p. 1097–102.

Kirschke H. Cathepsin S. In: Barrett AJ, Rawlings ND, Woessner JF, editors. Handbook of proteolytic enzymes. 2nd ed. London: Elsevier; 2004c. p. 1104–7.

Kornfeld S. Structure and function of the mannose 6-phosphate/insulinlike growth factor II receptors. Annu Rev Biochem. 1992;61:307–30.

Kornfeld S, Mellman I. The biogenesis of lysosomes. Annu Rev Cell Biol. 1989;5:483–525.

Krueger S, Kalinski T, Hundertmark T, Wex T, Kuster D, Peitz U, Ebert M, Nagler DK, Kellner U, Malfertheiner P, Naumann M, Rocken C, Roessner A. Up-regulation of cathepsin X in Helicobacter pylori gastritis and gastric cancer. J Pathol. 2005;207:32–42.

Lemansky P, Brix K, Herzog V. Subcellular distribution, secretion, and posttranslational modifications of clusterin in thyrocytes. Exp Cell Res. 1999;251(1):147–55.

Lenarcic B, Turk V. Thyroglobulin type-1 domains in equistatin inhibit both papain-like cysteine proteinases and cathepsin D. J Biol Chem. 1999;274(2):563–6.

Lenarcic B, Ritonja A, Strukelj B, Turk B, Turk V. Equistatin, a new inhibitor of cysteine proteinases from Actinia equina, is structurally related to thyroglobulin type-1 domain. J Biol Chem. 1997;272(21):13899–903.

Li Z, Hou WS, Escalante-Torres CR, Gelb BD, Bromme D. Collagenase activity of cathepsin K depends on complex formation with chondroitin sulfate. J Biol Chem. 2002;277(32):28669–76.

Li Z, Yasuda Y, Li W, Bogyo M, Katz N, Gordon RE, Fields GB, Brömme D. Regulation of collagenase activities of human cathepsins by glycosaminoglycans. J Biol Chem. 2004;279(7):5470–9.

Liaudet-Coopman E, Beaujouin M, Derocq D, Garcia M, Glondu-Lassis M, Laurent-Matha V, Prebois C, Rochefort H, Vignon F. Cathepsin D: newly discovered functions of a long-standing aspartic protease in cancer and apoptosis. Cancer Lett. 2006;237(2):167–79.

Linke M, Herzog V, Brix K. Trafficking of lysosomal cathepsin B-green fluorescent protein to the surface of thyroid epithelial cells involves the endosomal/lysosomal compartment. J Cell Sci. 2002a;115(Pt 24):4877–89.

Linke M, Jordans S, Mach L, Herzog V, Brix K. Thyroid stimulating hormone upregulates secretion of cathepsin B from thyroid epithelial cells. Biol Chem. 2002b;383(5):773–84.

Luke CJ, Pak SC, Askew YS, Naviglia TL, Askew DJ, Nobar SM, Vetica AC, Long OS, Watkins SC, Stolz DB, Barstead RJ, Moulder GL, Brömme D, Silverman GA. An intracellular serpin regulates necrosis by inhibiting the induction and sequelae of lysosomal injury. Cell. 2007;130(6):1108–19.

Machleidt W, Borchart U, Fritz H, Brzin J, Ritonja A, Turk V. Protein inhibitors of cysteine proteinases. II. Primary structure of stefin, a cytosolic protein inhibitor of cysteine proteinases from human polymorphonuclear granulocytes. Hoppe Seylers Z Physiol Chem. 1983;364(11):1481–6.

Mason RW, Gal S, Gottesman MM. The identification of the major excreted protein (MEP) from a transformed mouse fibroblast cell line as a catalytically active precursor form of cathepsin L. Biochem J. 1987;248(2):449–54.

Mayer K, Iolyeva ME, Meyer-Grahle U, Brix K. Intestine-specific expression of green fluorescent protein-tagged cathepsin B: proof-of-principle experiments. Biol Chem. 2008;389(8):1085–96.

Mehtani S, Gong Q, Panella J, Subbiah S, Peffley DM, Frankfater A. In vivo expression of an alternatively spliced human tumor message that encodes a truncated form of cathepsin B. Subcellular distribution of the truncated enzyme in COS cells. J Biol Chem. 1998;273(21):13236–44.
Menard R, Sulea T. Cathepsin X. In: Barrett AJ, Rawlings ND, Woessner JF, editors. Handbook of proteolytic enzymes. 2nd ed. London: Elsevier; 2004. p. 1113–6.
Meulendyke KA, Wurth MA, McCann RO, Dutch RE. Endocytosis plays a critical role in proteolytic processing of the Hendra virus fusion protein. J Virol. 2005;79(20):12643–9.
Millet JK, Whittaker GR. Host cell proteases: critical determinants of coronavirus tropism and pathogenesis. Virus Res. 2015;202:120–34.
Mohamed MM, Sloane BF. Cysteine cathepsins: multifunctional enzymes in cancer. Nat Rev Cancer. 2006;6(10):764–75.
Mort JS. Cathepsin B. In: Barrett AJ, Rawlings ND, Woessner JF, editors. Handbook of proteolytic enzymes. 2nd ed. London: Elsevier; 2004. p. 1079–86.
Mort JS, Buttle DJ. Cathepsin B. Int J Biochem Cell Biol. 1997;29(5):715–20.
Moyle G, Gazzard B. Current knowledge and future prospects for the use of HIV protease inhibitors. Drugs. 1996;51(5):701–12.
Müntener K, Zwicky R, Csucs G, Rohrer J, Baici A. Exon skipping of cathepsin B: mitochondrial targeting of a lysosomal peptidase provokes cell death. J Biol Chem. 2004;279(39):41012–7.
Naganawa Y, Itoh K, Shimmoto M, Kamei S, Takiguchi K, Doi H, Sakuraba H. Stable expression of protective protein/cathepsin A-green fluorescent protein fusion genes in a fibroblastic cell line from a galactosialidosis patient. Model system for revealing the intracellular transport of normal and mutated lysosomal enzymes. Biochem J. 1999;340(Pt 2):467–74.
Nielsen R, Courtoy PJ, Jacobsen C, Dom G, Lima WR, Jadot M, Willnow TE, Devuyst O, Christensen EI. Endocytosis provides a major alternative pathway for lysosomal biogenesis in kidney proximal tubular cells. Proc Natl Acad Sci U S A. 2007;104(13):5407–12.
Novinec M, Lenarčič B, Baici A. Clusterin is a specific stabilizer and liberator of extracellular cathepsin K. FEBS Lett. 2012;586(7):1062–6.
Obermajer N, Magister S, Kopitar AN, Tepes B, Ihan A, Kos J. Cathepsin X prevents an effective immune response against Helicobacter pylori infection. Eur J Cell Biol. 2009;88:461–71.
Ong PC, McGowan S, Pearce MC, Irving JA, Kan WT, Grigoryev SA, Turk B, Silverman GA, Brix K, Bottomley SP, Whisstock JC, Pike RN. DNA accelerates the inhibition of human cathepsin V by serpins. J Biol Chem. 2007;282(51):36980–6.
Pager CT, Dutch RE. Cathepsin L is involved in proteolytic processing of the Hendra virus fusion protein. J Virol. 2005;79(20):12714–20.
Pager CT, Craft WWJ, Patch J, Dutch RE. A mature and fusogenic form of the Nipah virus fusion protein requires proteolytic processing by cathepsin L. Virology. 2006;346(2):251–7.
Peloille S, Esnard A, Dacheux JL, Guillou F, Gauthier F, Esnard F. Interactions between ovine cathepsin L, cystatin C and alpha 2-macroglobulin. Potential role in the genital tract. Eur J Biochem. 1997;244(1):140–6.
Platt MO, Evans D, Keegan PM, McNamara L, Parker IK, Roberts LM, Caulk AW, Gleason RLJ, Seifu D, Amogne W, Penny C. Low-cost method to monitor patient adherence to HIV antiretroviral therapy using multiplex Cathepsin zymography. Mol Biotechnol. 2016;58(1):56–64.
Pohlmann R, Boeker MW, von Figura K. The two mannose 6-phosphate receptors transport distinct complements of lysosomal proteins. J Biol Chem. 1995;270(45):27311–8.
Poller W, Willnow TE, Hilpert J, Herz J. Differential recognition of alpha 1-antitrypsin-elastase and alpha 1-antichymotrypsin-cathepsin G complexes by the low density lipoprotein receptor-related protein. J Biol Chem. 1995;270(6):2841–5.
Pshezhetsky AV. Lysosomal carboxypeptidase A. In: Barrett AJ, Rawlings ND, Woessner JF, editors. Handbook of proteolytic enzymes. 2nd ed. London: Elsevier; 2004. p. 1923–9.
Rachner TD, Khosla S, Hofbauer LC. Osteoporosis: now and the future. Lancet. 2011;377(9773):1276–87.
Rawlings ND. Protease families, evolution and mechanism of action. In: Brix K, Stöcker W, editors. Proteases: structure and function. Berlin: Life Sciences, springer.com; 2013. p. 1–36.

Rawlings ND, Barrett AJ, Finn R. Twenty years of the MEROPS database of proteolytic enzymes, their substrates and inhibitors. Nucleic Acids Res. 2016;44(D1):D343–50.
Rawn SM, Cross JC. The evolution, regulation, and function of placenta-specific genes. Annu Rev Cell Dev Biol. 2008;24:159–81.
Reiser J, Adair B, Reinheckel T. Specialized roles for cysteine cathepsins in health and disease. J Clin Invest. 2010;120(10):3421–31.
Rivera LE, Colon K, Cantres-Rosario YM, Zenon FM, Melendez LM. Macrophage derived cystatin B/cathepsin B in HIV replication and neuropathogenesis. Curr HIV Res. 2014;12(2):111–20.
Rochefort H, Garcia M, Glondu M, Laurent V, Liaudet E, Rey JM, Roger P. Cathepsin D in breast cancer: mechanisms and clinical applications, a 1999 overview. Clin Chim Acta. 2000;291(2):157–70.
Rock KL, Shen L. Cross-presentation: underlying mechanisms and role in immune surveillance. Immunol Rev. 2005;207:166–83.
Rubin H. Systemic effects of cancer: role of multiple proteases and their toxic peptide products. Med Sci Monit. 2005;11(7):RA221–8.
Sadaghiani AM, Verhelst SH, Gocheva V, Hill K, Majerova E, Stinson S, Joyce JA, Bogyo M. Design, synthesis, and evaluation of in vivo potency and selectivity of epoxysuccinyl-based inhibitors of papain-family cysteine proteases. Chem Biol. 2007;14(5):499–511.
Saftig P, Hunziker E, Wehmeyer O, Jones S, Boyde A, Rommerskirch W, Moritz JD, Schu P, von Figura K. Impaired osteoclastic bone resorption leads to osteopetrosis in cathepsin-K-deficient mice. Proc Natl Acad Sci U S A. 1998;95(23):13453–8.
Salpeter SJ, Blum G. Ready, set, cleave: proteases in action. Chem Biol. 2013;20(2):137–8.
Salvesen GS. Cathepsin G. In: Barrett AJ, Rawlings ND, Woessner JF, editors. Handbook of proteolytic enzymes. 2nd ed. London: Elsevier; 2004. p. 1524–6.
Sanman LE, Bogyo M. Activity-based profiling of proteases. Annu Rev Biochem. 2014;83:249–73.
Scharf JG, Braulke T. The role of the IGF axis in hepatocarcinogenesis. Horm Metab Res. 2003;35(11–12):685–93.
Schaschke N, Assfalg-Machleidt I, Machleidt W, Moroder L. Substrate/propeptide-derived endo-epoxysuccinyl peptides as highly potent and selective cathepsin B inhibitors. FEBS Lett. 1998;421(1):80–2.
Schechter I, Berger A. On the active site of proteases. 3. Mapping the active site of papain; specific peptide inhibitors of papain. Biochem Biophys Res Commun. 1968;32(5):898–902.
Schilling K, Körner A, Sehmisch S, Kreusch A, Kleint R, Benedix Y, Schlabrakowski A, Wiederanders B. Selectivity of propeptide-enzyme interaction in cathepsin L-like cysteine proteases. Biol Chem. 2009;390(2):167–74.
Siintola E, Partanen S, Strömme P, Haapanen A, Haltia M, Maehlen J, Lehesjoki AE, Tyynelä J. Cathepsin D deficiency underlies congenital human neuronal ceroid-lipofuscinosis. Brain. 2006;129(Pt 6):1438–45.
Silverman GA, Bird PI, Carrell RW, Church FC, Coughlin PB, Gettins PG, Irving JA, Lomas DA, Luke CJ, Moyer RW, Pemberton PA, Remold-O'Donnell E, Salvesen GS, Travis J, Whisstock JC. The serpins are an expanding superfamily of structurally similar but functionally diverse proteins. Evolution, mechanism of inhibition, novel functions, and a revised nomenclature. J Biol Chem. 2001;276(36):33293–6.
Simmons G, Zmora P, Gierer S, Heurich A, Pöhlmann S. Proteolytic activation of the SARS-coronavirus spike protein: cutting enzymes at the cutting edge of antiviral research. Antiviral Res. 2013;100(3):605–14.
Sloane BF, Moin K, Krepela E, Rozhin J. Cathepsin B and its endogenous inhibitors: the role in tumor malignancy. Cancer Metastasis Rev. 1990;9(4):333–52.
Sloane BF, List K, Fingleton B, Matrisian L. Proteases in cancer—significance for invasion and metastasis. In: Brix K, Stöcker W, editors. Proteases: structure and function. Berlin: Life Sciences, springer.com; 2013. p. TBD.
Spiess E, Brüning A, Gack S, Ulbricht B, Spring H, Trefz G, Ebert W. Cathepsin B activity in human lung tumor cell lines: ultrastructural localization, pH sensitivity, and inhibitor status at the cellular level. J Histochem Cytochem. 1994;42(7):917–29.

Sun L, Liu S, Chen ZJ. SnapShot: pathways of antiviral innate immunity. Cell. 2010;140:436. https://doi.org/10.1016/j.cell.2010.01.041.
Tamhane T, Wolters BK, Illukkumbura R, Maelandsmo GM, Haugen MH, Brix K. Construction of a plasmid coding for green fluorescent protein tagged cathepsin L and data on expression in colorectal carcinoma cells. Data Brief. 2015;5:468–75.
Tamhane T, Lllukkumbura R, Lu S, Maelandsmo GM, Haugen MH, Brix K. Nuclear cathepsin L activity is required for cell cycle progression of colorectal carcinoma cells. Biochimie. 2016;122:208–18.
Tedelind S, Poliakova K, Valeta A, Hunegnaw R, Yemanaberhan EL, Heldin NE, Kurebayashi J, Weber E, Kopitar-Jerala N, Turk B, Bogyo M, Brix K. Nuclear cysteine cathepsin variants in thyroid carcinoma cells. Biol Chem. 2010;391(8):923–35.
Tepel C, Bromme D, Herzog V, Brix K. Cathepsin K in thyroid epithelial cells: sequence, localization and possible function in extracellular proteolysis of thyroglobulin. J Cell Sci. 2000;113(Pt 24):4487–98.
Tholen M, Hillebrand LE, Tholen S, Sedelmeier O, Arnold SJ, Reinheckel T. Out-of-frame start codons prevent translation of truncated nucleo-cytosolic cathepsin L in vivo. Nat Commun. 2014;5:4931.
Tolosa E, Li W, Yasuda Y, Wienhold W, Denzin LK, Lautwein A, Driessen C, Schnorrer P, Weber E, Stevanovic S, Kurek R, Melms A, Bromme D. Cathepsin V is involved in the degradation of invariant chain in human thymus and is overexpressed in myasthenia gravis. J Clin Invest. 2003;112:517–26.
Tong TR. SARS coronavirus anti-infectives. Recent Pat Antiinfect Drug Discov. 2006;1(3):297–308.
Toomes C, James J, Wood AJ, Wu CL, McCormick D, Lench N, Hewitt C, Moynihan L, Roberts E, Woods CG, Markham A, Wong M, Widmer R, Ghaffar KA, Pemberton M, Hussein IR, Temtamy SA, Davies R, Read AP, Sloan P, Dixon MJ, Thakker NS. Loss-of-function mutations in the cathepsin C gene result in periodontal disease and palmoplantar keratosis. Nat Genet. 1999;23:421–4.
Travis J. Structure, function, and control of neutrophil proteinases. Am J Med. 1988;84(6A):37–42.
Turk V, Bode W. The cystatins: protein inhibitors of cysteine proteinases. FEBS Lett. 1991;285(2):213–9.
Turk B, Turk V. Lysosomes as "suicide bags" in cell death: myth or reality? J Biol Chem. 2009;284(33):21783–7.
Turk B, Turk D, Turk V. Lysosomal cysteine proteases: more than scavengers. Biochim Biophys Acta. 2000;1477(1–2):98–111.
Turk B, Turk D, Dolenc I, Turk V. Dipeptidyl-peptidase I. In: Barrett AJ, Rawlings ND, Woessner JF, editors. Handbook of proteolytic enzymes. 2nd ed. London: Elsevier; 2004. p. 1192–6.
Turk V, Stoka V, Turk D. Cystatins: biochemical and structural properties, and medical relevance. Front Biosci. 2008;(13):5406–20.
Velasco G, Lopez-Otin C. Cathepsin O. In: Barrett AJ, Rawlings ND, Woessner JF, editors. Handbook of proteolytic enzymes. 2nd ed. London: Elsevier; 2004. p. 1102–3.
Vogt C, Eickmann M, Diederich S, Moll M, Maisner A. Endocytosis of the Nipah virus glycoproteins. J Virol. 2005;79(6):3865–72.
von Figura K. Molecular recognition and targeting of lysosomal proteins. Curr Opin Cell Biol. 1991;3(4):642–6.
Weber E, Barbulescu E, Medek R, Reinheckel T, Sameni M, Anbalagan A, Moin K, Sloane BF. Cathepsin B-deficient mice as source of monoclonal anti-cathepsin B antibodies. Biol Chem. 2015;396(3):277–81.
Weis M, Maisner A. Nipah virus fusion protein: importance of the cytoplasmic tail for endosomal trafficking and bioactivity. Eur J Cell Biol. 2015;94(7–9):316–22.
Weiss-Sadan T, Gotsman I, Blum G. Cysteine proteases in atherosclerosis. FEBS J. 2017. https://doi.org/10.1111/febs.14043. (Epub ahead of print).
Whisstock JC, Silverman GA, Bird PI, Bottomley SP, Kaiserman D, Luke CJ, Pak SC, Reichhart JM, Huntington JA. Serpins flex their muscle: II. Structural insights into target peptidase recognition, polymerization, and transport functions. J Biol Chem. 2010;285(32):24307–12.

Willnow TE, Moehring JM, Inocencio NM, Moehring TJ, Herz J. The low-density-lipoprotein receptor-related protein (LRP) is processed by furin in vivo and in vitro. Biochem J. 1996;313(Pt 1):71–6.

Zeeuwen PL, Cheng T, Schalkwijk J. The biology of cystatin M/E and its cognate target proteases. J Invest Dermatol. 2009;129(6):1327–38.

Zhou XY, Morreau H, Rottier R, Davis D, Bonten E, Gillemans N, Wenger D, Grosveld FG, Doherty P, Suzuki K, Grosveld GC, D'Azzo A. Mouse model for the lysosomal disorder galactosialidosis and correction of the phenotype with overexpressing erythroid precursor cells. Genes Dev. 1995;9:2623–34.

Zou F, Schmon M, Sienczyk M, Grzywa R, Palesch D, Boehm BO, Sun ZL, Watts C, Schirmbeck R, Burster T. Application of a novel highly sensitive activity-based probe for detection of cathepsin G. Anal Biochem. 2012;421(2):667–72.

Zwicky R, Müntener K, Csucs G, Goldring MB, Baici A. Exploring the role of 5′ alternative splicing and of the 3′-untranslated region of cathepsin B mRNA. Biol Chem. 2003;384(7):1007–18.

Part III

Protease Inhibitors

The Antiviral Potential of Host Protease Inhibitors

11

Torsten Steinmetzer and Kornelia Hardes

Abstract

The replication of numerous pathogenic viruses depends on host proteases, which therefore emerged as potential antiviral drug targets. In some cases, e.g., for influenza viruses, their function during the viral propagation cycle is relatively well understood, where they cleave and activate viral surface glycoproteins. For other viruses, e.g., Ebola virus, the function of host proteases during replication is still not clear. Host proteases may also contribute to the pathogenicity of virus infection by activating proinflammatory cytokines. For some coronaviruses, human proteases can also serve in a nonproteolytical fashion simply as receptors for virus entry. However, blocking of such protein-protein contacts is challenging, because receptor surfaces are often flat and difficult to address with small molecules. In contrast, many proteases possess well-defined binding pockets. Therefore, they can be considered as well-druggable targets, especially, if they are extracellularly active. The number of their experimental crystal structures is steadily increasing, which is an important prerequisite for a rational structure-based inhibitor design using computational chemistry tools in combination with classical medicinal chemistry approaches. Moreover, host proteases can be considered as stable targets, and their inhibition should prevent rapid resistance developments, which is often observed when addressing viral proteins. Otherwise, the inhibition of host proteases can also affect normal physiological processes leading to a higher probability of side effects and a narrow therapeutic window. Therefore, they should be preferably used in combination therapies with additional antiviral drugs. This strategy should provide a stronger antiviral efficacy, allow to use lower drug doses, and minimize side effects. Despite numerous experimental findings on their antiviral activity, no small-molecule inhibitors of host proteases have been approved for the treatment of virus infections, so far.

T. Steinmetzer (✉) · K. Hardes
Institute of Pharmaceutical Chemistry, Philipps University, Marburg, Germany
e-mail: steinmetzer@uni-marburg.de

E. Böttcher-Friebertshäuser et al. (eds.), *Activation of Viruses by Host Proteases*,
https://doi.org/10.1007/978-3-319-75474-1_11

Abbreviations

Agm	Agmatine
BBI	Bowman-Birk inhibitor
CCHFV	Crimean-Congo hemorrhagic fever virus
CMK	Chloromethyl ketone
Dec	Decanoyl
EboV	Ebola virus
GP	Glycoprotein
HA	Hemagglutinin
HAT	Human airway trypsin-like peptidase
HIV	Human immunodeficiency virus
HPAIV	Highly pathogenic avian influenza virus
HPIV	Human parainfluenza virus
hTyr	Homotyrosin
IAV	Influenza A virus
IBV	Influenza B virus
LASV	Lassa virus
LBS	Lysine binding site
LCMV	Lymphocytic choriomeningitis virus
LPAIV	Low pathogenic avian influenza virus
MERS-CoV	Middle East respiratory syndrome coronavirus
PAR-2	Protease-activated receptor 2
PC	Proprotein convertase
Phac	Phenylacetyl
PPI	Protein-protein interaction
SARS-CoV	Severe acute respiratory syndrome coronavirus
S-protein	Spike protein
TLSP	Trypsin-like serine protease
TTSP	Type II transmembrane serine protease

11.1 Introduction

At present, 588 human proteases are listed in the degradome database (Quesada et al. 2009), which can be further divided into five classes based on their catalytic mechanism. The majority belongs to the family of metallo (192 members), serine (184 enzymes), and cysteine proteases (164 members); in addition, 27 threonine and 21 aspartyl proteases are known, so far. Proteases are well-druggable targets and numerous small-molecule protease inhibitors have been approved in the past. The more than ten inhibitors of the angiotensin-converting enzyme (ACE) (Turk 2006), as well as the neprilysin (enkephalinase) inhibitor prodrugs sacubitril (Howell and Cameron 2016) and racecadotril (acetorphan) target Zn^{2+}-dependent metalloproteases. The ACE blockers and sacubitril are suitable for long-term usage as antihypertensive drugs, whereas racecadotril is approved as an antidiarrheal drug. Despite limited use, treatment of high blood pressure is also possible with aliskiren, a small-molecule inhibitor of the aspartate protease renin (Wood et al. 2003). Numerous synthetic inhibitors of the trypsin-like serine proteases (TLSP) thrombin (argatroban, dabigatran etexilate, bivalirudin) and factor Xa (rivaroxaban, apixaban, edoxaban, betrixaban) can be used as anticoagulants (Straub et al. 2011). Except argatroban and bivalirudin, all of these clotting protease inhibitors are orally available and suited for long-term use in the prevention of stroke, e.g., in patients suffering from atrial fibrillation. Meanwhile, more than ten gliptins have been approved in various countries. The gliptins are a class of oral hypoglycemic drugs for the treatment of diabetes mellitus type 2 targeting the serine protease dipeptidyl peptidase 4, thereby reducing the degradation of incretin hormones leading to an enhanced insulin secretion (Scheen 2015). The proteasome inhibitors bortezomib (Adams 2004) and carfilzomib (Kortuem and Stewart 2013) are used in patients with multiple myeloma and the first approved drugs targeting threonine proteases. Meanwhile, two additional proteasome inhibitors, ixazomib and oprozomib, obtained orphan drug status (Manasanch and Orlowski 2017). Besides addressing human proteases, numerous inhibitors of the aspartyl protease of HIV (Ghosh et al. 2016) and the NS3/4A serine protease of the hepatitis C virus (McCauley and Rudd 2016) are on the market. With few exceptions, most of these inhibitors are routinely used in combination with other drugs and not as single agents. Despite large efforts, no cysteine protease inhibitor has been approved, so far. One of the most advanced inhibitors of the papain-like bone-degrading protease cathepsin K is the nitrile derivative odanacatib, which was developed for the treatment of osteoporosis. Based on a clinical phase III trial, a high efficacy with increasing bone mineral density and reduced risk of fractures was initially reported, as well as a good safety profile (Chapurlat 2015). However, its further development was stopped at the end of 2016 due to a slightly increased risk of stroke (Mullard 2016). Other cathepsin K inhibitors, like the nitrile balicatib, failed in phase II due to complications with skin fibrosis (Brömme et al. 2016; Runger et al. 2012). Despite the lack of approved cysteine protease inhibitors, many other examples confirm the suitability of at least some proteases as excellent drug targets.

This should also apply to numerous host proteases, which are involved at various steps during the propagation cycle of certain viruses. The inhibition of host enzymes could be advantageous compared to the classical addressing of viral targets due to a low risk for rapid drug resistances. However, side effects may occur by targeting host enzymes, which are required for normal physiological processes. Moreover, most host proteases belong to families of structurally closely related enzymes, and therefore it might be challenging to address a single target without affecting other family members. In order to avoid side effects and a narrow therapeutic window, it is therefore advisable to develop such host protease inhibitors mainly for combination therapies, which should enable the use of lower drug doses.

In the following sections, numerous examples for the antiviral activity of inhibitors mainly addressing human serine proteases will be provided. In addition, structural aspects of the protease-inhibitor complexes will be discussed. Although first approved inhibitors of ACE and thrombin have been discovered long before the target structures have been determined, all subsequent successful developments in the field of protease inhibitors were strongly supported by the availability of crystal structures, which is an important prerequisite for a rational structure-based drug design.

11.2 Serine Proteases as Antiviral Targets

11.2.1 Trypsin-Like Serine Proteases

The majority of the human serine proteases belongs to the subfamily S1A, possessing a chymotrypsin-like folding pattern, and to the family S8 with the two subfamilies S8A and S8B exhibiting a subtilisin- or kexin-like folding (Rawlings et al. 2014). First studies on the limited proteolysis and essential maturation of viral glycoproteins by host proteases were performed on influenza viruses. It could be demonstrated that the cleavage of the hemagglutinin (HA) precursor could be blocked by the broad-spectrum serine protease inhibitor diisopropylfluorophosphate (DFP) (Klenk and Rott 1973). It was found that the poor infectivity of influenza A viruses (IAV) grown in cultures of chick embryo cells could be strongly increased after treatment with exogenous trypsin, which cleaves substrates after basic residues like arginine or lysine (Klenk et al. 1975). At the same time, comparable results were described for the trypsin-catalyzed activation of HA from influenza B viruses (Lazarowitz and Choppin 1975). Although the digestive protease trypsin is not found in the respiratory tract, these initial studies suggested that other trypsin-like airway proteases should be involved in HA maturation. For instance, plasmin efficiently activated the HA of the A/WSN virions but failed to cleave the influenza B HA (Lazarowitz and Choppin 1975). A detailed analysis of the substrate sequences revealed that HAs of human and other mammalian influenza viruses as well as HAs of low pathogenic avian influenza viruses (LPAIV) are cleaved after a single arginine residue before a constant P1′-P3′ Gly-Leu-Phe segment by trypsin-like serine proteases. In contrast, HAs of high pathogenic avian influenza viruses (HPAIV) are activated by furin-like proprotein convertases (PCs) at inserted multibasic sequences containing additional arginine or lysine residues in the adjacent non-primed positions (Garten and Klenk 2008). In addition to the HA activation of HPAIV, many other viruses depend on the correct cleavage of their surface glycoproteins by furin-like PCs (basic PCs) or by the neutral PC SKI-1 (Klenk and Garten 1994; Pasquato et al. 2013). Consequently, different inhibitor structures depending on the specific virus strains are required to target the appropriate activating protease.

The human genome encodes for approximately 70 different trypsin-like serine proteases, which cleave after a single basic residue, preferably after arginine in P1 position (Schechter and Berger 1967). Few of them, like matriptase or TMPSS13 (MSPL), strongly prefer substrates with additional basic residues in the non-primed region close to the cleavage site, e.g., in P4 and/or P3 position. The full-length enzymes can strongly differ in their molecular weights due to the presence of specific protein domains. However, all of them possess a relatively similar catalytic domain of approximately 225–230 amino acids with a strong structural homology to chymotrypsin of the subfamily S1A of serine proteases. This facilitates a common numbering of the residues within the catalytic domain with respect to chymotrypsin(ogen) used throughout this chapter. So far, crystal structures are available for ~25 different trypsin-like serine proteases. In most cases, the structures have only been determined for their catalytic domains, and their knowledge is normally sufficient for a rational structure-based design of active-site-directed inhibitors. The protease domain of the trypsin-like serine proteases consists of two six-stranded barrel domains, held together by several transdomain straps. The residues Ser195, His57, and Asp102 of the catalytic triad are located at the junction between these two barrels. All trypsin-like serine proteases contain a negatively charged aspartate residue at the bottom of their S1 pocket responsible for accepting substrates and inhibitors with basic P1 residues.

Data from virus-infected cell cultures suggested that different secreted trypsin-like serine proteases are involved in the HA activation of human IAV and LPAIV depending on the host and specific virus strain. A protease named tryptase Clara was isolated from Clara cells of rat airway epithelium (Kido et al. 1992). Furthermore, miniplasmin purified from rat lungs was described as potential HA activator (Murakami et al. 2001). Miniplasmin is a degraded version of plasmin, only comprising the kringle 5 and protease domain and lacking the N-terminal kringle domains 1–4 (Al-Horani and Desai 2014). However, plasmin cannot efficiently activate the HAs with monobasic cleavage sites found in the presently circulating influenza strains, because it prefers substrates with bulky P2 residues like Phe, Tyr, and Trp (Swedberg and Harris 2011). In embryonated chicken eggs, a blood clotting factor Xa-like protease was found to activate HA (Gotoh et al. 1990). Again, it seems unlikely that human factor Xa could be involved, which is a very specific protease with a strong preference for substrates with glycine in P2 position due to the presence of a bulky Tyr99 on top of its S2 binding pocket. This special structural feature limits the access of substrates with larger P2 side chains. So far, only prothrombin and eventually the protease-activated receptor 2 (PAR-2) are known as well-cleavable natural factor Xa substrates (Oe et al. 2016); both possess a glycine in P2 position. Moreover, in rats and mice, cellular ectopic trypsins due to severe influenza infections and cytokine storms were detected as HA-activating proteases (Kido et al. 2012; Pan et al. 2011; Wang et al. 2010). An unusual extrapancreatic trypsin expression was also found in some human tumors (Paju et al. 2001; Sorsa et al. 1997), raising speculations that this could also happen as a consequence of excessive inflammation during viral infections. Porcine tryptases are additional candidates (Chen et al. 2000; Sato et al. 2003), although human lung tryptase has failed to activate the HA of the 1918 IAV (Stevens et al. 2004). Among the 15 kallikrein-related peptidases (KLKs) (Goettig et al. 2010), 12 of them possess a trypsin-like substrate specificity, and a potential HA activation of seasonal IAV was observed for KLK5 and KLK12 (Hamilton and Whittaker 2013).

Additional candidates have been found among the type II transmembrane serine proteases (TTSPs), which comprise 17 different enzymes (Antalis et al. 2011; Garten et al. 2015). Most of them show a restricted tissue-specific expression pattern; an exception is matriptase, which is ubiquitously found in epithelial layers of most tissues. In 2006, the TTSPs TMPRSS2 (epitheliasin) and human airway trypsin-like peptidase (HAT or TMPRSS11D) were identified as activating enzymes of monobasic HAs of seasonal IAV when overexpressed in MDCK cells (Böttcher-Friebertshäuser et al. 2010; Böttcher et al. 2006). A potential HA cleavage of the 1918 IAV was also described by TMPRSS4 (Bertram et al. 2010; Chaipan et al. 2009). The virus spread is eventually promoted by additional TTSPs, like DESC1 (Zmora et al. 2014) or hepsin, whereas no HA processing was found for TMPRSS3 and TMPRSS6 (also named as matriptase-2) (Bertram et al. 2010). In contrast to these TTSPs involved in the HA activation of LPAIV, a strong substrate preference for HAs from HPAIV with Lys as P4 residue has been identified for TMPRSS13 (MSPL) (Okumura et al. 2010). This is a striking difference to the substrate profile of the PC furin, which strongly prefers arginine in P4 position (Rockwell et al. 2002). Independent studies with knockout mice revealed that the HA of the seasonal H1N1 virions including the 2009 pandemic virus is dominantly activated by TMPRSS2 (Hatesuer et al. 2013; Sakai et al. 2014; Tarnow et al. 2014), whereas TMPRSS2 alone is not sufficient for the maturation of the HA from H3N2 subtypes. A recent study showed that only the combined TMPRSS2$^{-/-}$ and TMPRSS4$^{-/-}$ knockout mice reduced H3N2 spread and signs of infection in lung, suggesting that both proteases are involved (Kuhn et al. 2016).

11.2.1.1 Inhibitors for the Treatment of Human Influenza Virus Infections

Proteinaceous Inhibitors

The concept of HA cleavage inhibition as realistic antiviral strategy was supported by studies with the broad-spectrum inhibitor aprotinin (**1**) in mice (Ovcharenko and Zhirnov 1994; Zhirnov et al. 1982a). The 58-mer Kunitz-type inhibitor, also known as BPTI or Trasylol®, is isolated from bovine lungs and relatively well tolerated in animals and humans. It inhibits numerous trypsin-like serine proteases such as trypsin, plasmin, matriptase, and plasma kallikrein (Ascenzi et al. 2003); its complex with trypsin is shown in Fig. 11.1.

Trasylol was intravenously used for many years to reduce perioperative bleeding during open heart surgery but was withdrawn from market in 2008 when it became clear that it was associated with increased numbers of deaths as compared to treatment with standard antifibrinolytics (Fergusson et al. 2008). As nonhuman protein, it may cause hypersensitive reactions during cardiac surgery, although the risk for anaphylactic reactions at first-time exposure is very low (<0.1%) but might increase to ~2.7% after reexposure (Levy and Adkinson 2008). However, the application of aprotinin against influenza in form of an aerosol at significantly lower doses should minimize side effects (Zhirnov et al. 2011). In addition, an aprotinin variant with stronger kallikrein and similar plasmin inhibition has been developed, which exhibits a reduced immunogenicity in chimpanzees (Apeler et al. 2004). Antiviral potency against influenza was recently also demonstrated with the endogenous proteinaceous inhibitor hepatocyte growth factor activator inhibitor 2 (HAI-2) (Hamilton et al. 2014). HAI-2 and HAI-1 are type 1 transmembrane proteins both containing two Kunitz domains, which inhibit numerous trypsin-like serine proteases including

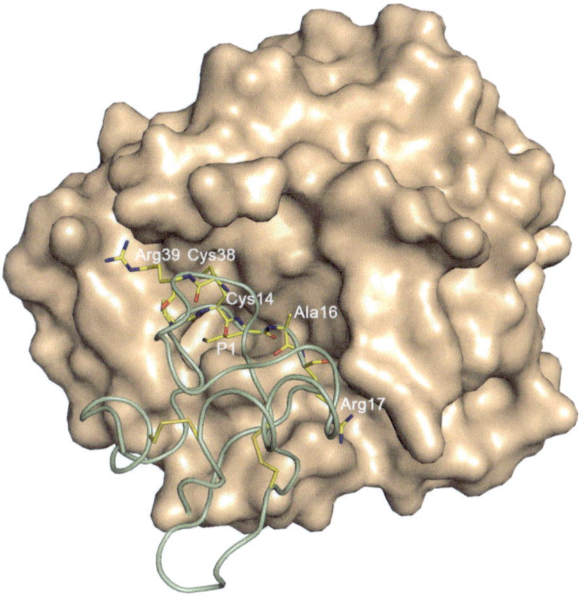

Fig. 11.1 Structure of aprotinin (**1**) in complex with trypsin (PDB, 3FP6). Trypsin is shown with its surface in beige, aprotinin as cartoon in green. Only the aprotinin residues in direct contact to the trypsin surface (Pro13-Cys-Lys-Ala-Arg17 and Cys38-Arg39) are shown as sticks with carbon atoms in yellow, as well as the three disulfide bridges. The P1 residue Lys15 binds into the S1 pocket of trypsin

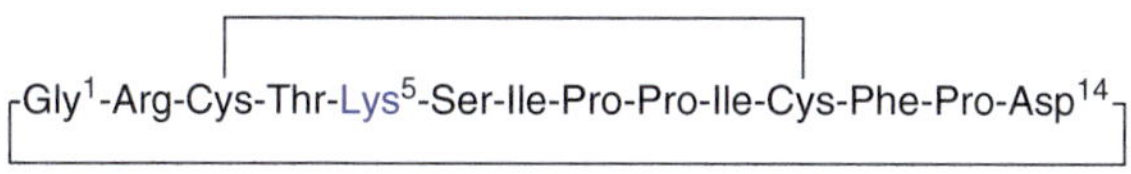

SFTI-1 (**2**)

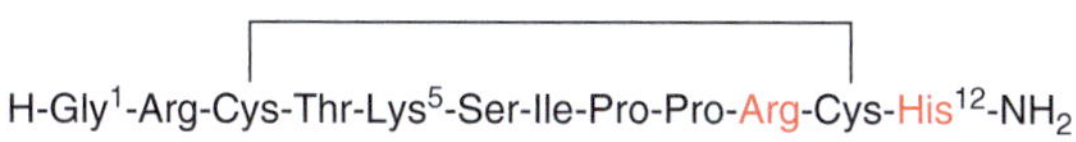

(**3**)

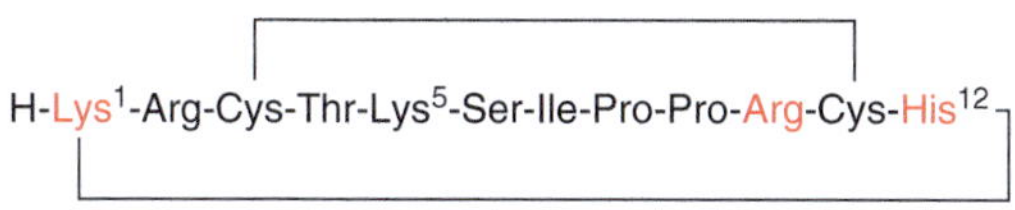

(**4**)

Fig. 11.2 Structure of the bicyclic 14-mer SFTI-1 (**2**) and of its optimized truncated monocyclic (**3**) and bicyclic (**4**) analogues

hepatocyte growth factor activator, matriptase (Szabo et al. 2008), hepsin, trypsin, and prostasin (Liu et al. 2017). Sunflower trypsin inhibitor SFTI-1 (**2**), a bicyclic 14-mer, provides also an excellent scaffold for the design of serine protease inhibitors (Luckett et al. 1999) (Fig. 11.2).

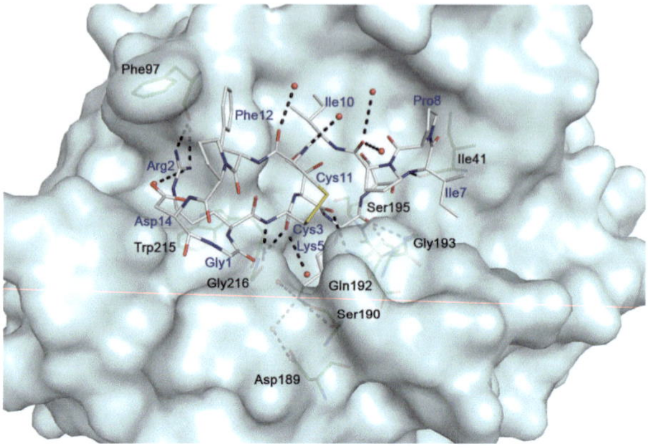

Fig. 11.3 Crystal structure of SFTI-1 (**2**) in complex with matriptase (PDB, 3P8F). Matriptase is shown with its transparent surface in cyan; important residues involved in H-bonds are shown as sticks with carbon atoms in green. SFTI-1 is presented as sticks with carbons in white, water molecules as red balls, and polar contacts as dashed lines in black. For better differentiation, matriptase residues are labeled in black and selected inhibitor residues in blue

It is one of the smallest naturally occurring peptidic protease inhibitors and comprises a typical Bowman-Birk motif of serine protease inhibitors (BBIs) with its characteristic 9-mer disulfide-bridged loop containing Lys5 as P1 residue (Li et al. 2007; McBride et al. 2002). The inhibitor is further stabilized by an additional head-to-tail backbone cyclization between the N-terminal Gly1 and C-terminal Asp14 residues. SFTI-1 inhibits trypsin and matriptase with K_i values of 100 pM (Luckett et al. 1999) and 100 nM (Long et al. 2001), respectively. After initial crystallization in complex with trypsin (Luckett et al. 1999), its binding mode has been determined also in the active site of matriptase (Fig. 11.3) (Yuan et al. 2011).

In contrast to the trypsin complex, the Lys5 amino group makes only a water-bridged interaction to Asp189 at the bottom of the S1 pocket of matriptase but forms no direct contact to its carboxyl group. It also interacts with the side chain and carbonyl oxygens of the adjacent Ser190 and with a second water molecule (Fig. 11.3). Numerous analogues of SFTI-1 have been made. For instance, its conversion to the more simple truncated monocyclic analogue **3** provided a K_i value of 6.2 nM against matriptase (Fittler et al. 2013). The bicyclic analogue **4**, containing a side-chain-to-tail cyclization, is even more potent with an inhibition constant of 2.6 nM (Fittler et al. 2014). However, despite this excellent potency, none of the SFTI analogues has been tested for antiviral activity.

Because of their high molecular weight aprotinin, HAI-1, HAI-2, or SFTI analogues should mainly inhibit extracellular HA processing, and it is rather unlikely that substantial amounts of these compounds can enter the cell and block the TTSPs in the secretory pathway. Despite this limitation, an aerosol formulation of aprotinin has been developed and approved in Russia for the treatment of mild-to-moderate influenza infections (Zhirnov et al. 2011). It was described that in case of mammalian IAV and LPAIV with monobasic HAs, progeny particles are assembled, which still contain some uncleaved HAs and therefore require further activation by extracellular enzymes sensitive to aprotinin. In contrast to these previous reports, more recent findings reveal that the monobasic HAs are processed in intracellular compartments without a further need for extracellular processing (Böttcher-Friebertshäuser et al. 2010, 2013). This suggests that small amounts of aprotinin can reach intracellular targets and achieve an antiviral efficacy (Zhirnov et al. 2011).

S1 Binders

Although the therapeutic use of peptide drugs (Vlieghe et al. 2010) and biologics including recombinant enzymes, hormones, and antibodies is steadily increasing and meanwhile well established, a treatment with synthetic, small-molecule drugs offers significant advantages. They can be produced at low costs in large quantities and offer the possibility for oral administration with a negligible risk of allergic reactions. In initial studies a weak antiviral effect after treatment with ECA (**5**, ε-aminocaproic acid, Fig. 11.4) was observed in mice infected by IAV including the plasmin-dependent WSN strain (Zhirnov et al. 1982a, b). Due to its small size, ECA (**5**) cannot bind efficiently to the active site of trypsin-like proteases and reduces mainly the activation of plasminogen by blocking the lysine binding site on the kringle domains of plasminogen (Al-Horani and Desai 2014). A pronounced protective effect was also observed, when IAV-infected mice were treated with a cocktail of 4-(2-aminoethyl)benzenesulfonylfluoride (**6**, AEBSF or Pefabloc®SC) and *p*-aminobenzamidine (**7**, *p*-AB) prior to infection. Both compounds are relatively unspecific protease inhibitors and can only occupy the S1 pocket of trypsin-like serine proteases. AEBSF **6** leads to a covalent sulfonylation of the active-site Ser195 side chain (second-order inactivation constant k_2/K_i for trypsin 30 M^{-1} s^{-1}), while *p*-AB acts as a reversible competitive inhibitor with K_i values in the two- and three-digit micromolar range depending on the specific target (Stürzebecher et al. 2001).

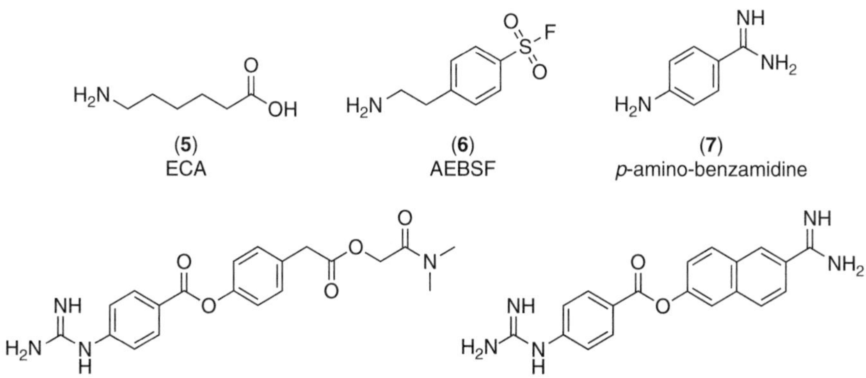

Fig. 11.4 Structures of the lysine analogue ECA and of numerous broad-spectrum inhibitors of trypsin-like serine proteases mainly targeting the S1-binding pocket

Although AEBSF **6** is more stable in buffer than the related phenylmethane sulfonylfluoride (PMSF), it is also susceptible to hydrolysis in aqueous media (Powers et al. 2002) and not suitable for further development. Derivatives of the classical active-site titrant *p*-nitrophenyl-*p*-guanidinobenzoate (Chase and Shaw 1967) including compounds like camostat **8** (FOY-305) and nafamostat **9** (FUT-175) (Fujii and Hitomi 1981) inhibit numerous trypsin-like serine proteases by covalent acylation of their active-site Ser195, providing relatively stable acyl-enzyme complexes. These compounds are active against numerous monobasic IAV and IBV strains (Hosoya et al. 1993; Lee et al. 1996; Someya et al. 1990) but suffer from instability of their ester bond in aqueous media and the circulation ($t_{1/2}$ <1 min after intravenous application of camostat in man (Midgley et al. 1994)). A very short half-life is also reported for nafamostat, although this could be an advantage for certain applications. Notably, nafamostat has been widely used as anticoagulant for hemodialysis patients with a tendency to bleed (Han et al. 2011).

Substrate Analogue Inhibitors

Although an antiviral efficacy was demonstrated for ECA **5** and the S1 ligands **6–9**, a stronger potency can be achieved with compounds also addressing adjacent binding sites. Trypsin-like host proteases possess relatively well-defined S2 and S3/S4 regions, often named proximal and distal binding pockets, respectively. Therefore, substrate analogue S3/S4-S2-segments were coupled with C-terminal decarboxylated P1 arginine mimetics or with stabilized non-cleavable P1-P1′ scaffolds, e.g., arginyl-ketone moieties or other warheads targeting the active-site serine.

A well-suited P1 mimetic to target trypsin- and furin-like PCs is the 4-amidinobenzylamide group (4-Amba). The 4-Amba anchor makes numerous polar interactions to residues at the bottom and exit of the S1 pocket (Fig. 11.5) but exhibits no preference for special trypsin-like serine proteases. Within this inhibitor type, an improved selectivity for individual family members can partially achieved only via their P2, P3, and/or P4 residues. All substrate analogue structures strictly require a P2 amino acid in L-configuration, whereas a D-configured P3-residue is often preferred. Its side chain can address the characteristic distal S3/4 binding pocket above Trp215, which is found and mostly well-defined in all trypsin-like serine proteases. In contrast, the side chain of a natural L-configured P3 residue is directed toward the solvent and makes fewer binding contributions. Interestingly, the backbone of the P3 residue is involved in a short antiparallel β-sheet interaction with a highly conserved glycine in position 216 of the protease domain, irrespective of an L- or D-configuration of the P3 amino acid. So far, no crystal structures of TTSPs involved in HA cleavage, like TMPRSS2, TMPRSS4, or HAT, are available. However, countless crystal structures of this inhibitor type in complex with numerous other trypsin-like serine proteases have been elucidated. As an example, Fig. 11.5 shows the binding mode of the substrate analogue inhibitor H-D-hTyr-Ala-4-Amba (**10**) (Fig. 11.6) in complex with trypsin, indicating the characteristic binding pockets (Maiwald et al. 2016). This inhibitor type could be refined following a well-established prodrug strategy that has been used for the development of the orally available thrombin inhibitor ximelagatran (Gustafsson et al. 2004).

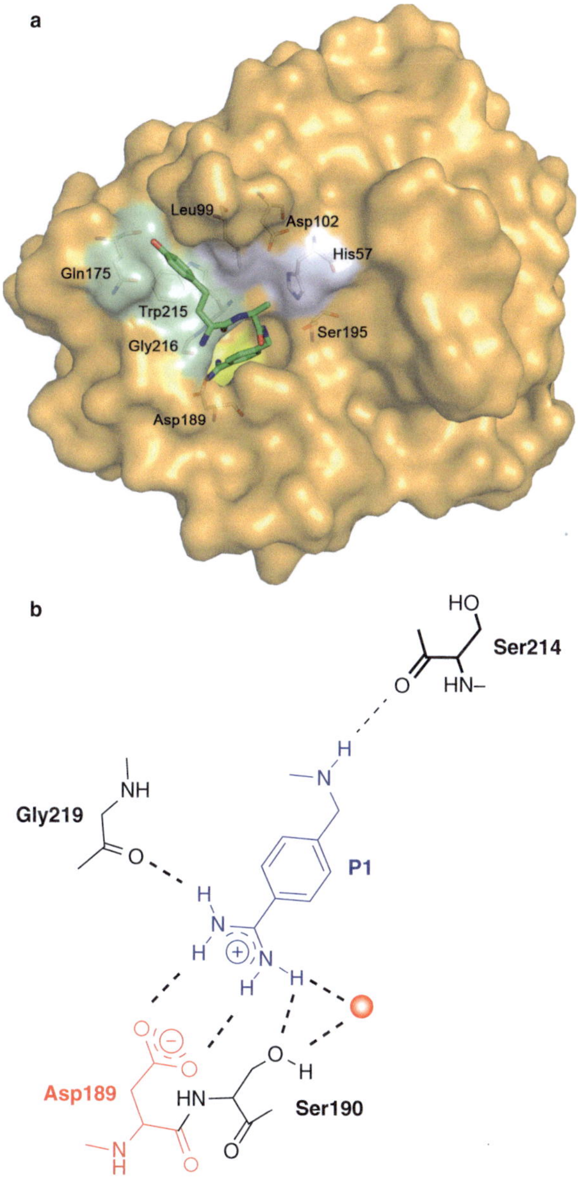

Fig. 11.5 Crystal structure of the substrate analogue inhibitor H-D-homoTyr-Ala-4-Amba (**10**) in complex with trypsin (PDB, 4MTB) (Maiwald et al. 2016). (**a**) Overall structure of the complex indicating the characteristic binding pockets similarly found in all trypsin-like serine proteases (S1 pocket with Asp189 at the bottom in yellow, S2 pocket below His57 and Leu99 in light blue, distal S3/4 pocket above Trp215 in light green). The trypsin residues of the catalytic triad (Ser195, His57, and Asp102) and some additional residues within the active site are labeled; water molecules are omitted. (**b**) Polar interactions of the 4-Amba residue in the S1 pocket of trypsin, a conserved water molecule found in nearly all crystal structures with arginine or benzamidines in P1 position, is shown as red ball. Most family members contain a Ser190 as shown for trypsin or an Ala190 (e.g., in HAT, thrombin, factor Xa) that makes an H-bond less

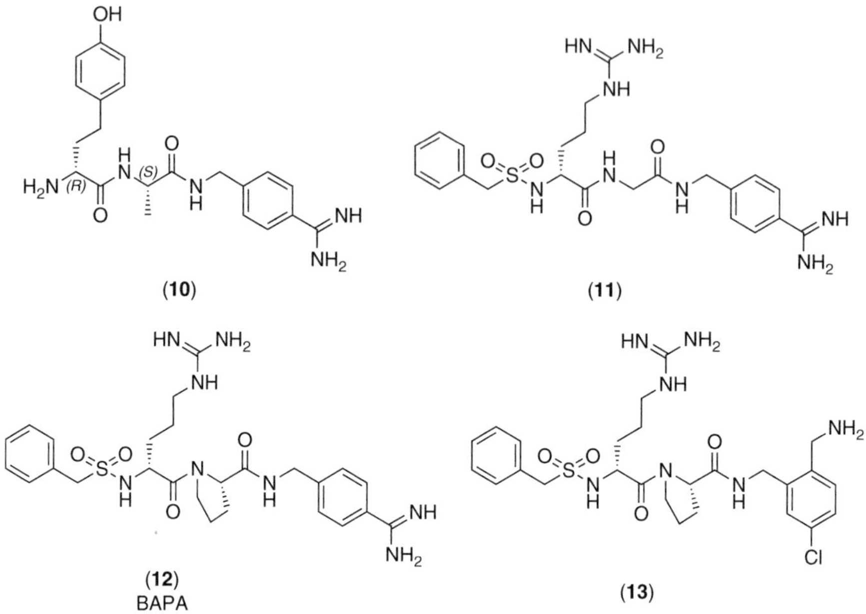

Fig. 11.6 Structures of substrate analogue inhibitors against trypsin-like serine proteases containing decarboxylated arginine mimetics like 4-amidinobenzylamide or 2-aminomethyl-5-chlorobenzylamide in P1 position (Maiwald et al. 2016; Schweinitz et al. 2006)

The conversion of the P1-amidine (pK_a = 11.5) into a hydroxyamidine (pK_a = 5.2) reduces the strongly basic character of the 4-Amba group and improves the bioavailability of the inhibitors. An enzyme system found in the liver and other organs (kidney, lung, brain, and intestine) consisting of cytochrome b5, a NADH-dependent cytochrome b5 reductase, and a P450 enzyme is involved in the reduction of the N-hydroxylated compounds (Clement 2002).

Due to a similar architecture of the S1 pocket in most trypsin-like serine proteases, it is challenging to achieve a pronounced selectivity for individual proteases within this inhibitor scaffold, although compound **10** is a relatively selective matriptase inhibitor (K_i = 26 nM) and shows reduced affinity against the clotting proteases thrombin (K_i = 300 nM) and factor Xa (K_i = 570 nM) (Maiwald et al. 2016). The potency of this inhibitor type can be further enhanced by elongation with a suitable P4 residue, preferably a sulfonyl group, whereas one of its oxygen atoms interacts with the amide NH of Gly219. For inhibitors containing a D-configured P3 residue, the N-terminal coupling of a benzylsulfonyl group induces a turn-like inhibitor backbone conformation when bound to the target, in which the N-terminal benzyl group binds in a shallow subpocket adjacent to the S1 site and comes in close van der Waals contact to the P1 phenyl ring (Fig. 11.7) above the Cys191-Cys220 disulfide bridge. Inhibitor **11** is a somewhat selective fXa inhibitor (K_i = 3.5 nM (Schweinitz et al. 2006)) due to the presence of glycine in P2 position but inhibits also trypsin and matriptase with inhibition constants of 10 nM and 130 nM, respectively.

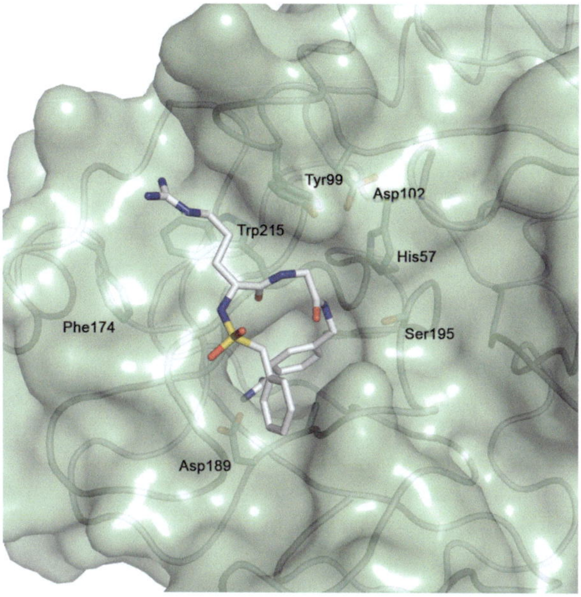

Fig. 11.7 Crystal structure of the substrate analogue inhibitor Bzls-D-Arg-Gly-4-Amba (**11**) in complex with a trypsin variant containing Leu99Tyr and Gly174Phe mutations (PDB, 3PMJ) (Tziridis et al. 2014). Only few side chains of characteristic residues within the active site of this trypsin mutant are shown and labeled. The inhibitor adopts a turn-like backbone conformation; the P3 D-Arg side chain binds into the distal S3/4 pocket above Trp215 and is probably involved in cation-π interactions to the aromatic residues Tyr99, Phe174, and Trp215

The replacement of glycine by proline in P2 position leads to the relatively nonselective inhibitor BAPA (**12**), which possesses one- or two-digit nanomolar potencies against the clotting proteases thrombin, fXa, plasma kallikrein (PK), fibrinolytic plasmin, and the TTSPs HAT, TMPRSS2, and matriptase (Hellstern et al. 2007; Maiwald et al. 2016; Sielaff et al. 2011a). A significant antiviral effect was found in BAPA-treated cell cultures infected with numerous monobasic IAV strains (Böttcher-Friebertshäuser et al. 2010, 2012; Böttcher et al. 2009; Sielaff et al. 2011a). Many analogues of this inhibitor scaffold with different P2 and/or P3 residues have been tested but were less effective than BAPA (data not published). It seems that compounds with a broader target spectrum exhibit an advantageous antiviral profile. This tendency confirms the promising results obtained for the relatively nonselective inhibitor aprotinin. Otherwise, less selective inhibitors that also target numerous clotting proteases might suffer from a narrow safety profile due to potential bleeding complications after i.v. treatment. Such side effects might be reduced, when the compounds will be inhaled, as described for aprotinin. Notably, BAPA (**12**) was well tolerated in a bleomycin-induced fibrosis model in mice when it was applied as aerosol via a microsprayer (data not published).

In principle, it should be possible to replace the C-terminal 4-Amba residue in these substrate analogue inhibitors. Numerous alternative P1 groups are known from the field of thrombin and fXa inhibitors (Straub et al. 2011). Besides basic groups, many chloro-substituted aromatic P1 residues as used, for example, in inhibitor **13** have been developed (Sisay et al. 2010). The chlorine addresses a highly conserved Tyr228 side chain found at the back of the S1 pocket, whereas the aminomethyl group comes out of the S1 pocket and does not bind to Asp189. It provides an enhanced inhibitory potency against thrombin and fXa but leads to a reduced inhibition of HAT, matriptase, plasmin, and of several other trypsin-like serine proteases. However, only weak anti-influenza activity was found for compound **13** suggesting that the relevant HA-cleaving host proteases are not susceptible to such chloro-substituted aromatic P1 structures.

Covalent Substrate Analogue Inhibitors

Another strategy is the design of covalent inhibitors (Singh et al. 2011). Covalent bonds can be irreversibly formed, e.g., when serine proteases are inactivated by chloromethyl ketones (CMK) (Powers et al. 2002). Although such CMKs are useful tools for initial biochemical experiments and structure analysis, they are not suitable for drug development due to stability and selectivity problems. A more eligible approach is the incorporation of electrophilic warheads, which also form covalent but reversible bonds to their targets. This can be achieved by coupling of substrate-analogue P3-P2 or P4-P2 segments with P1 boroarginine and arginal groups or with P1-P1′-arginyl ketones. These electrophilic groups can be attacked by the active-site serine providing covalent boronate ester-, hemiacetal-, or hemiketal-like complexes. The formed covalent bond significantly improves the affinity of the inhibitors. In the 1990s, a few arginal and arginyl-ketone-derived thrombin inhibitors reached preclinical and early clinical development (Steinmetzer et al. 2001), which was stopped due to the discovery of more suited non-covalent inhibitors, like melagatran containing the 4-Amba residue. The arginal derivative **14** acylated with the N-terminal Abz-fluorophore (Fig. 11.8) was described as an inhibitor of HAT ($K_i = 54$ nM) (Wysocka et al. 2010). CVS-3983 (**15**), a second aldehyde derivative containing an unusual dialkylated P3 residue, inhibits matriptase with an inhibition constant of 3.3 nM and reduced prostate cancer growth in a xenograft mice model (Galkin et al. 2004). However, none of these aldehydes was tested for antiviral potential. The reuse of reversible arginyl ketone inhibitors against TTSPs has started in 2012 with compound **16** (IN-1) (Fig. 11.8). Although it is a much stronger matriptase inhibitor ($K_i = 11$ pM), it potently inhibits additional trypsin-like serine proteases like matriptase-2, hepsin, HAT, and trypsin with K_i values <10 nM (Colombo et al. 2012). Compound **16** reduces breast cancer progression in mice (Zoratti et al. 2015) and inhibits H1N1 IAV propagation in human Calu-3 cells with an EC_{50} of 5.6 μM (Beaulieu et al. 2013). It was postulated that the antiviral efficacy is due to matriptase inhibition. However, since the inhibitor concentrations used were well above the K_i values against related proteases, it is conceivable that the inhibition of additional targets contributes to the observed antiviral effects. Numerous analogues with modified P4-P2 peptide segments like compound **17** (K_i value for matriptase 0.92 nM) targeting also matriptase-2, hepsin, and hepatocyte growth factor activator (HGFA) have been prepared for tumor treatment or regulation of iron overload but were not further tested with influenza viruses, so far (Duchene et al. 2014; Han et al. 2014, 2016).

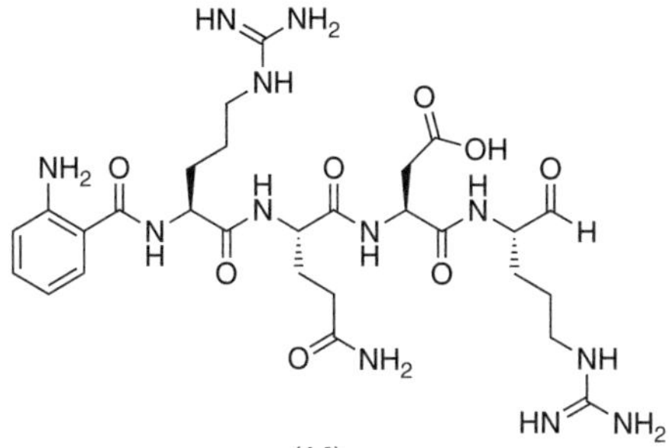

(14)

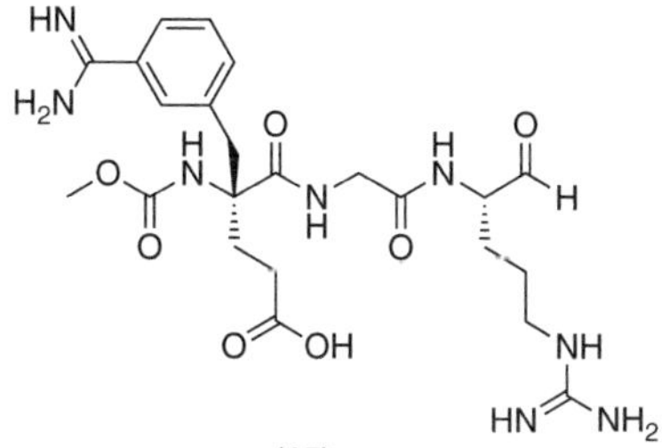

(15)
CVS-3983

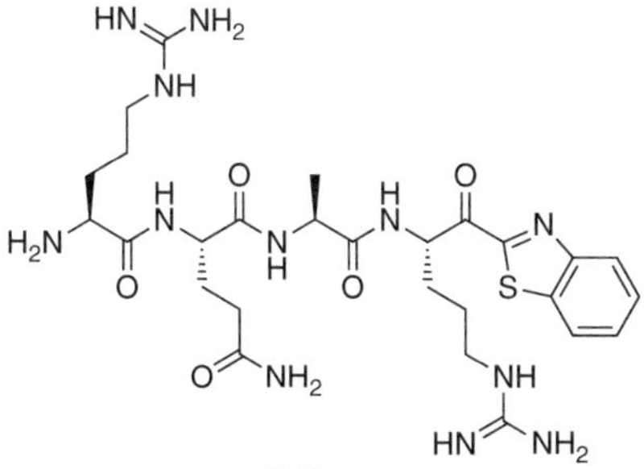

(16)
IN-1

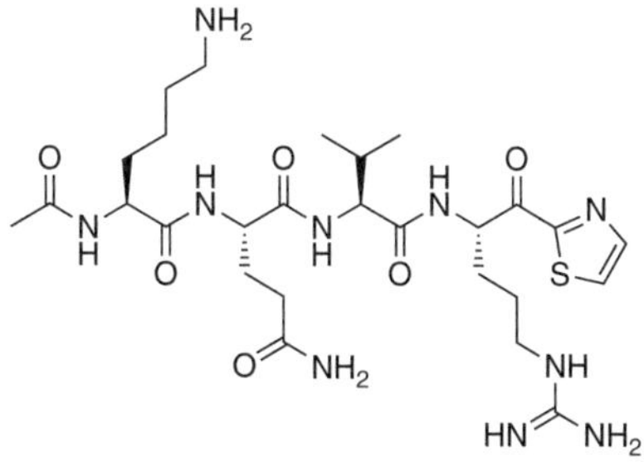

(17)

Fig. 11.8 Structures of peptidic arginal-, arginyl-ketobenzothiazole, and ketothiazole inhibitors (Colombo et al. 2012; Galkin et al. 2004; Han et al. 2016; Wysocka et al. 2010)

Sulfonylated 3-Amidinophenylalanine Derivatives

Suitable scaffolds for non-covalent inhibitors of trypsin-like serine proteases are the tertiary amides of sulfonylated 3-amidinophenylalanines, which were originally developed for the inhibition of trypsin, thrombin, or uPA and further used for the design of matriptase inhibitors (Steinmetzer et al. 2006; Stürzebecher et al. 1997). With some exception, this class of compounds provides relatively broad-spectrum inhibitors targeting numerous trypsin-like serine proteases. The uPA inhibitor **18** (Stürzebecher et al. 1999) (WX-UK1, Fig. 11.9) in combination with the 5-fluorouracil prodrug capecitabine reached clinical development for tumor therapy (Setyono-Han et al. 2005). Moreover, an orally available formulation in form of its hydroxyamidino prodrug mesupron (**19**) has also reached clinical phase II studies (indication pancreas and breast cancer). This suggests that this 3-amidinophenylalanine inhibitor type is suitable for drug development. Otherwise, compound **18** is only a moderate inhibitor of numerous trypsin-like serine proteases including matriptase with K_i values close to 0.5 μM. Its sterically demanding and rigid triisopropylphenylsulfonyl (Tips) group cannot fully occupy the distal binding pocket above Trp215 in the usual way (Setyono-Han et al. 2005; Stürzebecher et al. 1999). A significantly improved and selective matriptase inhibition (K_i = 3.8 nM) was achieved with inhibitor **20**. Its N-terminal β-alanyl-amide targets the distal S3/4 pocket and is most likely involved in cation-π interactions to Trp215 and Phe99 of matriptase (Steinmetzer et al. 2006). However, the tribasic and strongly polar

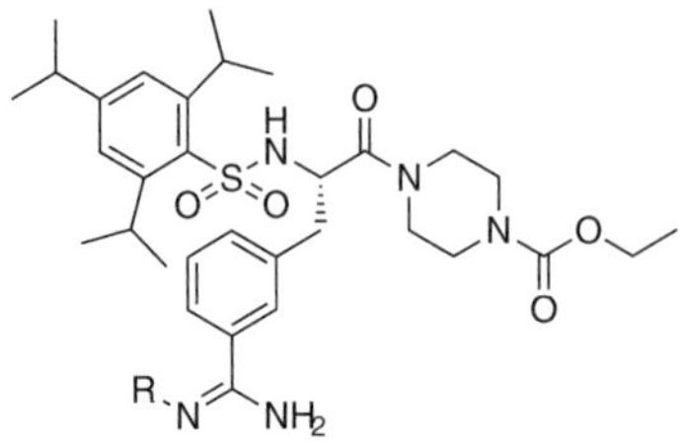

(**18**) R=H (WX-UK1)
(**19**) R=OH (mesupron)

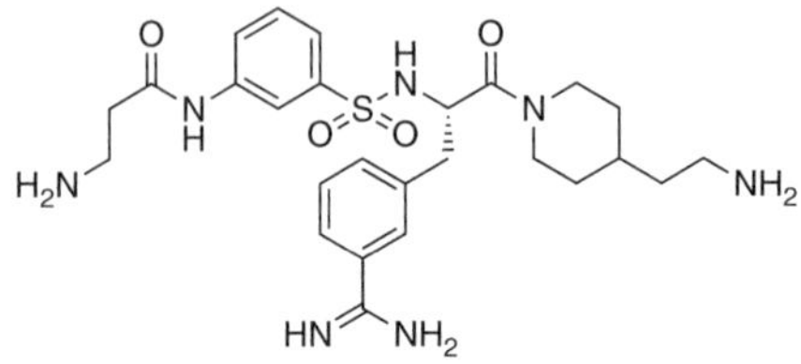

(**20**)

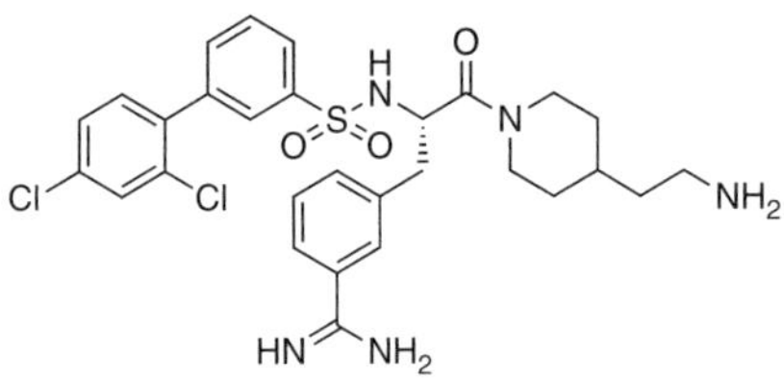

(**21**)
MI-432

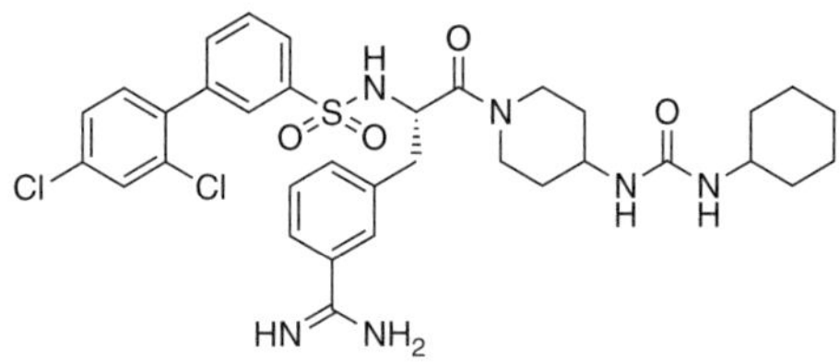

(**22**)
MI-462

Fig. 11.9 Structures of tertiary amides of sulfonylated 3-amidinophenylalanines (Hammami et al. 2012; Steinmetzer et al. 2006; Stürzebecher et al. 1999). The hydroxyamidino prodrug strategy of mesupron (**19**) could also be adapted for other benzamidine inhibitors

character limits the membrane permeability and bioavailability of this compound. Therefore, the number of basic groups was stepwise reduced and provided compound **21** (MI-432) containing an N-terminal substituted biphenyl-3-sulfonyl group. This analogue inhibits matriptase and TMPRSS2 with similar inhibition constants of 2 nM and 0.9 nM, respectively (Hammami et al. 2012; Meyer et al. 2013). It is also a relatively potent thrombin inhibitor (K_i = 20 nM) but poor inhibitor of HAT (K_i = 1.7 μM). Replacement of its C-terminal amino group by an urea portion provided the monobasic analogue **22** (MI-462) (Hammami et al. 2012) (K_i = 5.1 nM for matriptase and 630 nM for thrombin). Compounds **20–22** (Fig. 11.9) reduced the propagation of various LPAIV strains with monobasic HA cleavage sites in infected cells (Baron et al. 2013; Meyer et al. 2013). For instance, treatment with 50 μM of compound **21** inhibited the multiple cycle replication of IAV H3N2 and H1N1 strains in Calu-3 cells (Meyer et al. 2013), in which TMPRSS2 was identified as the major HA-cleaving protease (Böttcher-Friebertshäuser et al. 2011).

So far, no crystal structure of TMPRSS2 is available in the protein data bank. Therefore, we have generated a homology model of TMPRSS2 in complex with inhibitor **21** using SWISS-MODEL (http://swissmodel.expasy.org/) (Biasini et al. 2014). The model was superimposed with the crystal structure of inhibitor **21** in complex with thrombin (PDB, 4E7R (Hammami et al. 2012)) and subsequent deletion of the thrombin structure. The inhibitor was energy minimized in the active site of TMPRSS2 using the software Molecular Operating Environment (MOE) (2016). The model shows the expected polar contacts known from numerous crystal structures of sulfonylated 3-amidinophenylalanines in complex with trypsin (Renatus et al. 1998), uPA (Zeslawska et al. 2000), and matriptase (Steinmetzer et al. 2006)

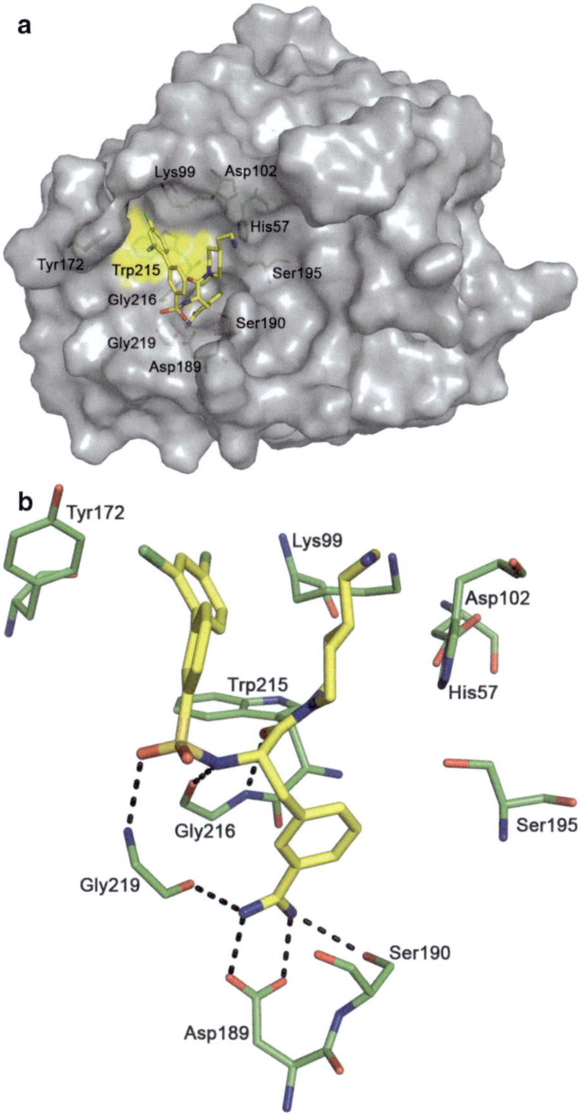

Fig. 11.10 Model of TMPRSS2 in complex with inhibitor **21**. (**a**) Structure of inhibitor **21** (stick model with yellow carbon atoms, nitrogen in blue, oxygen in red, chlorine in green) bound into the active site of TMPRSS2, shown with its surface in gray. The 2,4-dichloro-biphenyl-3-sulfonyl residue is directed toward the distal S3/4 pocket above Trp215 indicated by a yellow surface. (**b**) Polar interactions of inhibitor **21** in the active site of TMPRSS2 are shown as dashed lines in black; important TMPRSS2 residues are shown with carbon atoms in green

(Fig. 11.10). The benzamidine binds to Asp189, Ser190, and Gly219 at the bottom of the S1 pocket, whereas the backbone amide NH and carbonyl oxygen of the P1 3-amidinophenylalanine make a short antiparallel β-sheet-like interaction with Gly216, and one sulfonyl oxygen binds to the NH of Gly219. It should be noted that in complexes with the substrate analogue inhibitors **10–17**, this Gly216 is addressed by the backbone NH and carbonyl of the P3 residue, which indicates a completely different binding mode of this inhibitor type.

Nonpeptidic Small-Molecule Inhibitors

Drug development is strongly focused on the design of nonpeptidic, small-molecule inhibitors. Usually, they are more stable than peptidic inhibitors, possess longer half-lives, and offer better chances to be orally bioavailable. Hundreds of highly potent nonpeptidic inhibitors targeting the clotting proteases are available, whereas only few TTSP inhibitors of this type have been described, so far. Most of the work was done in the field of matriptase inhibitors (Fig. 11.11). The bis-benzamidine **23** has a K_i value of 208 nM (Enyedy et al. 2001); and an enhanced potency ($K_i = 40$ nM) was described for analogue **24** (Goswami et al. 2013). The tribasic compounds **25** and **26** show improved matriptase inhibition with K_i values of 10 nM and 3 nM, respectively (Goswami et al. 2014). A less rigid core segment was used for the design of the tribasic inhibitor **27**, which inhibits matriptase and matriptase-2 with K_i values of 38 nM and 3.6 μM (Furtmann et al. 2016). A group from Korea reported the discovery of the 2-hydroxydiarylamide derivatives **28** and **29** as TMPRSS4 inhibitors with IC_{50} values of 12 μM and 6 μM (Kang et al. 2013), respectively. Surprisingly, also the mucolytic cough suppressant bromhexine **30** was described as TMPRSS2 inhibitor ($IC_{50} = 0.75$ μM).

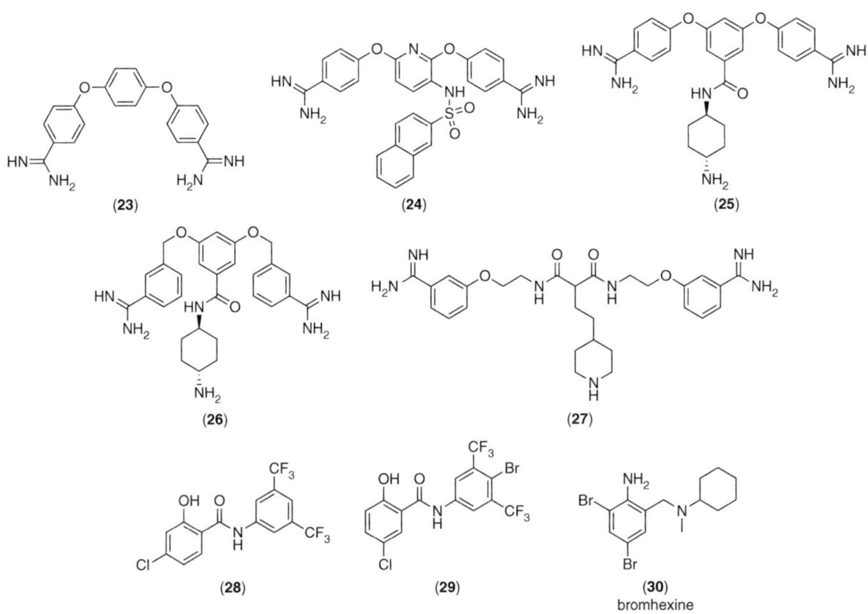

Fig. 11.11 Structures of nonpeptidic inhibitors of the TTSPs matriptase (**23-27**) (Enyedy et al. 2001; Furtmann et al. 2016; Goswami et al. 2013, 2014), TMPRSS4 (**28**, **29**) (Kang et al. 2013), and TMPRSS2 (**30**) (Lucas et al. 2014)

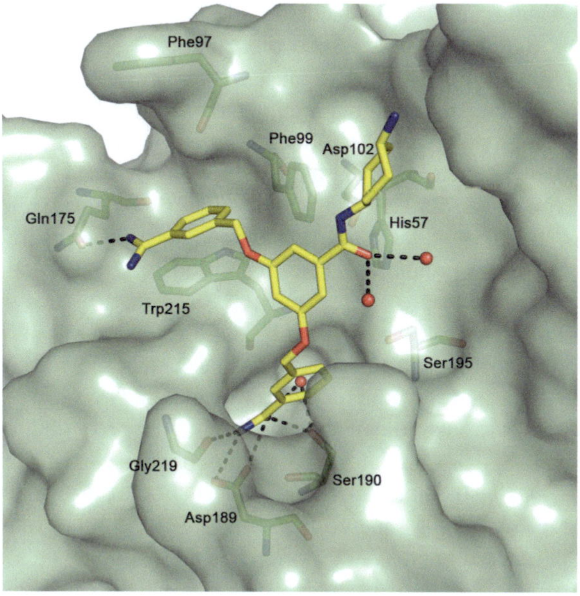

Fig. 11.12 Crystal structure of inhibitor **26** shown as sticks with carbon atoms in yellow (oxygen in red, nitrogen in blue) in complex with matriptase, presented with a transparent surface in green (PDB, 4O97) (Goswami et al. 2014). Important matriptase residues within the active site are shown with carbon atoms in green and are labeled. Water molecules in direct contact to the inhibitor are provided as red spheres, polar contacts as dashed lines in black

The binding mode of a few bis-benzamidine and several tribasic analogues in complex with matriptase was determined by crystal structure analysis. The complex with inhibitor **26** is shown in Fig. 11.12. One benzamidine occupies the S1 site, and the second targets the distal S3/4 pocket above Trp215, whereas the aminocyclohexyl group is directed toward the 60-loop of matriptase but not involved in direct polar contacts. Due to the lack of experimental structures of TMPRSS2 and TMPRSS4, the binding mode of compounds **28–30** cannot be trustfully predicted. Most of these nonpeptidic inhibitors were developed to address the TTSPs as targets for tumor treatment (Goswami et al. 2015; Kang et al. 2013; Lucas et al. 2014), although they might be also suitable for other indications.

Besides influenza viruses, other viral pathogens are also suitable targets for inhibitors of trypsin-like serine proteases (TLSPs). A few examples will be presented in the following paragraph.

11.2.1.2 TLSP Inhibitors for Treatment of Paramyxo- and Coronavirus Infections

Paramyxoviridae contain two surface glycoproteins, the receptor-binding protein (HN, H, G) and the fusion protein F. In contrast to the *Orthomyxoviridae*, where fusion of the viral membrane with host cell membrane occurs in the endosome, the activated F protein of many paramyxoviruses induces fusion of the virus envelope and the plasma membrane. The F_0 precursor of human parainfluenza virus 1 (HPIV-1) is cleaved at a monobasic sequence typical for TLSPs (DNPQTR↓FFGAV) (Diederich and Maisner 2007). HPIV-1 causes respiratory infections, such as croup, especially among young children. In principle, some of the above-described inhibitors could also be suitable for the treatment of HPIV infections. With other paramyxoviruses, such as measles virus and mumps virus, F_0 is activated by furin, whereas F_0 of Nipah virus is cleaved by cathepsin L. Consequently, furin or cathepsin L inhibitors should be suitable for the treatment of these virus infections.

The spike (S) surface protein of respiratory coronaviruses (CoV) is also synthesized as inactive precursor protein that has to be activated by host proteases. The N-terminal S1 unit of the cleaved protein binds to receptors on host cells, and the C-terminal S2 unit enables the fusion of the viral membrane with host cell membranes. Both functions are essential for CoV propagation. Original studies suggested that the endosomal cysteine protease cathepsin L is solely required for spike activation and subsequent SARS-CoV infectivity (Simmons et al. 2005). Notably, a cathepsin dependency was also found for the enveloped Ebola virus of the *Filoviridae* family (Chandran et al. 2005) and Nipah virus (Diederich and Maisner 2007), as described above. Later, numerous groups have found that in the presence of cathepsin L inhibitors, the S-protein of SARS-CoV and MERS-CoV is activated by TMPRSS2 (Gierer et al. 2013; Matsuyama et al. 2010; Shulla et al. 2011; Zmora et al. 2014). Consequently, inhibition of SARS-CoV growth in Calu-3 airway epithelial cells was achieved by a combination treatment with the broad-spectrum serine protease inhibitor camostat (**8**) (Fig. 11.4) and the cathepsin inhibitor (23,25) trans-epoxysuccinyl-L-leucylamido-3-methyl-butane ethyl ester (EST) (Kawase et al. 2012). Based on these observations, it is suggested that SARS-CoV enters the host cells via two distinct pathways, one using TTSPs like TMPRSS2 and a second using the endosomal cathepsins L and/or B for spike activation. Interestingly, a recent study with a fresh clinical isolate of the human CoV 229E revealed a clear preference for host cell entry via TMPRSS2, whereas after 20 passages in HeLa cells the cathepsin L pathway became more important (Shirato et al. 2017). However, the cell culture virus showed a reduced ability for replication suggesting that the endosomal pathway is disadvantageous for HCoV-229E infection in humans. Based on these results, the authors suggested to target TMPRSS2 rather than endosomal cathepsins in CoV infections (Shirato et al. 2017). A similar tendency was previously found in an animal model of SARS-CoV infection, where viral spread and pathogenesis were only prevented by the TLSP inhibitor camostat and not by broad-spectrum vinylsulfone-type cysteine protease inhibitors targeting cathepsins L and B (Zhou et al. 2015). However, the authors argue that their new vinylsulfone inhibitors might be excellent lead structures for the development of inhibitors of Ebola virus entry.

11.2.2 Proprotein Convertases

The family of proprotein convertases comprises nine calcium-dependent serine endoproteases (furin, PC1, PC2, PC4, PC5, PACE4, PC7, SKI-1/S1P, and PCSK9). These enzymes play an important role in the maintenance of cell homeostasis by activation or deactivation of proteins including prohormones as well as proforms of transcription factors, membrane receptors, and extracellular membrane proteins (Seidah and Prat 2012; Thomas 2002). Besides their physiological role, PCs are also involved in the activation of viral proteins and bacterial toxins. For example, several enveloped viruses depend on the cleavage of their surface glycoproteins by PCs to gain fusion capacity, which is required for virus propagation (Pasquato et al. 2013; Thomas 2002). All PCs possess a multidomain structure comprising a signal peptide, an N-terminal prodomain, followed by a catalytic domain, where the catalytic triad Ser-His-Asp is located, and a P-domain (Fig. 11.13) (Henrich et al. 2005).

Four PCs (furin, PC5B, PC7, and SKI-1/S1P) possess a transmembrane and C-terminal cytoplasmatic domain, which anchors them to cellular membranes. Furthermore, furin, PC5B, and SKI-1/S1P can be shed and released in a soluble

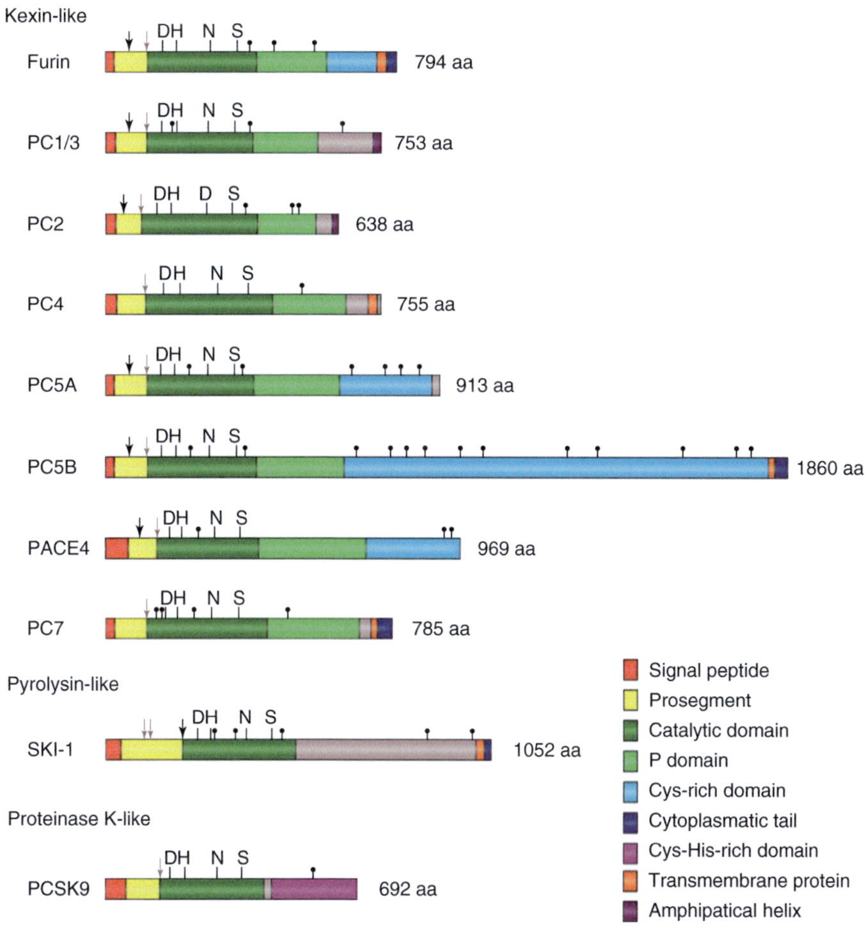

Fig. 11.13 Schematic structure of the human PCs. Residues belonging to the catalytic triad (S, H, and D) and oxyanion hole (N or in case of PC2 D) are indicated. N-glycosylation (⫯) and prodomain processing sites are highlighted with an arrow (primary cleavage site with a gray and secondary with a black arrow) (Seidah and Prat 2012)

form into the extracellular space. PC1 and PC2 are maintained in dense-core granules, whereas the remaining PC4, PC5A, PACE4, and PCSK9 are secreted. Furin and the related six PCs PC1, PC2, PC4, PC5, PACE4, and PC7 recognize multibasic cleavage sequences and, therefore, are also known as basic PCs or furin-like PCs. The catalytic domains of these enzymes show more than 50% identity (Thomas 2002). Due to the similar cleavage sites, a redundant behavior of these enzymes was found in overexpression experiments as well as in vitro studies. Despite overlapping consensus sequences and a high sequence homology, minor modifications in the recognition sequence as well as in their cellular localization lead to specific cleavages by the different PCs. In contrast, SKI-1/S1P cleaves after the consensus cleavage site (K/R)-X-(V/L/I)-Z↓, where Z is any amino acid except Val, Pro, Cys, Glu, or Asp and the spacer X is preferably a basic residue (Seidah 2013). After an autocatalytic cleavage at VFAQ↓S, PCSK9 forms a proteolytically inactive complex with its prosegment. So far, no other PCSK9 substrates are known.

Knockout of PCs in mice reveals their physiological significance especially during embryogenesis and contributes to the identification of specific PC substrates (Creemers and Khatib 2008; Seidah and Prat 2012; Taylor et al. 2003). Due to severe malformations, knockout of furin or SKI-1/S1P in mice leads to embryonic death, whereas PC5-deficient mice die at birth. For PACE4, a lethality of 25% was observed in knockout mice at embryonic day 14. In contrast, PC1- and PC2-deficient mice are viable but have several neuroendocrine peptide processing defects, and PC7 knockout mice show a loss of anxiety. The knockout of PC4 leads to infertility especially in male mice, whereas PCSK9 deficiency leads to lower plasma cholesterol levels.

The maturation of the PCs requires an autoproteolytic cleavage (Seidah and Prat 2012; Thomas 2002). In case of furin and furin-like PCs, two cleavages are needed to gain full enzymatic activity. After the removal of the signal peptide in the endoplasmic reticulum, a first cleavage leads to a conformational change of the enzyme, which is then the latent form. Enzymatic activity is obtained after a second cleavage in the prosegment, which leads to a release of the prosegment. The cleavages depend on the pH in the respective organelles. With exception of PC2 and SKI-1/S1P, the prosegments act as inhibitors of their respective enzyme. In case of SKI-1/S1P, three cleavages are required for full enzymatic activity. The subcellular localization of the PCs differs. Furin, PC5B, and PC7 have sorting signals in their cytosolic tails, which mediate recycling between the TGN and the cell surface. SKI-1/S1P shows also a broad distribution and is found in the ER, Golgi apparatus, endosomes, and lysosomes. In contrast, PC1 and PC2 are primarily found in dense-core vesicles of the secretory pathway, and PC4 is localized only in the plasma membrane of male and female germ cells. Furin, PC7, and SKI-1/S1P are ubiquitously distributed, and PC5 and PACE4 are widely distributed. In contrast, the expression of PC1 and PC2 is limited to neural and endocrine cells. PCSK9 can be found predominantly in the liver, intestine, and kidney.

The reported inhibitors against PCs can be categorized into various groups including macromolecular compounds, pure peptides, peptidomimetics, as well as nonpeptidic compounds, which will be described in the following sections.

11.2.2.1 Inhibitors of Basic Proprotein Convertases

Protein-Based Inhibitors

A common approach for protease inhibitor development is the optimization of natural inhibitors by mutation of their inhibitory recognition loops. The α1-antitrypsin Portland (α1-PDX) is a bioengineered serpin-type inhibitor containing the furin-adapted sequence R355-I-P-R358 instead of A355-I-P-M358 in its inhibitory loop. α1-PDX inhibits furin with a K_i value of 0.6 nM in a slow tight-binding manner and is supposed to act as suicide substrate, yielding an inactive enzyme (Jean et al. 1998). It also inhibits PC1 (K_i = 260 nM) and PC5 (K_i = 2.3 nM) but has reduced potency against PC2, PC7, and PACE4 (K_i >1000 nM). Its expression in cells blocked the processing of HIV gp160 as well as measles virus fusion protein and, thus, inhibited virus spread (Anderson et al. 1993; Watanabe et al. 1995). Based on the reactive loop of α1-PDX, numerous mini-PDX peptides have been prepared. These acyclic- or disulfide-bridged cyclic 30-mers inhibit furin with IC_{50} values of 731 nM and 569 nM, respectively (Basak and Lotfipour 2005).

Turkey ovomucoid third domain (OMTKY3) belongs to the family of Kazal-type inhibitors and normally inactivates serine proteases of the S1A fold that prefer a neutral P1 residue. Exchange of A15-C-T-L18 to R15-C-L-R18 in its reactive site loop leads to a moderate furin inhibitor with an association constant K_a of 1.1×10^7 M^{-1} (Lu et al. 1993), which roughly corresponds to a reciprocal dissociation equilibrium constant of ~90 nM.

Like OMTKY3, inter-alpha-inhibitor protein (IαIp) is known to be a potent serine protease inhibitor, e.g., against trypsin, chymotrypsin, or acrosin. IαIp was first isolated from human plasma and is a multicomponent complex, consisting of two heavy and one light chain, called bikunin, which are linked via a chondroitin linker. Bikunin possesses two protease domains of the Kunitz-type that are likely to inhibit furin, because treatment with IαIp provided a significant protection against anthrax toxin in cell culture studies and in mice (Opal et al. 2005). Eglin C was originally isolated from the leech *Hirudo medicinalis* and belongs to the potato I inhibitor family. It inhibits several serine proteases, e.g., subtilisin, human leukocyte elastase, or cathepsin G. Insertion of a multibasic recognition site by mutation of P42-V-T-L45 to R42-V-K-R45 resulted in a strong furin inhibitor with a K_i value of 1.6 nM (Komiyama and Fuller 2000; Liu et al. 2004).

Additional furin inhibitors were designed by mutation of the homotetrameric glycoprotein α_2-macroglobulin (α_2-M), which is found in high concentrations in human blood. It is a potent broad-spectrum protease inhibitor with a unique inhibition mechanism. After protease mediated cleavage in the so-called bait region, the internal S-esters hydrolyze and trigger a conformational change of α_2-M. The protease is enclosed by α_2-M and sterically shielded from its substrates (Barrett and Starkey 1973). Replacement of its original G683-F-Y-E-S-D688 sequence by R683-S-K-R-S-L688 yielded a potent furin inhibitor, which blocked the processing of von Willebrand factor, TGF-β1, and HIV-1 gp160 (Van Rompaey et al. 1997).

The 45-kDa proteinase inhibitor 8 (PI8) was the first reported furin inhibitor, which is not a serpin reactive-site mutant. PI8 belongs to the group of ovalbumin-type

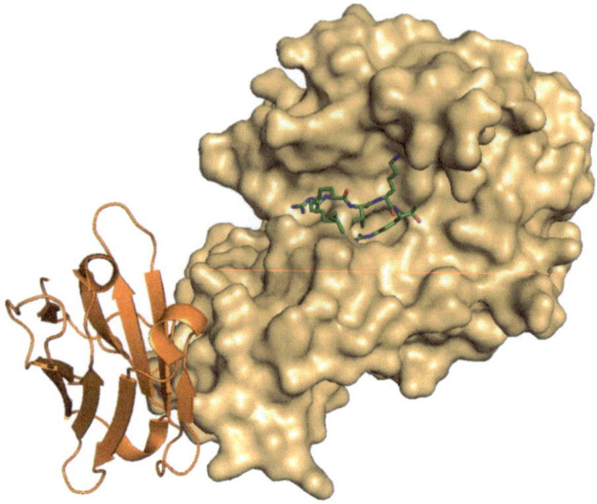

Fig. 11.14 Structure of human furin (shown with its surface in beige) in complex with the irreversible active-site inhibitor Dec-Arg-Val-Lys-Arg-CMK (stick model with carbons in green, nitrogen in blue, and oxygen in red) and with the furin inhibitory nanobody Nb14, shown in cartoon style in orange (PDB, 5JMO) (Dahms et al. 2016b). Although the nanobody binds far away from the active site, it blocks the binding of larger proteinaceous substrates

serpins, containing two furin recognition sequences within its R336-N-S-R-C-S-R342 segment. The inhibition of furin by PI8 consists of two steps, starting with the rapid formation of a loose complex and followed by the slow isomerization to a stable complex. The overall K_i value for recombinant and soluble furin in vitro is 53.8 pM (Dahlen et al. 1998).

Recently, several specific nanobodies against furin have been described (Zhu et al. 2012). These nanobodies are dromedary-derived single-domain antigen-binding fragments. A crystal structure of one of these nanobodies in complex with furin and an additional active-site inhibitor reveals that they do not directly bind to the active site (Fig. 11.14) (Dahms et al. 2016b). Nevertheless, some of these nanobodies inhibit furin-mediated cleavage of diphtheria toxin by a noncompetitive mechanism (K_i values of approximately 25 μM) and the activation of anthrax toxin (Zhu et al. 2012). Obviously, they block the cleavage of larger proteinaceous substrates by steric hindrance.

Peptide-Based Inhibitors

The second group of inhibitors comprises various peptides and peptidomimetics. Due to their relatively high molecular weight, negligible oral bioavailability, and stability problems, these compounds are not the first choice for drug development. Nevertheless, numerous injectable peptides are used in therapeutic applications (Vlieghe et al. 2010). The first potent PC inhibitors were the irreversible substrate analogue chloromethyl ketones (CMK), which covalently bind to the enzyme (Garten et al. 1989; Hallenberger et al. 1992). The most widely used basic PC

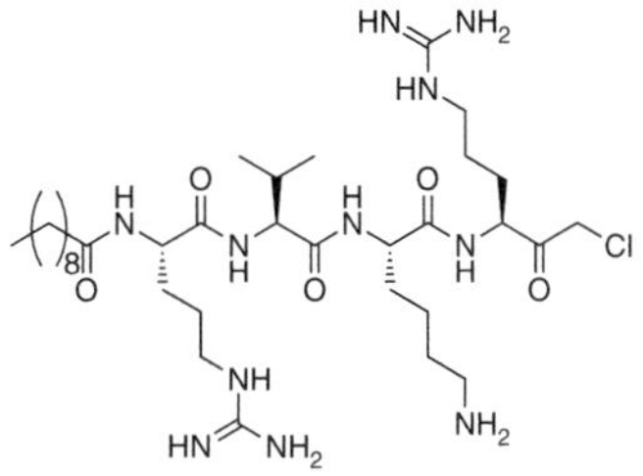

(31)

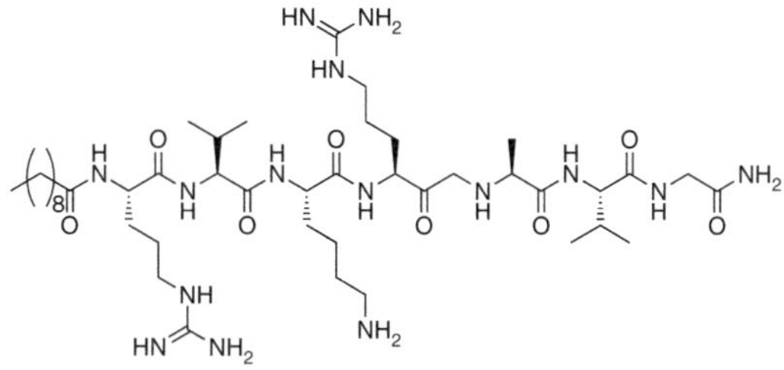

(32)
K_i (furin) 3.4 nM

Fig. 11.15 Structures of substrate analogue arginyl-ketone-derived furin inhibitors. The CMK derivative **31** is a covalent irreversible inhibitor, whereas the ketomethylene compound **32** binds by a covalent reversible mechanism (Angliker 1995; Garten et al. 1994)

inhibitor is the commercially available derivative Dec-Arg-Val-Lys-Arg-CMK (**31**, Fig. 11.15), which targets all of the known furin-like PCs in the nanomolar range (Garten et al. 1994; Jean et al. 1998). In cell culture, it blocks the furin-catalyzed cleavage of HIV gp160 into gp120 and gp41. Furthermore, the first crystal structure of furin was obtained when it was complexed with this inhibitor (Henrich et al. 2003). However, CMKs are not suited for further development. In vivo studies with the thrombin inhibitor D-Phe-Pro-Arg-CMK revealed a very short half-life <5 min (Hanson and Harker 1988). The same P4-P1 segment of the CMK **31** served for the design of analogues containing pseudo peptide bonds, like the ketomethylene inhibitor **32** (K_i value of 3.4 nM against furin, Fig. 11.15) (Angliker 1995).

So far, only few endogenous PC inhibitors are known. A prominent role plays autoinhibition by prodomains (Zhong et al. 1999). Furin-like PCs are synthesized as zymogen and are activated by autocatalytic cleavage within the prodomain. The prosegment acts as an intramolecular chaperon, needed for the correct folding, regulation of enzymatic activity, and transport within the secretory pathway. A moderate K_i value of 156 nM was determined for the complete 83-mer prodomain of furin. Furthermore, this inhibitor reduced the proliferation, migration, and invasion of cancer cells (Basak et al. 2010). Several truncated derivatives have been synthesized. The most potent one, the 24-mer DYYHFWHRGVTKRSLSPHRPRHSR, inhibits furin with an inhibition constant of 0.9 μM. Moreover, some peptides derived from the prodomain of PC1 inhibit furin in the same range (Basak and Lazure 2003).

A combinatorial peptide library containing approximately 52 million hexapeptides was scanned to identify PC1 and PC2 inhibitors. For instance, Ac-Leu-Leu-Arg-Val-Lys-Arg-NH_2 inhibits PC1 and PC2 with K_i values of 3.2 and 360 nM, respectively, whereas it is only a moderate furin inhibitor with an inhibition constant of 1.4 μM. On the other hand, the unprotected analogue H-Leu-Leu-Arg-Val-Lys-Arg-OH has a stronger furin affinity (0.42 μM) but reduced potency against PC2 (3.4 μM) (Cameron et al. 2000).

We have developed a new peptidomimetic lead structure for basic PCs by incorporation of decarboxylated arginine derivatives in P1 position. Starting from the

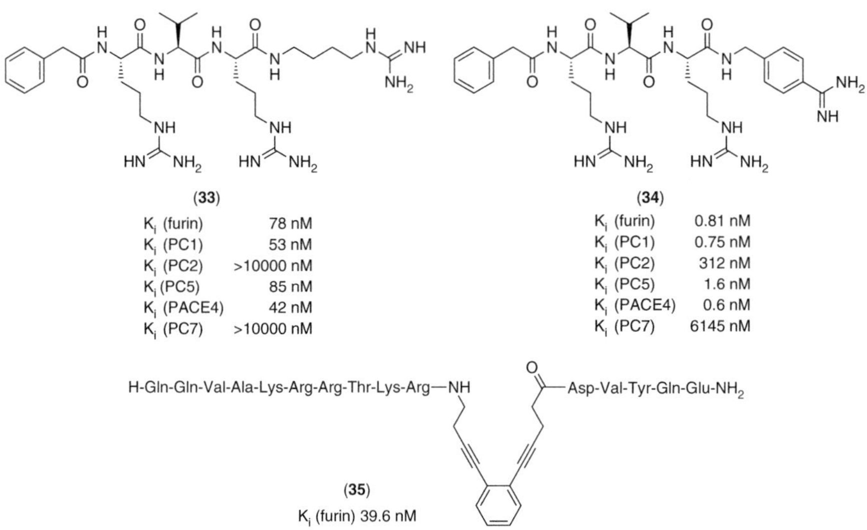

Fig. 11.16 Peptidomimetic PC inhibitors (Basak et al. 2009; Becker et al. 2010; Cameron 2000)

agmatine compound Phac-Arg-Val-Arg-4-Agm (**33**, K_i = 78 nM, Fig. 11.16) or the equipotent noragmatine analogue, the approximately 100-fold more potent 4-Amba inhibitor **34** has been prepared (Fig. 11.16). It inhibits furin with a K_i value of 0.81 nM and PC1, PACE4, and PC5 with similar efficacy but has reduced potency against PC2 and PC7 (Becker et al. 2010) and negligible affinity against the trypsin-like serine proteases thrombin, factor Xa, and plasmin. Further modification of the P3 position by *tert*-leucine or penicillamine provided inhibitors with enhanced potencies (Becker et al. 2011; Hardes et al. 2015). A considerably improved affinity was achieved by incorporation of basic P5 residues (Becker et al. 2012). The most potent analogue **36** (MI-1148, Fig. 11.17) contains a *para*-guanidinomethyl substitution at the P5 phenyl ring and inhibits furin with a K_i value of 5.5 pM (Hardes et al. 2015). These inhibitors have a similar selectivity profile, as described for analogue **34**, and also inhibit PC4 in the low picomolar range. To the best of our knowledge, these compounds are the most potent synthetic inhibitors of furin-like PCs. Some of these compounds enabled the determination of crystal structures in complex with human furin (Fig. 11.17) (Dahms et al. 2014, 2016a; Hardes et al. 2015). A nearly identical binding mode was found for an analogous inhibitor containing the P5 guanidinomethyl substitution in meta-position (Dahms et al. 2014). The antiviral efficacy of inhibitor **36** and/or of its P3 Val analogue MI-701 (Becker et al. 2012; Hardes et al. 2015) was extensively studied in cell culture infected with numerous furin-dependent viruses. A significant antiviral effect was found with HPAIV H5N1 and H7N1 strains (Hardes et al. 2015; Lu et al. 2015) and with canine distemper virus (CDV), which belongs to the *Paramyxoviridae* and is closely related to measles virus (Hardes et al. 2015). Moreover, both inhibitors reduced the replication of Semliki Forest and chikungunya virus, both belonging to the alphaviruses,

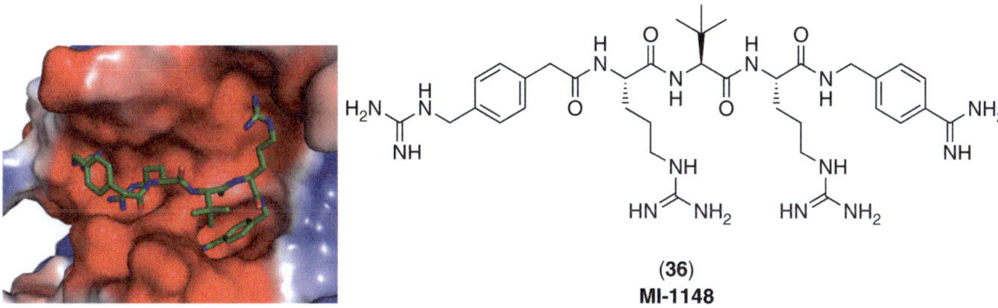

(36)
MI-1148
K_i (furin) 5.5 pM

Fig. 11.17 Structure of inhibitor MI-1148 (stick model with carbons in green, nitrogen in blue, and oxygen in red) in complex with human furin (PDB, 4RYD) (Hardes et al. 2015). Furin is shown in surface representation colored by its electrostatic potential (negatively charged regions are shown in red and positive areas in blue)

where furin cleaves the precursor protein p62 in the TGN (Hardes et al. 2017). An antiviral effect was also found against flaviviruses like West Nile and dengue-2 virus (Kouretova et al. 2017) and, for MI-701, also against Borna disease virus (Lennartz et al. 2016). Furthermore, both compounds strongly protected cells against anthrax, Shiga, and diphtheria toxin (Becker et al. 2012; Hardes et al. 2015).

Furthermore, inhibitor **35** (Fig. 11.16) containing a β-turn inducing enediynyl amino acid moiety was prepared. The amino and carboxyl groups of this unusual amino acid were coupled to peptide sequences around the cleavage sites within the prodomain of furin. This compound inhibits furin with a K_i value of 40 nM and blocks the cleavage of a fluorogenic peptide derived from the spike protein of human SARS coronavirus with an IC_{50} value of 193 nM (Basak et al. 2009).

In addition, the HA cleavage site of a H5 influenza virus was used as scaffold for the development of peptidic inhibitors (Shiryaev et al. 2007). The best compound TPRARRRKKRT-NH_2 (**37**, Table 11.1) inhibits furin with a K_i value of 23 nM, whereas other PCs are less affected. Further optimization was achieved with peptide (**38**), which inhibits furin, PC5, and PACE4 in the low nanomolar range, but has reduced potency against PC7 (K_i = 490 nM) (Remacle et al. 2010).

The inhibitory potency against furin could be further enhanced with poly-arginine derivatives, found by a positional scanning of combinatorial L- and D-hexapeptide libraries (Cameron et al. 2000). Hexa-D-arginine (**39**, Table 11.1) inhibits furin and PC5 with K_i values around 200 nM in the same range, whereas it is less active against PC1 and PC7. The elongated analogue nona-D-arginine-amide (D9R-amide, compound **40**, Table 11.1) possesses a significantly improved inhibition constant of 1.3 nM. In contrast to the analogous L-peptide, which was cleaved by furin, the D-configured D9R-amide was found to be fully stable (Cameron et al. 2000; Kacprzak et al. 2004).

Another attempt to synthesize potent PC inhibitors is based on the monocyclic sunflower trypsin inhibitor SFTI-1 (Fig. 11.2) (Fittler et al. 2015). Therefore, the SFTI-1 backbone was used as a starting point for several modifications like the implementation of a furin cleavage motif and truncation of the inhibitor. The most

Table 11.1 Peptidic inhibitors of basic PCs (Cameron et al. 2000; Fittler et al. 2015; Fugere et al. 2007; Kwiatkowska et al. 2016; Remacle et al. 2010; Shiryaev et al. 2007)

		K_i value (nM)					
No.	Compound	Furin	PC1	PC4	PC5	PACE4	PC7
37	TPRARRRKKRT-amide	23		441	232	162	152
38	Acetyl-RARRRKKRT-amide	6.5			2.6	2.6	490
39	H-(DArg)$_6$-OH	106[a] 265[b]	13 000		206	580	1875
40	H-(DArg)$_9$-amide	1.3			19		81
41	KRCKKSIPPICF-amide	0.49					
42	Acetyl-LLLRVKR-amide	400				18	

[a]Cameron et al. (Cameron et al. 2000)
[b]Fugere et al. (Fugere et al. 2007)

potent compound (**41**) (Table 11.1) of this series inhibits furin in the low nanomolar range (K_i = 0.49 nM), whereas matriptase-1 is only poorly (K_i = 560 nM) and trypsin not affected.

Moreover, peptidomimetic compounds containing a multi-Leu motif have been described (Levesque et al. 2012). The aim of this modification was the discrimination between furin and PACE4. The most selective derivative of this series (**42**) inhibits PACE4 with a K_i value of 18 nM, whereas furin was 22-fold less affected. Further modification with polyethylene glycols of different length could improve the selectivity profile, whereas incorporation of 4-amidinobenzylamide or 2,3-dehydroagmatine as P1 residue resulted in less selective PACE4 inhibitors compared to furin (Kwiatkowska et al. 2016).

Small-Molecule Inhibitors

Small molecules are the most promising candidates for drug development, because they have a better chance for oral bioavailability and should be better suited to address intracellular targets. Nevertheless, only a few nonpeptidic PC inhibitors have been reported; most of them possess moderate potency. The only efficacious derivatives were obtained within a series of guanidinylated 2,5-dideoxystreptamine derivatives; the most potent analogue **43** (Fig. 11.19) inhibits furin and PC5 with inhibition constants of 6 and 4 nM, respectively, whereas its affinity against PACE4 and PC7 is reduced (Jiao et al. 2006). Derivative **44** (1n) enabled the determination of a crystal structure in complex with human furin (Fig. 11.18) (Dahms et al. 2017). Surprisingly, two inhibitor molecules bind close to the active-site cleft. The first molecule (with carbon atoms in green) binds to the S4 and S2 sites of furin, without addressing its S1 pocket. The binding of the second molecule (shown with carbon atoms in yellow), which does not form strong polar contacts to furin, could be an artifact caused by the unusually high inhibitor concentrations used during crystallization.

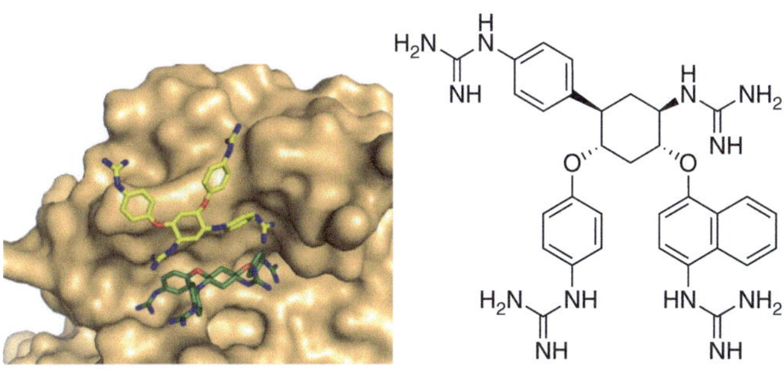

(44)
1n

K_i (furin)	46 nM
K_i (PC5B)	21 nM
K_i (PACE4)	58 nM
K_i (PC7)	1100 nM

Fig. 11.18 Crystal structure of the simultaneous binding of two 2,5-dideoxystreptamine derivatives **44** (**1n**, shown as stick model) in complex with human furin (PDB, 5MIM) (Dahms et al. 2017). Surprisingly, the S1 binding pocket is not addressed

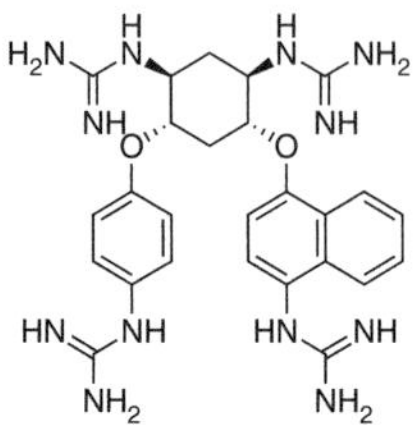

(43)

K_i (furin) 6 nM
K_i (PC5B) 4 nM
K_i (PACE4) 25 nM
K_i (PC7) 415 nM

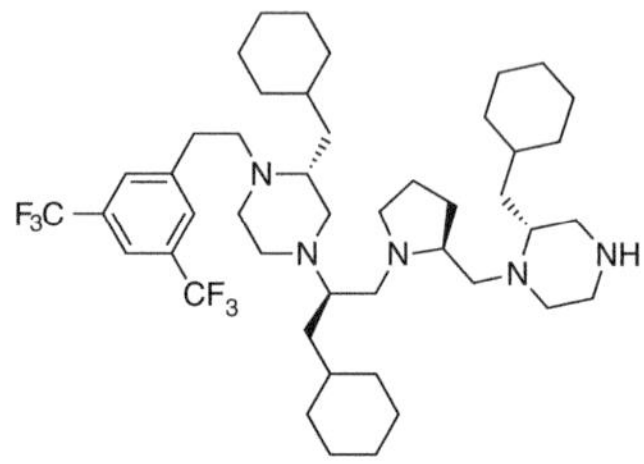

(45)

K_i (furin) > 25000 nM
K_i (PC1) > 40000 nM
K_i (PC2) 540 nM

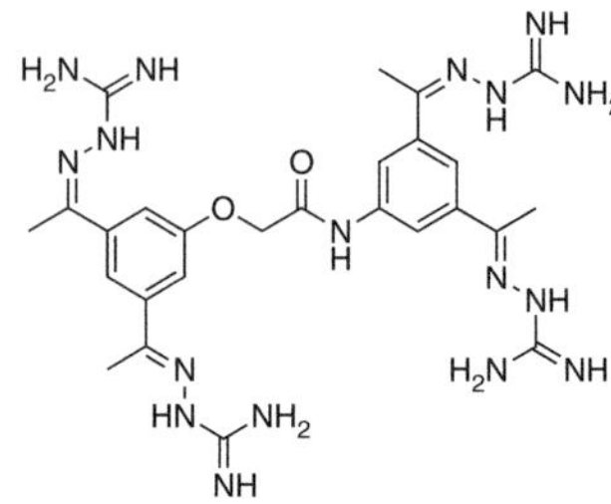

(46)

K_i (furin) 580 nM

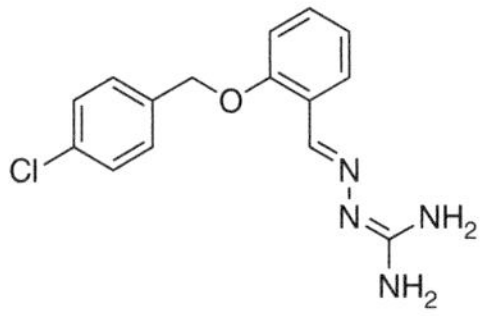

(47)

K_i (furin) 11800 nM

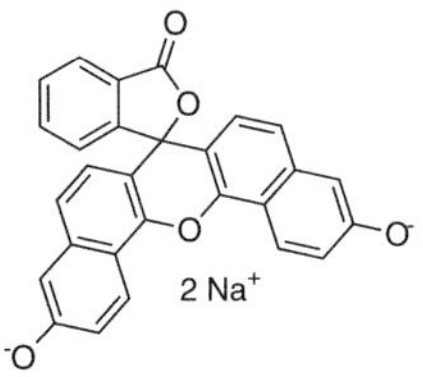

(48)

K_i (furin) 12000 nM

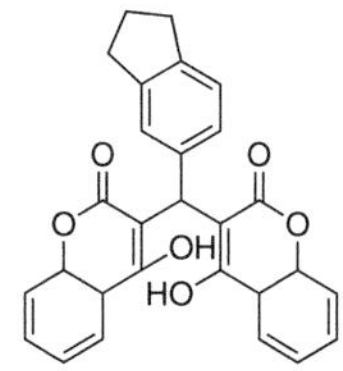

(49)

K_i (furin) 1000 nM
K_i (PC5) 4000 nM
K_i (PACE4) 13000 nM
K_i (PC7) 79000 nM

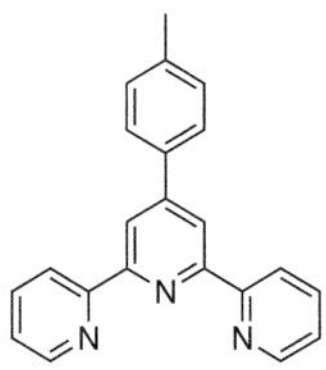

(50)

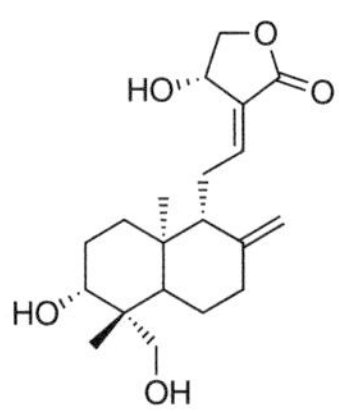

(51)

K_i (furin) 1000 μM
K_i (PC1) 750 μM

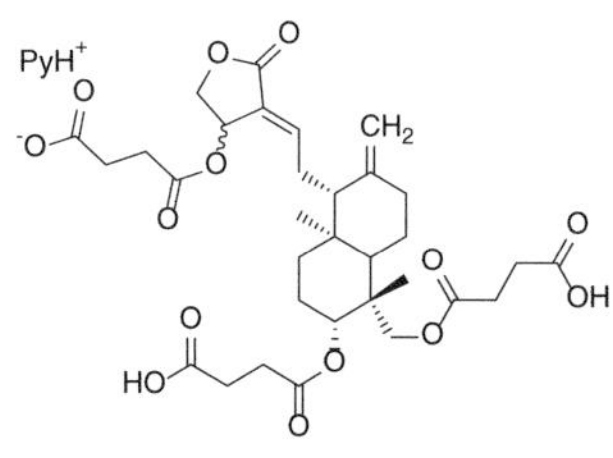

(52)

K_i (furin) 2600 nM
K_i (PC1) 21900 nM
K_i (PC7) 26000 nM

Fig. 11.19 Known small-molecule inhibitors of PCs (Basak et al. 1999; Coppola et al. 2008; Jiao et al. 2006; Komiyama et al. 2009; Kowalska et al. 2009; Podsiadlo et al. 2004; Sielaff et al. 2011b)

Reduction of peptide bonds by treatment of their amide precursors with borane-tetrahydrofuran provided a series of multibasic piperazine and pyrrolidine containing derivatives, like **45** (Fig. 11.19) with submicromolar affinity against PC2 and weak affinity against PC1 and furin (Kowalska et al. 2009). Among a series of weakly basic guanylhydrazones, several compounds like derivative **46** were prepared, which inhibit furin with inhibition constants close to 0.5 μM (Sielaff et al. 2011b). Even, the mono-guanylhydrazone derivative (**47**) still showed a significant furin inhibition with a K_i value of 11.8 μM (Komiyama et al. 2009). Compound **47** also inhibits PC5 and PACE4 in the same range, whereas a reduced potency was observed against PC7. Despite the poor in vitro potency of the cell-permeable naphthofluorescein derivative B3 (**48**) with a K_i value of 12 μM against furin, it was able to inhibit the furin-catalyzed activation of the remodeling protease proMT1-MMP, leading to decreased MMP-2 activation and cell motility of CHO cells. Treatment with B3 also reduced the invasiveness of human fibrosarcoma cells (HT1080) (Coppola et al. 2008).

Furthermore, furin-inhibiting dicoumarol derivatives could be identified by HTS. Some of these compounds protected cells against furin-activated anthrax toxin and inhibited proMT1-MMP processing. Compound **49** noncompetitively binds to furin with an inhibition constant of 1 μM (Komiyama et al. 2009). Recently, a series of zinc and copper ion chelate complexes was described to inhibit furin in the micromolar range; the structure of the most suitable chelate ligand TTP (**50**) is shown in Fig. 11.19. The authors speculated that the active-site histidine might be coordinated by the zinc or copper ions. Interestingly, the solvated Zn^{2+} was less potent than its chelated form, whereby the free chelate ligands did not affect furin (Podsiadlo et al. 2004).

One approach to find new nonpeptidic furin inhibitors is the testing of compounds from natural sources. An example of this attempt is the screening of the chemical constituents of the medicinally used plant *Andrographis paniculata*. Derived from the major component andrographolide **51**, several semisynthetic compounds have been tested. The most potent derivative is the andrographolide-trisuccinate pyridinium salt **52** (Fig. 11.19) with a K_i value of 2.6 μM (Basak et al. 1999).

11.2.2.2 SKI-1/S1P Inhibitors

Peptide-Based Inhibitors

In analogy to the CMK inhibitors against furin-related PCs, irreversible inhibitors for SKI-1/S1P were prepared (Pasquato et al. 2006). The substrate analogue backbone was derived from the cleavage site of the Lassa virus (LASV) glycoprotein IYISRRLL. Based on this sequence, a 4-mer (Dec-RRLL-CMK; **53**, Fig. 11.20), a 6-mer (Dec-ISRRLL-CMK), and a 7-mer (Dec-YISRRLL-CMK) were synthesized. Interestingly, the 4-mer shows in vitro a 250-fold higher potency compared to the longer derivatives, whereas in cell culture experiments, all inhibitors are almost equipotent. The CMK derivatives were able to block the infection of lymphocytic choriomeningitis virus (LCMV) and chimeras of LCMV containing the LASV glycoprotein.

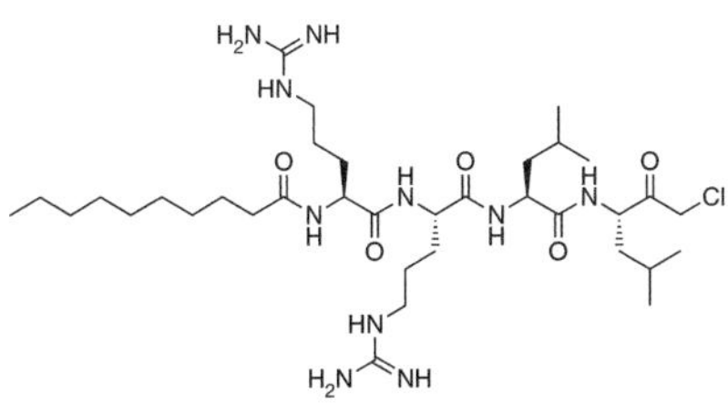

(**53**)
IC_{50} 9 nM

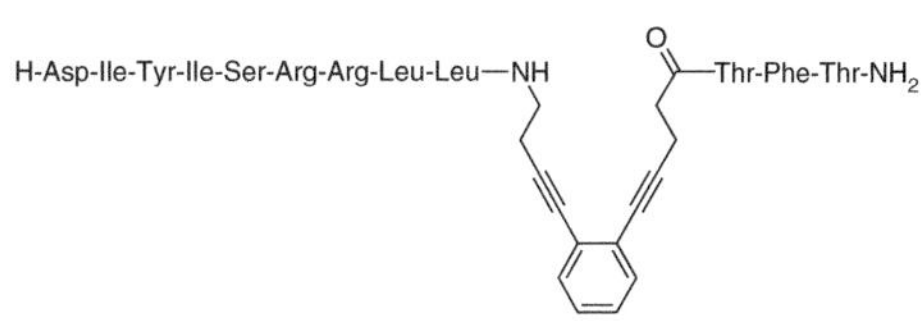

(**54**)
K_i 0.82 μM

Fig. 11.20 Known peptidic inhibitors of SKI-1/S1P (Basak et al. 2015; Pasquato et al. 2006)

Furthermore, enediynyl peptides were synthesized based on the amino acid sequence of cleavage sites of known substrate sequences like the LASV glycoprotein and the prodomain of SKI-1/S1P (Basak et al. 2015). The most potent compound (**54**, Fig. 11.20) inhibits SKI-1/S1P with a K_i value of 0.82 μM.

Protein-Based Inhibitors

The concept of mutation and thereby optimization of natural serine protease inhibitors for a new target was also applied for the development of SKI-1/S1P inhibitors (Maisa et al. 2009; Pullikotil et al. 2004). Instead of the multibasic furin cleavage site, the SKI-1/S1P-specific motif (R/K)-X-X-(L/T)↓ (Seidah and Chretien 1999) was introduced. The overexpression of the α1-AT RRVL variant was found to inhibit both CCHFV (Crimean-Congo hemorrhagic fever virus) and LASV glycoprotein maturation. Blocking of arenavirus GPC processing had a strong antiviral effect in suppressing cell-to-cell spread and formation of viral particles. Micromolar concentrations of the wild-type prosegment of SKI-1/S1P were needed to inhibit the enzyme in vitro. A prosegment with the amino acid exchange R134E was the most potent mutant inhibiting the cleavage of CCHFV preGC (Pullikotil et al. 2004).

Small-Molecule Inhibitors

With the role of SKI-1/S1P in the cholesterol and fatty acid synthesis in mind, a high-throughput screen was conducted to identify lead structures lowering plasma cholesterol and triglycerides (Hay et al. 2007). Based on screening results, improved derivatives were prepared. The most potent compound from this study (**55**, Fig. 11.21) inhibits SKI-1/S1P with an IC_{50} value of 8 nM. However, the less potent analogue PF-429242 (**56**) was used for further evaluation because of its lower molecular weight and lipophilicity. PF-429242 is a reversible, competitive inhibitor of SKI-1/S1P with an IC_{50} value of 170 nM. Furthermore, it is highly selective for SKI-1/S1P; other serine proteases like trypsin, plasmin, kallikrein, or furin are not inhibited by this compound. However, PF-429242 suffers from rapid clearance and poor oral bioavailability in rats. Inhibition of SKI-1/S1P by PF-429242 blocks the cleavage of the glycoprotein of the old-world arenaviruses LASV and LCMV, resulting in reduced virus replication in infected cell cultures (Urata et al. 2011). Interestingly, no escape variants could be detected during or after the treatment with this SKI-1/S1P inhibitor. These studies were extended to several new-world arenaviruses like Junin or Guanarito virus, where PF-429242 also efficiently blocked glycoprotein processing and virus production (Pasquato et al. 2012).

A set of nonpeptidic isocoumarinyl sulfone derivatives was tested against SKI-1/S1P. However, only one compound (**57**, Fig. 11.21) showed a weak inhibition of SKI-1/S1P (K_i = 255 μM) (Basak et al. 2015).

Several known serine protease inhibitors, among them AEBSF (4-(2-aminoethyl)-benzene sulfonylfluoride) and *p*-aminobenzamidine (Fig. 11.4) as well as inhibitors of furin or furin-related PCs like Dec-RVKR-CMK, were tested against SKI-1/S1P. A significant inhibition was only found for DCI (3,4-dichloroisocoumarin, compound **58**, Fig. 11.21), which exhibited a slow irreversible binding mode with an apparent inhibition constant of 6.8 μM (Bodvard et al. 2007).

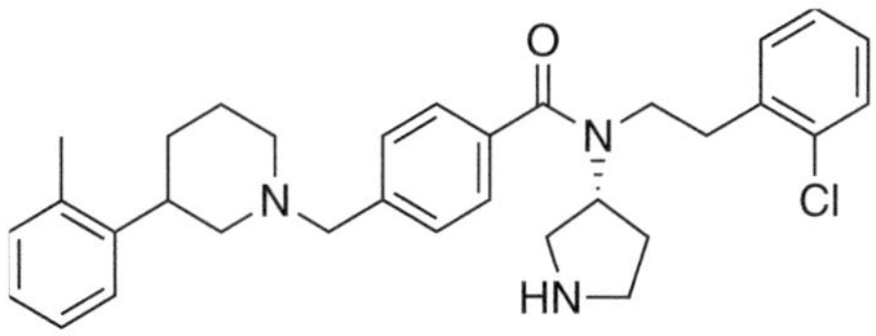

(55)
IC_{50} 8 nM

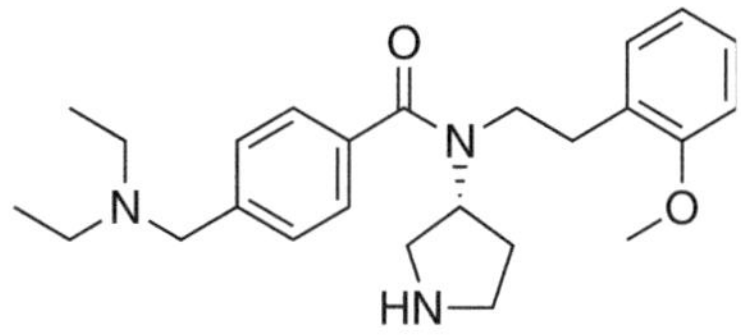

(56)
PF-429242
IC_{50} 170 nM

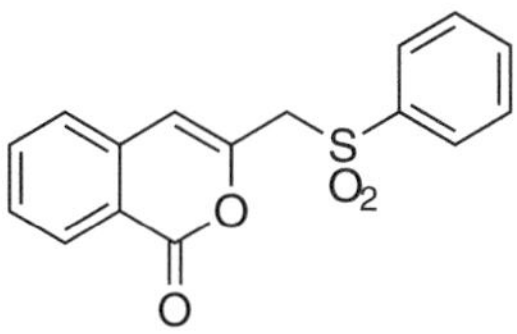

(57)
K_i 255 μM

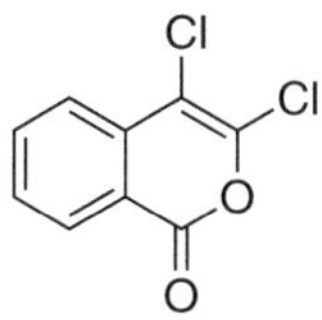

(58)
$K_{i(app)}$ 6.8 μM

Fig. 11.21 Known small-molecule inhibitors of SKI-1/S1P (Basak et al. 2015; Bodvard et al. 2007; Hay et al. 2007)

11.3 Miscellaneous Host Proteases as Antiviral Targets

Numerous viral surface proteins are cleaved by endosomal cysteine proteases. The papain-like cathepsin L activates the Ebola virus (EboV) glycoprotein (GP) (Chandran et al. 2005) or Nipah virus F protein (Diederich et al. 2008), and despite contradictory reports, cathepsins might be also involved in the GP activation of some coronaviruses (Zhou et al. 2015). In contrast to the majority of serine protease inhibitors, most cysteine protease inhibitors achieve potency by a covalent modification of their target enzymes, which can be reversible or irreversible (Siklos et al. 2015). Only a few basic examples of typical cysteine protease inhibitors will be provided in this paragraph. Classical warheads leading to a relatively specific irreversible inhibition of cysteine proteases are the epoxysuccinates, vinylsulfones, and allylsulfones, whereas chloromethyl and diazomethyl ketones react with both the cysteine and serine proteases. The archetypal epoxysuccinate E-64 (**59**) (Fig. 11.22) was isolated from *Aspergillus japonicus* and forms a thioether bond with the active-site cysteine (Hanada et al. 1978). It is a nonselective inhibitor of nearly all cysteine cathepsins, with the exception of cathepsin C (Turk et al. 2012). Many analogues of E-64 have been prepared by replacement of its leucyl-agmatine segment. The more hydrophobic ethyl ester prodrug E-64d (**60**) reached clinical phase III trials (Satoyoshi 1992). Peptidic vinylsulfones represent an additional class of widely used cysteine protease inhibitors, including numerous cathepsins. The active-site cysteine attacks the Michael acceptor like vinylsulfone and becomes irreversibly

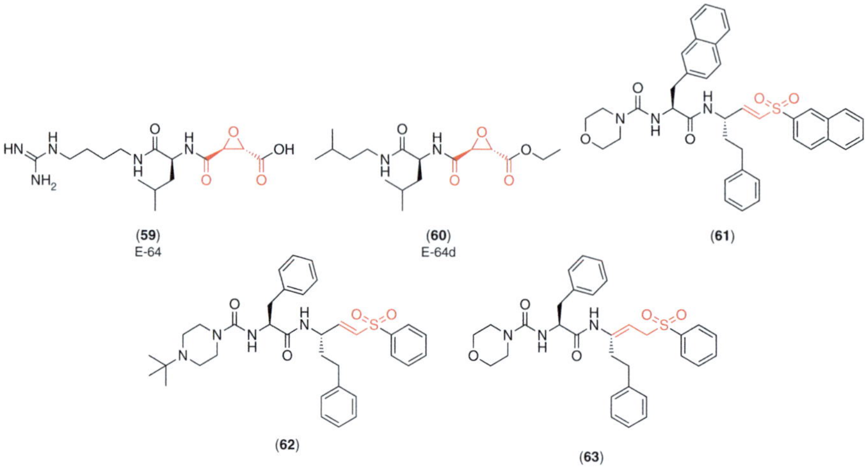

Fig. 11.22 Structures of covalent irreversible cysteine protease inhibitors, which make an alkylation of the active-site cysteine. The reactive epoxysuccinate, vinylsulfone, and allylsulfone segments are colored in red (Fennell et al. 2013; Palmer et al. 1995; Siklos et al. 2015; Zhou et al. 2015)

alkylated. For instance, cathepsins L and B are very rapidly inactivated by inhibitor **61** with second-order rate constants (k_{inact}/K_i) of $9.2 \times 10^6\ M^{-1}s^{-1}$ and $4.2 \times 10^5\ M^{-1}s^{-1}$, respectively (Palmer et al. 1995). A considerable inhibition of EboV GP and SARS-CoV S-protein cleavage was observed for close analogues in cell culture, like compound **62**. However, this compound was completely inactive in a lethal SARS-CoV mouse model (Zhou et al. 2015). A considerably slower inactivation of cathepsin B ($k_{obs}/[I] = 9.2 \times 10^6\ M^{-1}s^{-1}$) and other cysteine proteases was found by the allylsulfone **63** and numerous analogues, indicating a reduced electrophilicity of the allylsulfone segment (Fennell et al. 2013). Peptidic chloromethyl or diazomethyl ketones (Powers et al. 2002) are also very effective in vitro cysteine protease inhibitors. However, they suffer from instability in vivo and often from low selectivity.

Similar as done in the field of serine protease inhibitors, α-keto-derived cysteine protease inhibitors have been prepared, including α-ketoacids (**64**), α-ketoesters (**65**), and α-ketoamides (**66**) (Fig. 11.23). The ketone moiety of these analogues is attacked by the active-site cysteine leading to a covalent but reversible thiohemiketal complex. Additional examples are summarized in a recently published review (Siklos et al. 2015). Probably the most promising warhead for the development of covalent reversible cysteine protease inhibitors is the nitrile group. Nitriles react with the active-site cysteine forming a thioimidate adduct. A myriad of peptidic nitrile inhibitors have been prepared (Frizler et al. 2010), for instance, compound **67** inhibits cathepsins L and B with IC_{50} values of 20 nM and 1.8 nM (Greenspan et al. 2001). A potency considerably enhanced by four to five orders of magnitude could be achieved by the replacement of the Cα carbon in P1 position with nitrogen. For instance, the resulting azanitrile **68** binds to cathepsin L with a K_i value of 74 pM and possesses a similar picomolar potency against cathepsins S and L

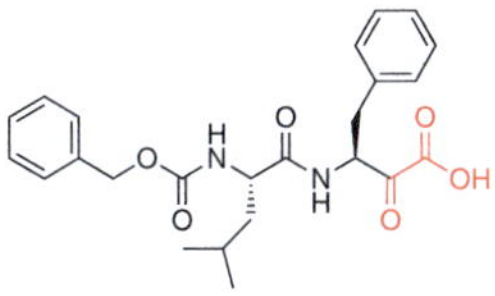

(64)

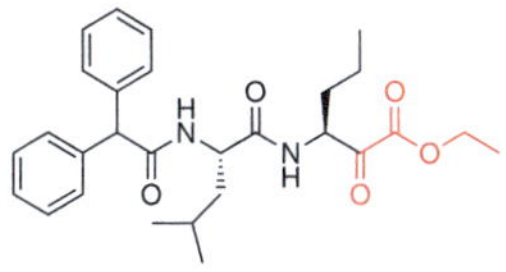

(65)

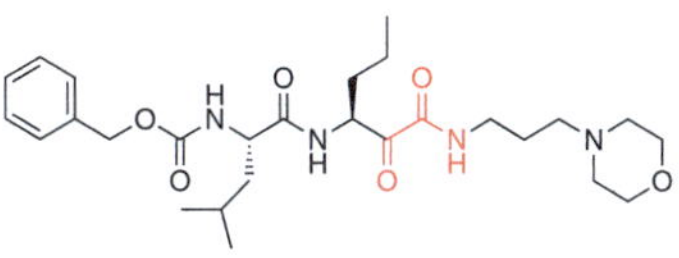

(66)

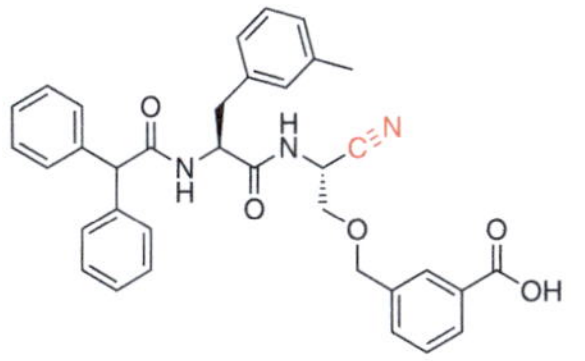

(67)

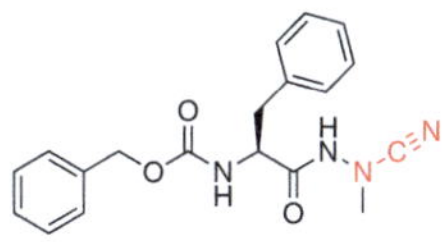

(68)

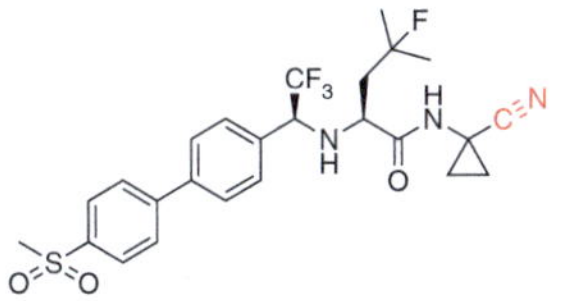

(69)

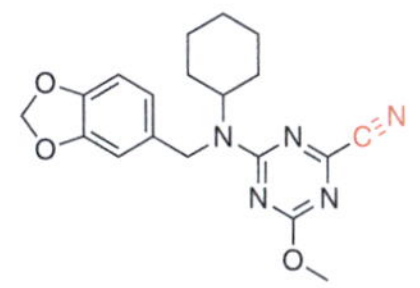

(70)

Fig. 11.23 Structures of covalent reversible cysteine protease inhibitors containing α-ketoacids, α-ketoesters, α-ketoamides, nitrile and azanitrile warheads, which are colored in red (Siklos et al. 2015)

(Löser et al. 2008). The cathepsin K inhibitor odanacatib **69** reached clinical phase III development, but its development was recently cancelled due to side effects. Moreover, completely nonpeptidic nitrile derivatives have been developed, such as compound **70** that inhibits cathepsin L with a K_i value of 2 nM (Fig. 11.23).

Numerous cathepsin inhibitors have been prepared in the past. However, despite a few proof of concept studies, their deeper characterization and optimization as antivirals are limited, so far. They have been mainly tested in nonviral applications, e.g., for cancer therapy, osteoporosis, and rheumatoid arthritis or in neurodegenerative diseases (Siklos et al. 2015; Turk et al. 2012).

Zinc-dependent host metalloproteases from the MMP (matrix metalloproteases) or the ADAM (a disintegrin and metalloprotease) families might also be involved in the entry and fusion of certain viruses, as recently described for a neurovirulent murine CoV strain (Phillips et al. 2017). Moreover, a considerable amount of the EboV GP is shed by the metalloprotease TACE (ADAM17) (Dolnik et al. 2004). The soluble GP activates dendritic cells and macrophages and causes the release of pro- and anti-inflammatory cytokines and affects vascular permeability. The dysregulated inflammatory host response seems to contribute to the high virus pathogenicity (Escudero-Perez et al. 2014). These results suggest that inhibitors of metalloproteases may have antiviral activity. Although more than 50 clinical trials with metalloprotease inhibitors for the treatment of various cancers failed, their broad anti-inflammatory potential has aroused new interest (Vandenbroucke and Libert 2014).

11.4 Host Proteases as Receptors for Virus Entry

A few respiratory viruses use membrane-bound host proteases independent from their proteolytic activity as entry receptors. A surface region far away from the active site of the ubiquitously expressed serine protease dipeptidyl peptidase 4 (DPP4, also called CD26) serves as human cellular receptor of MERS-CoV (Raj et al. 2013). Consequently, no antiviral effect could be observed after treatment of MERS-CoV-infected cells with active-site-directed DPP4 inhibitors. Moreover, the SARS-CoV and HCoV-NL63 use angiotensin-converting enzyme 2 (ACE2) (Li et al. 2003; Wu et al. 2009) and aminopeptidase N (Yeager et al. 1992) as human receptors. Both proteases do not show any sequence or structural similarity with DPP4 (Wang et al. 2013). The crystal structure of the receptor-binding domain of the MERS- and SARS-CoV S-protein in complex with DPP4 (Lu et al. 2013; Wang et al. 2013) and ACE2 (Li et al. 2005) has been determined, respectively. The complexes reveal typical protein-protein interactions (PPI). The receptor region on DPP4 and ACE2 are relatively flat missing deep binding pockets normally found in the active site of proteases. Although no examples are known so far, it should be possible to inhibit the entry of these virions by blocking the described PPIs with suitable ligands.

Conclusion

So far, only inhibitors addressing viral proteases have been approved for the treatment of certain virus infections. A huge arsenal of excellent inhibitors against host proteases has been developed in the past for treatment of chronic diseases, such as hypertension, diabetes, risk of thrombosis, inflammatory ailments, and cancer, but only few of them reached the clinic. Despite loss of patent protection, many of these failed inhibitors or their analogues could still be suitable for short-term treatment of acute life-threatening infectious diseases, without being hampered by side effects that might develop after long-term application. One of the most important prerequisites for successful drug development is the identification of a valid target. For some virus infections, the relevant host proteases have been identified, in other cases there are still uncertainties, and further basic research on target identification is needed. Since proteases usually belong to families of similar enzymes which substitute each other, a broad-spectrum inhibitor could be tolerable or even advantageous for the special treatment of infectious disease, although selective drugs are usually preferred for most applications to minimize side effects. Ideally, host protease inhibitors should be used in combination with additional drugs. This strategy should improve the antiviral efficacy and allow the use of reduced concentrations, thereby minimizing side effects. The development of effective and tolerable host protease inhibitors will hopefully expand the arsenal of antiviral drugs in the future.

References

Adams J. The proteasome: a suitable antineoplastic target. Nat Rev Cancer. 2004;4:349–60.

Al-Horani RA, Desai UR. Recent advances on plasmin inhibitors for the treatment of fibrinolysis-related disorders. Med Res Rev. 2014;34:1168–216.

Anderson ED, Thomas L, Hayflick JS, Thomas G. Inhibition of HIV-1 gp160-dependent membrane fusion by a furin-directed alpha 1-antitrypsin variant. J Biol Chem. 1993;268:24887–91.

Angliker H. Synthesis of tight binding inhibitors and their action on the proprotein-processing enzyme furin. J Med Chem. 1995;38:4014–8.

Antalis TM, Bugge TH, Wu Q. Membrane-anchored serine proteases in health and disease. Prog Mol Biol Transl Sci. 2011;99:1–50.

Apeler H, Peters J, Schröder W, Schneider KH, Lemm G, Hinz V, Rossouw GJ, Dembowsky K. Expression, purification, biochemical and pharmacological characterization of a recombinant aprotinin variant. Arzneimittelforschung. 2004;54:483–97.

Ascenzi P, Bocedi A, Bolognesi M, Spallarossa A, Coletta M, De Cristofaro R, Menegatti E. The bovine basic pancreatic trypsin inhibitor (Kunitz inhibitor): a milestone protein. Curr Protein Pept Sci. 2003;4:231–51.

Baron J, Tarnow C, Mayoli-Nüssle D, Schilling E, Meyer D, Hammami M, Schwalm F, Steinmetzer T, Guan Y, Garten W, et al. Matriptase, HAT, and TMPRSS2 activate the hemagglutinin of H9N2 influenza A viruses. J Virol. 2013;87:1811–20.

Barrett AJ, Starkey PM. The interaction of alpha 2-macroglobulin with proteinases. Characteristics and specificity of the reaction, and a hypothesis concerning its molecular mechanism. Biochem J. 1973;133:709–24.

Basak A, Lazure C. Synthetic peptides derived from the prosegments of proprotein convertase 1/3 and furin are potent inhibitors of both enzymes. Biochem J. 2003;373:231–9.

Basak A, Lotfipour F. Modulating furin activity with designed mini-PDX peptides: synthesis and in vitro kinetic evaluation. FEBS Lett. 2005;579:4813–21.

Basak A, Cooper S, Roberge AG, Banik UK, Chrétien M, Seidah NG. Inhibition of proprotein convertases-1, -7 and furin by diterpines of Andrographis paniculata and their succinoyl esters. Biochem J. 1999;338(Pt 1):107–13.

Basak A, Khatib AM, Mohottalage D, Basak S, Kolajova M, Bag SS. A novel enediynyl peptide inhibitor of furin that blocks processing of proPDGF-A, B and proVEGF-C. PLoS One. 2009;4:e7700.

Basak A, Chen A, Scamuffa N, Mohottalage D, Basak S, Khatib AM. Blockade of furin activity and furin-induced tumor cells malignant phenotypes by the chemically synthesized human furin prodomain. Curr Med Chem. 2010;17:2214–21.

Basak A, Goswami M, Rajkumar A, Mitra T, Majumdar S, O'Reilly P, Bdour HM, Trudeau VL, Basak A. Enediynyl peptides and iso-coumarinyl methyl sulfones as inhibitors of proprotein convertases PCSK8/SKI-1/S1P and PCSK4/PC4: Design, synthesis and biological evaluations. Bioorg Med Chem Lett. 2015;25:2225–37.

Beaulieu A, Gravel É, Cloutier A, Marois I, Colombo É, Désilets A, Verreault C, Leduc R, Marsault É, Richter MV. Matriptase proteolytically activates influenza virus and promotes multicycle replication in the human airway epithelium. J Virol. 2013;87:4237–51.

Becker GL, Sielaff F, Than ME, Lindberg I, Routhier S, Day R, Lu Y, Garten W, Steinmetzer T. Potent inhibitors of furin and furin-like proprotein convertases containing decarboxylated P1 arginine mimetics. J Med Chem. 2010;53:1067–75.

Becker GL, Hardes K, Steinmetzer T. New substrate analogue furin inhibitors derived from 4-amidinobenzylamide. Bioorg Med Chem Lett. 2011;21:4695–7.

Becker GL, Lu Y, Hardes K, Strehlow B, Levesque C, Lindberg I, Sandvig K, Bakowsky U, Day R, Garten W, et al. Highly potent inhibitors of proprotein convertase furin as potential drugs for treatment of infectious diseases. J Biol Chem. 2012;287:21992–2003.

Bertram S, Glowacka I, Blazejewska P, Soilleux E, Allen P, Danisch S, Steffen I, Choi SY, Park Y, Schneider H, et al. TMPRSS2 and TMPRSS4 facilitate trypsin-independent spread of influenza virus in Caco-2 cells. J Virol. 2010;84:10016–25.
Biasini M, Bienert S, Waterhouse A, Arnold K, Studer G, Schmidt T, Kiefer F, Gallo Cassarino T, Bertoni M, Bordoli L, et al. SWISS-MODEL: modelling protein tertiary and quaternary structure using evolutionary information. Nucleic Acids Res. 2014;42:W252–8.
Bodvard K, Mohlin J, Knecht W. Recombinant expression, purification, and kinetic and inhibitor characterisation of human site-1-protease. Protein Expr Purif. 2007;51:308–19.
Böttcher E, Matrosovich T, Beyerle M, Klenk HD, Garten W, Matrosovich M. Proteolytic activation of influenza viruses by serine proteases TMPRSS2 and HAT from human airway epithelium. J Virol. 2006;80:9896–8.
Böttcher E, Freuer C, Steinmetzer T, Klenk HD, Garten W. MDCK cells that express proteases TMPRSS2 and HAT provide a cell system to propagate influenza viruses in the absence of trypsin and to study cleavage of HA and its inhibition. Vaccine. 2009;27:6324–9.
Böttcher-Friebertshäuser E, Freuer C, Sielaff F, Schmidt S, Eickmann M, Uhlendorff J, Steinmetzer T, Klenk HD, Garten W. Cleavage of influenza virus hemagglutinin by airway proteases TMPRSS2 and HAT differs in subcellular localization and susceptibility to protease inhibitors. J Virol. 2010;84:5605–14.
Böttcher-Friebertshäuser E, Stein DA, Klenk HD, Garten W. Inhibition of Influenza virus infection in human airway cell cultures by an antisense peptide-conjugated morpholino oligomer targeting the hemagglutinin-activating protease TMPRSS2. J Virol. 2011;85:1554–62.
Böttcher-Friebertshäuser E, Lu Y, Meyer D, Sielaff F, Steinmetzer T, Klenk HD, Garten W. Hemagglutinin activating host cell proteases provide promising drug targets for the treatment of influenza A and B virus infections. Vaccine. 2012;30:7374–80.
Böttcher-Friebertshäuser E, Klenk HD, Garten W. Activation of influenza viruses by proteases from host cells and bacteria in the human airway epithelium. Pathog Dis. 2013;69:87–100.
Brömme D, Panwar P, Turan S. Cathepsin K osteoporosis trials, pycnodysostosis and mouse deficiency models: commonalities and differences. Expert Opin Drug Discovery. 2016;11:457–72.
Cameron A, Appel J, Houghten RA, Lindberg I. Polyarginines are potent furin inhibitors. J Biol Chem. 2000;275:36741–9.
Chaipan C, Kobasa D, Bertram S, Glowacka I, Steffen I, Tsegaye TS, Takeda M, Bugge TH, Kim S, Park Y, et al. Proteolytic activation of the 1918 influenza virus hemagglutinin. J Virol. 2009;83:3200–11.
Chandran K, Sullivan NJ, Felbor U, Whelan SP, Cunningham JM. Endosomal proteolysis of the Ebola virus glycoprotein is necessary for infection. Science. 2005;308:1643–5.
Chapurlat RD. Odanacatib: a review of its potential in the management of osteoporosis in postmenopausal women. Ther Adv Musculoskelet Dis. 2015;7:103–9.
Chase TJ, Shaw E. p-Nitrophenyl-p'-guanidinobenzoate HCl: a new active site titrant for trypsin. Biochem Biophys Res Commun. 1967;29:508–14.
Chen Y, Shiota M, Ohuchi M, Towatari T, Tashiro J, Murakami M, Yano M, Yang B, Kido H. Mast cell tryptase from pig lungs triggers infection by pneumotropic Sendai and influenza A viruses. Purification and characterization. Eur J Biochem. 2000;267:3189–97.
Clement B. Reduction of N-hydroxylated compounds: amidoximes (N-hydroxyamidines) as prodrugs of amidines. Drug Metab Rev. 2002;34:565–79.
Colombo É, Désilets A, Dûchene D, Chagnon F, Najmanovich R, Leduc R, Marsault É. Design and synthesis of potent, selective inhibitors of matriptase. ACS Med Chem Lett. 2012;3:530–4.
Coppola JM, Bhojani MS, Ross BD, Rehemtulla A. A small-molecule furin inhibitor inhibits cancer cell motility and invasiveness. Neoplasia. 2008;10:363–70.
Creemers JW, Khatib AM. Knock-out mouse models of proprotein convertases: unique functions or redundancy? Front Biosci. 2008;13:4960–71.
Dahlen JR, Jean F, Thomas G, Foster DC, Kisiel W. Inhibition of soluble recombinant furin by human proteinase inhibitor 8. J Biol Chem. 1998;273:1851–4.
Dahms SO, Hardes K, Becker GL, Steinmetzer T, Brandstetter H, Than ME. X-ray structures of human furin in complex with competitive inhibitors. ACS Chem Biol. 2014;9:1113–8.

Dahms SO, Arciniega M, Steinmetzer T, Huber R, Than ME. Structure of the unliganded form of the proprotein convertase furin suggests activation by a substrate-induced mechanism. Proc Natl Acad Sci U S A. 2016a;113:11196–201.

Dahms SO, Creemers JW, Schaub Y, Bourenkov GP, Zogg T, Brandstetter H, Than ME. The structure of a furin-antibody complex explains non-competitive inhibition by steric exclusion of substrate conformers. Sci Rep. 2016b;6:34303.

Dahms SO, Jiao GS, Than ME. Structural studies revealed active site distortions of human Furin by a small molecule inhibitor. ACS Chem Biol. 2017;12:1211–6.

Diederich S, Maisner A. Molecular characteristics of the Nipah virus glycoproteins. Ann N Y Acad Sci. 2007;1102:39–50.

Diederich S, Thiel L, Maisner A. Role of endocytosis and cathepsin-mediated activation in Nipah virus entry. Virology. 2008;375:391–400.

Dolnik O, Volchkova V, Garten W, Carbonnelle C, Becker S, Kahnt J, Stroher U, Klenk HD, Volchkov V. Ectodomain shedding of the glycoprotein GP of Ebola virus. EMBO J. 2004;23:2175–84.

Dûchene D, Colombo É, Désilets A, Boudreault PL, Leduc R, Marsault É, Najmanovich R. Analysis of subpocket selectivity and identification of potent selective inhibitors for matriptase and matriptase-2. J Med Chem. 2014;57:10198–204.

Enyedy IJ, Lee SL, Kuo AH, Dickson RB, Lin CY, Wang S. Structure-based approach for the discovery of bis-benzamidines as novel inhibitors of matriptase. J Med Chem. 2001;44:1349–55.

Escudero-Pérez B, Volchkova VA, Dolnik O, Lawrence P, Volchkov VE. Shed GP of Ebola virus triggers immune activation and increased vascular permeability. PLoS Pathog. 2014;10:e1004509.

Fennell BD, Warren JM, Chung KK, Main HL, Arend AB, Tochowicz A, Gotz MG. Optimization of peptidyl allyl sulfones as clan CA cysteine protease inhibitors. J Enzyme Inhib Med Chem. 2013;28:468–78.

Fergusson DA, Hebert PC, Mazer CD, Fremes S, MacAdams C, Murkin JM, Teoh K, Duke PC, Arellano R, Blajchman MA, et al. A comparison of aprotinin and lysine analogues in high-risk cardiac surgery. N Engl J Med. 2008;358:2319–31.

Fittler H, Avrutina O, Glotzbach B, Empting M, Kolmar H. Combinatorial tuning of peptidic drug candidates: high-affinity matriptase inhibitors through incremental structure-guided optimization. Org Biomol Chem. 2013;11:1848–57.

Fittler H, Avrutina O, Empting M, Kolmar H. Potent inhibitors of human matriptase-1 based on the scaffold of sunflower trypsin inhibitor. J Pept Sci. 2014;20:415–20.

Fittler H, Depp A, Avrutina O, Dahms SO, Than ME, Empting M, Kolmar H. Engineering a constrained peptidic scaffold towards potent and selective Furin inhibitors. ChemBioChem. 2015;16:2441–4.

Frizler M, Stirnberg M, Sisay MT, Gütschow M. Development of nitrile-based peptidic inhibitors of cysteine cathepsins. Curr Top Med Chem. 2010;10:294–322.

Fugere M, Appel J, Houghten RA, Lindberg I, Day R. Short polybasic peptide sequences are potent inhibitors of PC5/6 and PC7: use of positional scanning-synthetic peptide combinatorial libraries as a tool for the optimization of inhibitory sequences. Mol Pharmacol. 2007;71:323–32.

Fujii S, Hitomi Y. New synthetic inhibitors of C1r, C1 esterase, thrombin, plasmin, kallikrein and trypsin. Biochim Biophys Acta. 1981;661:342–5.

Furtmann N, Häussler D, Scheidt T, Stirnberg M, Steinmetzer T, Bajorath J, Gütschow M. Limiting the number of potential binding modes by introducing symmetry into ligands: structure-based design of inhibitors for trypsin-like serine proteases. ChemMedChem. 2016;22:610–25.

Galkin AV, Mullen L, Fox WD, Brown J, Duncan D, Moreno O, Madison EL, Agus DB. CVS-3983, a selective matriptase inhibitor, suppresses the growth of androgen independent prostate tumor xenografts. Prostate. 2004;61:228–35.

Garten W, Klenk HD. Cleavage activation of the influenza virus hemagglutinin and its role in pathogenesis. In: Klenk HD, Matrosovich M, Stech J, editors. Avian Influenza. Basel: Karger; 2008. p. 156–67.

Garten W, Stieneke A, Shaw E, Wikström P, Klenk HD. Inhibition of proteolytic activation of influenza virus hemagglutinin by specific peptidyl chloroalkyl ketones. Virology. 1989;172:25–31.

Garten W, Hallenberger S, Ortmann D, Schäfer W, Vey M, Angliker H, Shaw E, Klenk HD. Processing of viral glycoproteins by the subtilisin-like endoprotease furin and its inhibition by specific peptidylchloroalkylketones. Biochimie. 1994;76:217–25.

Garten W, Braden C, Arendt A, Peitsch C, Baron J, Lu Y, Pawletko K, Hardes K, Steinmetzer T, Böttcher-Friebertshäuser E. Influenza virus activating host proteases: identification, localization and inhibitors as potential therapeutics. Eur J Cell Biol. 2015;94:375–83.

Ghosh AK, Osswald HL, Prato G. Recent progress in the development of HIV-1 protease inhibitors for the treatment of HIV/AIDS. J Med Chem. 2016;59:5172–208.

Gierer S, Bertram S, Kaup F, Wrensch F, Heurich A, Kramer-Kuhl A, Welsch K, Winkler M, Meyer B, Drosten C, et al. The spike protein of the emerging betacoronavirus EMC uses a novel coronavirus receptor for entry, can be activated by TMPRSS2, and is targeted by neutralizing antibodies. J Virol. 2013;87:5502–11.

Goettig P, Magdolen V, Brandstetter H. Natural and synthetic inhibitors of kallikrein-related peptidases (KLKs). Biochimie. 2010;92:1546–67.

Goswami R, Mukherjee S, Wohlfahrt G, Ghadiyaram C, Nagaraj J, Chandra BR, Sistla RK, Satyam LK, Samiulla DS, Moilanen A, et al. Discovery of Pyridyl Bis(oxy)dibenzimidamide derivatives as selective Matriptase inhibitors. ACS Med Chem Lett. 2013;4:1152–7.

Goswami R, Mukherjee S, Ghadiyaram C, Wohlfahrt G, Sistla RK, Nagaraj J, Satyam LK, Subbarao K, Palakurthy RK, Gopinath S, et al. Structure-guided discovery of 1,3,5 tri-substituted benzenes as potent and selective matriptase inhibitors exhibiting in vivo antitumor efficacy. Bioorg Med Chem. 2014;22:3187–203.

Goswami R, Wohlfahrt G, Mukherjee S, Ghadiyaram C, Nagaraj J, Satyam LK, Subbarao K, Gopinath S, Krishnamurthy NR, Subramanya HS, et al. Discovery of O-(3-carbamimidoylphenyl)-l-serine amides as matriptase inhibitors using a fragment-linking approach. Bioorg Med Chem Lett. 2015;25:616–20.

Gotoh B, Ogasawara T, Toyoda T, Inocencio NM, Hamaguchi M, Nagai Y. An endoprotease homologous to the blood clotting factor X as a determinant of viral tropism in chick embryo. EMBO J. 1990;9:4189–95.

Greenspan PD, Clark KL, Tommasi RA, Cowen SD, McQuire LW, Farley DL, van Duzer JH, Goldberg RL, Zhou H, Du Z, et al. Identification of dipeptidyl nitriles as potent and selective inhibitors of cathepsin B through structure-based drug design. J Med Chem. 2001;44:4524–34.

Gustafsson D, Bylund R, Antonsson T, Nilsson I, Nyström JE, Eriksson U, Bredberg U, Teger-Nilsson AC. A new oral anticoagulant: the 50-year challenge. Nat Rev Drug Discov. 2004;3:649–59.

Hallenberger S, Bosch V, Angliker H, Shaw E, Klenk HD, Garten W. Inhibition of furin-mediated cleavage activation of HIV-1 glycoprotein gp160. Nature. 1992;360:358–61.

Hamilton BS, Whittaker GR. Cleavage activation of human-adapted influenza virus subtypes by kallikrein-related peptidases 5 and 12. J Biol Chem. 2013;288:17399–407.

Hamilton BS, Chung C, Cyphers SY, Rinaldi VD, Marcano VC, Whittaker GR. Inhibition of influenza virus infection and hemagglutinin cleavage by the protease inhibitor HAI-2. Biochem Biophys Res Commun. 2014;450:1070–5.

Hammami M, Rühmann E, Maurer E, Heine A, Gütschow M, Klebe G, Steinmetzer T. New 3-amidinophenylalanine-derived inhibitors of matriptase. Med Chem Commun. 2012;3:807–13.

Han SJ, Kim HS, Kim KI, Whang SM, Hong KS, Lee WK, Lee SH. Use of Nafamostat Mesilate as an anticoagulant during extracorporeal membrane oxygenation. J Korean Med Sci. 2011;26:945–50.

Han Z, Harris PK, Jones DE, Chugani R, Kim T, Agarwal M, Shen W, Wildman SA, Janetka JW. Inhibitors of HGFA, Matriptase, and Hepsin Serine Proteases: a nonkinase strategy to block cell signaling in cancer. ACS Med Chem Lett. 2014;5:1219–24.

Han Z, Harris PK, Karmakar P, Kim T, Owusu BY, Wildman SA, Klampfer L, Janetka JW. alpha-Ketobenzothiazole Serine Protease inhibitors of aberrant HGF/c-MET and MSP/RON Kinase pathway signaling in cancer. ChemMedChem. 2016;11:585–99.

Hanada K, Tamai M, Yamagishi M, Ohmura S, Sawada J, Tanaka I. Isolation and characterization of E–64, a new thiol protease inhibitor. Agric Biol Chem. 1978;42:523–8.

Hanson SR, Harker LA. Interruption of acute platelet-dependent thrombosis by the synthetic anti-thrombin D-phenylalanyl-L-prolyl-L-arginyl chloromethyl ketone. Proc Natl Acad Sci U S A. 1988;85:3184–8.

Hardes K, Becker GL, Lu Y, Dahms SO, Köhler S, Beyer W, Sandvig K, Yamamoto H, Lindberg I, Walz L, et al. Novel Furin inhibitors with potent anti-infectious activity. ChemMedChem. 2015;10:1218–31.

Hardes K, Ivanova T, Thaa B, McInerney GM, Klokk TI, Sandvig K, Künzel S, Lindberg I, Steinmetzer T. Elongated and shortened peptidomimetic inhibitors of the proprotein convertase furin. ChemMedChem. 2017;12:613–20.

Hatesuer B, Bertram S, Mehnert N, Bahgat MM, Nelson PS, Pöhlman S, Schughart K. Tmprss2 is essential for influenza H1N1 virus pathogenesis in mice. PLoS Pathog. 2013;9:e1003774.

Hay BA, Abrams B, Zumbrunn AY, Valentine JJ, Warren LC, Petras SF, Shelly LD, Xia A, Varghese AH, Hawkins JL, et al. Aminopyrrolidineamide inhibitors of site-1 protease. Bioorg Med Chem Lett. 2007;17:4411–4.

Hellstern P, Stürzebecher U, Wuchold B, Haubelt H, Seyfert UT, Bauer M, Vogt A, Stürzebecher J. Preservation of in vitro function of platelets stored in the presence of a synthetic dual inhibitor of factor Xa and thrombin. J Thromb Haemost. 2007;5:2119–26.

Henrich S, Cameron A, Bourenkov GP, Kiefersauer R, Huber R, Lindberg I, Bode W, Than ME. The crystal structure of the proprotein processing proteinase furin explains its stringent specificity. Nat Struct Biol. 2003;10:520–6.

Henrich S, Lindberg I, Bode W, Than ME. Proprotein convertase models based on the crystal structures of furin and kexin: explanation of their specificity. J Mol Biol. 2005;345:211–27.

Hosoya M, Shigeta S, Ishii T, Suzuki H, De Clercq E. Comparative inhibitory effects of various nucleoside and nonnucleoside analogues on replication of influenza virus types A and B in vitro and in ovo. J Infect Dis. 1993;168:641–6.

Howell EH, Cameron SJ. Neprilysin inhibition: a brief review of past pharmacological strategies for heart failure treatment and future directions. Cardiol J. 2016;23:591–8.

Jean F, Stella K, Thomas L, Liu G, Xiang Y, Reason AJ, Thomas G. alpha1-Antitrypsin Portland, a bioengineered serpin highly selective for furin: application as an antipathogenic agent. Proc Natl Acad Sci U S A. 1998;95:7293–8.

Jiao GS, Cregar L, Wang J, Millis SZ, Tang C, O'Malley S, Johnson AT, Sareth S, Larson J, Thomas G. Synthetic small molecule furin inhibitors derived from 2,5-dideoxystreptamine. Proc Natl Acad Sci U S A. 2006;103:19707–12.

Kacprzak MM, Peinado JR, Than ME, Appel J, Henrich S, Lipkind G, Houghten RA, Bode W, Lindberg I. Inhibition of furin by polyarginine-containing peptides: nanomolar inhibition by nona-D-arginine. J Biol Chem. 2004;279:36788–94.

Kang S, Min HJ, Kang MS, Jung MG, Kim S. Discovery of novel 2-hydroxydiarylamide derivatives as TMPRSS4 inhibitors. Bioorg Med Chem Lett. 2013;23:1748–51.

Kawase M, Shirato K, van der Hoek L, Taguchi F, Matsuyama S. Simultaneous treatment of human bronchial epithelial cells with serine and cysteine protease inhibitors prevents severe acute respiratory syndrome coronavirus entry. J Virol. 2012;86:6537–45.

Kido H, Yokogoshi Y, Sakai K, Tashiro M, Kishino Y, Fukutomi A, Katunuma N. Isolation and characterization of a novel trypsin-like protease found in rat bronchiolar epithelial Clara cells. A possible activator of the viral fusion glycoprotein. J Biol Chem. 1992;267:13573–9.

Kido H, Okumura Y, Takahashi E, Pan HY, Wang S, Yao D, Yao M, Chida J, Yano M. Role of host cellular proteases in the pathogenesis of influenza and influenza-induced multiple organ failure. Biochim Biophys Acta. 2012;1824:186–94.

Klenk HD, Garten W. Host cell proteases controlling virus pathogenicity. Trends Microbiol. 1994;2:39–43.

Klenk HD, Rott R. Formation of influenza virus proteins. J Virol. 1973;11:823–31.

Klenk HD, Rott R, Orlich M, Blodorn J. Activation of influenza A viruses by trypsin treatment. Virology. 1975;68:426–39.

Komiyama T, Fuller RS. Engineered eglin c variants inhibit yeast and human proprotein processing proteases, Kex2 and furin. Biochemistry. 2000;39:15156–65.

Komiyama T, Coppola JM, Larsen MJ, van Dort ME, Ross BD, Day R, Rehemtulla A, Fuller RS. Inhibition of furin/proprotein convertase-catalyzed surface and intracellular processing by small molecules. J Biol Chem. 2009;284:15729–38.

Kortuem KM, Stewart AK. Carfilzomib. Blood. 2013;121:893–7.

Kouretova J, Hammamy MZ, Epp A, Hardes K, Kallis S, Zhang L, Hilgenfeld R, Bartenschlager R, Steinmetzer T. Effects of NS2B-NS3 protease and furin inhibition on West Nile and Dengue virus replication. J Enzyme Inhib Med Chem. 2017;32:712–21.

Kowalska D, Liu J, Appel JR, Ozawa A, Nefzi A, Mackin RB, Houghten RA, Lindberg I. Synthetic small-molecule prohormone convertase 2 inhibitors. Mol Pharmacol. 2009;75:617–25.

Kühn N, Bergmann S, Kosterke N, Lambertz RL, Keppner A, van den Brand JM, Pöhlmann S, Weiß S, Hummler E, Hatesuer B, et al. The proteolytic activation of (H3N2) Influenza A virus hemagglutinin is facilitated by different type II Transmembrane serine proteases. J Virol. 2016;90:4298–307.

Kwiatkowska A, Couture F, Levesque C, Ly K, Beauchemin S, Desjardins R, Neugebauer W, Dory YL, Day R. Novel insights into structure-activity relationships of N-terminally modified PACE4 inhibitors. ChemMedChem. 2016;11:289–301.

Lazarowitz SG, Choppin PW. Enhancement of the infectivity of influenza A and B viruses by proteolytic cleavage of the hemagglutinin polypeptide. Virology. 1975;68:440–54.

Lee MG, Kim KH, Park KY, Kim JS. Evaluation of anti-influenza effects of camostat in mice infected with non-adapted human influenza viruses. Arch Virol. 1996;141:1979–89.

Lennartz F, Bayer K, Czerwonka N, Lu Y, Kehr K, Hirz M, Steinmetzer T, Garten W, Herden C. Surface glycoprotein of Borna disease virus mediates virus spread from cell to cell. Cell Microbiol. 2016;18:340–54.

Levesque C, Fugère M, Kwiatkowska A, Couture F, Desjardins R, Routhier S, Moussette P, Prahl A, Lammek B, Appel JR, et al. The Multi-Leu peptide inhibitor discriminates between PACE4 and furin and exhibits antiproliferative effects on prostate cancer cells. J Med Chem. 2012;55:10501–11.

Levy JH, Adkinson NFJ. Anaphylaxis during cardiac surgery: implications for clinicians. Anesth Analg. 2008;106:392–403.

Li W, Moore MJ, Vasilieva N, Sui J, Wong SK, Berne MA, Somasundaran M, Sullivan JL, Luzuriaga K, Greenough TC, et al. Angiotensin-converting enzyme 2 is a functional receptor for the SARS coronavirus. Nature. 2003;426:450–4.

Li F, Li W, Farzan M, Harrison SC. Structure of SARS coronavirus spike receptor-binding domain complexed with receptor. Science. 2005;309:1864–8.

Li P, Jiang S, Lee SL, Lin CY, Johnson MD, Dickson RB, Michejda CJ, Roller PP. Design and synthesis of novel and potent inhibitors of the type II transmembrane serine protease, matriptase, based upon the sunflower trypsin inhibitor-1. J Med Chem. 2007;50:5976–83.

Liu ZX, Fei H, Chi CW. Two engineered eglin c mutants potently and selectively inhibiting kexin or furin. FEBS Lett. 2004;556:116–20.

Liu M, Yuan C, Jensen JK, Zhao B, Jiang Y, Jiang L, Huang M. The crystal structure of a multidomain protease inhibitor (HAI-1) reveals the mechanism of its auto-inhibition. J Biol Chem. 2017;292:8412–23.

Long YQ, Lee SL, Lin CY, Enyedy IJ, Wang S, Li P, Dickson RB, Roller PP. Synthesis and evaluation of the sunflower derived trypsin inhibitor as a potent inhibitor of the type II transmembrane serine protease, matriptase. Bioorg Med Chem Lett. 2001;11:2515–9.

Löser R, Frizler M, Schilling K, Gütschow M. Azadipeptide nitriles: highly potent and proteolytically stable inhibitors of papain-like cysteine proteases. Angew Chem Int Ed Engl. 2008;47:4331–4.

Lu W, Zhang W, Molloy SS, Thomas G, Ryan K, Chiang Y, Anderson S, Laskowski M Jr. Arg15-Lys17-Arg18 turkey ovomucoid third domain inhibits human furin. J Biol Chem. 1993;268:14583–5.

Lu G, Hu Y, Wang Q, Qi J, Gao F, Li Y, Zhang Y, Zhang W, Yuan Y, Bao J, et al. Molecular basis of binding between novel human coronavirus MERS-CoV and its receptor CD26. Nature. 2013;500:227–31.

Lu Y, Hardes K, Dahms SO, Böttcher-Friebertshäuser E, Steinmetzer T, Than ME, Klenk HD, Garten W. Peptidomimetic furin inhibitor MI-701 in combination with oseltamivir and ribavirin efficiently blocks propagation of highly pathogenic avian influenza viruses and delays high level oseltamivir resistance in MDCK cells. Antiviral Res. 2015;120:89–100.
Lucas JM, Heinlein C, Kim T, Hernandez SA, Malik MS, True LD, Morrissey C, Corey E, Montgomery B, Mostaghel E, et al. The androgen-regulated protease TMPRSS2 activates a proteolytic cascade involving components of the tumor microenvironment and promotes prostate cancer metastasis. Cancer Discov. 2014;4:1310–25.
Luckett S, Garcia RS, Barker JJ, Konarev AV, Shewry PR, Clarke AR, Brady RL. High-resolution structure of a potent, cyclic proteinase inhibitor from sunflower seeds. J Mol Biol. 1999;290:525–33.
Maisa A, Stroher U, Klenk HD, Garten W, Strecker T. Inhibition of Lassa virus glycoprotein cleavage and multicycle replication by site 1 protease-adapted alpha(1)-antitrypsin variants. PLoS Negl Trop Dis. 2009;3:e446.
Maiwald A, Hammami M, Wagner S, Heine A, Klebe G, Steinmetzer T. Changing the selectivity profile—from substrate analog inhibitors of thrombin and factor Xa to potent matriptase inhibitors. J Enzyme Inhib Med Chem. 2016;31:89–97.
Manasanch EE, Orlowski RZ. Proteasome inhibitors in cancer therapy. Nat Rev Clin Oncol. 2017;14:417–33.
Matsuyama S, Nagata N, Shirato K, Kawase M, Takeda M, Taguchi F. Efficient activation of the severe acute respiratory syndrome coronavirus spike protein by the transmembrane protease TMPRSS2. J Virol. 2010;84:12658–64.
McBride JD, Watson EM, Brauer AB, Jaulent AM, Leatherbarrow RJ. Peptide mimics of the Bowman-Birk inhibitor reactive site loop. Biopolymers. 2002;66:79–92.
McCauley JA, Rudd MT. Hepatitis C virus NS3/4a protease inhibitors. Curr Opin Pharmacol. 2016;30:84–92.
Meyer D, Sielaff F, Hammami M, Böttcher-Friebertshäuser E, Garten W, Steinmetzer T. Identification of the first synthetic inhibitors of the type II transmembrane serine protease TMPRSS2 suitable for inhibition of influenza virus activation. Biochem J. 2013;452:331–43.
Midgley I, Hood AJ, Proctor P, Chasseaud LF, Irons SR, Cheng KN, Brindley CJ, Bonn R. Metabolic fate of 14C-camostat mesylate in man, rat and dog after intravenous administration. Xenobiotica. 1994;24:79–92.
Molecular Operating Environment (MOE). C.C.G.I., 1010 Sherbooke St. West, Suite #910, Montreal, QC, Canada, H3A 2R7, 2016.
Mullard A. Merck &Co. drops osteoporosis drug odanacatib. Nat Rev Drug Discov. 2016;15:669.
Murakami M, Towatari T, Ohuchi M, Shiota M, Akao M, Okumura Y, Parry MA, Kido H. Mini-plasmin found in the epithelial cells of bronchioles triggers infection by broad-spectrum influenza A viruses and Sendai virus. Eur J Biochem. 2001;268:2847–55.
Oe Y, Hayashi S, Fushima T, Sato E, Kisu K, Sato H, Ito S, Takahashi N. Coagulation Factor Xa and Protease-Activated Receptor 2 as novel therapeutic targets for diabetic nephropathy. Arterioscler Thromb Vasc Biol. 2016;36:1525–33.
Okumura Y, Takahashi E, Yano M, Ohuchi M, Daidoji T, Nakaya T, Böttcher E, Garten W, Klenk HD, Kido H. Novel type II transmembrane serine proteases, MSPL and TMPRSS13, Proteolytically activate membrane fusion activity of the hemagglutinin of highly pathogenic avian influenza viruses and induce their multicycle replication. J Virol. 2010;84:5089–96.
Opal SM, Artenstein AW, Cristofaro PA, Jhung JW, Palardy JE, Parejo NA, Lim YP. Inter-alpha-inhibitor proteins are endogenous furin inhibitors and provide protection against experimental anthrax intoxication. Infect Immun. 2005;73:5101–5.
Ovcharenko AV, Zhirnov OP. Aprotinin aerosol treatment of influenza and paramyxovirus bronchopneumonia of mice. Antiviral Res. 1994;23:107–18.
Paju A, Sorsa T, Tervahartiala T, Koivunen E, Haglund C, Leminen A, Wahlström T, Salo T, Stenman UH. The levels of trypsinogen isoenzymes in ovarian tumour cyst fluids are associated with promatrix metalloproteinase-9 but not promatrix metalloproteinase-2 activation. Br J Cancer. 2001;84:1363–71.

Palmer JT, Rasnick D, Klaus JL, Brömme D. Vinyl sulfones as mechanism-based cysteine protease inhibitors. J Med Chem. 1995;38:3193–6.
Pan HY, Yamada H, Chida J, Wang S, Yano M, Yao M, Zhu J, Kido H. Up-regulation of ectopic trypsins in the myocardium by influenza A virus infection triggers acute myocarditis. Cardiovasc Res. 2011;89:595–603.
Pasquato A, Pullikotil P, Asselin MC, Vacatello M, Paolillo L, Ghezzo F, Basso F, Di Bello C, Dettin M, Seidah NG. The proprotein convertase SKI-1/S1P. In vitro analysis of Lassa virus glycoprotein-derived substrates and ex vivo validation of irreversible peptide inhibitors. J Biol Chem. 2006;281:23471–81.
Pasquato A, Rochat C, Burri DJ, Pasqual G, de la Torre JC, Kunz S. Evaluation of the anti-arenaviral activity of the subtilisin kexin isozyme-1/site-1 protease inhibitor PF-429242. Virology. 2012;423:14–22.
Pasquato A, Ramos da Palma J, Galan C, Seidah NG, Kunz S. Viral envelope glycoprotein processing by proprotein convertases. Antiviral Res. 2013;99:49–60.
Phillips JM, Gallagher T, Weiss SR. Neurovirulent Murine Coronavirus JHM.SD uses cellular zinc metalloproteases for virus entry and cell-cell fusion. J Virol. 2017;91. e01564–16.
Podsiadlo P, Komiyama T, Fuller RS, Blum O. Furin inhibition by compounds of copper and zinc. J Biol Chem. 2004;279:36219–27.
Powers JC, Asgian JL, Ekici OD, James KE. Irreversible inhibitors of serine, cysteine, and threonine proteases. Chem Rev. 2002;102:4639–750.
Pullikotil P, Vincent M, Nichol ST, Seidah NG. Development of protein-based inhibitors of the proprotein of convertase SKI-1/S1P: processing of SREBP-2, ATF6, and a viral glycoprotein. J Biol Chem. 2004;279:17338–47.
Quesada V, Ordónez GR, Sáñchez LM, Puente XS, López-Otin C. The Degradome database: mammalian proteases and diseases of proteolysis. Nucleic Acids Res. 2009;37:D239–43.
Raj VS, Mou H, Smits SL, Dekkers DH, Muller MA, Dijkman R, Muth D, Demmers JA, Zaki A, Fouchier RA, et al. Dipeptidyl peptidase 4 is a functional receptor for the emerging human coronavirus-EMC. Nature. 2013;495:251–4.
Rawlings ND, Waller M, Barrett AJ, Bateman A. MEROPS: the database of proteolytic enzymes, their substrates and inhibitors. Nucleic Acids Res. 2014;42:D503–9.
Remacle AG, Gawlik K, Golubkov VS, Cadwell GW, Liddington RC, Cieplak P, Millis SZ, Desjardins R, Routhier S, Yuan XW, et al. Selective and potent furin inhibitors protect cells from anthrax without significant toxicity. Int J Biochem Cell Biol. 2010;42:987–95.
Renatus M, Bode W, Huber R, Stürzebecher J, Stubbs MT. Structural and functional analyses of benzamidine-based inhibitors in complex with trypsin: implications for the inhibition of factor Xa, tPA, and urokinase. J Med Chem. 1998;41:5445–56.
Rockwell NC, Krysan DJ, Komiyama T, Fuller RS. Precursor processing by kex2/furin proteases. Chem Rev. 2002;102:4525–48.
Runger TM, Adami S, Benhamou CL, Czerwinski E, Farrerons J, Kendler DL, Mindeholm L, Realdi G, Roux C, Smith V. Morphea-like skin reactions in patients treated with the cathepsin K inhibitor balicatib. J Am Acad Dermatol. 2012;66:e89–96.
Sakai K, Ami Y, Tahara M, Kubota T, Anraku M, Abe M, Nakajima N, Sekizuka T, Shirato K, Suzaki Y, et al. The host protease TMPRSS2 plays a major role in in vivo replication of emerging H7N9 and seasonal influenza viruses. J Virol. 2014;88:5608–16.
Sato M, Yoshida S, Iida K, Tomozawa T, Kido H, Yamashita M. A novel influenza A virus activating enzyme from porcine lung: purification and characterization. Biol Chem. 2003;384:219–27.
Satoyoshi E. Therapeutic trials on progressive muscular dystrophy. Intern Med. 1992;31:841–6.
Schechter I, Berger A. On the size of the active site in proteases. I. Papain. Biochem Biophys Res Commun. 1967;27:157–62.
Scheen AJ. A review of gliptins for 2014. Expert Opin Pharmacother. 2015;16:43–62.
Schweinitz A, Stürzebecher A, Stürzebecher U, Schuster O, Stürzebecher J, Steinmetzer T. New substrate analogue inhibitors of factor Xa containing 4-amidinobenzylamide as P1 residue: part 1. Med Chem. 2006;2:349–61.

Seidah NG. Chapter 721—Site-1 Protease A2—Rawlings, Neil D. In: Salvesen G, editor. Handbook of proteolytic enzymes. New York: Academic Press; 2013, p. 3265–70.
Seidah NG, Chrétien M. Proprotein and prohormone convertases: a family of subtilases generating diverse bioactive polypeptides. Brain Res. 1999;848:45–62.
Seidah NG, Prat A. The biology and therapeutic targeting of the proprotein convertases. Nat Rev Drug Discov. 2012;11:367–83.
Setyono-Han B, Stürzebecher J, Schmalix WA, Muehlenweg B, Sieuwerts AM, Timmermans M, Magdolen V, Schmitt M, Klijn JG, Foekens JA. Suppression of rat breast cancer metastasis and reduction of primary tumour growth by the small synthetic urokinase inhibitor WX-UK1. Thromb Haemost. 2005;93:779–86.
Shirato K, Kanou K, Kawase M, Matsuyama S. Clinical isolates of human Coronavirus 229E bypass the endosome for cell entry. J Virol. 2017;91. e01387–16.
Shiryaev SA, Remacle AG, Ratnikov BI, Nelson NA, Savinov AY, Wei G, Bottini M, Rega MF, Parent A, Desjardins R, et al. Targeting host cell furin proprotein convertases as a therapeutic strategy against bacterial toxins and viral pathogens. J Biol Chem. 2007;282:20847–53.
Shulla A, Heald-Sargent T, Subramanya G, Zhao J, Perlman S, Gallagher T. A transmembrane serine protease is linked to the severe acute respiratory syndrome coronavirus receptor and activates virus entry. J Virol. 2011;85:873–82.
Sielaff F, Böttcher-Friebertshäuser E, Meyer D, Saupe SM, Volk IM, Garten W, Steinmetzer T. Development of substrate analogue inhibitors for the human airway trypsin-like protease HAT. Bioorg Med Chem Lett. 2011a;21:4860–4.
Sielaff F, Than ME, Bevec D, Lindberg I, Steinmetzer T. New furin inhibitors based on weakly basic amidinohydrazones. Bioorg Med Chem Lett. 2011b;21:836–40.
Siklos M, BenAissa M, Thatcher GR. Cysteine proteases as therapeutic targets: does selectivity matter? A systematic review of calpain and cathepsin inhibitors. Acta Pharmaceut Sin B. 2015;5:506–19.
Simmons G, Gosalia DN, Rennekamp AJ, Reeves JD, Diamond SL, Bates P. Inhibitors of cathepsin L prevent severe acute respiratory syndrome coronavirus entry. Proc Natl Acad Sci U S A. 2005;102:11876–81.
Singh J, Petter RC, Baillie TA, Whitty A. The resurgence of covalent drugs. Nat Rev Drug Discov. 2011;10:307–17.
Sisay MT, Steinmetzer T, Stirnberg M, Maurer E, Hammami M, Bajorath J, Gütschow M. Identification of the first low-molecular-weight inhibitors of matriptase-2. J Med Chem. 2010;53:5523–35.
Someya A, Tanaka N, Okuyama A. Inhibition of influenza virus A/WSN replication by a trypsin inhibitor, 6-amidino-2-naphthyl p-guanidinobenzoate. Biochem Biophys Res Commun. 1990;169:148–52.
Sorsa T, Salo T, Koivunen E, Tyynela J, Konttinen YT, Bergmann U, Tuuttila A, Niemi E, Teronen O, Heikkila P, et al. Activation of type IV procollagenases by human tumor-associated trypsin-2. J Biol Chem. 1997;272:21067–74.
Steinmetzer T, Hauptmann J, Stürzebecher J. Advances in the development of thrombin inhibitors. Expert Opin Investig Drugs. 2001;10:845–64.
Steinmetzer T, Schweinitz A, Stürzebecher A, Dönnecke D, Uhland K, Schuster O, Steinmetzer P, Müller F, Friedrich R, Than ME, et al. Secondary amides of sulfonylated 3-amidinophenylalanine. New potent and selective inhibitors of matriptase. J Med Chem. 2006;49:4116–26.
Stevens J, Corper AL, Basler CF, Taubenberger JK, Palese P, Wilson IA. Structure of the uncleaved human H1 hemagglutinin from the extinct 1918 influenza virus. Science. 2004;303: 1866–70.
Straub A, Roehrig S, Hillisch A. Oral, direct thrombin and factor Xa inhibitors: the replacement for warfarin, leeches, and pig intestines? Angew Chem Int Ed Engl. 2011;50:4574–90.
Stürzebecher J, Prasa D, Hauptmann J, Vieweg H, Wikström P. Synthesis and structure-activity relationships of potent thrombin inhibitors: piperazides of 3-amidinophenylalanine. J Med Chem. 1997;40:3091–9.

Stürzebecher J, Vieweg H, Steinmetzer T, Schweinitz A, Stubbs MT, Renatus M, Wikström P. 3-Amidinophenylalanine-based inhibitors of urokinase. Bioorg Med Chem Lett. 1999;9:3147–52.
Stürzebecher J, Schweinitz A, Schmalix WA, Wikström P. Synthetic urokinase inhibitors as potential anti-invasive drugs. IDrugs. 2001;4:677–83.
Swedberg JE, Harris JM. Plasmin substrate binding site cooperativity guides the design of potent peptide aldehyde inhibitors. Biochemistry. 2011;50:8454–62.
Szabo R, Hobson JP, List K, Molinolo A, Lin CY, Bugge TH. Potent inhibition and global co-localization implicate the transmembrane Kunitz-type serine protease inhibitor hepatocyte growth factor activator inhibitor-2 in the regulation of epithelial matriptase activity. J Biol Chem. 2008;283:29495–504.
Tarnow C, Engels G, Arendt A, Schwalm F, Sediri H, Preuss A, Nelson PS, Garten W, Klenk HD, Gabriel G, et al. TMPRSS2 is a host factor that is essential for pneumotropism and pathogenicity of H7N9 influenza A virus in mice. J Virol. 2014;88:4744–51.
Taylor NA, Van De Ven WJ, Creemers JW. Curbing activation: proprotein convertases in homeostasis and pathology. FASEB J. 2003;17:1215–27.
Thomas G. Furin at the cutting edge: from protein traffic to embryogenesis and disease. Nat Rev Mol Cell Biol. 2002;3:753–66.
Turk B. Targeting proteases: successes, failures and future prospects. Nat Rev Drug Discov. 2006;5:785–99.
Turk V, Stoka V, Vasiljeva O, Renko M, Sun T, Turk B, Turk D. Cysteine cathepsins: from structure, function and regulation to new frontiers. Biochim Biophys Acta. 2012;1824:68–88.
Tziridis A, Rauh D, Neumann P, Kolenko P, Menzel A, Bräuer U, Ursel C, Steinmetzer P, Stürzebecher J, Schweinitz A, et al. Correlating structure and ligand affinity in drug discovery: a cautionary tale involving second shell residues. Biol Chem. 2014;395:891–903.
Urata S, Yun N, Pasquato A, Paessler S, Kunz S, de la Torre JC. Antiviral activity of a small-molecule inhibitor of arenavirus glycoprotein processing by the cellular site 1 protease. J Virol. 2011;85:795–803.
Van Rompaey L, Ayoubi T, Van De Ven W, Marynen P. Inhibition of intracellular proteolytic processing of soluble proproteins by an engineered alpha 2-macroglobulin containing a furin recognition sequence in the bait region. Biochem J. 1997;326(Pt 2):507–14.
Vandenbroucke RE, Libert C. Is there new hope for therapeutic matrix metalloproteinase inhibition? Nat Rev Drug Discov. 2014;13:904–27.
Vlieghe P, Lisowski V, Martinez J, Khrestchatisky M. Synthetic therapeutic peptides: science and market. Drug Discov Today. 2010;15:40–56.
Wang S, Le TQ, Kurihara N, Chida J, Cisse Y, Yano M, Kido H. Influenza virus-cytokine-protease cycle in the pathogenesis of vascular hyperpermeability in severe influenza. J Infect Dis. 2010;202:991–1001.
Wang N, Shi X, Jiang L, Zhang S, Wang D, Tong P, Guo D, Fu L, Cui Y, Liu X, et al. Structure of MERS-CoV spike receptor-binding domain complexed with human receptor DPP4. Cell Res. 2013;23:986–93.
Watanabe M, Hirano A, Stenglein S, Nelson J, Thomas G, Wong TC. Engineered serine protease inhibitor prevents furin-catalyzed activation of the fusion glycoprotein and production of infectious measles virus. J Virol. 1995;69:3206–10.
Wood JM, Maibaum J, Rahuel J, Grutter MG, Cohen NC, Rasetti V, Ruger H, Goschke R, Stutz S, Fuhrer W, et al. Structure-based design of aliskiren, a novel orally effective renin inhibitor. Biochem Biophys Res Commun. 2003;308:698–705.
Wu K, Li W, Peng G, Li F. Crystal structure of NL63 respiratory coronavirus receptor-binding domain complexed with its human receptor. Proc Natl Acad Sci U S A. 2009;106:19970–4.
Wysocka M, Spichalska B, Lesner A, Jaros M, Brzozowski K, Legowska A, Rolka K. Substrate specificity and inhibitory study of human airway trypsin-like protease. Bioorg Med Chem. 2010;18:5504–9.
Yeager CL, Ashmun RA, Williams RK, Cardellichio CB, Shapiro LH, Look AT, Holmes KV. Human aminopeptidase N is a receptor for human coronavirus 229E. Nature. 1992;357:420–2.

Yuan C, Chen L, Meehan EJ, Daly N, Craik DJ, Huang M, Ngo JC. Structure of catalytic domain of Matriptase in complex with Sunflower trypsin inhibitor-1. BMC Struct Biol. 2011;11:30.
Zeslawska E, Schweinitz A, Karcher A, Sondermann P, Sperl S, Stürzebecher J, Jacob U. Crystals of the urokinase type plasminogen activator variant beta(c)-uPAin complex with small molecule inhibitors open the way towards structure-based drug design. J Mol Biol. 2000;301:465–75.
Zhirnov OP, Ovcharenko AV, Bukrinskaya AG. Protective effect of protease inhibitors in influenza virus infected animals. Arch Virol. 1982a;73:263–72.
Zhirnov OP, Ovcharenko AV, Bukrinskaya AG. Proteolytic activation of influenza WSN virus in cultured cells is performed by homologous plasma enzymes. J Gen Virol. 1982b;63:469–74.
Zhirnov OP, Klenk HD, Wright PF. Aprotinin and similar protease inhibitors as drugs against influenza. Antiviral Res. 2011;92:27–36.
Zhong M, Munzer JS, Basak A, Benjannet S, Mowla SJ, Decroly E, Chretien M, Seidah NG. The prosegments of furin and PC7 as potent inhibitors of proprotein convertases. In vitro and ex vivo assessment of their efficacy and selectivity. J Biol Chem. 1999;274:33913–20.
Zhou Y, Vedantham P, Lu K, Agudelo J, Carrion R Jr, Nunneley JW, Barnard D, Pöhlmann S, McKerrow JH, Renslo AR, et al. Protease inhibitors targeting coronavirus and filovirus entry. Antiviral Res. 2015;116:76–84.
Zhu J, Declercq J, Roucourt B, Ghassabeh GH, Meulemans S, Kinne J, David G, Vermorken AJ, Van de Ven WJ, Lindberg I, et al. Generation and characterization of non-competitive furin-inhibiting nanobodies. Biochem J. 2012;448:73–82.
Zmora P, Blazejewska P, Moldenhauer AS, Welsch K, Nehlmeier I, Wu Q, Schneider H, Pöhlmann S, Bertram S. DESC1 and MSPL activate influenza A viruses and emerging coronaviruses for host cell entry. J Virol. 2014;88:12087–97.
Zoratti GL, Tanabe LM, Varela FA, Murray AS, Bergum C, Colombo E, Lang JE, Molinolo AA, Leduc R, Marsault E, et al. Targeting matriptase in breast cancer abrogates tumour progression via impairment of stromal-epithelial growth factor signalling. Nat Commun. 2015;6:6776.

Index

E. Böttcher-Friebertshäuser et al. (eds.), *Activation of Viruses by Host Proteases*,
https://doi.org/10.1007/978-3-319-75474-1

U

V

W

Y

Z

MIX
Papier aus verantwortungsvollen Quellen
Paper from responsible sources
FSC® C105338

If you have any concerns about our products,
you can contact us on
ProductSafety@springernature.com

In case Publisher is established outside the EU,
the EU authorized representative is:
Springer Nature Customer Service Center GmbH
Europaplatz 3, 69115 Heidelberg, Germany

Printed by Libri Plureos GmbH
in Hamburg, Germany